IET CIRCUITS, DEVICES AND SYSTEMS SERIES 2

Series Editors: Dr D.G. Haigh
Dr R.S. Soin

Analogue
IC Design
The Current-Mode
Approach

Other volumes in this series:

Analogue
IC Design
The Current-Mode
Approach

Edited by C. Toumazou,
F.J. Lidgey and D.G. Haigh

The Institution of Engineering and Technology

List of Contents

Foreword

This volume arose from an idea by Chris Toumazou to attempt to publish the notes for a Workshop on "Current-Mode Analogue IC Design", organised by the editors as part of the 1990 IEEE International Symposium on Circuits and Systems held at New Orleans on 30th April 1990.

The timescale of five months from contacting the authors to publication seemed daunting but the energy and enthusiasm shown by Chris and John and the eminence of the authors they had lined-up meant that the project had to go ahead. The publishers helped by allowing a very short time for printing the manuscript.

In the event, the editors have produced a volume which far transcends the original workshop and constitutes a "Bible" for modern integrated circuit design techniques. It is an excellent complement to the first volume in the Circuits and Systems Series entitled "GaAs Technology and its Impact on Circuits and Systems" and will be of benefit to students, both undergraduate and postgraduate, and researchers and circuit designers in academia and the Electronics Industry.

David Haigh
Randeep Singh Soin
London, March 1990

List of Contributors

P.E Allen
School of Electrical Engineering
Georgia Institute of Technology,
Atlanta Georgia
USA

S. Bibyk
S.T Dupuie
M. Ismail
Department of Electrical Engineering
The Ohio State University
205 Dreese Laboratory
USA

D. F. Bowers
Precision Monolithic Inc.
Santa Clara
California
USA

S. J. Daubert
AT & T Bell Laboratories
Murray Hill
New Jersey
USA

B. Gilbert
Analog Devices Semiconductor Inc.
Beaverton
Oregon
USA

D. G. Haigh
Department of Electronic and Electrical
Engineering
University College
London
UK.

J. B. Hughes
Philips Research Laboratories
Redhill
Surrey
UK

F. J. Lidgey
School of Engineering
Oxford Polytechnic
Oxford
UK

D. G. Nairn
C. A. T. Salama
H. W. Singor
A. S. Sedra
Department of Electrical Engineering
University of Toronto
Toronto Ontario
Canada

G. W. Roberts
Department of Electrical Engineering
McGill University
Montreal Quebec
Canada

R. Schaumann
Electrical Engineering Department
Portland State University
Portland Oregon
USA

E. Seevinck
Philips Research Laboratories
Eindhoven
The Netherlands

M. A. Tan
Department of Electrical Engineering
Bilkent University
Maltepe
Ankara
Turkey

C. Toumazou
Department of Electrical Engineering
Imperial College of Science
Technology & Medicine
London
UK

D. Vallancourt
Department of Electrical
Engineering and Center for
Telecommunications Research
Columbia University
New York
USA

E. Vittoz
Centre Suisse D'Electronique et
de Microtechnique S.A.
Recherche et Developpement
Neuchatel
Switzerland

D. C. Wadsworth
Phototronics Co.
Regd. Box 977
Manotick
Ontario
Canada

G. Wegmann
Swiss Federal Institute
of Technology
Lausanne
Switzerland

Chapter 1

Introduction

Chris Toumazou, John Lidgey and David Haigh

1.1 Analogue IC Design

Analogue Integrated Circuit Design is becoming increasingly important with growing opportunities. The emergence of IC's incorporating mixed analogue and digital functions on a single chip has led to an advanced level of analogue design .

Analogue IC design has traditionally been hampered by process technology, generally because technology has been optimized for digital circuits. This has resulted in an apparent " design time Syndrome", where a single IC may contain only 20% analogue functions which take 80% of the design time. However this situation is now changing as we are experiencing the development of a new generation of " technology specific" analogue design techniques. Furthermore as new and more mature device technologies such as true complementary silicon bipolar junction (BJT), mixed Silicon bipolar and complimentary metal oxide semiconductor devices (BiCMOS) and Gallium Arsenide (GaAs) are becoming available they bring with them the requirement for novel analogue design, methods, techniques and CAD tools necessary for the successful development and exploitation of these technologies for the future market place. Coupled with these technological improvements are the ever shrinking feature size of devices on IC's and the consequential reduction of power supply voltages which has fuelled the creation of "alternative" analogue design techniques.

State-of-the-art analogue integrated circuit design is receiving a tremendous boost from the development and application of current-mode processing, which is rapidly superceeding traditional approaches based on voltage-mode designs. There are many advantages to be gained from a wider view of analogue signal processing embracing current-mode techniques. This book draws together contributions from the world's most eminent analogue IC designers to provide, for the first time, a comprehensive text devoted to this important and exciting new area of analogue electronics.

1.2 The Current-Mode Approach

Analogue design has historically been viewed as a voltage dominated form of signal processing. This has been apparent in analogue IC design where

possibilities of microwave and optical system design being transferred to the realms of the analogue IC design arena.

Furthermore, with maturing CMOS VLSI, BiCMOS, GaAs, and true complementary bipolar technology, current-mode analogue design techniques will play an important role in successfully exploiting these technologies in the analogue domain. As a consequence many of the early current-mode circuit techniques are enjoying a reinassance and a new generation of current-mode building blocks and systems are being developed, as indicated in Figure 1.1 and described throughout this book.

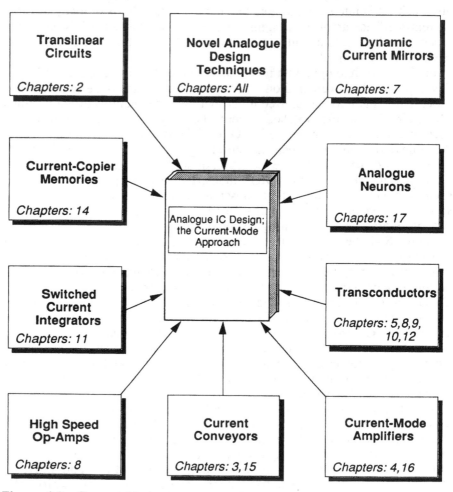

Figure 1.2   Current-Mode Analogue Building blocks.

1.4 New Analogue Building-Blocks

Early circuit design principles and techniques for current-mode processing, such as the translinear circuit principal introduced by Barrie Gilbert in 1972, are becoming powerful tools for the development of high performance analogue circuits and systems. A further consequence of the development of current-mode analogue signal processing has been the emergence of new analogue building-blocks ranging from the current-conveyor and current-feedback op-amps through to sampled-data current circuits, such as dynamic current-mirrors and analogue neural networks. Figure 1.2 shows the range of these building-blocks that are to be found in this text and within each block the Chapter reference number is shown. For example, the current conveyor is an extremely powerful analogue building block, combining voltage and current-mode capability. It has proved to be functionally flexible and versatile, rapidly gaining acceptance as a practical device with a wide range of high performance circuit and system applications. The recent introduction of a commercially integrated circuit current-conveyor is reported and is indeed very timely and welcome.

1.5 Current-Mode Systems

The maturity of current-mode analogue signal processing is seen from the development of systems based on the current-mode approach. A wide variety of systems applications described herein is shown in Figure 1.3, covering important areas such as continuous-time and sampled-data filters through general analogue interfacing, A/D and D/A converters to current-mode neural networks. The area is still in its infancy and over the next five years or so, we will see more and more system developments structured from current-mode analogue circuits and sub-systems.

1.6 Review of the Chapters

This book comprises three main sections, these being circuit level, sub-system level and system level design. Appropriately the book begins in Chapter 2 with a tutorial on 'Current-mode circuits from a translinear viewpoint', by Barrie Gilbert, the inventor of the translinear circuit principle. Since it's inception in 1972, the 'translinear circuit principle' has promoted the conception of a variety of elegant non-linear circuits, aided their understanding and simplified their analysis. The principle is reviewed and applications ranging from multipliers/dividers to trignometric function generators are described. With the recent emergence of truly complementary BJTs this important circuit principle will become confirmed as a powerful tool in analogue circuit design. The prospects of heterojunction devices which have the basic operation of a BJT but draw no base current (infinite beta) is described in the chapter. The applications that lie ahead for such a device are phenomenal, and Barrie Gilbert portrays his views on the possibilties of this conceptually "ideal" semiconductor

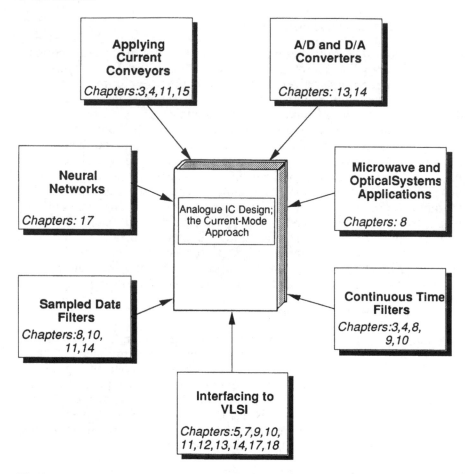

Figure 1.3   Current-Mode System Applications.

device! Further developments of the translinear circuit principle into the MOS domain are reported later in Chapter 12.

At the time of it's introduction in 1968 the advantages of the current-conveyor were not fully apparent, particularly since the monolithic op-amp was well established. All too often the electronics engineer tends to think in voltage terms rather than current and it is only now that the analogue designer is realizing the full benefits of the current-mode approach. The current-conveyor is essentially a voltage/current-mode hybrid circuit and consequently is an extremely versatile analogue building-block. This is confirmed by the wealth of literature on the current-conveyor which is the topic of Chapter 3, entitled 'Current-conveyor theory and practice', by Adel Sedra, the co-inventor of the current-conveyor, and Gordon Roberts. Although the current-conveyor is more than 20 years old, it is "a circuit idea whose time has come".

Chapter 4, by Chris Toumazou and John Lidgey, describes amplifier development work towards a universal current-mode analogue amplifier. Beginning with adapted voltage-mode op-amps configured to provide current-mode performance, based on the technique of supply-current sensing, they show through applications the advantages of this approach in many areas of analogue circuit and system design. The latter part of this Chapter concentrates on the development of novel architecture analogue amplifiers, including an operational floating current-conveyor through to a fully differential input/output current-mode op-amp.

Another important current-mode analogue building-block is the transconductor. Chapter 5 by Scott Dupiue and Mohammed Ismail describes the detailed development of a variety of high frequency CMOS transconductors. Aspects of modelling, linearity, distortion, noise, high frequency performance, signal handling and power consumption are design issues addressed in this Chapter. The authors stress the value of the transconductor as an ideal interface between voltage and current-mode circuits in VLSI.

The current-mirror is a ubiquitous building-block for all semiconductor technologies. Practical aspects of BJT current-mirrors are explored by Barrie Gilbert in Chapter 6. Often encountered though not usually well documented from a practical viewpoint, the author reviews common and not-so-common current-mirrors producing a comprehensive treatise on the subject. The emphasis of Chapter 6 is on the importance of selecting a particular current-mirror to suit a specific application and ground rules are developed for achieving the desired performance goals.

Unfortunately, offsets and noise performance of CMOS current-mirrors are significantly poorer than BJTs and a new generation of dynamic current-mirrors, which overcomes the majority of these inaccuracies, is presented by Eric Vittoz and George Wegmann in Chapter 7. The technique is based on a simple idea of a switched transistor MOS transistor giving an output which is a one-to-one discontinuous copy of the input current. The evolution and design of the dynamic current-mirror is described and the authors show how it can be extended to a variety of different circuits to create very precise analogue CMOS building-blocks.

Gallium Arsenide technology is maturing rapidly with the achievement of higher yields, lower costs and faster turnrounds. Current developments in modern wideband communication systems are creating a firm need for circuit and system components realized in this technology. In Chapter 8 Chris Toumazou and David Haigh present the design and development of GaAs analogue building-blocks, such as current-mirrors, transconductors and operational amplifiers for applications in precision filters, microwave and optical communications. As in other technologies, the current-mode approach is essential to fully exploit the high speed potential of GaAs technology.

Mixed-analogue and digital VLSI requires the development of analogue continuous-time filters. Current-based transconductance-capacitance (gm-

architecture the long-tail pair has been abandoned for a complementary common-emitter and common-base stage and the quoted slew-rate of these amplifiers is some two orders of magnitude higher. Furthermore current-feedback op-amps exhibit an almost constant bandwidth irrespective of closed-loop gain, a feature of many of the current-mode amplifier designs described throughout this text. The current feedback op-amp architecture is a partial step towards a true current-mode op-amp, but it has been designed in such a way that it can be used in an almost identical feedback application to that of the conventional op-amp. Chapter 16 also describes the design of commercially available high performance precision instrumentation amplifiers again employing active current feedback techniques to improve performance such as CMRR.

A new exciting development is the electronic circuit implementation of analogue neural networks. Neural networks mimic the functionality of the human brain and to date much work has been devoted to signal processing applications such as pattern recognition. In Chapter 17, Steven Bibyk and Mohammed Ismail describe the development of novel neural building-blocks for analog MOS VLSI. The emphasis is again on the use of current-mode techniques to exploit VLSI technology, resulting in optimum neural network designs. Performance features of current-mode neural networks include increased dynamic range and frequency performance and improved linearity. Floating gate technology is a maturing technology which will have significant impact on the future of analogue design. Such technology is suited to applications requiring adjustment of a transistor parameter such as threshold voltage with high resolution in minimal chip area. One of the most attractive applications is for non-volatile floating gate on-chip analog memories for neural networks and examples of these circuits are described.

1.7 Future current-mode analogue IC Design

Chapter 18 concludes the book with Phillip Allen reviewing the future of analogue integrated circuit design. Topics ranging from present status of analogue IC design and technological and design trends through to design automation and testing are discussed. The Chapter emphasises the important role that technology will play in the future of analogue IC design and design methodologies and techniques. It is clear that future of anlogue signal processing is "very dynamic" and current-mode analogue IC design will be catylitic in a decade where we are witnessing tremendous improvements in process technologies. It is only now becoming possible with such improvements in technology that current-mode analogue techniques are providing dramatic benefits in practical circuits and systems. This book has attempted to capture these developments and should stimulate interest in helping to make "Analog IC Design; the current-mode approach" more accessible to those in the electronics industry.

Chapter 2

Current-mode Circuits From A Translinear Viewpoint: A Tutorial

Barrie Gilbert

2.1 Introduction

When the word "translinear" was coined in 1975 [1] it was with a very specific purpose in mind. A novel class of circuits had gradually emerged following the discovery of a few special cases in 1967 and first reported at the 1968 ISSCC [2]. These circuits could only be accurately realized using *monolithic* BJTs, since isothermal operation and tight matching of device geometry and doping profiles were essential. Hence, although such circuits might have been conceivable in the early '60s, they certainly would not have been realizable in the days of discrete devices. These circuits were based on the remarkable fact that the trans̲conductance of a BJT is li̲near̲ly proportional to its collector current, hence the term *trans-linear*. Further, the word carried with it the connation of "lying somewhere between familiar home territories of the linear circuit and the formidable terrains of the nonlinear". Indeed, in addition to their well-known application in multipliers and other nonlinear circuits, translinear concepts are found embedded in many contemporary linear integrated circuits. The most familiar example is the current mirror; the classical four-transistor Class-AB output stage of almost any op-amp can be viewed in translinear terms; the current-conveyor, used in the recently-rediscovered current-feedback amplifier, is yet another example.

While the term "translinear" has this more general application, the inputs and outputs of what will be called "strict-TL" circuits are entirely in the form of currents, and no voltages other than the junction voltages are involved. The circuit function in such cases is essentially independent of the overall magnitude of the operating currents, that is, the "bias" level; rather, the behavior is invariably a consequence of *current ratios* within the circuit. In fact, in this Chapter extensive use will be made of the idea of a *modulation index* to describe a signal. This is a dimensionless variable, usually in the range 0 to 1 or -1 to +1, which when multiplied by some bias current yields the actual signal. Thus, within the same physical circuit, a given function can often be achieved at bias levels from nanoamps, at which point operation is slow and accuracy may deteriorate due to second-order device effects, up to milliamps, where the circuit speed is as fast as any possible circuit topology, because voltage swings are at an absolute minimum. Furthermore, the function invariably remains exact up to the limits of the available bias range

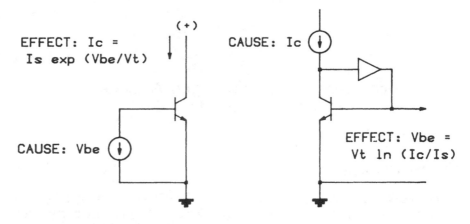

EFFECT: Ic =
Is exp (Vbe/Vt)

CAUSE: Vbe

(+)

CAUSE: Ic

EFFECT: Vbe =
Vt ln (Ic/Is)

Figure 2.1 Two views of BJT biasing: in (a) the applied V_{BE} generates I_C; in (b) V_{BE} is forced to adjust to accommodate the demanded I_C.

and, most significantly, is *fundamentally insensitive* to variations in temperature over the full range of operation possible with silicon.

2.2   General Principles

This Chapter will describe many examples of translinear circuits, which come as close to true current-mode operation as any circuit can.

At the outset, however, it must be stated that the term "current-mode circuit" has no rigorous meaning. The behavior of electrical networks is always the result of an interplay between voltage and current. Choice of impedance level is frequently crucial to successful implementation of a signal-processing function. In designing with the silicon bipolar junction transistor (BJT), biasing plays an important role in setting impedance levels. The small-signal view of circuit design requires that strongly nonlinear devices be represented by linearized models, to allow well-known linear mathematical methods to be used in analysis. In this traditional view, the BJT is often viewed as a current-controlled current-source, in which the finite common-emitter current gain, beta, is regarded as a key parameter in determining circuit behavior. It is often assumed to be more or less constant with collector current, I_C, so presenting the transistor as essentially linear in nature. This "beta-view" diminishes the emphasis on the far more important role of V_{BE}, which is often taught as being only an approximate characteristic of the device: the words "about 700mV for a typical silicon transistor" are to be found in too many text-books.

In fact, the relationship between I_C and V_{BE} is at the very heart of the BJT. Figure 2.1 shows that this relationship can be viewed in two equivalent ways; in (a) the base-emitter junction of the transistor is driven by a voltage, V_{BE}, resulting in a collector current, I_C, which is an exponential function of V_{BE}; in (b) the transistor is operated in a reciprocal fashion, by forcing I_C and

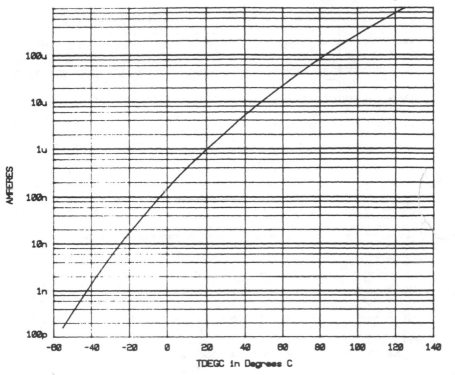

Figure 2.2 For a BJT having an Is of 1E-16A a fixed V_{BE} of 600mV causes I_C to range from 100pA to 1mA over -55°C to +125°C.

observing the resulting V_{BE}. In the first case

$$I_C = I_S(T)\, exp\, (V_{BE}/V_T) \tag{2.1a}$$

where V_T has the customary meaning (the thermal voltage, kT/q, which evaluates to 25.85mV at 300K) and I_S is the saturation current. Eq 2.1a omits the "-1" which should follow V_{BE}/V_T in the exponential argument, since for most practical purposes it can be safely ignored. It should be included in calculations involving high-temperature, low-current operation.

It is, perhaps, the notorious temperature sensitivity of the saturation current, I_S, and the resulting unpredictability of I_C when driven in this manner, that has led to the general belief that reliance on the relationship V_{BE}/I_C is to be avoided in biasing the BJT. It increases by about 9.5% per degree Celsius, and thus alters by a factor of about ten million over the -55°C to +125°C temperature range. I_C, being directly proportional to I_S, varies by the same amount if V_{BE} is held at a fixed value; Figure 2.2, plotted for a typical BJT, shows a 600mV bias will result in an I_C of anything from 100pA to 1mA for $I_S = 10^{-16}$A at 300K. As might be expected from its very small value, I_S cannot be measured directly. It is certainly not a "leakage" current,

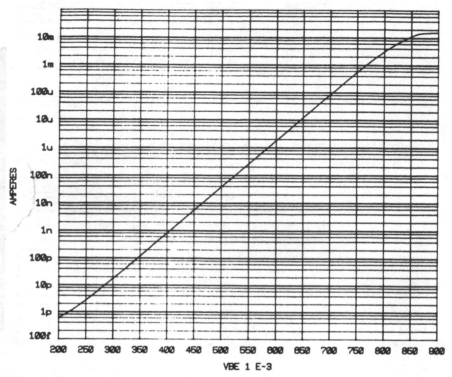

Figure 2.3   The key to the translinear principle: the very exact logarithmic relationship between I_C and V_{BE}, shown here for T=27°C.

to be measured by reverse-biasing either the collector or emitter junctions. It is a property of the base region of the transistor, a *scaling* parameter arising out of a multiplicity of process-related quantities, including the doping levels, and several fundamental constants, including the band-gap voltage E_g, as well as temperature. It is almost, but not exactly, proportional to emitter area. In practice, I_S is inferred through direct measurement of the V_{BE} at a reference value of $I_C = I_R$ and known temperature $T = T_R$.

In the reciprocal case, the circuit "output" (V_{BE}) is only a mild function of temperature:

$$V_{BE} = V_T \, ln \, \{I_C/I_S(T)\} \tag{2.1b}$$

Incidentally, the "amplifier", represented by the triangle in Figure 2.1b, is included to remind us that we wish to force I_C, and not allow the base current to introduce an error in I_C. This function is often adequately supplied by a simple emitter-follower, but an op-amp connected as a voltage-follower might also be used here.

The value of the relationship shown in alternative forms in Eqs 2.1a and 2.1b can be summed up by a plot of $log(I_C)$ versus V_{BE}, as shown in Figure

2.3. The most obvious feature of this plot is its extraordinary linearity. By differentiating Eq 2.1a we find

$$\frac{\delta I_C}{\delta V_{BE}} = g_m = \frac{I_C}{V_T}$$

(2.2)

that is, the transconductance of an ideal BJT is a linear function of its collector current.

While this last property of the BJT is widely known to be useful in circuit design, it is pivotal to the *translinear* view of the BJT, and from this starting point numerous, sometimes startling, consequences arise. Although much of the literature devoted to BJT circuit design makes adequate reference to this aspect of the device, it is rarely presented as the *key* to comprehending the general behavior of the vast canvas of circuits based on the bipolar transistor, in which both voltages and currents play an equally important role.

In this view of the transistor, the role of beta is secondary; indeed, as we have already seen, we would rather have no base current at all. Similarly, all other artifacts of real devices are not *essential* to the nature of the BJT. Certainly, they need to be accounted for in a comprehensive device model, but it cannot be said that they ever serve any *useful* function. Thus, base resistance only generates noise in amplifiers, or slows down the charging of the base (so reducing the bandwidth of linear circuits, or introducing delay in digital circuits such as ECL) or results in a g_m lower than predicted by Eq 2.2; the finite Early voltage only puts a limit on the maximum voltage gain that a single-stage amplifier can exhibit, and so on. Of course, junction capacitances are always unwelcome.

2.2.1 *Translinear Loops and Networks*

While we have stressed the importance of considering both the current and voltage levels in an optimal view of BJT circuits, there nevertheless exists an extensive class of BJT circuits whose function depends primarily on the use of *currents* as signals or functional variables, and which can be designed and understood, certainly at a basic level, using methods in which *voltages need not be considered at all*. It is circuits of this sort that are often meant by the use of the term "current-mode". These circuits are comprised almost exclusively of BJTs arranged in one or more *closed loops of junctions*. Note here that not all translinear circuits involve such closed loops. Many translinear circuits simply invoke BJT behavior to realize some exact linear or nonlinear function. For example, the familiar differential-pair depends on translinearity to provide a g_m which is well defined; in a more complex case, this same behavior can be invoked to generate a very accurate sine function [3], and so on. Such circuits can be referred to as "translinear networks", which is a more general term, and includes the more specialized

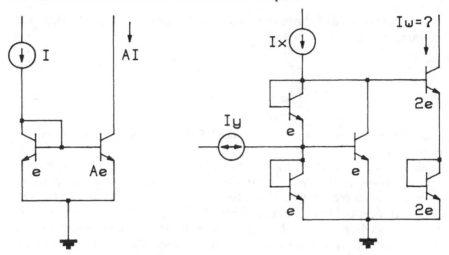

Figure 2.4 Two simple examples of translinear loops: the behavior of the current mirror (a) is apparent, but what is to be made of (b)?

circuits employing translinear loops. To distinguish between these two general classes, we will use the abbreviation "TL" to uniquely refer to a translinear circuit involving one or more closed loops of junctions. It is useful, at least in the present context, to refer to the remaining types of translinear networks as "TN" circuits.

While not usually viewed in this light, a simple current-mirror, Figure 2.4a, is a TL circuit, employing a single loop. It can be viewed as a combination of the "current-in/voltage-out" sub-circuit of Figure 1b driving the "voltage-in/current-out" sub-circuit of Figure 1a. Of course, we don't generally think of the current mirror in such detail: its form is so commonplace that most readers will not only "feel" that this is a current-mode configuration, in which the voltage arising across the common base-emitter port is of little interest, but intuitively recognize that the output current will be scaled by the emitter area, A.

But what is to be made of the circuit shown in Figure 2.4b? It looks a bit like an over-grown current mirror, that somehow needs two inputs, Ix and Iy, to develop its single output Iw. Or, how about the behavior of a much more familiar circuit, the diode bridge, shown in Figure 2.5? Given the three drive currents Ia, Ib and Ic, how is one to calculate the four diode currents? It is the main purpose of this Chapter to show that the analysis of such circuits becomes easy when the *translinear principle*, which specifically applies to TL circuits, is invoked. We will return to the circuit of Figure 2.1b, fully-equipped to analyze it, after discussing this concept, which could be said to be to nonlinear circuits based on junctions what KCL and KVL are to circuits based on linear elements.

This powerful concept has promoted the conception of numerous, often quite complex, nonlinear circuits and made the analysis of their behavior a simple, routine matter. By contrast, traditional analyses using full circuit

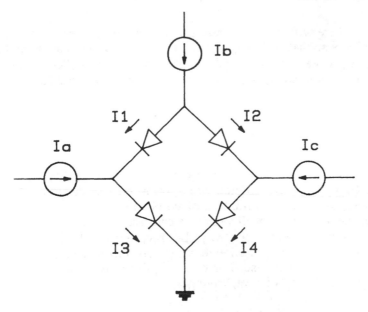

Figure 2.5 A diode bridge can be viewed as a translinear loop: given Ia, Ib and Ic, what are the four diode currents?

equations, relating voltages and currents, are cumbersome and needlessly burdened with factors which do not appear at all in the final result. This includes the strongly temperature-dependent terms V_T (increasing by about 0.33% for every 1°C increase in temperature) and the saturation current I_S (which, as we've seen, varies wildly with temperature). Indeed, one of the special merits of TL circuits is that their behavior in numerous instances is not dependent on the actual value of the signal currents at all, but primarily on their *ratios,* as will soon become apparent. Stated differently, they exhibit the same basic function whether biased at extremely low currents (nanoamps) or fairly high currents (many milliamps - depending on the details of the process on which the transistors are fabricated, and their geometry). It is interesting to note that many of the current-mode circuits in Chapters 3, 4, 15 and 16 exploit indirectly the translinear circuit principle; particularly the more recent bipolar current-conveyoy realisation.

2.2.2 *The Translinear Principle*

In a closed loop containing *n* PN junctions, such as shown in Figure 2.5, where the junctions are biased into forward conduction by some means, not shown in the Figure, the junction voltages, V_{Fk}, must sum to zero:

$$\sum_{k=1}^{k=n} V_{Fk} = 0$$

(2.3)

The junctions here will usually represent the base-emitter terminals of each BJT in the loop, so that each V_{Fk} is actually the V_{BE} of a transistor and the currents shown in each junction will usually represent the collector current, I_{Ck}. Accordingly, we can replace V_{Fk} by the value given in Eq 2.1b:

$$\sum_{k=1}^{k=n} V_T \ln \frac{I_{Ck}}{I_{Sk}} = 0$$

(2.4)

where the inclusion of a separate I_S for each junction includes the possibility that the junctions may have different areas, or even be made from different device types (for example, a mixture of NPN and PNP transistors). The thermal voltage V_T appears in all terms; we can assume that it is equal for all junctions. In fact, this may not always be true, and in TL circuits of the highest accuracy some compensation may be needed to account for the slight departure from this ideal case. For most applications, however, we can safely assume that the V_T's cancel, leaving

$$\sum_{k=1}^{k=n} \ln \frac{I_{Ck}}{I_{Sk}} = 0$$

(2.4a)

Noting that the summation of a series of logarithmic terms can be written as a product, and that $\ln(1)=0$, we can rewrite Eq 2.4 as

$$\prod_{k=1}^{k=n} \frac{I_{Ck}}{I_{Sk}} = 1$$

(2.5)

Now, any practical circuit will operate with $I_C/I_S \gg 1$. For example, even at a collector current as low as 100pA, this ratio will be typically 10,000. Thus, for the product to remain unity while maintaining sensible operating currents, two fundamental conditions must be met:

1) There must be an even number of junctions (at least two) in a TL loop.
2) There must be an equal number of clockwise-facing (CW) and counterclockwise-facing (CCW) junctions.

The ordering of the junctions is not important at this stage, and, in practice, the variety of possible orderings allows for interesting topological variants to be devised in the synthesis of TL circuits. It was noted above that one could mix junctions of different types. It will be apparent that when this is done they need to appear in opposing pairs, since their saturation currents may have significantly different temperature behavior. For now, we will assume that all the devices are of the same type, differing, if at all, only in emitter area.

Given the need for symmetry, Eq. 2.5 can be stated in a slightly different way:

$$\prod_{CW} \frac{I_{Ck}}{I_{Sk}} = \prod_{CCW} \frac{I_{Ck}}{I_{Sk}}$$

(2.6)

In a properly-implemented design, all the saturation currents in a TL loop will be proportional to the emitter areas. Because of the different nature of carrier injection at the emitter edge compared to that in the planar region directly under the emitter, I_S only scales approximately when the area is altered by changing the length or width. Therefore it is good TL practice to realize emitter area ratios by repeating a fixed geometry an integral number of times. In the examples presented in this Chapter, the small letter 'e' beside an emitter implies the use of a "unit" emitter (the absolute size is often unimportant); multiple emitters are shown as, for example, "2e".

With these precautions in mind, we can take the next step in the derivation of a useful theory, which is to replace I_{Sk} in Eq. 2.6 by factors of the form $A_k J_{Sk}$, where A_k is the emitter area and J_{Sk} is the geometry-independent *saturation current-density*. Since this latter factor can be assumed to equally weight the LHS and RHS of the equation (and it can be arranged to satisfy this requirement, even when mixed junction types are used, by appropriate pairing) we are left with

$$\prod_{CW} \frac{I_{Ck}}{A_k} = \prod_{CCW} \frac{I_{Ck}}{A_k}$$

(2.7)

Now, recognizing that the ratios I_{Ck}/A_k are just the emitter current densities we can write the translinear principle in its most succinct and compact form:

$$\prod_{CW} J = \prod_{CCW} J$$

(2.8)

This is the translinear principle, which can be stated in this way:

IN A CLOSED LOOP CONTAINING AN EVEN NUMBER OF FORWARD BIASED JUNCTIONS, ARRANGED SO THAT THERE ARE AN EQUAL NUMBER OF CLOCKWISE-FACING AND COUNTERCLOCKWISE- FACING POLARITIES, THE PRODUCT OF THE CURRENT DENSITIES IN THE CLOCKWISE DIRECTION IS EQUAL TO THE PRODUCT OF THE CURRENT DENSITIES IN THE COUNTERCLOCKWISE DIRECTION.

As we shall see, the application of the translinear principle (abbreviated "TLP") permits a rapid appreciation of the behavior of even quite complicated circuits and, while direct synthesis of a given function is often possible, analysis is so simple that one can often find a satisfactory form more quickly by trying out a variety of promising candidates. Seevinck [4, 4a] has very ably addressed the topic of TL synthesis, but its likely that most circuits in use today arose out of a more heuristic approach.

2.2.3 *Use of Emitter Area Ratios*

The ratio in emitter area between pairs of devices in a TL circuit is very important. Often, it is altered deliberately to effect some desired outcome. An example might be the current mirror (Figure 2.1a) where the area ratio A can be used to scale the output current. Deliberate use of emitter area ratios is often helpful in reducing, or even eliminating, errors (most often, distortion) due to finite junction resistance. As we shall see, in our discussion of RMS-DC conversion, area ratios can advantageously be used to improve accuracy for signals of high crest-factor, while leaving the scale-factor unchanged.

Modifying Eq. 2.7 to consolidate the emitter areas into a composite term:

$$\prod_{CW} \frac{1}{A_k} \prod_{CW} I_{Ck} = \prod_{CCW} \frac{1}{A_k} \prod_{CCW} I_{Ck}$$

(2.9)

hence

$$\prod_{CW} J = \lambda \prod_{CCW} J$$

(2.10)

where λ is called the "area-ratio factor":

$$\lambda = \frac{\displaystyle\prod_{CW} A_k}{\displaystyle\prod_{CCW} A_k}$$

(2.11)

For example, consider the TL loop shown in Figure 2.6, and let the junction areas have the same numbering as the currents. We have A_1, A_4, A_6 and A_8 in the CW direction and A_2, A_3, A_5 and A_7 in the CCW direction. Thus,

$$\lambda = \frac{A_1 A_4 A_6 A_8}{A_2 A_3 A_5 A_7}$$

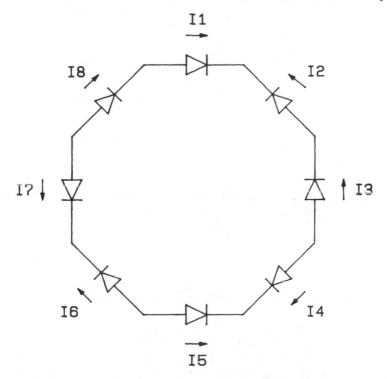

Figure 2.6 The basic idea behind every TL circuit: a closed loop of forward-biased junctions results in exact, temperature-stable behavior.

Note that it is often desired that λ be as close to unity as possible, but this does not require that the emitter areas all be equal. For example, let A_1=21, A_4=2, A_6=5, A_8=8, A_2=3, A_3=7, A_5=20 and A_8=4 - all different values, yet λ remains unity. On occasions, this kind of juggling of areas is sometimes useful in minimizing errors arising from ohmic resistance. It will also be apparent that very large scaling factors - many thousands - can be realized from a combination of practically-achievable areas.

2.2.4 V_{BE} Mismatch

Frequently, unintentional errors in emitter area ratios (most usually referred to as "V_{BE} mismatch" when this ratio is supposed to be unity) occur in the implementation. These arise from local variations in junction doping and in the photolithographic delineation of the emitter opening. They can also be caused by thermal gradients on the chip. Since V_{BE} varies by about - 2mV/°C, and 2mV is roughly equivalent to an emitter area ratio of 1.08 (that is, exp(2/26) at 300K) only a small variation in temperature can cause significant errors in TL circuits. Even well-executed designs have been known to fall prey to this problem, particularly when a power-driving

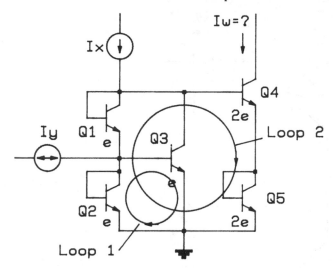

Figure 2.7   Vector difference circuit of Figure 2.4b partitioned into two TL loops for analysis; emitter areas are indicated by 'e', '2e', etc.

output amplifier is included on the monolithic chip.

Not only can heat from the output stage cause a fixed thermal gradient which disrupts operation of a TL "core" and can in many cases directly introduce distortion, but fluctuation in this gradient as the output varies can further degrade performance. This effect is widely known in op-amp practice, where it is referred to as "thermal feedback". It can lower the low-frequency open-loop gain and cause serious distortion. For this reason, high-accuracy monolithic circuits invariably use a highly symmetrical layout and often arrange critical pairs of transistors as cross-connected quads. This practice is even more important in TL circuits. For example, in the well-known six-transistor multiplier, a residual V_{BE} mismatch of only 5μV will result in about 0.01% parabolic nonlinearity (generating even-order distortion). Finally, stresses in the silicon can cause the V_{BE} of adjacent transistors to differ. These stresses can arise from inadequate assembly techniques, or from thermal effects.

It is for these reasons that one occasionally meets with some skepticism about the accuracy potential of TL circuits. That the practical problems can be surmounted is evidenced by the many high-precision circuits which are now commercially available. The latest monolithic analog multipliers, for example, provide total harmonic distortion levels of -80dB or better, compared to the -40dB of the earliest attempts at using these ideas.

2.2.5   *Some Examples of TL Analysis; Multiple Loops*

We are now ready to solve the two circuit problems posed earlier, beginning with the mysterious circuit, presented in the introduction, and redrawn in Figure 2.7. It is apparent that there are actually *two* overlapping TL loops.

So far we have assumed that the TL network contains only a single loop. In fact, many circuits involve two or more overlapping loops. The theory does not require any modification to address this possibility. It is only necessary to use the appropriate value of current for the device or devices contained in more than one loop.

The minor loop (1) is trivial; it can be seen from inspection that the currents in Q2 and Q3 are essentially the same; it is a current mirror. (At this stage, we are not ready to consider second-order effects, which, in this case, are predominantly the increased collector current in Q3, arising from having a higher collector voltage than Q2, and the errors due to finite beta. These are discussed later.)

Applying TLP to Loop 2

$$\frac{I_{C4}}{2} \frac{I_{C5}}{2} = IC_1 IC_2$$

$$\text{(CW)} \quad \text{(CCW)}$$

Now,

$$I_{C2} = Iy + I_{C1} \quad \text{and} \quad I_{C1} = Ix - I_{C3} \quad (\text{or } Ix - I_{C2}),$$

so

$$2I_{C2} = Ix + Iy$$

$$I_{C1}I_{C2} = \left\{ 1x - \frac{Ix + Iy}{2} \right\} \left\{ \frac{Ix + Iy}{2} \right\}$$

$$= \frac{Ix^2 - Iy^2}{4}$$

But

$$I_{C4}I_{C5} \quad (= Iw^2) = 4 I_{C1}I_{C2}$$

and therefore

$$Iw = \sqrt{(Ix^2 - Iy^2)}$$

By the way, this vector-difference example, like many TL circuits, arose out of the sort of "what-if" approach mentioned earlier. Its practical value is somewhat limited by the fact Ix must be unipolar, and vector operations often require both inputs to have either sign. Also, Ix in the negative direction must not exceed Iy (it can in the positive direction). In spite of these shortcomings, the circuit is useful, and at this point provides a good example of the ease with which the function can be rapidly ascertained by application of TLP. Later, we will look at some other vector-operation circuits of more practical value.   Note here that the subscript "w" will be reserved for outputs.

The second example posed earlier was that of the diode bridge. We shall assume here that the four diodes have equal geometry and behave ideally, having a junction voltage related to forward current by the classical PN junction equation (Eq. 2.1). In fact, two-terminal diodes do not behave so nicely: they have considerable series resistance. Diode-connected transistors, on the other hand, neatly circumvent that problem, since most of their resistance arises in the base circuit, where only a small fraction of the forward current (I_F/β) flows.

We will make use of the bridge example to introduce another idea which finds wide utility in TL circuits, namely, the use of a temporary dimensionless variable, in the range of 0 to 1, or alternatively -1 to +1, to represent some unknown currents. This variable can be viewed as a "modulation index" acting on a bias current; the convention used throughout this Chapter is that lower-case variables lie in the range 0 to 1 while upper-case variables lie in the range -1 to +1. While on the topic of conventions, note that collector, base and emitter currents will be denoted by the use of upper-case-subscripted variables, I_C, I_B and I_E respectively, while bias and signal variables will be denoted by non-subscripted lower-case modifiers, such as Ix, Ia, Ie, I2, and so on.

In Figure 2.8a the variable x temporarily describes the current in the two upper diodes, which sum to Ib. The value of the currents in the lower diodes follow directly.

Applying TLP

$$(1 - x)\, Ib\, \{\ Ic + (1 - x)\, Ib\} \ = \ x\, Ib\, (\, Ia + Ib\,)$$

$$\text{(CW)} \qquad\qquad\qquad \text{(CCW)}$$

(From now on, the reader will not be reminded of which current products are clockwise and which are counterclockwise. However, we will continue to use the convention that CW is on the LHS of the equations and CCW is on the RHS.)

Solving the loop equation

$$x = \frac{Ib + Ic}{Ia + 2Ib + Ic}$$

The complete solution for the diode currents I1, I2, I3 and I4 follows directly.   We can check the result with a specific example, using Ia=1mA, Ib=2mA and Ic=3mA. The obtained solution is shown in Figure 2.8b: if correct, the product of the CW currents should equal the product of the counterclockwise currents. It does: it is 2.8125 mA^2 in both directions.

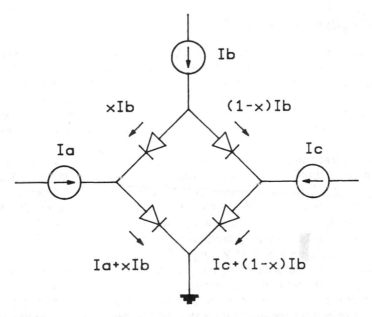

Figure 2.8a The approach to the diode bridge problem: introduce a temporary unknown *x* then apply the TL principle to the loop.

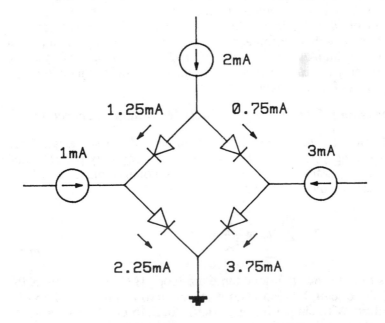

Figure 2.8b The solution when Ia=1mA, Ib=2mA and Ic=3mA; note that the ratio 1.25mA/0.75mA is equal to the ratio 3.75mA/Figure 2.25mA.

2.2.6   The Ratiometric Viewpoint

Now, the concept of variables embedded in chained products of current, with dimension of $(Amps)^n$, where n is the number of junctions in either direction, may be a bit disconcerting. An alternative, and probably more satisfying, way of thinking about TL loop behavior is in terms of the independent current *ratios* between junctions taken in opposing pairs; these are always dimensionless and often within an order of magnitude of unity (0.1 to 10).

Indeed, the great value of TL circuits stems from this *ratiometric* behavior, which, it should be stressed, is fundamentally independent of the general magnitude of the bias currents (see Figure 2.3), or the process on which the devices are made (with appropriate reservations about matching device structures), or the operating temperature (since the V_T factors canceled in Eq 2.4 and the basic I_S factors in Eq. 2.6). This expectation has been proven to be reliable over a wide range of circumstances.

This viewpoint does not require a change in formulation, but only affects the way in which the circuit is understood. Thus, in the example just analyzed, rather than verifying the result by calculating the total product in CW and CCW directions we might consider the current ratios in the upper and lower pairs. These ratios must be the same, and while the bridge example may be somewhat prosaic, we shall later see a striking similarity in the TL form of the most common multiplier cells, where a similar consideration of ratios is invaluable in our approach to these circuits. Following through with this approach, we find that both the upper and lower ratios are 5/2 (1.25mA/0.75mA and 3.75mA/2.25mA respectively). The reader will quickly realize the value of identifying opposing junctions and thinking in terms of their current ratios.

2.2.7   Extension of TL Theory to Include Voltage Generators

In basic TL circuits, the ratiometric behavior just discussed arises from setting the net loop voltage to zero. However, it is certainly possible, and occasionally even useful, to include one or more voltage sources in a loop, when the governing equation becomes

$$\sum_{k=1}^{k=n} V_T \ln \frac{I_{Ck}}{I_{Sk}} = V_L$$

(2.4b)

where V_L is the net voltage inserted into the loop. This radically alters the behavior of the circuit, but can readily be cast into a form that shows its potential utility. With simple manipulation, Eq. 2.4b can be shown to be equivalent to the modification of the emitter-area-factor, λ, to

$$\lambda' = \lambda \exp(V_L/V_T)$$

(2.12)

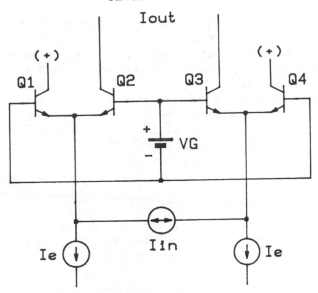

Figure 2.9 A common AGC cell can be viewed as a translinear network; it also depends on the predictable logarithmic junction law.

It follows that if a temperature-stable effect is desired, V_L must be PTAT (proportional to absolute temperature). One use of such a voltage is to null out the "V_{BE} mismatch" in a TL loop where this would degrade accuracy. Modern analog multipliers [5] use this technique to achieve much lower even-order distortion than obtainable by other techniques, which include the careful cross-quadding of critical pairs, and in some cases taking this to a much deeper level of interdigitation. Both approaches have their merits. The modified-λ approach, of course, requires the use of some form of active trimming technique, as well as the means to measure distortion during this step of manufacture. The "correct-as-fabricated" approach (using heavy interdigitation of transistors to minimize statistical fluctuations in the effective value of λ) has the advantage of requiring no trimming (at least, to maintain low distortion and offsets in a multiplier) and is much more immune to on-chip thermal gradients and stress-induced errors (particularly those due to post-packaging stress). On the other hand, the latter approach consumes much more chip area, often requires the use of double-level metal to maintain the required low interconnection resistance, and may result in reduced bandwidth, due to the much higher parasitic capacitances and lowered current densities.

Incidentally, the fact that the trimming voltage may often be quite small (V_{BE} mismatches are typically less than $250\mu V$ in a modern process, using appropriate layout techniques, which is just another way of saying that the emitter area ratio of critical pairs can be held to within ±1%) it is often possible to use a short length of metallic interconnect as an approximately-PTAT resistor, thus generating the needed correction voltage when driven by a temperature-stable current. Aluminum, the most commonly-used interconnection material, has a temperature coefficient of about 3900ppm/°C, which makes it "almost right"; in fact, a graph of resistance versus temperature shows an almost linear form which extrapolates to zero at a temperature of about 30K.

While beyond the scope of this short treatment, it should be mentioned that there are many general translinear circuits in which the inclusion of one or more voltage generators is essential to their operation. By way of illustration, Figure 2.9 shows a widely-used automatic gain-control cell, which can be viewed as two translinear loops: Q1, Q2 and V_G on the left, and Q3, Q4 and V_G on the right. For $V_G=0$ (and assuming equal emitter areas) the factor λ is 1, so half of the input current Iin appears at the output collectors, that is, the current-gain is -6dB. As V_G is varied, the effective λ is also varied (Eq. 2.12) and the gain can be raised (though only by 6dB at most, being about -0.1dB when V_G is +115mV at an operating temperature of 300K) or lowered. In general, the current-gain is given by

$$\frac{\text{Iout}}{\text{Iin}} = \frac{\exp{(V_G/V_T)}}{\exp{(V_G/V_T)} + 1}$$

Clearly, for large negative values of V_G (when $exp\ (V_G/V_T) \ll 1$) the loss in decibels becomes an almost linear function of V_G:

$$\text{Loss (dB)} \approx -\ 8686\ V_G\ /\ V_T$$

This is within 1dB of the exact result for $V_G/V_T \geq 2$. However, a precise relationship between gain and V_G is rarely important in AGC circuits, which generally operate in a closed-loop fashion to restore the final detected output to some desired reference level.

The general field of translinear circuits - viewed as all circuits in which the basic "translinearity" of a BJT is invoked, to varying degrees of precision is, of course, immense. Since it is not possible to discuss these in any depth here, most of the examples in this Chapter will be concerned with "strict TL" circuits, that is, ones in which no additional voltage sources are introduced. However, an occasional "TN" circuit will be included to illustrate the wider utility of translinear approach to BJT design.

2.2.8 *Effect of Other Device Non-Idealities*

We have already dealt with the most troublesome of the non-ideal aspects of the BJT in TL circuits, namely, V_{BE} mismatch. Before leaving the subject, it needs to be stressed that this popular term is useful in considering such behavioral aspects as op-amp input offset, where the two transistors in a typical BJT differential pair are carefully arranged to operate at the same current, but can be misleading in TL analysis. In the first place, pairs of devices may never operate at the same current. Secondly, it will by now be obvious that the actual V_{BE}'s are only of incidental interest in determining the overall function: both their absolute value and their "matching" will vary greatly with temperature, while the circuit behavior will be stably "locked-in" by the area-ratio factor, λ (or factors, in multiple-loop circuits).

It is well-known that V_{BE} is also affected by the collector-base voltage, V_{BC}, through base-width modulation (Early effect) and is increased by the finite ohmic resistances, most notably, the effective base resistance, which we will here just call R_B, which, for simplicity, will be blamed for all effects that are ohmic in nature and cause V_{BE} to be higher than predicted by Eq. 2.1b.

Consider first the effect of V_{BC}. The value of V_{BE} predicted by Eq. 2.1b usually assumes a V_{BC} of zero. (Incidentally, it need not: one can alter the definition of I_S to reflect any chosen V_{BC}, but any value other than zero would be arbitrary.) If we include the effect of base-width modulation, it is readily shown that the collector current is increased by the factor $(1 + V_{BC}/V_{AF})$, where V_{AF} is the forward Early voltage of the device. Thus, the basic equations need to be amended, to read

$$I_C = (1 + V_{BC} / V_{AF}) I_S(T) \, exp \, (V_{BE} / V_T) \qquad (2.1c)$$

and

$$V_{Be} = V_T \ln \frac{I_C}{I_S(T)(1 + V_{BC}/A_{AF})} \qquad (2.1d)$$

Now, for many high-frequency transistors, V_{AF} may be fairly low (in the range 5-50V) and the effect of collector bias voltage on TL circuits using such devices can be very noticeable. For example, even at the upper end of this range, a 1V "error" in the choice of collector bias could change I_C by 2%, and correspondingly more for devices at the lower end of the range.

However, this problem is not as severe as it might at first seem. First, devices are often biased in pairs, and collector-biasing effects cancel (remember, TL circuits are usually only sensitive to current ratios, not the absolute currents). Second, V_{AF} is essentially independent of temperature, so removing a possible source of operating variability. Third, it is usually fairly easy to find a place to connect the collector to minimize this type of error. In some cases, this might require the use of a cascode stage (when the collector is delivering an output) or a specially-provided bias line designed to keep V_{CB} at or near zero. This is not always true, and in circuits designed to provide the highest precision the choice of the correct collector biasing can be critical. For example, in a circuit comprising a mix of diode-connected devices (whose V_{CB} is always zero) and "active" devices (whose collector bias may be fixed, but whose base voltage varies with the signal, hence whose V_{CB} is a function of signal) distortion in amplifier or multiplier applications can arise. Sometimes, this problem can be dealt with by the addition of diodes in the collectors; their forward voltage closely tracks the V_{BE} of the transistor, thus keeping V_{CB} constant.

Note that in the examples presented here, collectors are sometimes shown as being connected to (+). This merely means "find a good place to anchor this collector". It does not mean "tie this collector to the positive supply voltage". It is unusual in strict-TL circuits to find the full supply voltage across devices.

Errors due to finite beta occur frequently in TL circuits, because the base current for one device must "rob" a driving source of some current to some other device. The simplest such example is in a standard current mirror, where the output device uses some of the input current, the rest of which biases the diode-connected device. Furthermore, the finite beta corresponds to a finite alpha, so that a current may be accurately established in the emitter circuit only to lose one or two percent of its value by the time it gets to the collector.

To simplify the presentation and the analyses, we will here often choose to ignore base current. Indeed, the ideal device for TL applications (and, one suspects, *all* applications!) would have the I_C/V_{BE} characteristics of the BJT and the zero-gate-current properties of the MOS FET. Such devices may one day be possible, using heterojunction techniques; even BJT's made with polysilicon emitters can provide greatly enhanced beta (and/or increased Early voltage) without sacrificing base resistance, although the concomitant increase in emitter contact resistance mitigates against their use in TL circuits.

In practice, many measures can be taken to deal with finite beta. One can, for example, insert driver stages (such as might be used to eliminate the same problem in the current mirror) or use base-current-cancellation techniques. In some fortuitous cases, base currents in a pair of output devices can be made to have the same *ratio* as the currents in the devices driving the nodes, resulting in essentially no errors (until the drive currents are eventually entirely absorbed by the output devices); an example of this will be provided. A further complication in the analysis of beta-related errors arises from the dependence of beta on temperature and on collector current. Fortunately, the latter is most serious at levels of current density where anomalies in V_{BE} arising from high-level injection pose a more serious threat to accuracy. A high quality BJT will exhibit a very flat beta over many decades of I_C.

Base resistance is often found to be the primary limitation to accuracy, since it does not introduce simple factors into the equations as we've just seen for base-width modulation. Rather, it results in a V_{BE} which is now a mixture of linear and logarithmic terms:

$$V_{BE} = V_T \, ln \, \{I_C/I_S(T)\} + I_C R_B/\beta \qquad (2.1e)$$

where β is the appropriate value for the current-gain, in the sense that it is a function of I_C as just noted, and it, too, is a function of V_{CB}; it is also a function of frequency. As evidence of the difficulties which R_B promises to introduce into TL analysis, note that we can no longer write a closed-form equation for $I_C(V_{BE})$.

The practical consequences of base resistance are quite capricious. For example, in four-transistor two-quadrant multiplier ("variable-gain") cells, R_B is a considerable nuisance, introducing odd-order distortion which varies with gain: it can be made to vanish at one value of gain, or sometimes at two (by judicious use of emitter area sizing, which also affects R_B) but is hard to eliminate completely. On the other hand, a close relative, the six-transistor four-quadrant multiplier cell can be designed to exhibit essentially zero odd-order distortion even in the presence of very large ohmic errors. Indeed, one can insert resistors in every emitter of such a multiplier, having voltage drops of several times V_T across them, without noticeably affecting distortion [6].

The subject of R_B-related errors is therefore not too readily quantified in general terms. It is complicated by the fact that its effective value is current-dependent (emitter crowding lowers R_B at high currents) and temperature-dependent (it typically increases by about 0.15%/°C). Analysis is complicated further by the role of beta, and *its* current- and temperature-dependence. We will endeavor to address the likely effects of base resistance in those cases where some clear solution can be suggested (as, for example, in the case of the RMS-DC converter core design).

It is likewise difficult to make general comments about the dynamic behavior of TL circuits, controlled by the forward base transit time τ_F and the junction capacitances, C_{JE}, C_{JC} and C_{JS}. Some apparently simple circuits (such as the translinear cross-quad) can exhibit exotic AC response under apparently benign operating circumstances. The reason for this behavior is never hard to identify (often much harder to fix) but it would be going well beyond the intent of this introductory material to attempt to include a full discussion of dynamic effects.

However, one point is worth making in this regard. The voltage swings in TL circuits are always quite small. They are "delta-V_{BE}'s", caused by the varying current ratios. Since these ratios are generally less than 10 for TL circuits intended for high speed use (such as current-mode amplifiers and multipliers) the signal voltages are rarely more than a few tens of millivolts. For example, in a multiplier driven to a full-scale (FS) modulation index of X=0.75 (a typical situation) the FS differential voltage across base pairs can be shown to be

$$\Delta V_{BE} = V_T \ln \frac{1 + X}{1 - X} = V_T \ln 7 = 50 \text{mV at } 300 \text{K}$$

It follows that about 1/40th as much displacement current flows in the capacitances as would for a typical 2V signal in a voltage-mode circuit. It is for this reason that very fast TL circuits can be made on standard junction-isolated processes, having high collector-substrate capacitance. Put another way, TL circuits are low-impedance circuits. In fact, they operate at fundamentally the lowest impedance possible in a bipolar circuit operating at a given bias level: $r_E = V_T/I_C$. If one wishes, this impedance, and hence any poles formed by it and the device capacitances, can be stabilized over temperature by choosing to make the bias currents PTAT. Usually this is not only unnecessary (since circuit speed is dominated by τ_F in most cases), but undesirable, since one of the benefits of TL circuits is that they can be operated close to their peak dynamic range without distortion, and it is therefore customary to choose bias currents which are only slightly more than the FS signal levels (and thus which should not vary with temperature). Also, in many cases, the operating currents act as scaling quantities (for example, the denominator current in a multiplier) and once again it is desirable to use temperature-stable currents.

The very small voltage swings in TL circuits also results in essentially complete freedom from the slew-rate limitations which often arise in voltage-mode circuits, when a node capacitance has insufficient current to charge it during a large-signal event. This is one of the reasons for wishing to process signals in the current-domain. If slew-rate limitations were to arise in a TL circuit, it would require the dual situation: a current would be limited by the branch *inductance* and the limited voltage available to alter this current. This may become a problem in microwave TL circuits, but to the author's knowledge it has not yet manifested itself.

2.3 Sqaring And Square-rooting

Squaring and square-rooting are basic functions in the amplitude domain, the need for which arises quite often. They are essential to the extraction of the exact RMS value of a signal, and in power measurement in general. It is possible to devise TN circuits, involving voltage-in current-out signals, to approximate the squaring function in two quadrants (that is, the output is of the same sign for either input). Figure 2.10 shows a circuit which provides an effective solution for input voltages Vin of up to ±150mV peak; larger inputs are accommodated by simply adding a resistor in series with the input. The use of the resistor divider formed by the two base resistors, R, results in one of the two outer transistors to conduct more heavily whether Vin swings either positive or negative. The form of the resulting collector currents is approximately a hyperbolic cosine. The output amplitude can be shown to be maximal (49.23% of the tail current Ie) when the area ratio A=6. However, the function accuracy at this input level is improved using A=10, when the output amplitude is still about 47% of Ie. The standing current in Q2 for Vin=0 is withdrawn by the collector bias current Ic. Using Ie=1.06mA and Ic=880μA, the output is zero for Vin=0 and 500μA for Vin=±150mV. The

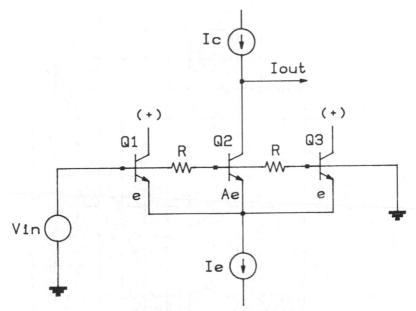

Figure 2.10 A simple signal squaring circuit; this is also translinear but not "TL", since there is no closed loop of junctions.

load circuit would preferably be set about a V_{BE} above ground, but because of the small voltage swings all collectors can be grounded. (In many AC applications, the signal could be capacitively-coupled to the bases, which could then be held slightly below ground.)

An initial disadvantage of this circuit in applications requiring accurate scaling is that the "detector efficiency" decreases with temperature because Vin is "scaled" by the thermal voltage V_T. This can be remedied by the use of aluminum resistors for R, to form the lower leg of a temperature-dependent attenuator, which converts the stable input voltage to one which is PTAT, and thus cancels the temperature-dependence of V_T. It can be shown that the required ratio for near-zero TC (recall, aluminum has a TC of slightly more than PTAT) is conveniently close to 10. If we choose to make R=2.5Ω (fairly readily done using standard metallization) and the upper leg of the attenuator 45Ω, we present a 50Ω overall input impedance and a peak input capacity of 1V RMS (+10dBm). Alternatively, an approximate solution to the scaling issue is to make Ie and Ic PTAT. This does not work in the same fundamentally-correct way as the PTAT attenuator, but for some applications, particularly over the narrower commercial temperature range of -25°C to +85°C, the compensation may be adequate. Figure 2.11 shows simulated results for the wider temperature range of -55°C to +125°C; the central curve is at 35°C. It can be seen that a tolerably-close fit to the square law is maintained and the output amplitude varies by about ±10% over this wide temperature range. Better amplitude compensation requires that Ie and Ic are "super-PTAT", that is, vary more rapidly with temperature.

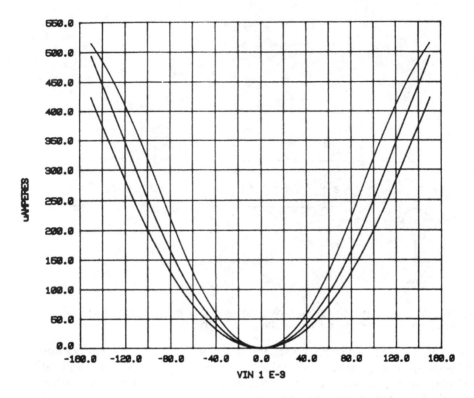

Figure 2.11 Output of the TN signal-squarer at -55°C (upper curve), +35°C and +125°C (lower) using PTAT tail current to stabilize amplitude.

2.3.1 Scaling

We cannot proceed very far in TL or TN design without confronting questions of scaling, as we've just seen. All nonlinear circuits raise this issue, since whenever a signal is squared, or two signals are multiplied, or some other strongly nonlinear operation is performed, the output has a higher dimension than the inputs, and must be restored to simple "Amps" or "Volts". In the case of the two-quadrant squaring circuit just described, we were a little bit casual about this issue: it came out the way it did through a sort of heuristic often employed in nonlinear design. This approach is particularly evident in the design of wide-band log-amps based on successive compression, or amplifier/limiter stages. Traditionally, these have been viewed as just a special kind of amplifier, but are more exactly defined by a nonlinear function of the form

$$Vout = Vy \ \log \frac{Vin}{Vx}$$

$$(2.13)$$

In fact, if it were not for the familiarity of the term "log-amp" all of the circuits performing this function ought more correctly to be called "logarithmic converters", to stress the importance of the scaling attributes. Notice in Eq 2.13 that there are two scaling voltages, Vx and Vy. The first is necessary because one cannot take the logarithm of a voltage, only a ratio. When Vin=Vx the output of the log-converter will be zero, so Vx is called the "intercept voltage" or sometimes (less precisely) the "log offset". The change in Vout for a given ratio change in Vin is controlled by the second voltage, which is called the "log slope"; in RF log-amps this may be defined in terms of so many Volts per decibel.

Now, if one examines the design of many extant log-amps, or consults the literature, it is clear that no direct path can be traced from the specified "log-offset" and "log-slope" to some identifiable *reference voltage* in the circuit which defines Vx and Vy. Because of the lack of rigor in this regard, these products have notoriously poor temperature stability and overall accuracy. Other questions arise when one gets serious about scaling. How can a log-amp accept bipolar input signals (usually, sinusoidal), since the logarithm of a negative argument does not have a real value? How does input waveform affect the behavior of a log-amp? This question arises because a demodulating amplifier (the kind most often used in RF and IF applications) includes an averaging filter.

This digression is only included here to point out the need for great diligence in pursuing the matter of scaling, not only in the more obvious nonlinear functions but in *every* case. To illustrate how one can take steps to put the scaling on a firmer foundation, let's take a closer look at the squarer just discussed. Its input is a voltage, but the output is a current. The complete function might then be written something like this:

$$\text{Iout} = \frac{\text{Vin}^2}{\text{Vref}} \frac{1}{\text{Rref}}$$

(2.14)

This certainly has the right dimensions, but where do Vref (an implied voltage reference) and Rref (a scaling resistance) come from? The answer is quite subtle. We can guess that Vref is actually traceable to the thermal voltage, V_T. We've already suggested the steps that might be taken to eliminate the temperature-proportional behavior of V_T (by using a PTAT attenuator), so this seems reasonable. Finding Rref is harder. Of course, it cannot be the voltage-dividing resistors, R; they simply form a ratio and can have any value over a wide range without affecting scaling. It must be traced to the resistor used in establishing the bias current, Ie, to which the output is clearly proportional. But if we did that, we must account for yet another voltage variable, the reference which sets up the bias currents in these resistors, and how could that fit into the equation?

The mistake in this case was to cast the transfer function in its most basic form, solely to achieve dimensional consistency. But a more careful consideration of the nature of the "V_T scaling" shows that we should have written

$$\text{Iout} = \left\{\frac{\text{Vin}}{V_T}\right\}^2 \frac{\text{Vref}}{\text{Rref}}$$

(2.14a)

Thus formulated, it is obvious that Vref/Rref is just the effective scaling current, generated by some reference (maybe an on-chip bandgap reference, or just some fraction of the supply voltage) and a resistor (probably in the emitter of some current-source transistor). Incidentally, it is worth noting that there are few, if any, monolithic elements available to generate an exact *current* reference; thus, it is necessary to first identify a voltage reference and divide this by a resistor. It follows immediately that the exact scaling of all voltage-in/current-out (that is, transconductance) elements requires the use of temperature-stable resistors.

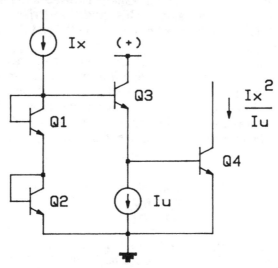

Figure 2.12 Simple one-quadrant TL squarer/divider; this circuit requires both the numerator and denominator inputs to be unipolar.

2.3.2 TL Squaring Circuits

The simplest one-quadrant squaring circuit in TL uses four transistors. It is shown in Figure 2.12. The TLP analysis is trivial: IbI4 = IaIa, so

$$\text{I4} = \text{Ia}^2/\text{Ib}. \ldots$$

. (2.15)

In the case of the TL squarer, the scaling is much more obvious: it is simply the supplied current Ib. Note that by fixing Ia we can implement a one-quadrant divider, with Ib as the input. This illustrates the fact that every "squarer" is actually a "squarer/divider" and this can be put to good use, as we shall see, in RMS-DC converters which make use of "implicit computation". The circuit of Figure 2.12 needs quite a lot of support circuitry, to convert input voltages, the most common currency in signal-processing systems, to currents, and in converting the output current back into the voltage domain. It is also prone to errors due to finite beta, particularly that of Q3 and Q4.

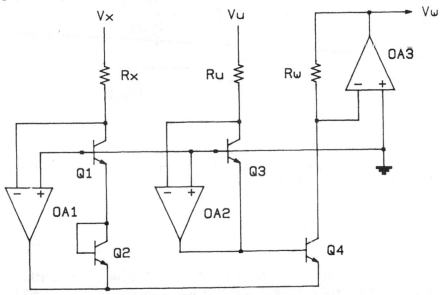

Figure 2.13 Essentially the same circuit as Figure 2.12, but having improved accuracy due to the use of op-amp to force collector currents.

By adding three op-amps, an accurate squarer can be realized (Figure 2.13). The improved accuracy arises from these features:

1) All devices operate at zero VCB.
2) The op-amps force collector currents in the input transistors (not emitter currents as in the circuit of Figure 2.12).
3) The op-amps also serve to provide very linear voltage-to-current conversion from inputs Vx and Vy and accurate current-to-voltage back to Vw.

Using devices of equal area (assumed throughout this Chapter unless otherwise stated) and applying a similar TLP analysis to that used in the last example, the transfer function is readily found to be

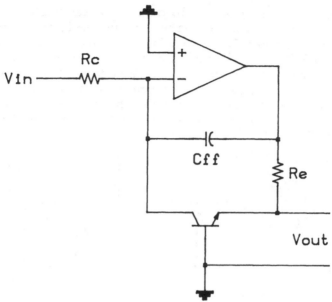

Figure 2.14 Method used to prevent HF instability when using a standard op-amps to force I_C ; this circuit can also be viewed as a log-amp.

$$Vw = \frac{Vx^2}{Vu} \frac{RuRw}{Rx^2}$$

(2.16)

This circuit is not quite practical as shown, at least, using off-the-shelf op-amps, which are invariably designed to be stable at closed-loop gains down to unity, that is, when the feedback from the output to the inverting input is 100%. In this case, however, there is further *voltage* gain supplied by the "grounded-base" transistors and their "load resistors", that is, by Q1/Q2 operating into Rx, and Q3 operating into Ru. This gain can be very large: we can quickly see in the case of the Ru/Q3 combination that it is Ru divided by the r_E of Q3, which in turn is V_T/Iu, or $V_T/(Vu/Ru)$. The feedback gain around OA2, therefore, is just Vu/V_T, which greatly exceeds unity for most practical values of Vu, thus destabilizing this op-amp. Note that a further problem arises in this regard due to the pole formed by Ru and the parasitic capacitance at the noninverting input of the op-amp.

The solution is simple: it is to insert an AC path around the transistor which prevents the feedback gain from becoming significantly greater than unity at any frequency. To effectively steer current at high frequencies away from the transistor a further resistor is added, as shown in Figure 2.14. The exact values depend on operating conditions. Re will typically be made about equal to Rc, so that the op-amp output will swing as far negatively as the input goes positive; Cff will be chosen to have a corner frequency with Re of no higher than one-fourth the unity-gain frequency of the op-amp. Values of Re=10kΩ and Cff=100pF are commonly used with 1MHz op-amps. Of course, in a fully integrated design, these op-amps would be specifically designed to support this rather special set of requirements.

Incidentally, this fragment of the squarer/divider turns out to be another familiar TN circuit, one which provides an output which is a logarithmic function of its input. It needs further refinement to be useful (both of the temperature-dependent factors V_T and $I_S(T)$ appear in its transfer function) and we will not pursue the matter in this Chapter. We should also note that "current-feedback" (transimpedance or "TZ") amplifiers (see Chapter 16) are particularly well suited to this application. Since they respond to the current at their (virtual short-circuit) non-inverting input node (rather than to the voltage at the open-circuit input of a classical op-amp) no gain results from the inclusion of the transistor in the feedback path. Re is still needed to form the dominant pole of the closed-loop response, as is always the case for TZ amplifiers, but Cff can be omitted. The result of this is not simply the saving of a cheap capacitor. The bandwidth (either as a stand-alone logarithmic converter or as part of a squarer/divider) is considerably extended, because the signal current continues to flow in the transistor (and is not shunted by Cff) up to the full bandwidth of the op-amp. Typical TZ op-amps provide bandwidths of from 30MHz to 200MHz.

2.3.3 RMS-DC Conversion

The squarer/divider can be used as the basis of an RMS-DC converter. Figure 2.15 shows how this is done. First, since the TL core operates in only one quadrant, the absolute value of the input Vx is generated. This is not difficult to implement; however, the use of such a circuit reduces the bandwidth potential of the core, as all signal currents drop to zero twice every cycle of a periodic input (except in the special case of a squarewave input having fast transitions, in which case the absolute value is constant if it is amplitude-symmetric, or which fluctuates between two non-zero values if not). The change in base-emitter voltage between the peak signal current and "zero" current is relatively large, and introduces unwelcome consequences into the HF accuracy. Nevertheless, this principle was successfully used in the first monolithic RMS to be commercially available (Analog Devices AD536), which has become a classic of its kind.

The RMS algorithm is implemented by an "implicit" technique. Rather than explicitly generating Vx^2 in one circuit, performing the low-pass filtering and then extracting the square-root in a second circuit, the output Vw of a squarer/divider is averaged (by the low-pass filter formed by the C and R associated with OA3) and then returned to provide the denominator (formerly Vu, in Figure 2.13). Thus,

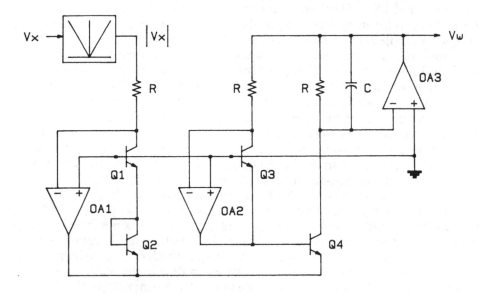

Figure 2.15   RMS-DC converter based on the squarer/divider; an absolute-value circuit provides the needed unipolar input; the output of the low-pass filter (Vw) is returned to the denominator input.

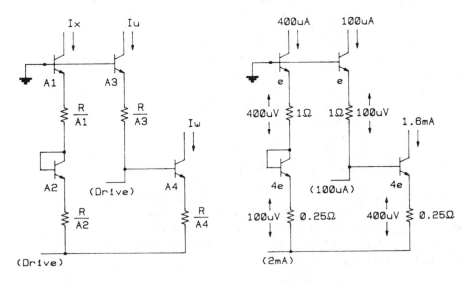

Figure 2.16   Figures for the analysis of errors in the RMS-DC converter arising from finite ohmic resistance, lumped into the emitter branches.

$$Vw = \frac{\text{Ave}\,(Vx^2)}{Vw}$$

So

$$Vw = \sqrt{\text{Ave}\,(Vx^2)}$$

This is not the place to discuss many of the interesting details in the design of an accurate RMS converter, but it does provide a good opportunity to examine how finite ohmic resistances typically affect accuracy in a TL circuit and how the errors may be reduced, in this case, by optimal choice of device geometries, which we will now do.

Figure 2.16a shows just the core and includes the ohmic resistances in the emitters, which, if beta is assumed to be independent of collector current, would just be R_B/β. While we may not know (or need to know) the actual value for a "unit" device, we can safely assume that it will scale in inverse proportion to emitter area, particularly if, as has been recommended earlier, the total area of any one transistor is formed by the use of several identical base-emitter regions connected in parallel. Of course, in practice, meticulous attention to balancing (or correctly ratioing) metallization resistances will also be needed.

In an RMS converter, the input (here Ix) can have peak values much greater than the RMS value of the signal; the ratio of these two quantities is called the "crest factor", denoted here by σ, which may have values of ten or even more. A DC or amplitude-symmetric squarewave input has a crest factor of 1, a sinusoid $\sqrt{2}$, and so on [7]. At high values of σ, the voltage errors due to the emitter resistances will cause the V_{BE} of Q1 and Q2 to be too high. On the other hand, Iu (being the quasi-DC RMS value) will be the smallest of all the currents in the core, and ohmic errors in Q3 will tend to be negligible, while the errors in the V_{BE} of Q4 will (in a core using all devices of the same geometry) be the largest, since its current peaks in proportion to σ^2, which might therefore be as much as 100 times the current in Q3.

It seems likely, from inspection of this situation, that some improvement might be possible by altering the emitter areas to cope with these conditions. In choosing these areas we would probably seek to maintain an overall scaling factor of unity, since there is no special merit to choosing any other scale. We might, for example, choose to make Q3 the smallest transistor (a unit emitter), and Q4 the largest, having ζ^2 unit emitters, where ζ is some high (but not necessarily the highest) value of crest factor to be handled by the circuit, say, 4. Q1 and Q2 would then have some intermediate size; to preserve unity-scaling, they would have an area of ζ units. In fact, that is not the optimal solution, as the following analysis shows.

Before proceeding, it will be helpful to note that the most severe test of accuracy at high crest factors for an RMS converter occurs when the input is a narrow unipolar pulse, since Q1/Q2 and Q4 then operate at the highest peak currents. As well as being the worst-case, it is also the simplest case to analyze, because *all* the error accrues during this high-current interval, that is, we do not have the complicating factor of post-core averaging, which "smears" the net error contribution of the quasi-DC output over an interval of variable error in the core. Furthermore, it is possible to generate very accurate pulses for test purposes, having exactly-known amplitude (referenced tightly to a precise DC voltage) and duty-cycle (by digital time-division), and an exact zero-value base-line.

To begin the analysis, we can write the equation for the *excess voltage* in the TL loop, that is, the voltage due just to ohmic effects:

$$\delta V = Ix \left\{ \frac{R}{A_1} + \frac{R}{A_2} \right\} - Iu \left\{ \frac{R}{A_3} \right\} - Iw \left\{ \frac{R}{A_4} \right\}$$

(2.17)

In the ideal case, of course, δV would always be zero, because all resistances would be zero. We desire here to find a solution which would make it zero *both for DC inputs and at some high crest value*, even when $R \ne 0$. Now, under dynamic conditions, the peak value of Ix at some general value of crest factor is σIu (from the definition of crest factor) and the peak value of Iw is $\sigma^2 Iu$ (from the basic TL core equations). Extracting IuR as a common factor, we get

$$\delta V = IuR \left\{ \frac{\sigma}{A_1} + \frac{\sigma}{A_2} - \frac{1}{A_3} - \frac{\sigma^2}{A_4} \right\}$$

(2.18)

For DC inputs $\sigma=1$, and Eq. 2.18 can be written

$$\delta V = IuR \left\{ \frac{A_1 + A_2}{A_1 A_2} - \frac{A_3 + A_4}{A_3 A_4} \right\}$$

(2.19)

This requires both that $A_1 A_2 = A_3 A_4$ and $A_1 + A_2 = A_3 + A_4$. Normalizing all areas to that of Q3, we can set $A_1 + A_3 = 1$ and $A_2 + A_4 = A$, which satisfies the unity-scaling requirement as well as making δV zero at DC. The excess voltage at general values of σ is then (from Eq. 2.18)

$$\delta V = IuR \left\{ \sigma(1 + \frac{1}{A}) - 1 - \frac{\sigma^2}{A} \right\}$$

(2.20)

This simple parabolic function has a second zero at $\sigma = A$, and a local maximum at $\sigma = (1+A)/2$, when $\delta V = IuR \{ (1+A)^2/4A - 1 \}$, becoming rapidly larger (negatively) above $\sigma = A$.

This is illustrated in Figure 2.16b, where we have used an emitter-area

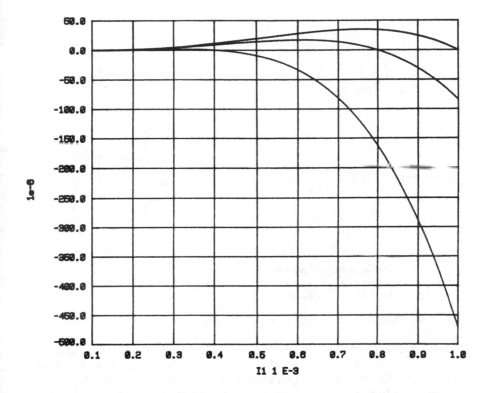

Figure 2.17 Errors in the RMS-DC converter for crest factors of 1 to 10, using A=4 (largest error at I1=1mA) A=8 and A=10.

ratio of 4, and a resistance for the unit emitter of 1Ω; for a pulse input having an RMS value of 100μA and a crest factor of 4, the peak input is 400μA, at which point the ohmic errors vanish. Figure 2.17 shows the actual error in the RMS output (not δV) for this circuit at values of σ from 1 to 10, using A=4, 8 and 10, holding the ideal RMS output at 100μA. The maximum error is -0.47% using A = 4, -0.08% using A = 8 and only +0.035% using A=10. A more complete discussion of this subject would mention the additional steps needed to achieve the results shown here. For example, the collector of Q4 is not, in a high accuracy design, connected directly to the non-inverting node of OA3 (Figure 2.15). To minimize base-width modulation effects, at the very least a diode should be inserted here. In fact, further steps can be taken to refine this core, which is capable of very high accuracy. There are also other interesting methods for RMS computation in TL form.

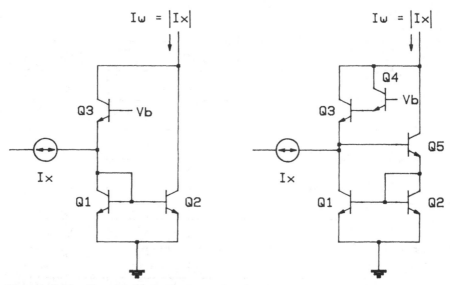

Figure 2.18   Simple absolute value circuits for use in current-mode applications: the beta errors in (a) are reduced by the use of Wilson mirror and Darlington cascode in (b).

2.3.4 Two-Quadrant TL Squarers

As mentioned earlier, many applications call for two-quadrant operation in squaring a signal, and a transconductance implementation was shown in Figure 2.10. Of course, one-quadrant circuits can be converted to two-quadrant operation by the addition of an absolute value (AV) circuit, as in the last example. In practice, this is most accurately achieved using op-amps, diodes and at least two resistors (the latter accurately defining the scale in the inverting direction). Such circuits can also be used to perform the voltage-to-current conversion needed to support a TL core. However, where the signal already exists in the form of a current it is possible to perform this function entirely in TL form, usually at higher operating speed.

Figure 2.18a shows a rudimentary, though perfectly practical, approach. When Ix is positive, it is handled by current-mirror Q1 and Q2; when negative, it is conveyed to the output node, presumed to be at least one V_{BE} above ground, by cascode transistor Q3. The bias voltage Vb is chosen to minimize the output error when Ix=0; this can also be one V_{BE} above ground, when the input node would vary from some value close to ground for Ix<0 to about a V_{BE} above ground for Ix>0. This circuit suffers from beta errors, which can be essentially eliminated by using a Darlington cascode and a Wilson mirror, as shown in Figure 2.18b. Seevinck describes more elaborate AV converters using TL methods [4].

The behavior of all AV circuits to high-speed signals of alternating polarity is less than entirely satisfactory, due to the need to switch transistors

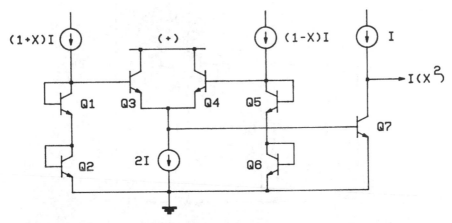

Figure 2.19   A two-quadrant TL squarer; the signal is represented by X, a dimensionless modulation index in the range -1 to +1.

fully off during one half cycle of the input. High-speed signal squaring, such as might be used for RF detection or power measurement, is preferably implemented through the use of circuits which are continuous through zero input. One possibility is shown in Figure 2.19. It comprises two overlapping TL loops, Q1-Q2-Q3-Q7 and Q5-Q6-Q4-Q7 and is driven by a pair of complementary currents (1+X)I and (1-X)I, which would in practice be obtained from a differential-output V/I converter [7], and requires a fixed bias current of 2I. In this circuit, all transistors remain active over the whole range of inputs, -1<X<+1. We shall assume for the moment that all devices have equal emitter areas.

Then, for the two loops

$$I_{C3}I_{C7} = I_{C1}I_{C2}$$

and

$$I_{C4}I_{C7} = I_{C5}I_{C6}.$$

Let C be a temporary variable, such that

$$I_{C3} = (1+C).2I$$

and

$$I_{C4} = (1-C).2I$$

It follows that

$$(1+C).2I.I_{C7} = (1+X)^2I^2 \ .. \ (a)$$

and

$$(1-C).2I.I_{C7} = (1-X)^2I^2 \ .. \ (b)$$

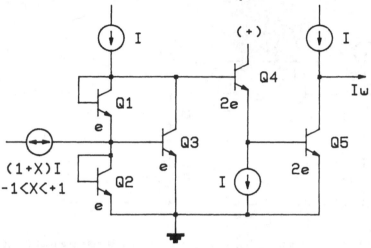

Figure 2.20 **This two-quadrant TL squarer does not require complementary drive currents; note similarity to vector difference circuit.**

By subtracting (b) from (a) and reducing, we find

$$I_{C7} \; = \; (1+X^2)I/2$$

Doubling the emitter area of Q7 and supplying a replication of the fixed current I to its collector, the output into a load, slightly above ground, is just X^2I.

   Many other variants of this form can be devised. Figure 2.20 shows a very slight alteration of the vector difference circuit, discussed at the beginning of this Chapter, which provides the squaring function, and has the advantage of requiring only a single bipolar input. The analysis is left to the reader: it is essentially the same as that given for the vector difference circuit, and shows that the output Iw is again X^2I. Notice the frequent recurrence of the form $(1+X)(1-X)=1-X^2$ in these topologies; we will have more to say about that later.

2.3.5   *Square-Rooting and Geometric-Mean*

Consider next the circuit shown in Figure 2.21. By now, the reader will have no difficulty in concluding, from little more than inspection, that

$$Iw \; = \; \sqrt{I_xI_y} \qquad\qquad (2.21)$$

   Both inputs must be positive (not even TL circuits can deal with imaginary currents!). If we choose to make one of the two inputs fixed, the circuit can be viewed as a square-rooting circuit; if both are variable, it is a geometric mean circuit. A potential hazard is the "β^2-loop" around Q1 and Q2: the product of the two beta poles can result in HF instability. The problem can be

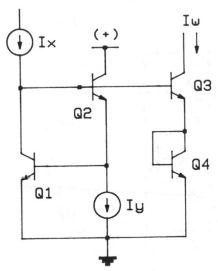

Figure 2.21 A TL square-rooting circuit; it can also be used to generate the geometric mean of Ix and Iy.

solved by the addition of a capacitor across the collector-base junction of Q1. This is satisfactory in using the circuit for low-speed applications, but better solutions exist to implement this TL form for fast signal square-rooting, such as might be required to implement geometry- and focus-correction algorithms in CRT display drivers.

2.3.6 *Vector Magnitude*

Closely allied to the squaring function is that of obtaining the vector magnitude of two or more variables. This, too, is very straightforward in TL form. Figure 2.22 shows what is probably the simplest circuit to do this. It will be recognized as yet another minor variation on the by-now familiar vector-difference circuit, and can be solved using the same analysis methods:

By inspection,

$$I_{C2} = I_{C3} = Iw - I_{C2} + Iy$$

whence

$$I_{C2} = (Iw + Iy)/2.$$

It follows that

$$I_{C1} = (Iw - Iy)/2$$

For the outer loop

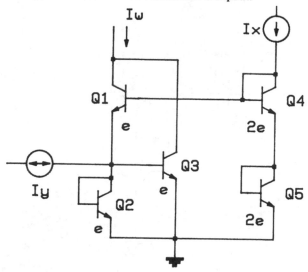

Figure 2.22 **A vector magnitude circuit, generating $Iw=\sqrt{(Ix^2+Iy^2)}$. Iy can be bipolar, but Ix must always flow in the direction shown.**

$$\frac{I_{C4}}{2}\frac{I_{C5}}{2} = I_{C1}I_{C2}$$

so

$$\frac{Ix^2}{4} = \frac{(Iw - Iy)}{2}\frac{(Iw + Iy)}{2}$$

whence

$$Iw = \sqrt{(Ix^2 + Iy^2)} \tag{2.22}$$

It will be apparent that the limitation on the magnitude of negative Iy (in similar circuits, limited to Ix) does not arise here, at least, in principle; in practice, Ix must supply the base current of Q1, so the circuit is beta-limited. Also note that Q1, Q2 and Q3 form the same $(1+X)(1-X)$ group mentioned above, and they also form essentially the absolute-value circuit of Figure 2.18a. Indeed, if Ix were zero (and the devices were those much-to-be-desired infinite-beta MOS-BJT's we yearn for!), we would expect the circuit to act simply in an AV mode, since

$$Iw = \sqrt{Ix^2} = |Ix|$$

An historically-earlier solution to the vector-magnitude is shown in Figure 2.23. This will be seen to have a completely symmetrical form. This is important because ohmic emitter resistances can seriously degrade

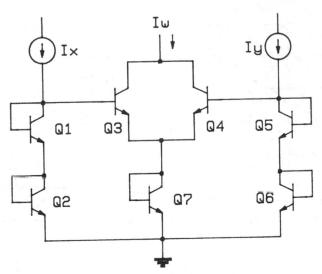

Figure 2.23 Another vector magnitude circuit. This has the advantage of being readily extended to higher powers and higher dimensions.

accuracy in all TL circuits, and, in the present context, it is more desirable to use topologies having a high degree of symmetry with regard to multiple inputs, thereby simplifying design and specification of performance, than to use possibly simpler circuits having grossly asymmetric performance. Thus, we now use seven transistors to do what was done by five in the last example, and no longer have the bipolar capability on the y-input. Nevertheless, the overall performance is satisfyingly predictable. Furthermore, the circuit can be easily improved, as we shall see.

The analysis proceeds along very similar lines to those used before, and for the sake of brevity will not be given here. However, TL circuits often reveal their general behavior by inspection, and we find that to be true here. We will assume for now that all devices have the same emitter area. When either Ix or Iy is zero, the circuit collapses into a current mirror, having more than the necessary number of devices, but a current mirror, nonetheless. For example, when Iy=0, the Ix input to Q1 and Q2 is just linearly replicated in Q3 and Q7; Q4, Q5 and Q6 become parasitic. This is consistent with the vector sum operation. When Ix=Iy, let's say, 1mA, we can see that the output can't be 2mA, because the TLP around either loop is satisfied:

$$2\,\text{mA}/2 \times 2\,\text{mA} \neq 1\,\text{mA} \times 1\,\text{mA}$$

$$I_{C3} \qquad I_{C7} \qquad I_{C1} \qquad I_{C2}$$

However, it is satisfied if the output is √2mA (the correct vector sum in this case):

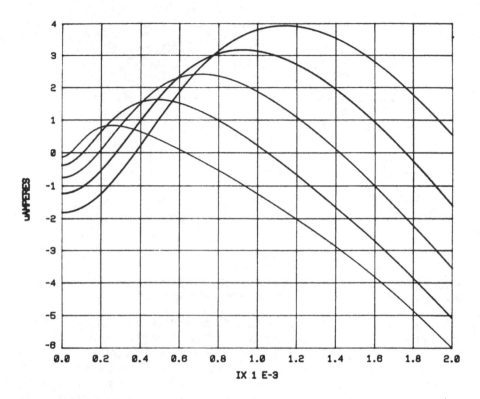

Figure 2.24 Typical errors for the vector magnitude circuit: see text for details. 1μA represents an error of 0.07%.

$$\sqrt{2}\,\text{mA}\,/\,2 \text{ x } \sqrt{2}\,\text{mA} \;\equiv\; 1\,\text{mA} \text{ x } 1\,\text{mA}$$

$$I_{C3} \qquad I_{C7} \qquad\quad I_{C1} \qquad\quad I_{C2}$$

The reader is urged to do the full analysis to verify that these are not just special cases.

Finite beta and ohmic resistance degrade accuracy, but in a fairly predictable way: as in a simple current mirror, a beta of 100 will introduce a scaling error of -2% when Ix and Iy either occur singly or are equal, so that's no surprise. Ohmic errors *tend* to cancel, because current-densities roughly track, but as might be expected there is something to be gained by playing around with emitter areas to minimize the peak errors over the full operating plane (Ix,Iy). To give some idea of this, Figure 2.24 shows a simulation of error current for an optimal circuit, having 10-emitter devices for Q1, Q2, Q5 and Q6, 7-emitter devices for Q3 and Q4 and a 15-emitter device for Q7. These transistors are presumed to have a current-independent beta of 100, very high Early voltage, and the unit device has an ohmic resistance referred to the emitter circuit of 5Ω (roughly, an R_B of 500Ω,

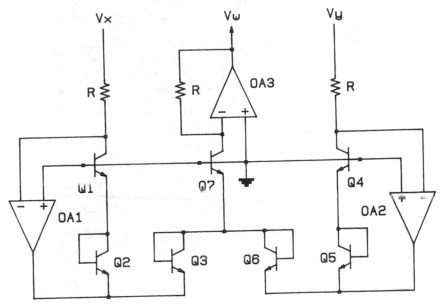

Figure 2.25 **The circuit of** **Figure 2.23 rearranged to allow the use of op-·** **amps to force collector currents and operate with voltage signals.**

typical of a minimum-geometry high-frequency transistor). Notice that the product of 7 and 15 (105) is 5% more than the product of 10 and 10, which closely compensates for the losses due to finite beta and the residual errors due to resistance. The Iy input has values of 0, 200μA, 400μA, 600μA, 800μA and 1mA, while the Ix input is swept from 0 to 2mA. The nominal "full-scale" output can be regarded as 1.4mA (Ix=Iy=1mA), so at this point on the plot, the error scaling is about 0.07% per μA.

The circuit can easily be expanded to handle any number of inputs J, by using J units like Q1, Q2, Q3, and/or to provide nth-rooting, by using n devices in each input branch and n-1 in the common central branch. In other words, this general TL form can perform the computation

$$Iw = \left\{ \sum_{k=1}^{k=J,} I_k^n \right\}^{1/n}$$

(2.23a)

and with very little extra complexity can be arranged to provide unequal "raising" and "rooting" powers:

$$Iw = \left\{ \sum_{k=1}^{k=J,} I_k^n \right\}^{1/m}$$

(2.23b)

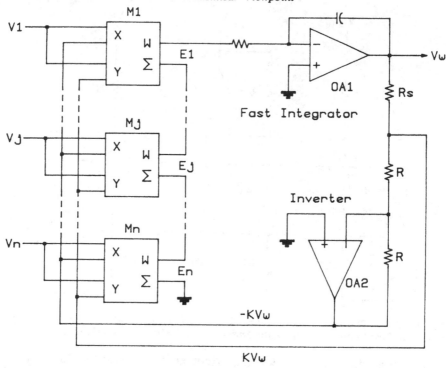

Figure 2.26   **A n-input vector magnitude circuit using complete multipliers and capable of accepting bipolar signal inputs.**

The appeal of these circuits is that they can, if required, operate at extremely high speed (sub-nanosecond) or at very low power levels (nanowatts, if necessary). It is not clear yet whether such operations are of value in neural networks, but the simplicity of the topologies and the possibility of operating at low power suggest that TL may find a role in this field. Figure 2.25 shows how the basic vector magnitude circuit can be rearranged for use with op-amps to improve accuracy and provide voltage-in/voltage-out operation; it can, of course, be extended along the lines just suggested.

This circuit still only allows unipolar operation; an interesting canonic form [8, see also 9,10] solving Eq. 2.23 for bipolar inputs, is shown in Figure 2.26. It uses complete voltage-in/voltage-out analog multipliers, which internally, of course, use TL circuits. The multipliers must provide differential input capabilities and the capability for "daisy-chaining", that is, the addition of the outputs by direct connection, as shown in the Figure. Many multipliers have these capabilities; in the past, their cost has been prohibitive. The availability of inexpensive 8-pin devices in plastic packages, such as the AD633, make this application practical.

The output of the j-th multiplier is

$$E_j = \frac{(V_j - KVw)(V_j + KVw)}{U} = \frac{V_j^2 - K^2 Vw^2}{U}$$

(2.24)

where U is the multiplier scaling voltage (10V in the case of the AD633). The outputs are summed and integrated, some time-constant τ, just large enough to ensure loop stability, to form the output Vw:

$$Vw = \frac{1}{U\tau} \int \sum_{j=1}^{j=n} \left\{ V_j^2 - K^2 Vw^2 \right\} dt$$

(2.25)

Since Vw is finite, the steady-state value of the integral must be zero:

$$\sum_{j=1}^{j=n} \left\{ V_j^2 - K^2 Vw^2 \right\} = 0$$

(2.26)

Hence

$$Vw = \frac{\sqrt{(V_1^2 + \ldots V_j^2 + \ldots V_n^2)}}{nK^2}$$

(2.27)

Finally, to maintain the overall scaling factor to unity, we need to set $nK^2=1$ which requires that

$$Rs = R(\sqrt{n} - 1)$$

(2.28)

Notice that the multiplier scaling factor does not enter into the final result, suggesting that the function could be less rigorously implemented in an ASIC realization. Alternatively, using multiplier/dividers, the denominator input could be connected to Vw, in which case the voltages Ek would be linearly related to the Vk's, resulting in a reduced dynamic range at the points, and thus improved accuracy. In high-speed applications, the AD834 can be used; this 500MHz monolithic multiplier has calibrated current outputs, allowing summation by simply connecting them in parallel.

2.4 Analog Multipliers And Dividers

We have arrived at the arch of this Chapter, now aware of the appealing simplicity of TL circuits, ready to review the most important commercial application of the principle, analog multipliers. A comprehensive history of the analog multiplier will be addressed in a future work. We note merely that the very first paper ever published about this form [2] leapfrogged from circuits capable of operation at only a few kilohertz, or megahertz at most, to 500MHz, without significant loss of accuracy. Since about 1970, virtually all analog multipliers and dividers have been based on translinear principles. The reason is very easily seen: unlike earlier methods, such as pulse-width-height (PHW) modulation, which utilizes the fact that the average value of a pulse is proportional to the *product* of its duty-cycle and its amplitude, the TL multiplier could be inexpensively fabricated in monolithic form (indeed, it absolutely required that medium to be fully developed), was as fast as any possible circuit (orders of magnitude faster than the PHW type, slowed by its averaging process) and could provide the complete function (no external capacitors were required).

During the early 70's many TL multipliers were introduced as "generic" products, most notably those from Motorola and Analog Devices, the former making available flexible and inexpensive building blocks, the latter pioneering the concept of the complete function, using thin-film (silicon-chromium) resistor technology and laser trimming to achieve very precise calibration. One or two monolithic products based on PHW were introduced in this period; they required considerable support components and only marginally exceeded the accuracy of the prevalent TL types, with drastically less bandwidth. The emphasis during the 80's was to work on refining the design and layout techniques to finally attain accuracy levels fully commensurate with the most exact alternatives. Concurrently, TL cells of all kinds, but particularly the four- and six-transistor units to be described here, began to find their way into numerous "embedded" applications; it is common, now, to find such cells in practically all monolithic communications circuits, performing modulation and demodulation at frequencies from audio to about 1GHz. The on-going development of very fast all-NPN processes and the more recent awareness of the value of fully-complementary processes continues to open up new uses for these cells, at ever higher speeds and accuracy. The TL principle has also been adapted to MOS usage (see Chapter 12); in sub-threshold operation, MOS transistors behave in a similar fashion to bipolar transistors, and their slow speed (milliseconds) in under these very low-current conditions may not always be a limitation. Again, one is bound to wonder about the potential for current-mode signal-processing in neural networks, where low speed and even low accuracy may be quite tolerable (see Chapter 17).

During the coming decade, it is likely that many of the traditional uses for analog multipliers will be displaced by various types of digital signal processing. On the other hand, the continued need to cope with very high-frequency signals (in the range 10MHz to 1GHz, in this context) assures a permanent place for TL techniques in the analog designer's repertoire. However, generic products will gradually be replaced by application-specific circuits, often embodying mixed-mode circuits (see Chapters 5 and 12) using embedded TL circuits alongside digital ones.

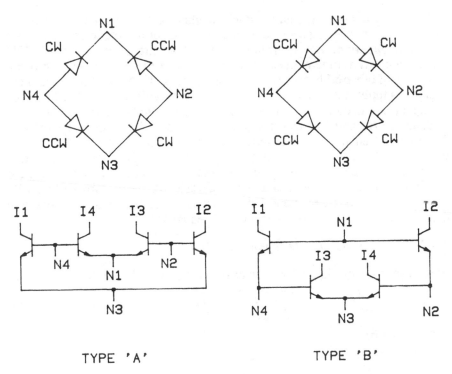

TYPE 'A'   TYPE 'B'

Figure 2.27   The two basic four-transistor TL forms; in type 'A', the junction polarities Alternate; in type 'B', they are Balanced.

2.4.1   Basic Four-Transistor Multiplier Cells

All TL multipliers are based on two basic cell types; these are shown in Figure 2.27, in both diode-ring and transistor-loop form, and are identified as type 'A', in which the letter can be thought as signifying the Alternating sequence of junctions CW-CCW-CW-CCW, and type 'B', which might be called the Balanced arrangement, where two opposing pairs of CCW-CW-arranged junctions are used, and where the sequence is CW-CW-CCW-CCW. These cells alone can be used to realize all multipliers (one-, two- and four-quadrant types) and dividers. They have already arisen many times in this Chapter in other guises, the most obvious cases being the TL squarer (Figure 2.12) and square-rooter (Figure 2.21), where they are of Type 'B' and of course, are basic elements of many multiple-loop circuits. This type is much more common; it's noteworthy that the only instance of type 'A' so far in this material has been the AGC cell (Figure 2.9), and that example is 'TN' but not 'TL'.

Figure 2.27 does not indicate how the currents are established in the junctions. In the transistor realizations, we can either force collector currents or emitter currents. Since these circuits operate only on current *ratios*, it may not seem to matter which we choose. This is not always true, as we shall see. In other respects, the relative merits of these two forms depend on the application. For example, type 'B' is attractive in an all-NPN monolithic environment because it can be driven by currents from the collectors of other NPN transistors ("from" nodes N2, N3 and N4 toward the negative supply line, with N1 anchored to some voltage point), while type 'A' appears to need one "outward" current (say, from N3) and two "inward" currents (to N2 and N4, with the presumption that the outer pair would be diode-connected and N3 anchored). Not obvious at this point is that the 'A' type can exhibit very low multiplication errors in the presence of finite betas, even when the operating currents in the outer pair is much lower than that in the inner pair (connected as just described). Before proceeding with the detailed analysis, note here that all transistors in the circuits discussed in this section are assumed to have nominally equal area.

2.4.2 The Beta-Immune Type 'A' Cell.

From inspection of Figure 2.27, the basic TL equation for the 'A' type cell is

$$I_1 I_3 \ = \ I_2 I_4 \tag{2.29a}$$

Equivalently

$$\frac{I_1}{I_2} = \frac{I_4}{I_3} \tag{2.29b}$$

It requires only a trivial rearrangement to see how any one of these currents can be called the "output" to realize a one-quadrant multiplier/divider, in which both numerator currents and the denominator current are unipolar. More commonly in signal-processing the need arises to handle at least one bipolar variable, that is, a two-quadrant multiplier is called for. The easiest way to provide this is shown in Figure 2.28, where the input signal is represented by the modulation index, X, the output by W and the 'gain-control' is provided by the current 2Iy. This circuit is well-known and its function intuitively obvious: the current gain is proportional to Iy. From the last equation we can surmise that the signal current in Q4 will increase with X, so identified this output as '+W'.

Applying TLP

$$(1+W)Iy.(1-X)Ix = (1-W)Iy.(1+X)Ix \tag{2.30}$$

so

$$W \equiv X$$

This is always an astonishing, almost unbelievable, result! It says that for all values of Ix and Iy, for any input signal right up to the limits

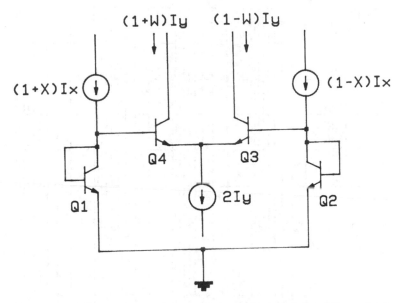

Figure 2.28 Type 'A' cell as a two-quadrant multiplier; the output modulation index W is identical to the input modulation index X, even when Iy is almost βIx - the 'beta-immunity' discussed in the text.

(from X=-1 to X=+1), for any transistor geometry and at any junction temperature, *the currents in the inner pair are always an exact, linear replication of those in the outer pair.* The dependence on Iy is seen by writing the full expression for the differential output current:

$$Iw = (1+W)Iy - (1-W)Iy = 2XIy \qquad (2.31)$$

Since the differential input signal is 2XIx, the current gain is just Iy/Ix. What is even more surprising is that this behavior is essentially preserved even when we take into account any errors due to beta. Let the symbol δ be used to represent either I_B/I_c, that is, $1/\beta$, or I_B/I_E, that is, $1/(\beta+1)$. This may seem to lack rigor, but the value will differ little whichever form is used, and we will sometimes switch definitions to suit circumstances. Just think of δ as being the "base current defect factor". From a pragmatic viewpoint, any circuit whose function appears to hang on the need for a counterbalancing of two such factors could hardly be called robust. In this case, we can expand Eq. 2.30 to include δ where it matters, that is, to the degree that the base currents of Q3 and Q4 rob some of the signal current from Q1 and Q2:

$$(1+W)Iy\{(1-X)Ix - \delta(1-W)Iy\} = (1-W)Iy\{(1+X)Ix - \delta(1+W)Iy\} \qquad (2.30)$$

so still

$$W \equiv X.$$

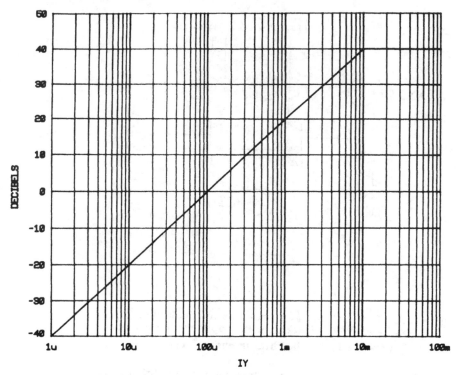

Figure 2.29 Simulation of the current gain of the cell shown in Figure 2.28 for Ix=100μA and β=100; it is exact right up to Iy=βIx.

This seems a bit fishy. Can this be true "for all values of Ix and Iy"?. Our experience with circuits tells us that there ought to be *some* penalty for finite beta. A basic current mirror, for example, having a nominal ratio of A (as defined by the emitter area ratio) actually exhibits a ratio which is lower, by the factor $(1-\delta)/(1+A\delta)$. Part of the answer lies in the fact that we have not yet finished the calculation, which is to show that the output current is

$$Iw = (1-\delta)(1+W)Iy - (1-\delta)(1-W)Iy = 2(1-\delta)XIy \qquad (2.31)$$

So the output *is* a bit smaller, but only by the factor $(1-\delta)$, which is of course, just alpha. The loss is the same as would be incurred by a signal passing through a cascode stage. This loss still appears to be quite independent of the gain Iy/Ix. To see how this can be possible, it's only necessary to consider how the ratio of the currents in Q1 and Q2 is affected by the base currents in Q4 and Q3: since the latter are exactly proportional to the same modulation index (X, or W) *this ratio is not affected.* This is the key. Only when each of the input currents is totally consumed by the bases of the inner pair does the circuit become unhinged: even then, it will obligingly deliver a current gain of beta (of course), although in practice, the de-biasing

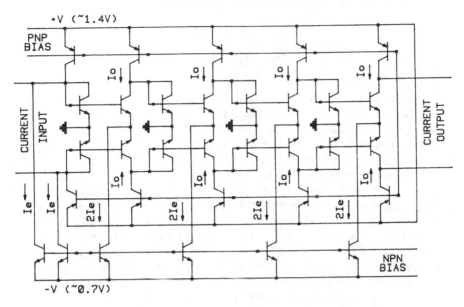

Figure 2.30 A variable-gain current-mode amplifier using four cascaded type 'A' cells; current-sources at left are only included to simplify the functional form of the gain, which increases very rapidly with Ie.

of Q1 and Q2, which hitherto behaved like clamping diodes to define the potentials at the base nodes, would cause some type of malfunction. It is a simple matter to add a diode at the Q3/Q4 emitter node to catch it under these conditions. Figure 2.29 shows a simulation result to support this prediction, using $\beta=100$, Ix=100μA and Iy = 1μA to 100mA. The gain exactly follows the ideal function within -0.1dB - the alpha error - until Iy=Ix, when it simply limits at +40dB, that is, $20\text{LOG}\beta$.

Real circuits actually work almost as well as this, but as stated earlier, it would be unwise to depend on close matching of betas: a modern high-yielding process can hold the beta-matching between devices which are adjacent on the die to within better than ±5%. On closer examination of circuit behavior it is found that base-current mismatches lead to parabolic nonlinearity, for reasons which are essentially the same as those for emitter-area mismatches. Nevertheless, generally good performance can be obtained for gains up to about $\beta/10$.

This cell has applications in variable-gain (or even fixed-gain) current-mode amplifiers, where this unique property is of value. Figure 2.30 shows one such topology, where another aspect of the circuit is put to good use, namely, its ability to be cascaded indefinitely without the accumulation of supply voltage. The current gain of this four-stage system can easily be calculated by inspection: every stage has a gain of Ie/(Io-Ie), so the overall gain is simply

$$G_i = \left\{ \frac{Ie}{Io - Ie} \right\}^4$$

(2.32)

The gain is unity when Ie = Io/2, and if we adopt the rule that we should not try to push the gain of each stage above $\beta/10$, say, 10, the maximum current gain of 80dB would be achieved when Ie=(10/11)Io. The close tracking required between Ie and Io may impose another limit on the gain per stage. However, notice that in AGC applications the NPN bias may be controlled by a closed-loop system, thus ameliorating the matching requirements. PNP transistors are shown for the current sources labeled Io; these can operate at very low collector bias voltages, so the whole amplifier can be powered by a +1.4V supply (or even lower). Also, because they only have to supply DC bias currents, split-collector lateral PNP transistors can advantageously be used, having much lower collector parasitic capacitances than a typical fully-complementary PNP and better balance between each pair of currents. The absolute value of Io affects the gain, but accuracy can be maintained even when the PNP betas are low by appropriate design techniques; this problem is made less severe by the fact that they all operate at an equal, fixed current. The negative supply needs to be even lower - less than one Vbe would suffice. Results for this type of amplifier were reported in [11].

Although we have digressed slightly from the topic of multipliers, it may be worth noting that the distinction between variable-gain amplifiers and two-quadrant multipliers is fuzzy. Also, there are relatively few situations in which one finds type 'A' TL cells.

One final example will be given here, and this shows how the basic cell, intrinsically only capable of one-quadrant operation (see Eq. 2.29), can be converted to four-quadrant operation by the use of multiple current drives. Figure 2.31 shows the circuit. The reader is invited to verify the function

$$Iw = \frac{(\pm Ix)(\pm Iy)}{Iu}$$

(2.33)

While the circuit is realizable, it is only included here as a curiosity, showing how the massive use of specially-scaled currents can alter the behavior of simple TL circuits in non-trivial ways. In practice, it suffers from the asymmetries mentioned earlier, due mainly to emitter resistance. There is no incentive in modern design to try to get so much function out of just four transistors, and in the next section, we'll find anyway that we can achieve better four-quadrant multiplication out of a type 'B' cell. Figure 2.31 is noteworthy for another reason, however, and this is the fact that only one of the four devices is actually used as a transistor! The remaining three are connected as diodes.

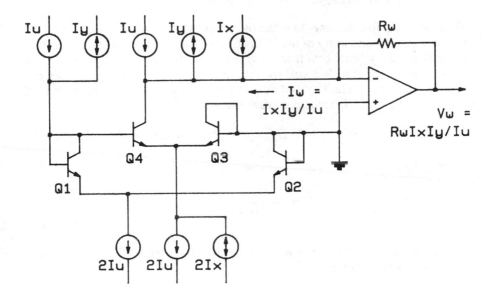

Figure 2.31   A translinear curiosity: a four-quadrant multiplier in which only one device (Q4) is used as a transistor; the rest are diodes.

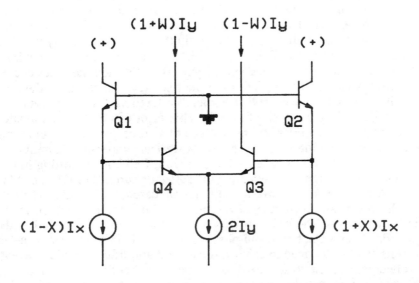

Figure 2.32   Type 'B' cell as a two-quadrant multiplier; this common form can be driven from further NPN transistors, but has large beta-errors.

2.4.3   The Beta-Allergic Type 'B' Cell

This section is so titled only to make the point that the 'B' cell does not enjoy the special beta-immunity of the 'A' cell; it is not highly allergic, just a little more affected by finite beta than its genetically stronger cousin. Figure 2.32 shows the basic form. We will assume that the basic TL analysis does not need to be restated and proceed directly to the equivalent equation derived for the 'A' cell, which accounted for the base currents:

$$(1+W)Iy\{(1-X)Ix + \delta(1+W)Iy\} = (1-W)Iy\{(1+X)Ix + \delta(1-W)Iy\} \quad (2.34)$$

so now

$$W = \frac{X}{1 + 2\delta Iy/Ix} \quad (2.35)$$

The output current error is worse than this, because of the further factor of alpha from the emitter circuit of Q3/Q4:

$$Iw = (1 - \delta)(1 + W)Iy - (1 - \delta)(1 - W)Iy = 2XIy \frac{1-\delta}{1 + 2\delta Iy/Ix} \quad (2.36)$$

It it clear that if we wish to operate this cell at high values of Iy/Ix some method for compensating for the gain errors caused by finite beta will be necessary. Note also that used as a two-quadrant multiplier element the linearity of response with regard to the y-input will be impaired by the non-constant denominator in Equation 2.36. Fortunately, in four-quadrant multipliers, which uses two overlapping type 'B' cells, nonlinearities caused by this mechanism in the two half-multiplier sections cancel; also, the ratio Iy/Ix will usually be unity, leaving a tolerable fixed scaling error, which is not a function of signal level, of $(1-3\delta)$. This error can be fully eliminated by arranging the multiplier scaling voltage to also be reduced by the same factor, thus ensuring lot-to-lot robustness as well as temperature stability.

Figure 2.33, sometimes called a "Gilbert Gain Cell", is interesting in that it re-uses the input signal, by summing the collector currents of Q1 and Q2 to the output. It can thus be viewed as a differential cascode with gain. The beta errors, however, are even more severe at any given gain, compared to the basic form, because the base currents of Q3 and Q4 are also added to the output, via Q1 and Q2, but in anti-phase. Figure 2.34 shows how a series of such cells can be cascaded to achieve broadband amplification with current-controlled gain. A useful feature of this scheme is that the operating bias at each stage is guaranteed to be exactly sufficient for the signal level at that stage, that is, the modulation index, X, is unaffected, but it acts on a progressively increasing bias level.

(1-X)(Ix+Iy)   (1+X)(Ix+Iy)

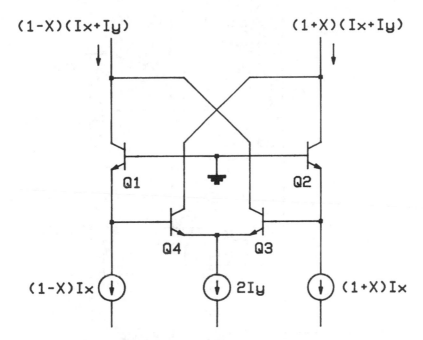

Figure 2.33 The "Gilbert Gain-Cell"; it provides a minimum gain of 1 and a maximum gain which is practically limited by beta to about 10.

The total current gain in this case is

$$G_i = \frac{1}{I_0} \sum_{k=0}^{k=n} I_k$$

(2.37)

This scheme has the advantage, in some applications, that gain remains linearly-dependent on any one of the input currents.

Finally, one further use for the gain cell is shown in Figure 2.35, where four-quadrant multiplication is performed in an economical and practical way. Here, the tail current to Q3 and Q4 varies from zero to 2Iy as Y varies over its peak range -1 to +1. When Y=-1, the differential output is just 2XIx; when Y=0 the current in Q3 and Q4 exactly cancels that in Q1 and Q2; when Y=+1, the output is -2XIx. This topology is not as accurate as the standard six-transistor circuit, and the errors due to ohmic emitter resistance are asymmetric. On the other hand, low-noise performance requires the use of low-resistance transistors anyway, and the fact that this type has only a single loop makes it easier to keep even-order distortion due to area mismatch low (by cross-quadding) or easier to trim (by the inclusion of just one PTAT voltage between the bases of Q1 and Q2.

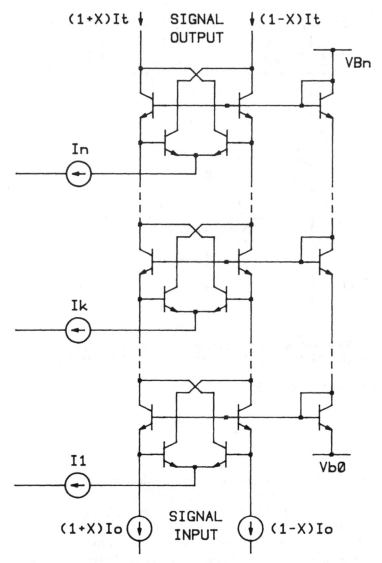

Figure 2.34 A variable-gain current-mode amplifier using four cascaded gain cells; the output bias current is the sum of all bias currents.

2.4.4 *Distortion Mechanisms*

The original paper on TL multiplier cells [6] dealt quite thoroughly with the topic of distortion. These are due primarily to emitter-area mismatches, which cause parabolic nonlinearities, thus, even-order harmonic distortion, and finite ohmic resistances, which introduce cubic nonlinearity, thus odd-order distortion.

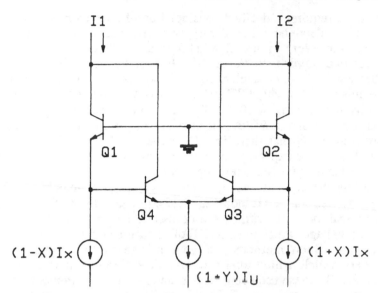

Figure 2.35 A four-quadrant multiplier based on the type 'B' cell; both X and Y can be bipolar.

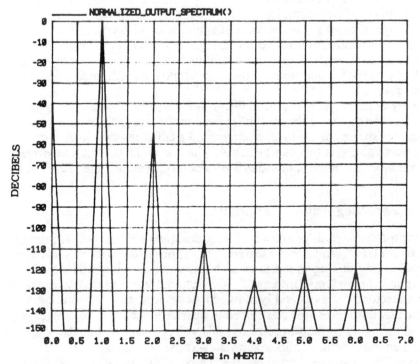

Figure 2.36 Demonstration of harmonic distortion caused by an emitter-area-mismatch of 1%; dominant feature is the -55dB 2nd harmonic.

The topic requires a detailed consideration of all factors, and will not be treated in any depth here. In order to give some idea of the magnitude of the problem, however, Figure 2.36 shows an FFT corresponding to the distortion in a simple type 'A' or 'B' cell when the net emitter-area-factor λ (see Section 2.2.3, in particular, Eq. 2.11) was set to 1.01, corresponding to a 'V_{BE} mismatch' of about 250μV, and driven by a sinusoidal input with a peak modulation index of ±0.75. The distortion is seen to be predominantly second-order, at about -55dB. It is also evident, from the -40dB component at zero-frequency, that there is a DC offset of 1% (of full-scale) at the output; this is to be expected, since a non-unity λ will cause the circuit to be imbalanced at its quiescent operating point.

Recalling the equivalence between λ and 'V_{BE} mismatch', or the total voltage inserted into a TL loop (Eq. 2.12), it's easy to see that this type of distortion can be dealt with by the deliberate introduction of a trimmable correction voltage, which must be PTAT to ensure stable compensation over temperature. Such methods are now in use in the highest-accuracy multipliers, which exhibit total harmonic distortion of less than -95dB (see Section 2.4.7). These use sophisticated automated measurement systems in manufacture, and realistic AC stimuli, to 'listen' for very small distortion products, and then laser trim appropriate resistors to null these residues.

A subtle source of distortion at high frequencies can arise when the drive currents become non-complementary; this is fully-equivalent to an area mismatch, and causes just the same magnitude of distortion. It is reduced only by meticulous attention to the design of the V/I converter. Over the years, these have evolved very considerably from the simple forms found in standard text-books, and some of these developments will be discussed in a future work.

Finally in this very brief look at distortion, we show Figure 2.37, which is an FFT of the output of the multiplier used in the last example, now having λ=1 but resistances of 1Ω in all emitters, and a tail current of 3mA instead of the correct value (for this case) of 2mA. There is no DC offset and virtually no even-order harmonics, but -52dB of second- harmonic and -92dB at the fifth.

2.4.5 The 'Log-Antilog' Multiplier

Bipolar transistor multipliers were being designed before the TL principle was formulated, and their design was approached from a different viewpoint, by thinking in terms of each transistor as a logarithmic or anti-logarithmic (exponential) element. Operational amplifiers were used to force collector currents Ix and Iy in two input devices, and the resulting V_{BE}s summed. A third transistor was forced to operate at a current Iu, and its V_{BE} was subtracted from the sum. Finally, this voltage was used as the V_{BE} of a fourth transistor, whose collector current was the output, say, Iw.

The operation was explained by an equation of the (non-rigorous) form

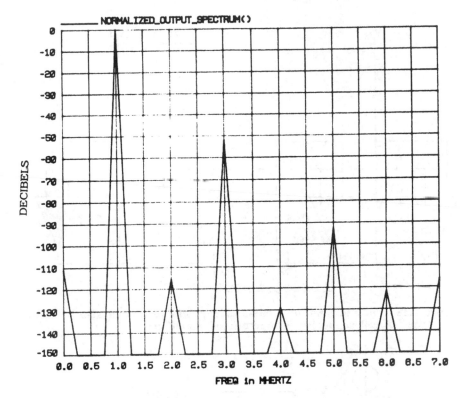

Figure 2.37 Harmonic distortion generated by ohmic resistance (see text for details); dominant feature is the -52dB 3rd harmonic.

$$\log Ix + \log Iy = \log Iu + \log Iw \qquad (2.38)$$

Of course, this is just another way of stating the TL principle, and there is absolutely no difference, in principle, between TL multipliers and 'log-antilog' types.

This can be seen by comparing a typical 'log-antilog' embodiment (Figure 2.38) with its TL reduction (Figure 2.39), using the same names for devices and currents. In the first case, the collector currents of Q1, Q2 and Q3 are accurately forced by op-amps OA1, OA2 and OA3, which also serve to generate virtual-ground nodes facilitating the conversion of the input voltages Vx, Vy and Vu to corresponding currents Ix, Iy, Iu. We have already seen the use of such techniques in Section 2.3.2. OA3 also supplies bias to Q4, whose collector current is simply $Iw = IxIy/Iu$, which is converted to an output voltage by Rw:

$$Vw = \frac{VxVy}{Vu} \frac{RuRW}{RxRy}$$

$$(2.39)$$

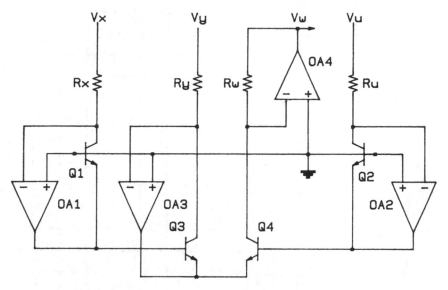

Figure 2.38 The "Log-Antilog" one-quadrant multiplier can also be viewed as a type 'B' cell supported by operational amplifiers.

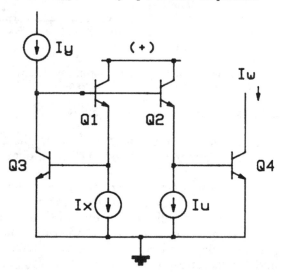

Figure 2.39 The TL reduction of the "Log-Antilog" multiplier.

Many commercial products based on this arrangement are available. With care, multiplication accuracies of 0.05% are achievable. However, operation is limited to one quadrant, and the use of operational amplifier techniques limits the speed of response. The minimal TL form, on the other hand, can be very fast, and, when signals already exist in the form of currents, it is very compact. An interesting modification to the log-antilog multiplier includes provisions for introducing a scaling constant between the LHS and RHS of Eq. 2.38, which allows signals (actually, signal ratios) to be raised to arbitrary powers; the AD538 from Analog Devices is a development of that idea.

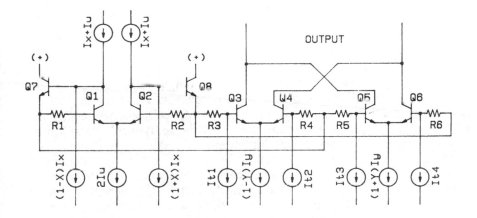

Figure 2.40   Rearrangement of the six-transistor four-quadrant multiplier core providing higher accuracy and multi-decade divider performance.

2.4.6   The Six-Transistor Multiplier Cell

Few readers of this book will be unaware of the six-transistor TL cell used for four-quadrant multiplication. Its economy of means and the simplicity with which it can be supported in a monolithic implementation has commended it for use in analog signal processing for over twenty years. Because of the wide availability of texts which quite ably discuss this circuit, there is no need to repeat the first-principles analysis here.

However, it is worth briefly reporting some recent developments, related to improving the accuracy of such multipliers while not sacrificing bandwidth. Figure 2.40 shows a core made up of two overlapping type 'A' cells Q1,Q2,Q3,Q4 and Q1,Q2,Q5,Q6. The common pair Q1,Q2 are driven at the collectors by bias currents Ix+Iu and by the X input signals (1-X)Ix and (1+X)Ix, and at the common emitter node by the current 2Iu, which as we shall see independently controls the denominator of the multiplication function. The means whereby the upper bias currents are made to exactly balance the lower bias currents in this section of the circuit need not concern us here; in practice, this balancing act is achieved by the use of a common-mode control loop which senses the voltage at the emitters of Q1,Q2 and holds it to a fixed value. Emitter followers Q7 and Q8 control the *differential* voltage between the bases to satisfy the requirement, enforced on the circuit by these currents, that

$$I_{C1} = (1+U)Iu$$

and

$$I_{C2} = (1-U)Iu$$

where

$$U = X\frac{Ix}{Iu}$$

(2.40)

Thus, for a given value of X, its effect on the output can be greater than in the standard multiplier core by choosing to make Iu less than Ix. In fact, because of the "beta-immunity" of the type 'A' cell, augmented by the additional current gain of the emitter followers, Iu can typically be as low as Ix/1000. This endows the cell with further utility as an accurate divider.

The currents It1 through It4 provide bias for the emitter followers and also provide a means whereby small correction voltages can be introduced to eliminate errors arising from emitter-area mismatch. This general structure has proven capable of distortion levels of -95dB with correspondingly small nonlinearities (as low as 0.01%), and demonstrate that the TL multiplier is capable of the highest accuracy.

2.5 The Translinear Cross-quad

We will now turn to a four-transistor cell which is not strict-TL in form (the loop is broken, making it a TN form) but which is very closely related and has so many uses, both by itself and embedded into TL circuits, that it is deserving mention in this context.

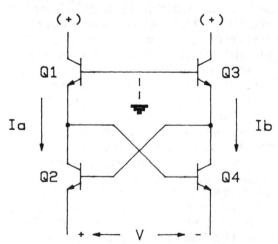

Figure 2.41 The translinear cross-quad; the voltage V remains very close to zero even for large ratios Ia/Ib.

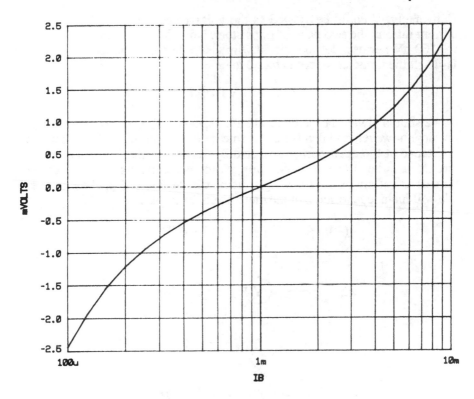

Figure 2.42 Demonstration of the low input voltage of the translinear cross-quad over a 100:1 range in Ib (Ia=1mA, β=100).

The generic cell is shown in Figure 2.41. In some way, a current Ia is established in Q1 and Q2 and Ib in Q3 and Q4. The voltage in the open port is

$$V = V_{BE4} + V_{BE1} - V_{BE3} - V_{BE2}$$

and since, in the ideal case where there are no base currents,

$$V_{BE1} = V_{BE2} \quad \text{and} \quad V_{BE3} = V_{BE4}$$

it follows that V is always zero, whatever values Ia and Ib may have! A real circuit can come quite close; it's easy to show that

$$V = V_T \ln \left\{ 1 + \delta \left(\frac{Ia}{Ib} - \frac{Ib}{Ia} \right) \right\} \sim V_T \delta \left(\frac{Ia}{Ib} - \frac{Ib}{Ia} \right)$$

$$(2.41)$$

For a ratio Ia/Ib=10 and β=100, V would be about 2.5mV; Figure 2.42 shows the form of V with Ia fixed and Ib swept from 100μA at 10mA.

Conversely, V could be viewed as the voltage required to establish a 10:1 current ratio in the two pairs of transistors. Note that this is much lower than the "60mV per decade" associated with a simple pair of junctions. If we let Ia=Ib=I, the input resistance is found to be

$$Rin \cong 4 \, VT \, / \, \beta \, I \qquad (2.42)$$

We should always be cautious when it appears we're getting something for nothing, however, and that is no less true here. The radical reduction in input resistance at the open port is obtained by 100% positive feedback, and the circuit is quite prone to oscillation if not correctly used. Nevertheless, numerous variations on this simple theme can be devised (and have been!). We will include here some of the more useful examples.

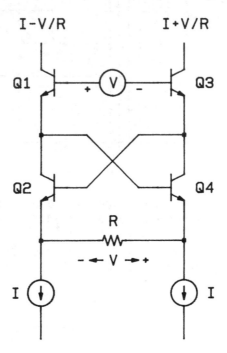

Figure 2.43 Caprio's quad; input V is accurately replicated across the resistor, providing linear V/I conversion for moderate input amplitudes.

2.5.1 *Caprio's Quad*

By breaking the loop again, at the bases of Q1 and Q3, and applying a voltage source as shown in Figure 2.43, we can arrange to replicate this voltage across the resistor, R, with very small errors, because of the V_{BE}-cancellation inherent in the quad. This arrangement was first published by Caprio [12] and it has enjoyed moderate success. It has two weaknesses, the first of which is that the voltage swing at the input cannot be very large

(±400mV at most) because the bias voltages for the collectors of Q2 and Q4 are only one V_{BE} at zero-signal, and these devices will saturate for large inputs, at which point the output polarity will invert. Second, the input resistance (at the bases of Q2 and Q3) is -R, which can lead to instabilities when the source is slightly reactive. Parasitic capacitances at the emitter nodes of Q2 and Q3 (such as attendant to the use of transistors for current sources I) can also cause instability.

These problems need not be fatal, and the circuit has utility as a voltage-to-current converter, allowing a high modulation index to be obtained at moderate signal levels without incurring significant nonlinearity. Headroom and beta-related errors can both be addressed by including emitter followers to drive the lower transistors, but precautions must then be taken to ensure HF stability.

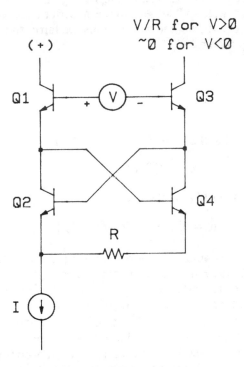

Figure 2.44 Half-wave rectifier derived from Caprio's quad; the form is readily converted to full-wave operation without full duplication.

2.5.2 Linear Half-Wave Rectifier

A trivial modification to Caprio's quad, the omission of one of the current sources, as shown in Figure 2.44, results in a half-wave transconductance rectifier. When the input voltage is more than about 50mV negative, the output is at its minimum value, which is simply I/β (from the base of Q2 thence via Q3). When the input voltage is zero, most of the bias current continues to flow in the Q1/Q2 branch, but the output will now be slightly larger. We can calculate the output for this condition quite readily. Let the currents in Q3 and Q4 be some xI, where $0 < x < 1$. In general, it can be shown that

$$V_T \ln \left\{ 1 + \delta \frac{(2x - 1)}{x(1 - x)} \right\} + xIR = V$$

$$(2.43)$$

The transcendental form of this equation precludes a forward solution, but we can map the function by letting x be the independent variable and calculating V. Of course, when δ is very small, the voltage across the resistor tracks the input exactly, which as well as enhancing the slope accuracy of the output, results in a very sharp "corner" to the rectifier transfer function. For the special case where V=0, we know that x must be fairly small, so the above equation can be simplified to

$$V_T \delta \frac{-1}{x} + xIR = 0$$

$$(2.44)$$

from which it follows that the zero-signal value of x is

$$x = \sqrt{(\delta V_T/IR)}$$

$$(2.45)$$

and the collector current of Q3 is therefore

$$Io = I\{ \delta + \sqrt{(\delta VT /IR)} \}$$

$$(2.46)$$

Figure 2.45 shows a simulation of the output for a circuit having I=1mA, R=500Ω and β=100. As predicted by the above analysis, the output starts at 10μA, becomes 34μA at V=0 and quickly becomes a linear function of V above this point, until, at V=+500mV, all of the available bias is steered to the output, which then limits at $I(1 - 2\delta)$, or $\alpha^2 I$. The slope of this output in the linear region is also reduced by the factor α^2. The additional line in this plot corresponds to the current in a perfect 500Ω transconductance stage, making the slope error of the circuit more apparent. As noted earlier, this circuit would generally not be used with inputs above about 400mV, to avoid saturation and fold-over.

2.5.3   PTAT (Delta-V_{BE}) Cell

The translinear quad finds common usage in biasing circuits as a "Delta-V_{BE}" cell, for generating voltages or currents which are PTAT. Figure 2.46 shows the simplest example. Its appeal lies in the fact that the magnitude of the

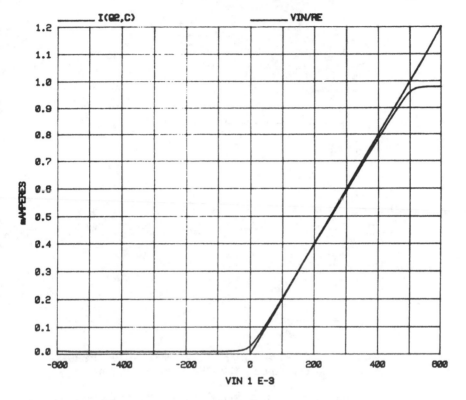

Figure 2.45 Simulated result for the half--wave rectifier; see text for details.

output is relatively immune to the bias current, I. Again, there are numerous variants of this simple circuit, the most common being to mirror the collector current of Q3 back to Q1, to almost eliminate dependence on I, which now is a very small start-up current. (It might be worth adding that this start-up provision is often excluded, not only in discussions of such circuits but in their monolithic implementation; many will start fortuitously, through leakage or transient currents, but it is good practice to consider carefully the start-up requirements over the full temperature and supply-voltage range.)

Consideration of the TN path shows that

$$I_{C3} = \frac{V_T}{R} \ln \frac{A_1 A_4}{A_2 A_3}$$

(2.47)

Usually, only the emitter area of Q4 is made large, $A_3=8$ being a common choice, when the delta-V_{BE} is theoretically 53.75mV at 300K. Making $A_1=8$ also will double this voltage, which in practice is invariably somewhat

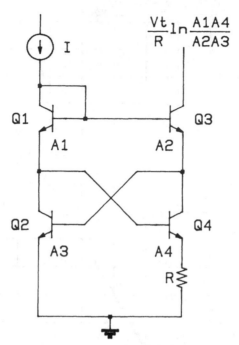

Figure 2.46   Delta-V_{BE} cell using the translinear cross-quad; the bias current I is not critical in determining the PTAT output current.

elevated, due to the finite beta and consequent voltages across base resistances R_B (or "ohmic emitter resistance", to be consistent with previous analyses in this Chapter; either view is acceptable).

A useful variant of this form has another resistor R in series with the emitter of Q1, and the areas are set to $A_1 = A_4 = A$ and $A_2 = A_3 = 1$. It can be shown that the sum of I_{C2} and I_{C4} (the total current into the *ground* node) is then PTAT and independent of the primary biasing current I, which is perhaps a little unexpected. Figure 2.47 shows a practical scheme based on this idea, providing regulation of about 1% in its 1mA output current for variations in supply voltage Vs from +4V to +12V. This improved regulation was achieved by slightly raising the resistor in the emitter of Q1. Figure 2.48 shows the simulated result. The replacement of the lower current mirror by a resistor allows a multiple of the delta-V_{BE} to be generated, which, when correctly proportioned, can provide a simple "band-gap" voltage reference, with the output at the base of Q4. Variants of this sort are endless.

Although we seem, at this point, to have strayed away from the theme of "current-mode circuits", we are nonetheless still very much concerned with the idea of *translinear* analysis, that is, an approach to BJT circuits in which the relationship between V_{BE} and I_C is of primary concern.

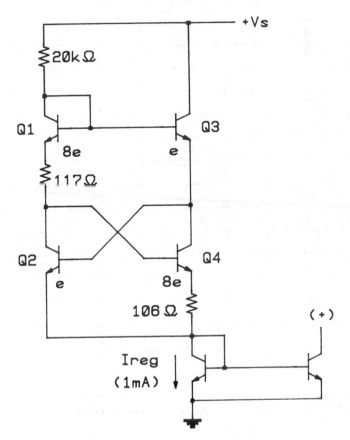

Figure 2.47 **A practical development of Figure Figure 2.46, in which the output Ireg is both PTAT and substantially independent of the supply voltage.**

2.5.4 Another Wideband Squarer

The translinear cross-quad can be used to generate a wideband virtual ground node in numerous ways. This makes it very easy to convert a voltage into a current with good linearity, in an analogous way to the use of the op-amps in earlier examples, but without the speed limitations of an op-amp. Figure 2.49 shows this, on the left hand-side of the schematic; we will explain the purpose of Q5 and Q6 in a moment. The emitter node of Q2 is the virtual ground, formed as explained at the beginning of this Section. The signal current i does not materially affect the voltage at this node, which is always within a few millivolts of ground. Thus, the current i is very precisely equal to V/R. With careful optimization of device geometries and operating conditions, this accuracy can be maintained up to frequencies of at least $f_T/10$.

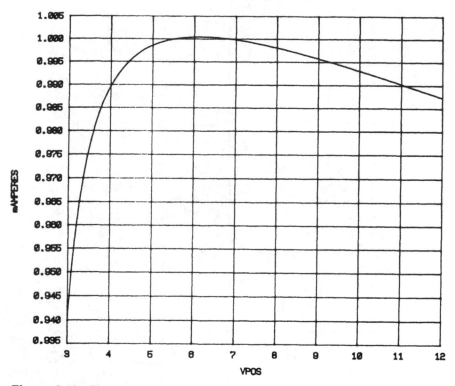

Figure 2.48   Simulated result for circuit of   Figure 2.47; Ireg is within 1% of its nominal value of 1mA for supply voltages from 4V to 11V.

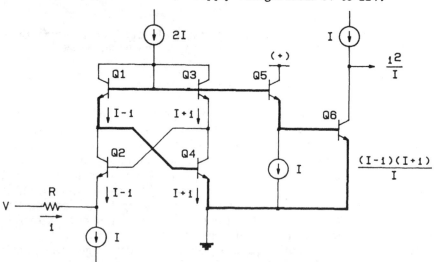

Figure 2.49   A two-quadrant squarer based on the translinear cross-quad, a mixed TN/TL circuit. The heavy line shows the main TL loop.

The currents in Q1 through Q4 must have the values shown. The significant aspect of this circuit is that Q1 and Q4 have a current product of the form (I-i)(I+i), which can be written as $I^2 - i^2$.

Now we can see that by simply adding Q5, operating at a fixed current, I, and Q6, generating the output, we have

$$I_{C6} = \frac{I^2 - i^2}{I} = I - \frac{i^2}{I}$$

(2.48)

By supplying the constant term by a further current source, the output is the two-quadrant square of the input. The main TL loop of this circuit is shown by the heavy line. The base current of Q5 is one of several sources of static error in this circuit; these, and dynamic errors, can be reduced to acceptable proportions using extensions of this idea.

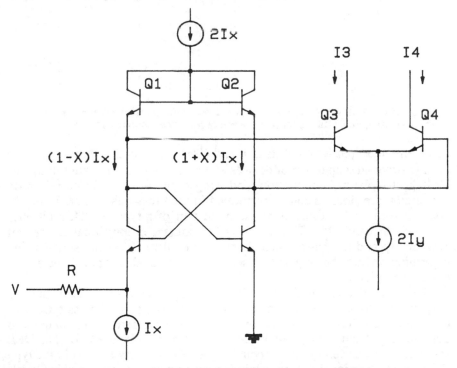

Figure 2.50 A two-quadrant multiplier based on the translinear cross-quad, another example of mixed TN/TL concepts.

2.5.5 *Another Two-Quadrant Multiplier*

A further application of the ubiquitous translinear cross-quad [15] is shown in Figure 2.50. Noting from our last example that the signal current becomes a

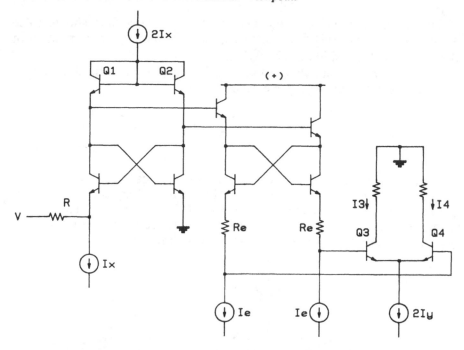

Figure 2.51   A development of Figure 2.50, this multiplier can provide high current-gains and operate between grounded sources and loads.

complementary pair of currents (I-i) and (I+i) suggests that it might also be used to drive an output pair of devices and realize a two-quadrant multiplier. Voltage-to-current conversion and single-sided to differential-mode conversion are simultaneously performed in the cross-quad, and Q1 and Q2 also perform double function as the input pair of a type 'B' TL multiplier cell with Q3 and Q4. The output is, of course, proportional to the tail current 2Iy and the current gain is proportional to Iy/Ix (the actual value depending on how the outputs are used). Thus, we again actually have a two-quadrant divider.

The input of this circuit is shown as being ground-referenced, but the outputs need to be biased above ground. By adding a second cross-quad, this one used as a differential emitter follower (DEF), we can arrange matters so that the output can drive grounded loads, as shown in Figure 2.51. The DEF stage provides current-gain, so reducing the beta-sensitivity seen in the type 'B' cell (see Section 2.4.3) to negligible proportions. It also provides level-shifting, through $2V_{BE}$, bringing the bases of the output transistors to a V_{BE} below ground, or slightly lower, if we include the common-mode voltage drop across the resistors Re. However, these resistors are included primarily to allow any parabolic nonlinearity due to emitter area mismatch (now due to the long and twisted TL path through all eight transistors) to be trimmed out, typically by differential adjustment of the current sources Ie. Note that the central cross-quad is essentially unaffected by mismatches in these currents, or by their absolute value. In fact, in actual monolithic realizations of these ideas, it is possible to omit them altogether, and bias the DEF solely by the base currents of Q3 and Q4, while maintaining excellent linearity! It should also be pointed out that the use of "ordinary" emitter followers here results in a new source of odd-order nonlinearity, inversely proportional to Ie, due to the same base currents causing small errors in the transference of the important differential voltage across the emitters of Q1 and Q2 to the bases of Q3 and Q4.

This circuit has the disadvantage of exhibiting higher noise than other forms, because of the R_B of the extra devices in the signal path. It is also quite difficult to ensure HF stability under all conditions of use. On the other hand, it can operate from very small inputs and provide high current gain, using small values of Ix.

2.6 Miscellaneous TL And TN Circuits

We conclude this chapter with a brief mention of some miscellaneous functions which can be realized in translinear loop (TL) form or simply through dependence on the logarithmic V_{BE} of the BJT, using what we have here called translinear networks (TN circuits).

Quite a large catalog of functions has been generated over the past twenty years, many of them being discovered rapidly in the late 60's and early 70's and left to gather dust. In this section we will only include those which are known to have practical utility. The reader is encouraged, however, to try out "what-if" experiments. We have said very little so far about the new possibilities opened up by the recently- available complementary processes, having PNP transistors which exhibit accurate translinear properties. A thorough discussion of these possibilities will have to wait until a later date, but this section will conclude with a brief mention of the topic.

Figure 2.52 shows one of the earliest extensions of the basic multiplier cells, which was called the "product/quotient" circuit [6] and the name stuck. It is readily shown that the current Io is not very important, as long as it is sufficient to support the output. The TL equation is

$$(1-a)Io \ Iu \ aIo = (1-a)Io \ Iy \ Ix$$

where a is a temporary variable in the range 0<a<1. It follows immediately that

$$Iw = aIo = \frac{IxIy}{Iu}$$

The utility of this circuit lies in the fact that it can generate this output using only currents which sink toward the negative supply, and can thus be supplied from the collectors of further NPN transistors. It will be immediately apparent that the concept can be extended almost without limit,

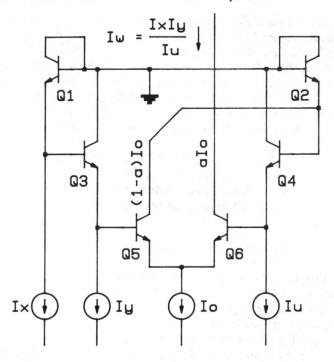

Figure 2.52   The "product-quotient" circuit; the bias current Io is only required to be larger than the output Iw.

to form the quotient of product sequences of the form

$$Iw = \frac{\prod_{k=1}^{n} I_k}{\prod_{j=1}^{n-1} Ij}$$

(2.49)

2.6.1   *Signal Normalization*

In many forms of pattern recognition, signals on multiple channels convey the useful information in the *ratios* of the elements in an array. For example, one type of speech recognition involves determining the power in a number of regions of the overall spectrum, and then performing feature extraction based on the relative levels in these sub-spectra. The absolute level of these signals should not have any direct bearing on the final outcome of classification. In this case, the signal can be thought of in terms of a single-column array, containing between 10 and 20 elements. The signal processing burden in speech recognition is fairly light, because of the low operating

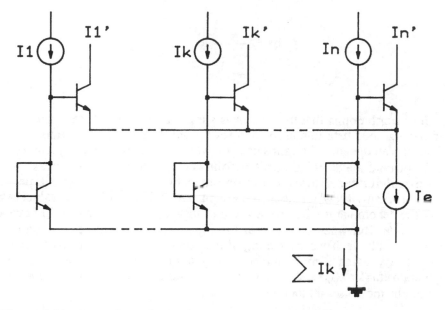

Figure 2.53 **An n-dimensional signal normalizer; the key idea here is that all ratios I1/I1', I2/I2' ... Ik/Ik' must be equal.**

speed required and the small size of the array. DSP techniques are now widely used to perform this function.

In machine vision, the signal may exists as a matrix, of perhaps 100 by 100 elements or even more. Once again, the features of interest (shapes) are embedded in the ratios of all these elements, not in their absolute magnitude. In this case, the computational burden may be quite severe, particularly when rapidly moving visual fields are to be analyzed. In such applications, there may be a role for simple analog circuits to perform the task of signal normalization, that is, to generate an output array in which each element is proportional to the corresponding element in the input array, divided by some factor which is a suitably-derived metric of the overall magnitude of the input, most simply, the largest element of the input array, or possibly the average of all elements.

The original idea to do this in TL form was published in 1968 [6] and later realized in monolithic form [13,14]. The basic circuit is shown in Figure 2.53. The n inputs are the currents I_k, and the outputs are I_k'. Applying TLP, it is readily shown that

$$\frac{I_1}{I_1'} = \ldots \frac{I_k}{I_k'} \ldots = \frac{I_n}{I_n'}$$

(2.50)

from which it follows that

$$I_k' = \frac{I_k}{\displaystyle\sum_{k=1}^{k=n} I_k} \, Ie$$

(2.51)

It is worth noting that this circuit is similar to the type 'A' multiplier cell in having good beta-immunity, and for the same reason (see section 2.4.2). Thus, it can operate with ratios of I_k'/I_k which can be nearly as high as beta. It is a trivial matter to add emitter-followers, if necessary, to increase this ratio still further. However, the output exhibits an undesirable type of pattern sensitivity: when all input elements are at their "full-scale" value, say I_{FS}, the denominator will have a value of nI_{FS}, so all the outputs will have a value Ie/n. But when only one of the inputs is present, the corresponding output will be Ie. Thus the scaling of the output array may vary by as much as $n{:}1$. By using a maximum-function circuit on this output array, to generate $Im{=}\text{MAX}(I_k)$ and continuously adjusting Ie to keep Im at some set-point, the more useful function

$$I_k' = \frac{I_k}{\text{MAX}(I_k)} \, I_{FS}$$

(2.52)

can be generated. This guarantees that at least one of the n outputs will have the full-scale value; all other elements in the output array will assume the same proportions as in the input array. Other types of scaling algorithms and their control loops have been devised [14].

The appeal of this circuit lies primarily in the prospect of performing massively parallel, truly concurrent signal processing. There is no real basic limit to how large n can be. The monolithic circuit described in [14] provided 16 inputs and outputs, but by bringing out the emitters of the array cell transistors, and one other control pin, the circuit could be expanded without limit. An LSI implementation might provide 64 inputs and outputs, with similar expansion facilities.

2.6.2   *Minimum and Maximum Functions*

We mentioned the need for the maximum function in the above circuit. In fact, it was implemented in that case as a negative-maximum-follower (NEGMAX), because the signal polarity at the output happened to be negative, and used a simple voltage-mode circuit based on substrate PNP transistors, which consume very little chip area and were adequately fast for that application. A POSMAX circuit is shown in Figure 2.54; of course, a NEGMAX circuit would reverse the polarity of the transistors and current sources. I1 and I2 would be chosen to minimize the input/output offset. The accuracy in the following mode is quite good when there is more than about

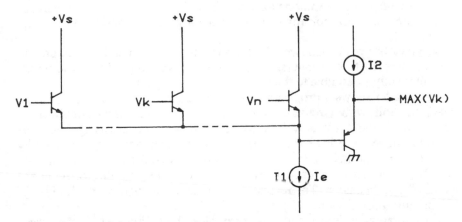

Figure 2.54 A voltage-mode maximum-follower; an even simpler embodiment would use diodes at the input and omit the emitter-follower.

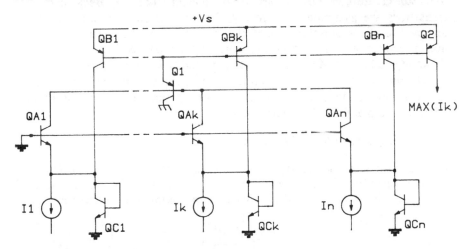

Figure 2.55 A new current-mode maximum-follower; the maximum current flows in all the PNPs QB, thus shutting off all except one NPN QA.

100mV between the inputs; if two inputs should have equal value, the output will be 18mV (V_T ln 2) too low; if all inputs happened to be equal, the output error would be $-V_T$ ln n. It is a straightforward matter to devise voltage-mode POSMAX, NEGMAX, POSMIN and NEGMIN circuits of much higher accuracy using a moderate amount of voltage gain.

These min/max circuits are really impostors in the present context, since they are not strongly dependent on translinearity. In fact, the input transistors could be just diodes, and the output emitter-follower only serves to level-shift. One could just as easily, and perhaps beneficially, use MOS input devices. While all this is true (and was the reason for omitting the more accurate voltage-mode min/max circuits here) it provides a suitable vantage point from which to look for ways of achieving these functions in current-mode.

Seevinck has published two TL forms [4] for minimum and maximum, but they are limited to two inputs, are asymmetric with regard to these inputs and not readily extendible to the general case of n inputs.

Figure 2.55 shows a proposed scheme which could be used as a n-input maximum-following circuit. It, too, is not really a translinear circuit: input devices QAk could just as well be MOS transistors; transistors QCk are just catching-diodes; the PNP transistors could be advantageously replaced by MOS (because of their "infinite current-gain"). Nevertheless, its novelty and possible value in an overall work on current-mode circuits provide the excuse for its inclusion. The reader is encouraged to invent some "truly-TL" n-input min-max circuits!

The operation is simple, and is best explained by letting $I_1=I_{max}$, the largest current. As soon as a small fraction of it flows in QA_1, the beta-boosting transistor Q1 turns on QB_1 to essentially conduct all of I_{max}. Because the PNP transistors are assumed to have equal emitter area, the same current is returned to all the input nodes. However, since it is the largest current, the non-maximum inputs are swallowed up by the mirrored I_{max} and the excess flows in catching diodes QC_k. There is probably something very interesting which could be done with these currents. An extra transistor Q2 delivers I_{max} into a grounded load; alternatively, it could be omitted and an output $n\,I_{max}$ could be taken at the +Vs supply point.

2.6.3   Trigonometric Functions

Considerable work has been put into finding strict-TL forms for synthesis of trigonometric functions. An early circuit for sine synthesis was included [1], reproduced in Figure 2.56. It produces an approximation, of the form

$$Iout = X\frac{1-X^2}{1+X^2}I$$

(2.53)

Here, $-1<X<+1$ corresponds to the angular range $-\pi$ to $+\pi$. The output amplitude is $0.3I$ and peak nonlinearity is 2.7%, which can be reduced to 0.4% with some adjustment to the center current. Figure 2.57 shows the simulated output of this circuit and its error. It is actually a special case of the generic differential form shown in Figure 2.58, which has the general equation

$$(1-A)^n(1+X)^m = (1+A)^n(1-X)^m$$

(2.54)

which solves to

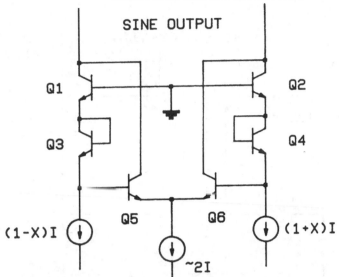

Figure 2.56 A TL sine shaper, providing an output of 30% of the tail current with a law-conformance accuracy of about Figure 2.7%.

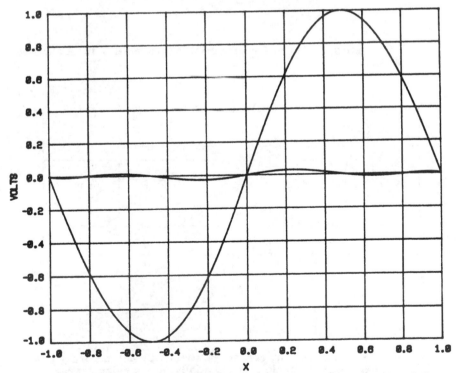

Figure 2.57 Simulated result for the TL sine shaper; the output and the error have been normalized.

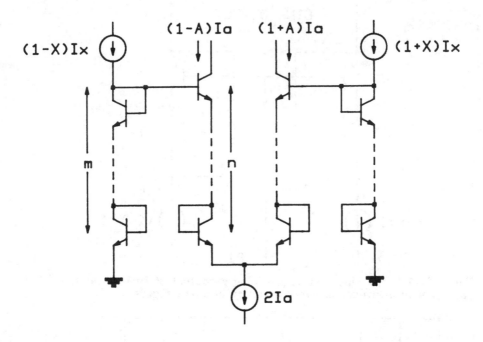

Figure 2.58 A generic TL cell for producing odd-symmetry functions (see text for details).

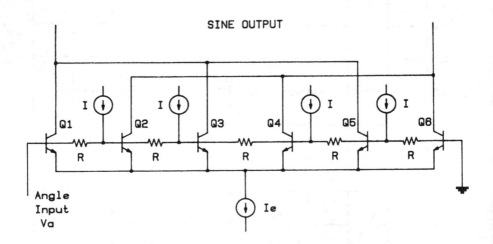

Figure 2.59 TN sine generator, capable of state-of-the-art accuracy and ±360° of angular range, extendible to essentially any angular range.

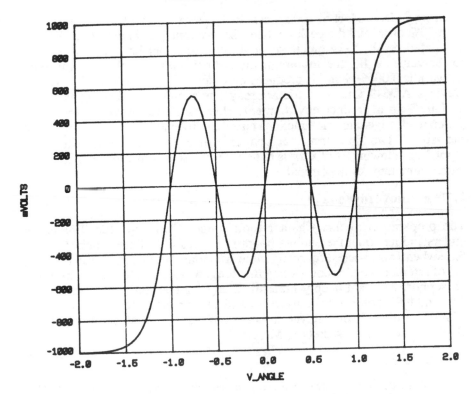

Figure 2.60 Simulation of the six-transistor TN sine circuit; see text for bias conditions.

$$A = \frac{Q^\sigma - 1}{Q^\sigma + 1}, \text{ where } Q = (1+X)/(1-X) \text{ and } \sigma = m/n$$

Extensions of this circuit, and an inverted form, have proven useful in compensating circuits with various types of odd-order nonlinearities.

A very accurate TN realization of the sine function is provided by networks of the form shown in Figure 2.59. A comprehensive study of such networks has been given in [3]. They can be viewed as "periodic transconductances", having an output which alternates in response to the angle voltage Va=V1-V2. When Va=0, the network is fully balanced; Q3 and Q4 equally share the tail current Ie and the differential output is zero. As Va increases, the base voltage of Q3 becomes higher than that of Q4 and a higher fraction of Ie is steered into it. When Va=2.5IR, a local maximum in I_{C3} occurs. This voltage corresponds to an angle of 90°. At Va=5IR, Q2 and Q3 equally share Ie, and so on. For very large values of Va, all of Ie is eventually steered into one output or the other. Of course, the output amplitude can be controlled by Ie.

Using an IR product of 100mV results in a sine amplitude of slightly more than 50% of Ie, and the peak nonlinearity is about 1%. Figure 2.60 shows a simulated result. Very much lower nonlinearity can be obtained using a lower value of IR; the minimum theoretical value for the six-transistor network is 0.006%, but this produces an output of only 20% Ie. The Analog Devices' AD639 Trigonometric Microsystem achieves nonlinearities as low as 0.02% in a product environment; this translinear *tour de force* can generate all trigonometric functions and their inverses [16] by making use of the ratio of two sine terms, each capable of operating with a differential input, thus allowing "90°" offsets to be introduced to convert sine to cosine in either numerator or denominator.

2.7 Acknowledgements

The preparation of this Chapter took longer than expected. For being understanding about the endless hours spent in front of the Macintosh at a time when I was needed for many more important reasons, for preparing the list of references and proof reading this lengthy text, and for never failing to be supportive, I am deeply grateful to my fiancee, Alicia. And for her outstanding support over many years and for numerous services associated with this project (including yet further proof reading!), I sincerely thank my always hard-working secretary, Monica.

2.8 References

[1] Gilbert, B., "Translinear circuits: a proposed classification", Electron.Lett., Vol. 11, pp. 14 - 16, 1975.

[2] Gilbert, B., "A dc-500 MHz amplifier/multiplier principle", Digest Tech. Papers, International Solid-States Circ. Conf., pp. 114 - 115, 1968.

[3] Gilbert, B., "A monolithic microsystem for analog synthesis of trigonometric functions and their inverses", IEEE JSSC, Vol. SC-17, No. 6, pp. 1179 - 1191, 1982.

[4] Seevinck, E., "Analysis and synthesis of translinear integrated circuits" Doctor of Science thesis, University of Pretoria, May, 1981.

[4a] Seevinck, E., "Analysis and synthesis of translinear integrated circuits", Elsevier, 1988.

[5] Gilbert, B., "A four-quadrant analog divider/multiplier with 0.01% distortion", Tech. Papers, International Solid-State Circ. Conf., pp. 248 - 249, 1983.

[6] Gilbert, B., "A precise four-quadrant multiplier with subnanosecond response", IEEE JSSC, Vol. SC-3, No. 4, pp. 365 - 373.

[7] Sheingold, D. H., (Ed.), Nonlinear circuits handbook, Analog Devices Inc., Norwood, MA 1976.

[8] Gilbert, B., "General technique for n-dimensional vector summation of bipolar signals", Electron. Lett., Vol. 12, No. 19, pp. 504 - 505, 1976.

[9]   Gilbert, B., "High-accuracy vector-difference and vector-sum circuits",Electron. Lett., Vol. 12, No. 11, pp. 293 - 294, 1976.

[10] Gilbert, B., "New analogue multiplier opens way to powerful function-synthesis", Microelectronics, Vol. 8, No. 1, pp. 26 - 36, 1976.

[11] Gilbert, B., "A new wide-band amplifier technique", IEEE JSSC, Vol. SC-3, No. 4, pp. 353 - 365, 1968.

[12] Caprio, R., "Precision differential voltage-current convertor", Electron.Lett., Vol. 9, pp. 147 - 148, 1973.

[13] Gilbert, B., "An analog array processor", Tech. Papers, International Solid-State Circ. Conf., pp. 286 - 287, 1984

[14] Gilbert, B., "A monolithic 16-channel analog array normalizer", IEEE JSSC, Vol. SC-19, No. 6, pp 956 - 963, 1984.

[15] Gilbert, B and Holloway, P., "Analog circuit techniques", Tech. Papers, International Solid-State Circ. Conf., pp. 200 - 201, 1980.

2.9   Bibliography

[1]   Paterson, W. L., "Multiplication and logarithmic conversion by operational amplifier-transistor circuits", Rev. Sci. Instr., Vol. 34, pp. 1311 - 1316, 1963.

[2]   Bruggeman, H., "Feedback stabilized four-quadrant analog multiplier", IEEE JSSC, Vol. SC-5, pp. 150 - 159, 1970.

[3]   Sansen, W.M.C., Meyer, R.G., "Distortion in bipolar transistor variable-gain amplifiers", IEEE JSSC, Vol. SC-8, pp. 275 - 282, 1973.

[4]   Traa, E., "An integrated function generator with two-dimensional electronic programming capability", IEEE JSSC, Vol. SC-10, pp. 458 - 463, 1975.

[5]   Ashok, S., "Translinear root-difference-of-squares circuit", Electron. Lett., Vol. 12, pp. 194 - 195, 1976.

[6]   Hamilton, D.J., Finch, K.B., "A single-ended current gain cell with AGC, low offset voltage, and large dynamic range", IEEE JSSC, Vol. SC-12, pp. 322 - 323, 1977.

[7]   Vittoz, E., Fellrath, J., "CMOS analog integrated circuits based on weak inversion operation", IEEE JSSC, Vol. SC-12, pp. 224 - 231, 1977.

[8]   Bahnas, Y. Z., Bloodworth, G.G., Brunnschweiler, A., "The noise properties of the linearized transconductance multiplier", IEEE JSSC, Vol. SC-12, pp. 580 - 584, 1977.

[9]   Huijsing, J.H., Lucas, P., de Bruin, B., "Monolithic analog multiplier-divider", IEEE JSSC, Vol. SC-17, pp. 9 - 15, 1982.

[10] Seevinck, E., Wassenaar, R.F., Wong, C.K., "A wideband technique for vector summation and rms-dc conversion", IEEE JSSC, Vol. SC-19, pp. 311 - 318, 1984.

Chapter 3

Current Conveyor Theory And Practice [1]

Adel S. Sedra and Gordon W Roberts

3.1 Introduction: The Current Conveyor Concept

A current conveyor is a four (possibly five) terminal device which when arranged with other electronic elements in specific circuit configurations can perform many useful analog signal processing functions [1, 2, 3]. In many ways the current conveyor simplifies circuit design in much the same manner as the conventional operational amplifier (op-amp). This stems largely from the fact that the current conveyor offers an alternative way of abstracting complex circuit functions, thus aiding in the creation of new and useful implementations. This together with the fact that the actual terminal behavior of the current conveyor, like the op-amp, approaches its ideal behavior quite closely. This, as most designers of op-amp circuits know, implies that one can design current conveyor circuits that work at levels that are quite close to their predicted theoretical performance. Hence, once the functionality of the current conveyor is understood, the reader should be able to design complex analog circuits using current conveyors rather easily.

At the time of the introduction of the current conveyor (1968) it wasn't clear what advantages the current conveyor offered over the conventional op-amp. Moreover, the electronics industry was just beginning to focus its efforts on the creation and application of the first generation of monolithic op-amps. Without clearly stated advantages, the electronics industry lacked the motivation to develop a monolithic current-conveyor realization. After all, the op-amp concept was entrenched in the minds of many analog circuit designers since the late 1940's and as far as IC manufacturers were concerned an op-amp market was already there to be tapped and expanded. It is only now that analog designers are discovering that the current conveyor offers several advantages over the conventional op-amp; specifically a current-conveyor circuit can provide a higher voltage gain over a larger signal bandwidth under small or large signal conditions than a corresponding op-

---

[1]This chapter is based in part on the material presented in a journal paper entitled " the current conveyor: history, progress and new results, " IEE Proceedings (Part G) April 1990.

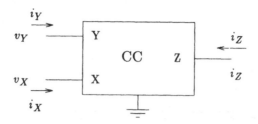

Figure 3.1: Black-box representation of the current conveyor.

amp circuit in effect a higher gain-bandwidth-product [4]. In addition, current conveyors have been extremely successful in the development of an instrumentation amplifier which does not depend critically on the matching of external components, instead depends only on the absolute value of a single component [5, 6, 7].

In this chapter we shall introduce the current conveyor concept in its various forms, demonstrate its various applications, and describe the design and experimental details of a monolithic current conveyor implemented in 5µm CMOS. In addition, we shall outline a method for converting specific types of active-RC circuits [2] into equivalent current-conveyor circuits. An important feature of this approach is that it preserves the sensitivities of the original active-RC circuit. Experimental and simulation examples will be given to illustrate this method.

3.2 The First Current Conveyor (CCI)

The current conveyor (CCI), as initially introduced, is a 3-port device whose black-box representation can be seen in Figure 3.1. The operation of this device is such that if a voltage is applied to input terminal Y, an equal potential will appear on the input terminal X. In a similar fashion, an input current I being forced into terminal X will result in an equal amount of current flowing into terminal Y. As well, the current I will be conveyed to output terminal Z such that terminal Z has the characteristics of a current source, of value I, with high output impedance. As can be seen, the potential of X, being set by that of Y, is independent of the current being forced into port X. Similarly, the current through input Y, being fixed by that of X, is independent of the voltage applied at Y. Thus the device exhibits a virtual short-circuit input characteristic at port X and a dual virtual open-circuit input characteristic at port Y.

In mathematical terms, the input-output characteristics of CCI can be described by the following hybrid equation

---

[2] Active-RC circuits are those that utilize voltage amplifiers together with RC elements.

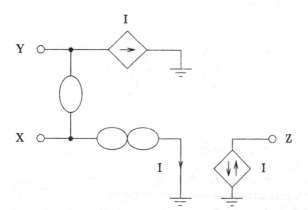

Figure 3.2: **Nullator-norator representation of CCI. The downward arrow in the controlled source at Z is for a CCII+, the upward arrow for a CCII-.**

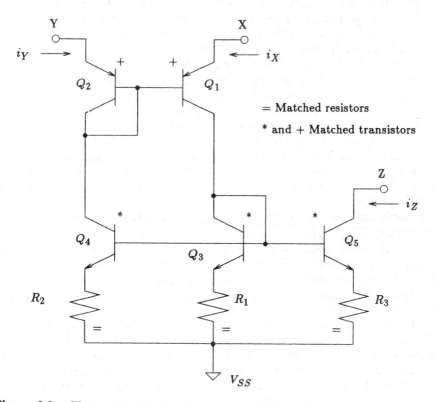

Figure 3.3: **First-order bipolar implementation of CCI.**

$$\begin{bmatrix} i_Y \\ v_X \\ i_Z \end{bmatrix} = \begin{bmatrix} 0 & 1 & 0 \\ 1 & 0 & 0 \\ 0 & \pm 1 & 0 \end{bmatrix} \begin{bmatrix} v_Y \\ i_X \\ v_Z \end{bmatrix}$$

(3.1)

where the variables represent total instantaneous quantities. Note the + sign applies for the CCI in which both Z and X flow into the conveyor, denoted CCI+. The - sign apply for the opposite polarity case, denoted CCI-. To visualize the interaction of the port voltages and currents described by the above matrix equation the nullator-norator [8] representation (commonly referred to as a nullor) shown in Figure 3.2 may be helpful. In this Figure 3.a single ellipse is used to represent the nullator element and two intersecting ellipses to represent the norator element. The nullator element has constitutive equations $V = 0$ and $I = 0$ whereas the norator has an arbitrary voltage-current relationship. Clearly, the nullator element is used to represent the virtual short circuit apparent between the X and Y terminals. Also included in this equivalent circuit are two dependent current sources. These are used to convey the current at port X to ports Y and Z.

A discrete first-order implementation of the current conveyor is depicted in Figure 3.3. Assuming that all marked transistors and resistors are matched and that all transistors have high current gain (both dc and incremental), it can be shown that the currents through transistors Q3 - Q5 are equal. This forces transistors Q1 and Q2 to have equal currents and thus equal V_{BE} drops. Thus X and Y track each other in both voltage and current. Operation of the circuit is independent of the absolute values of resistors and supply voltage as long as linear operation of the transistors is ensured throughout the operating range.

An early application for the current conveyor was its use as a wideband current measuring device [9], an alternative to the oscilloscope current probe based on the Hall effect. Although the current meter based on the current conveyor required that the measured circuit be broken to insert terminals X and Y of the conveyor, very impressive results were obtained. Input impedances less than one ohm and frequency range of operation extending from dc to 100 MHz were measured.

On closer observation of Figure 3.3, one sees that there will be slight differences in the currents carried by different transistors due to nonunity common-base current gains. However, these differences can be reduced, for instance, using more elaborate current mirrors. Also, the polarity of the output current at Z can be inverted by using an additional mirror stage. The

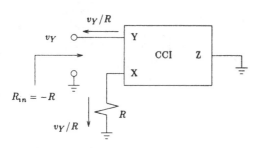

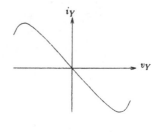

Short-Circuit Stable

(a)

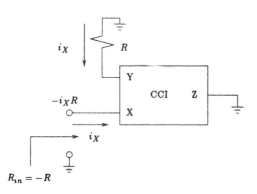

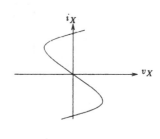

Open-Circuit Stable

(b)

Figure 3.4: CCI implementation of a negative impedance converter.

entire circuit can of course be inverted (pnp's replaced by npn's and vice versa, and the negative supply replaced by a positive supply).

Another early application of CCI was its use as a negative impedance converter (NIC). For this application terminal Z is grounded and the resistor to be converted is connected either between X and ground or between Y and ground. See the two circuits configurations displayed in Figure 3.4. If a resistor R is connected between X and ground, then looking into Y one sees a resistance -R that is short-circuit stable (similar to that observed in a tunnel diode). Alternatively, if R is connected between Y and ground then the input resistance at X is -R and is open-circuit stable (similar to that observed in a silicon controlled switch). It is interesting to note that a recently published paper describes an NIC using a CMOS implemented current conveyor [10]. Connecting two complementary current conveyors as in Figure 3.5 results in a class AB circuit capable of bipolar operation. A numberof variations on this latter circuit have been reported (see for example ref. 11).

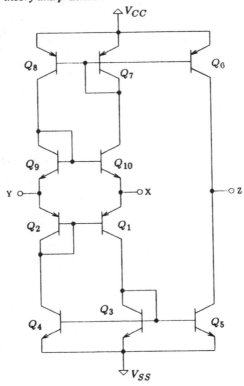

Figure 3.5: A class AB implementation of CCI.

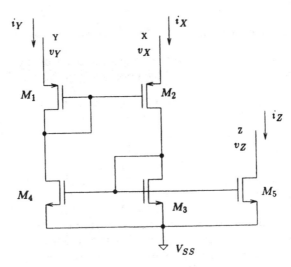

Figure 3.6: First-order CMOS implementation of CCI.

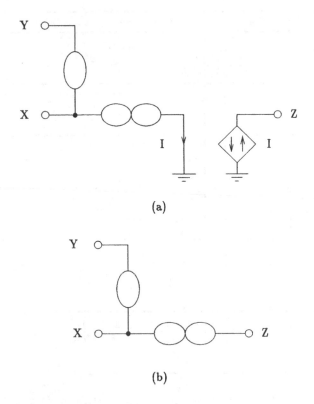

Figure 3.7: (a) Nullator - norator representation of a CCII. The downward arrow in the controlled source at Z is for a CCII+, the upward arrow for a CCII-. (b) A simplified representation of a CCII-.

A major problem that hindered the fabrication of the CCI in IC form in the 1960s is its use of high quality pnp devices. Since complementary devices are available in CMOS technology, it is easy to fabricate a CMOS current conveyor of the type shown in Figure 3.6. In fact a similar circuit has recently been reported by Temes and Ki [12] for application as a high-speed current buffer. Finally, we observe that the four-transistor cell (M1 to M4) in this circuit can be found as part of the biasing network of some high-performance CMOS op amps [13].

3.3 The Second Generation Current Conveyor (CCII)

To increase the versatility of the current conveyor, a second version in which no current flows in terminal Y, was introduced [2]. This building block has since proven to be more useful than CCI. Utilizing the same block diagram representation of Figure 3.1, CCII is described by

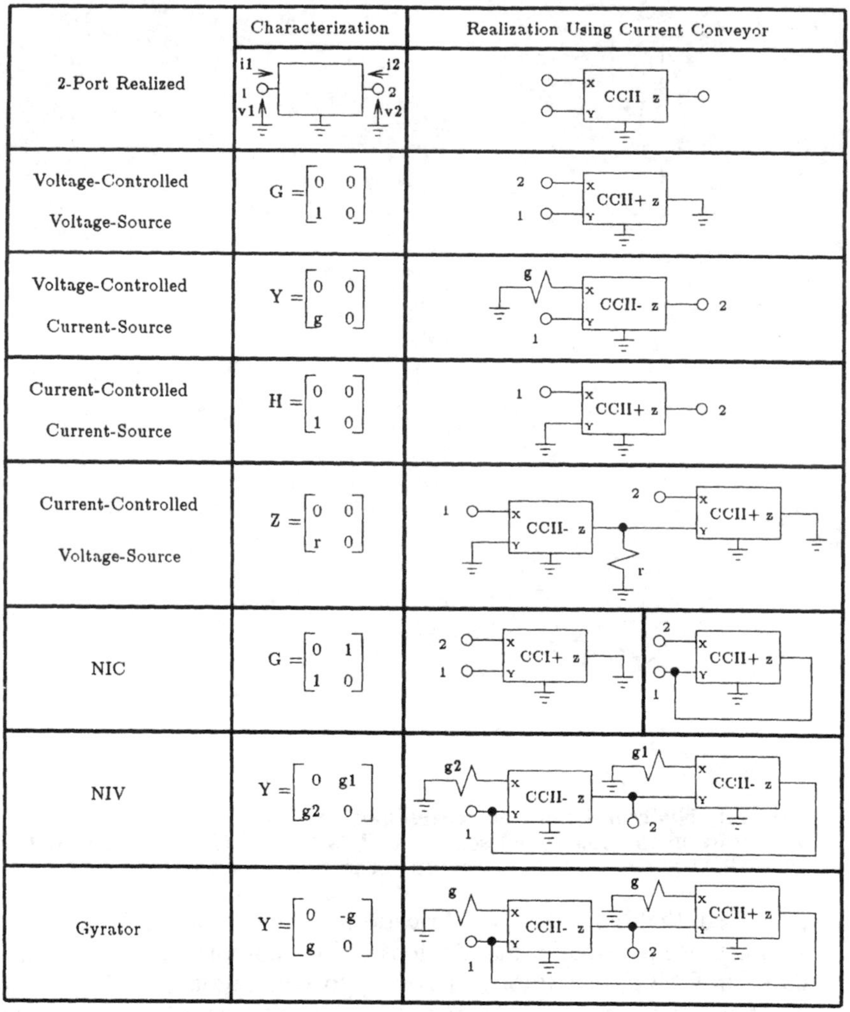

	Characterization	Realization Using Current Conveyor
2-Port Realized	$\begin{array}{c}i1 \rightarrow \quad \leftarrow i2\\ 1 \; \circ \quad \circ \; 2\\ v1 \uparrow \quad \quad \uparrow v2\end{array}$	
Voltage-Controlled Voltage-Source	$G = \begin{bmatrix} 0 & 0 \\ 1 & 0 \end{bmatrix}$	
Voltage-Controlled Current-Source	$Y = \begin{bmatrix} 0 & 0 \\ g & 0 \end{bmatrix}$	
Current-Controlled Current-Source	$H = \begin{bmatrix} 0 & 0 \\ 1 & 0 \end{bmatrix}$	
Current-Controlled Voltage-Source	$Z = \begin{bmatrix} 0 & 0 \\ r & 0 \end{bmatrix}$	
NIC	$G = \begin{bmatrix} 0 & 1 \\ 1 & 0 \end{bmatrix}$	
NIV	$Y = \begin{bmatrix} 0 & g1 \\ g2 & 0 \end{bmatrix}$	
Gyrator	$Y = \begin{bmatrix} 0 & -g \\ g & 0 \end{bmatrix}$	

Figure 3.8: **Applications of current conveyors to active network synthesis.**

$$\begin{bmatrix} i_Y \\ v_X \\ i_Z \end{bmatrix} = \begin{bmatrix} 0 & 0 & 0 \\ 1 & 0 & 0 \\ 0 & \pm 1 & 0 \end{bmatrix} \begin{bmatrix} v_Y \\ i_X \\ v_Z \end{bmatrix}$$

(3.2)

Thus, terminal Y exhibits an infinite input impedance. The voltage at X follows that applied to Y, thus X exhibits a zero input impedance. The current supplied to X is conveyed to the high-impedance output terminal Z where it is supplied with either positive polarity (in CCII+) or negative

Functional Element	Function	Realization Using Current Conveyor
Current Amplifier	$Io = (R1/R2)I1$	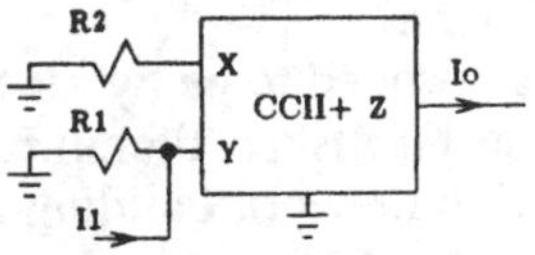
Currrent Differentiator	$Io = CR \, dI1/dt$	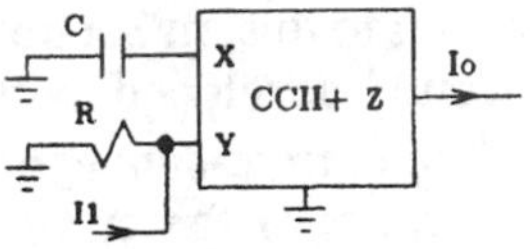
Current Integrator	$Io = 1/CR \int I1dt$	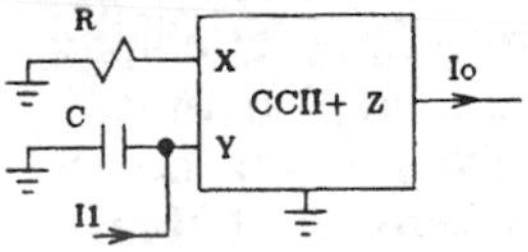
Current Summer	$Io = -\sum_{j}^{n} Ij$	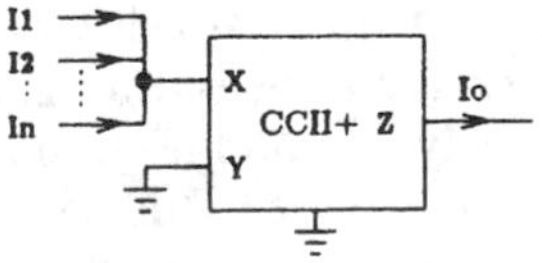
Weighted Current Summer	$Io = -\sum_{j}^{n} Ij \, Rj/R$	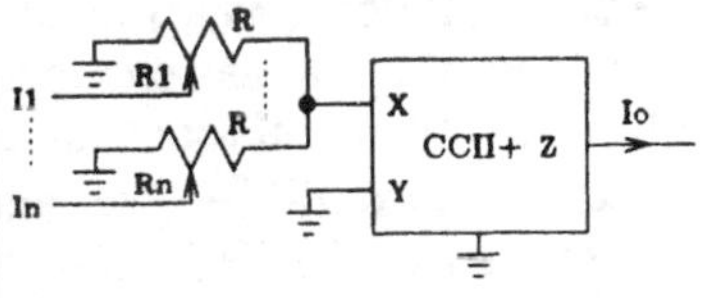

Figure 3.9: Applications of current conveyors to analog computation.

polarity (in CCII-). In terms of a nullor, the port behavior of the second generation current conveyor (positive or negative) can be depicted as shown in Figure 3.7(a). The similarity and difference between a CCI and CCII are clearly evident when one compares their equivalent circuits in Figures. 3.2 and 3.7(a). In the case of a CCII-, the dependent current source is redundant; current flowing into terminal X must flow out of terminal Z. Hence, the equivalent circuit of CCII- can be represented with a single nullor element as shown in Figure 3.7(b).

The first widely-available paper on CCII [2] illustrated its application in the realization of controlled sources, impedance converters, impedance inverters, gyrators, and various analog computation elements. In Figures 3.8 and 3.9 we summarise the results of this paper and present these circuits in table form for easy reference. The circuits presented in these two Figures are subdivided according to their usual roles as either active network elements or analog signal processing elements. A companion paper gave realizations for a number of nonlinear building blocks that had been postulated by Chua [14]. However, these circuits are not commonly used and therefore are not presented here.

CCII has proved to be by far the more useful of the two current conveyor types. The published literature provides CCII realizations for almost all known active network building blocks. A great deal of work has also been reported on the design of active-RC filters utilizing CCIIs [15, 16, 17, 18, 19, 20]. In addition to this, in Section 4 of this chapter, we shall outline a general method, recently developed by the authors, for converting active-RC circuits into equivalent current-conveyor circuits [21].

The inventors of the current conveyor knew that although they had a powerful building block it would have little impact unless it became available in IC form. They attempted to interest Canada's only semiconductor manufacturer at that time (Microsystems International Limited) in the current conveyor with no success. As an alternative, their attention was directed to devising CCII realizations utilizing the then emerging IC op amp. One such realization utilizing an IC op amp and an IC transistor array was reported at that time [22]. It was their view, however, that the op amp is not the most convenient building block for realizing CCII; the op amp is fundamentally a voltage-mode device while the current conveyor is a current-mode device. This view proved not entirely valid in light of the ingenious schemes suggested a number of years later [23, 24, 25].

Until the past half dozen years or so, few circuit realizations of CCII have been reported. The situation has changed dramatically with the appearance of a number of good implementations; some of these utilize op amps alone (for example ref. 26); others utilize IC op amps together with BJT IC arrays [23]; and others yet utilize CMOS technology resulting in fully integrated conveyors [27, 28]. It is also interesting to report that several monolithic bipolar realizations of the CCII- have been fabricated, although these have been labelled as monolithic nullor elements [5, 29]. Many of these realizations can be derived using the rationale outlined in the following.

3.3.1 A Basis For The Circuit Realization Of CCII

The CCII may be viewed as an ideal transistor, either bipolar or MOS. To illustrate this point, consider the NMOS transistor shown in Figure 3.10. If the transistor were ideal, its V_{GS} would approach zero. In such a case a voltage applied to the gate would result in an equal voltage at the source. While the gate terminal would approximate an open circuit (as the conveyor terminal Y), the source terminal would exhibit a zero input impedance (just as the conveyor terminal X). A current injected at the source would be

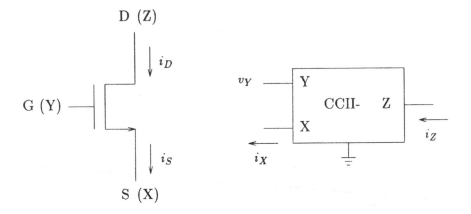

Figure 3.10: Comparison of CCII- and NMOS transistor.

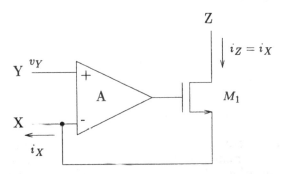

Figure 3.11: Negative current conveyor using a 'super-transistor'.

conveyed to the drain where the impedance level would be infinite (just as terminal Z in the conveyor). It follows that an ideal transistor behaves as a negative current conveyor (CCII-). This equivalence should also be obvious from the fact that the nullator-norator equivalent circuit of a transistor is that in Figure 3.7(b).

To create a more ideal transistor we place the NMOS transistor in the negative feedback loop of an op amp, as shown in Figure 3.11. The result is a CCII- with reasonably good performance. In this CCII- realization, however, current is restricted to flow out of terminal X. An alternative CCII- realization can be obtained by placing a PMOS transistor in the negative feedback loop of an op amp, in which case current will be restricted to flow into the X terminal. It follows that a CCII realization that allows bidirectional current flow can be obtained by placing a complementary pair of MOS transistors in the op amp feedback path as shown in Figure 3.12(a).

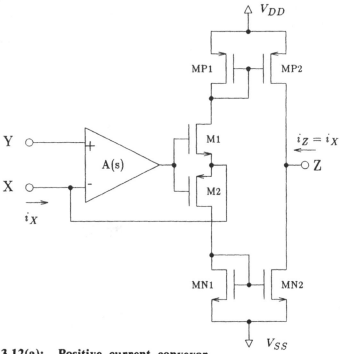

Figure 3.12(a):   Positive current conveyor.

Subsequently, this current is transferred to the output node via a complementary pair of current mirrors. Observe, however, that this circuit is now a CCII+ realization.   To obtain a CCII- circuit two additional complementary mirrors are required, as shown in Figure 3.12(b).

In the circuits of Figure 3.12 the M1 - M2 pair of transistors may be thought of as a class B output stage for the op amp. This circuit, therefore, is not much different than that used by Wilson [4] depicted in Figure 3.13 except that he senses the total supply current of a commercially available op amp (simply because he does not have access to the output stage of the op amp as would be available in a fully integrated conveyor realization).   We have recently fabricated a realization based on the circuits of Figure 3.12 with Northern Telecom's 5μm CMOS process. The design details and the measured performance of these monolithic current conveyor circuits will be discussed below.

3.3.2 A 5μm CMOS CCII Implementation

The current conveyor circuit relies upon the ability of the circuit to act as a voltage buffer between its inputs and upon the ability to convey current between two ports at extremely different impedance levels. Interestingly enough, these two attributes of a current conveyor can be realized

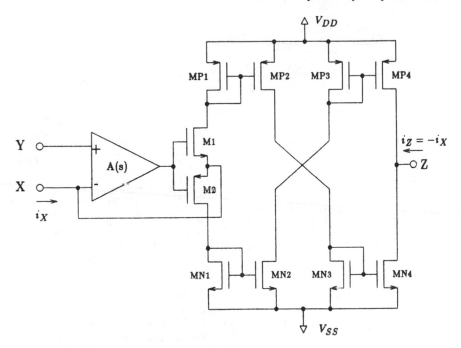

Figure 3.12(b):   Negative current conveyor.

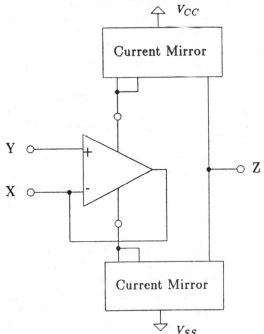

Figure 3.13:   Positive current conveyor implementation

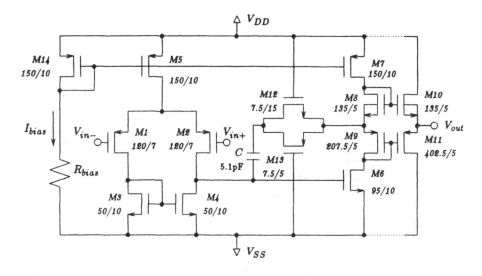

Figure 3.14: Circuit schematic of the CMOS op-amp.

independent of one another and later be combined to form the current conveyor. It is this approach that we have used in the design of our monolithic current conveyor. Specifically, we have fabricated a high gain op amp circuit with a class AB output buffer stage and several different types of current mirrors. Separate pins provide access to the input-output points of each of these building blocks allowing one to investigate various circuit combinations. Invariably the high frequency performance will suffer due to the additional stray capacitances at the pins but this is an acceptable cost when one considers the flexibility purchased.

The op amp topology is based on the popular two-stage CMOS op amp configuration [13] with an additional low output impedance stage. The circuit schematic of the CMOS op amp is shown in Figure 3.14. The drains of the transistors that make up the output buffer stage (M10 and M11) are not connected directly to the power rails but rather connected to external pins of the integrated circuit. This provides a possible means of sensing the current in the output stage, as required in the conveyor realizations of Figure 3.12. In fact, transistors M1 and M2 of the current conveyors depicted in Figure 3.12 are actually transistors M10 and M11 of the CMOS op amp.

The op amp circuit shown in Figure 3.14 utilizes two power supplies of $\pm$ 5V and is biased externally through a resistor to obtain a bias current of 25 μA. M1 and M2 with current source M5 form the differential input stage. M3 and M4 in the current mirror configuration converts the output differential signal to a single-ended signal. Transistor M6 along with biasing transistor M7 form the second stage. M8 and M9 provide a bias voltage for the output stage consisting o f M10 and M11. Devices M12 and M13

Table 3.1 Measured CMOS Op Amp Parameters

Parameters	Experimental Results (IB = 25 mA)	
Open loop gain (50 Hz)	72.7	dB
Output resistance	1051	O
3dB Frequency	400	Hz
Unity Gain Freq	2.2	MHz
Slew Rate (pos)	+2.26	V per msec
Slew Rate (neg)	-6.87	V per msec
Power Dissipation	1.21	mW
InputOffsetVoltage	17.4	mV
PSRR 100 Hz 10 KHz 500 KHz	68.3 47.8 13.1	dB dB dB

along with compensation capacitor C provide the internal compensation of the op amp. M14 with bias resistor R_{bias} supply the bias voltage for current sources M5 and M7.   In Table 3.1 we list a typical set of op amp characteristics measured from a small sample set of fabricated op amps.

The requirements on the current mirrors to be used within the current conveyors are: linear current gain, high output impedance, wide output voltage swing, small input bias voltage, and good high frequency response. The ability to satisfy some of the above requirements depends upon the type of current mirror chosen. There are basically five different types of current mirrors in CMOS technology [13]: (1) simple current mirror, (2) cascode or stacked current mirror; herein referred to as the stacked current mirror, (3) Wilson current mirror, (4) improved Wilson current mirror, and (5) cascode current mirror with improved biasing; herein referred to as a modified cascode current mirror. Prior to fabrication a SPICE investigation was performed to compare the various attributes of the five current mirrors above and thus we came to the conclusion that one must trade-off output impedance and current conveying accuracy for larger output voltage swing. The simple and stacked current mirrors typify this performance trade-off and will be the only mirrors considered for implementation in this work.

3.3.2.1 Positive Current Conveyor

The first circuit implemented is the positive current conveyor since the negative current conveyor can be considered an extension of the positive current conveyor. In the implementation, simple current mirrors were used

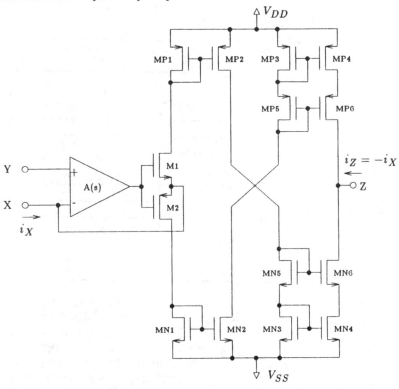

Figure 3.15: Negative current conveyor design using stacked current mirrors.

despite their low output impedance and poor current gain. The reason behind this choice follows from the allowable signal swing at the X terminal. The positive signal swing of the X terminal (refer to Figure 3.12a) is determined by the state of transistor M1 while the negative signal swing is determined by M2. As long as both transistors remain saturated, the output stage of the op amp will perform as expected. Thus the negative signal swing is restricted to $V_{ds\ sat2}$ above the negative input bias voltage of the n-channel current mirror while the positive signal swing is restricted to Vds sat1 below the positive input bias voltage of the p-channel current mirror. Since the supply lines are limited to ±5V, the use of current mirrors with large input bias voltages would severely restrict the signal swing at terminal X.

The circuit implementation of the positive current conveyor is as shown in Figure 3.12(a). The dimensions of the PMOS transistors of the current mirror, MP1 and MP2, are 365μm by 10μm while the dimensions of the NMOS transistors, MN1 and MN2, are 200μm by 10μm. The larger than minimum lengths of the transistors are used to provide an increased output impedance at terminal Z while the large W/L ratio of the transistors results in smaller input bias voltage requirements.

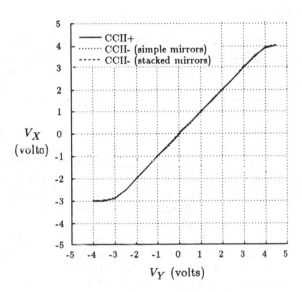

Figure 3.16: Measured large-signal voltage characteristics between terminals Y and X for the positive and negative current conveyor.

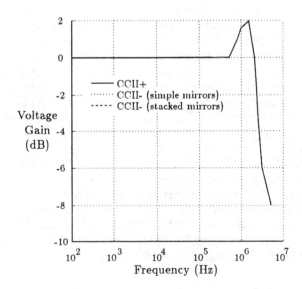

Figure 3.17: Measured small-signal magnitude response of the voltage gain between terminals X and Y.

3.3.2.2 Negative Current Conveyor

The negative current conveyor circuit can be derived from the positive current conveyor through the addition of two current mirrors. Unlike the current mirrors of the positive current conveyor, the additional current mirrors are not restricted to small input voltages. As such, either simple or stacked current mirrors can be used to achieve different current conveyor performances.

A negative current conveyor using simple current mirrors was illustrated in Figure 3.12(b). The dimensions of the transistors making up the simple current mirrors are identical to those used in the positive current conveyor. We depict a negative current conveyor using stacked current mirrors in Figure 3.15. The only difference between Figure 3.15 and Figure 3.12(b) is that the second pair of simple current mirrors have been replaced by a pair of stacked current mirrors. The stacked current mirrors are constructed of 342.5μm by 5μm PMOS transistors and 207.5μm by 5μm NMOS transistors. Reduction of the channel length is possible in the stacked current mirrors due to the high impedance nature of the circuit. The W/L ratios of the transistors within the stacked current mirrors were increased to reduce the input bias voltage as well as providing as large an output signal swing as possible.

3.3.2.3 Experimental Results

The current conveyors were constructed with the appropriate building blocks and measurements were taken to determine the small and large signal characteristics. Due to the building-block concept, the positive and negative current conveyors can physically be constructed with the same op amp and first pair of current mirrors. This provides the opportunity to compare the effects of using different current mirrors with additional variables introduced by employing different op amps and first pair of current mirrors. As anticipated, the voltage transfer characteristics between terminal Y and X are identical for all current conveyors since they utilize the same op amp. The large signal voltage transfer characteristic between terminal Y and X is shown in Figure 3.16.

The test points of Figure 3.16 were taken with terminal X loaded with a 10 MΩ probe and terminal Z grounded. Under this "open circuit" test condition, the maximum voltage difference between Y and X did not exceed 8 mV over the linear range. For the situation with X loaded, the large signal voltage gain remains unchanged provided the current supplied by terminal X does not exceed the current supplying capability of the op amp output stage. Thus, for linear circuit operation, the large-signal voltage gain between terminals Y and X is 1.0 V/V.

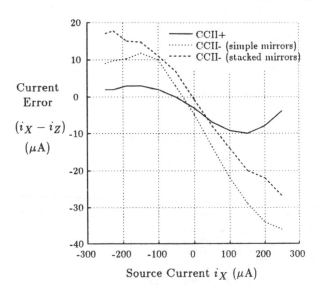

Figure 3.18: Measured large-signal current errors.

The small-signal magnitude response of the voltage gain from terminal Y to X as a function of frequency is shown in Figure 3.17. The same open circuit test conditions were used for this measurement as for the above large-signal voltage transfer characteristics. As expected the bandwidth of this voltage gain response corresponds with the unity-gain bandwidth of the op amp (see Table 3.1) and is the same for all three current conveyors investigated.

Other large-signal parameters related to the input terminals include the input resistance at terminal Y which was measured to be greater than 10 MΩ (the input resistance of the multimeter) while the large-signal input resistance at terminal X was found to be approximately 0.5 Ω.

The large-signal current gain between terminal X and Z (i_Z/i_X) was measured by loading X and Z with 10 kΩ resistances while applying a voltage source at terminal Y (ie. inverting amplifier configuration). In this manner, X and i_Z can be controlled and measured with voltage levels, although, these measurements will be affected by the output resistance at terminal Z. The results of the measurements are plotted in Figure 3.18 where the current difference i_X - i_Z is displayed as a function of i_X. If one excludes any offset current error, one can see from these results that CCII+ has the lowest current error, followed by CCII- (stacked mirrors) and then CCII- (simple mirrors).

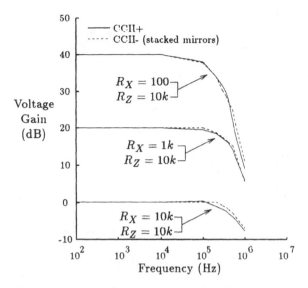

Figure 3.19: **Measured small-signal magnitude response of the voltage gain between terminals X and Z of an inverting amplifier with gain defined by R_Z/R_X.**

The small-signal magnitude response of the current gain from terminal X to Z as a function of frequency can be estimated from the voltage-gains displayed in Figure 3.19 for the above mentioned inverting amplifier test setup. Here we include only those measured results applicable to the positive current conveyor and the negative current conveyor with stacked mirrors. (The frequency response of the negative current conveyor with simple mirrors was not measured). It is evident from the various voltage-gain measurements that the current-gain bandwidth is not as large as should be expected, in fact it is somewhat less than the bandwidth of the voltage gain vY/vX (Figure 3.17). We attribute this loss of current-gain bandwidth to the parasitic poles created by the stray capacitance at the pins interconnecting the amplifier and the various current mirrors, and the poles created by the capacitance shunting the output resistance of the inverting amplifier. This was the price we paid for fabricating our conveyor circuits in building-block-form and for using a voltage gain to estimate a current gain. It is interesting to note that the effective unity-gain-bandwidth product exhibited by these current conveyor circuits is larger than the gain-bandwidth product of the op-amp employed in the realization. For example, the effective 3dB bandwidth of the inverting amplifier with a voltage gain of 100 is 100 kHz, hence the gain-bandwidth product is 10 MHz. This is 4.5 times greater than

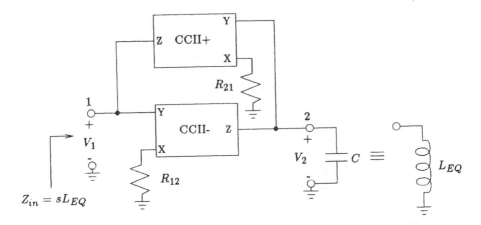

Figure 3.20: Grounded gyrator realization using current conveyors. A grounded inductor is seen looking into port 1 when port 2 is terminated with capacitor C.

the unity-gain-bandwidth of the op-amp. These results corroborate recent results reported by other researchers [30, 31].

The output impedance seen at the Z output of each current conveyor varies depending upon whether the current conveyor is sourcing or sinking current. Specifically, when the current conveyor is sourcing current the output impedance is less than when it is sinking current. This is attributed to the lower output resistance of p-channel devices as compared to their n-channel counterparts. When sourcing currents of approximately 100 μA, the output impedance of the positive current conveyor is approximately 280 kΩ, the output impedance of the negative current conveyor with simple current mirrors is approximately 270 kΩ and the output impedance of the negative current conveyor with stacked current mirrors is approximately 1.0 MΩ.

3.3.3   A 5th-Order High-Pass Filter Example

In the area of active filter design, inductor simulation has attracted considerable interest. The advantage of designing active filters by simulating the inductors of a passive LCR realization of the filter include low component sensitivities and the ability to utilize the extensive knowledge of LCR filter design. One of the methods in which an inductor can be simulated is through the use of a gyrator and a single capacitor. A gyrator can be realized by connecting together two current conveyors of opposite polarity as shown in Figure 3.20 (also illustrated in Figure 3.8). By terminating one of the ports with a grounded capacitor the impedance seen looking into the opposite port is that of a grounded inductor. Measurements have determined that the effective inductance seen looking into port 1 is $L_{EQ} \sim 0.85 C_2 R_{12} R_{21}$

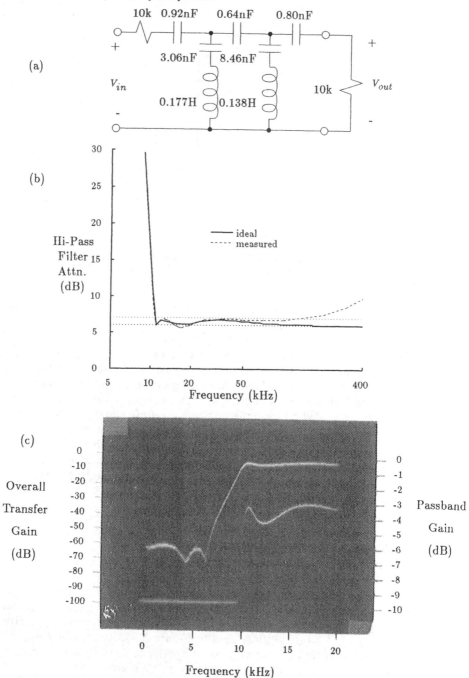

Figure 3.21: (a) **Fifth-order high-pass filter** (b) **passband frequency response** (c) **a photograph of the measured frequency response illustrating both passband and stopband operation.**

when constructed with our monolithic current conveyor. The deviation from the theoretical inductance value of $C_2 R_{12} R_{21}$ is due to the nonunity current gain of each current conveyor.

As an application of this simulated inductance, we have replaced the grounded inductors in the 5th-order highpass elliptic LC ladder prototype network displayed in Figure 3.21(a) with the active circuit shown in Figure 3.20.

Frequency and impedance scaling were used to give the high-pass filter a passband edge of 10kHz and a load impedance of 10kΩ. The circuit was designed to have a passband ripple of 1dB and a minimum stopband attenuation of 54dB. In Figure 3.21(b) the measured and theoretical passband response are plotted. It can be seen that near the passband edge the circuit response closely matches the ideal response. Also observable is the effect of the limited frequency response of the gyrator on the filter. Specifically, for frequencies greater than 100kHz, the filter passband response exceeds the 1dB attenuation margin indicated with the dotted lines. Figure 3.21(c) shows a photograph of the frequency response of the filter as displayed on an HP3580A Spectrum Analyzer. Observable are the two zeros in the filter stopband located at 4.8kHz and 6.8kHz. These values match reasonably closely to the locations of the ladder transmission zeros of 4.66kHz and 6.85kHz; however, the difference is enough to decrease the minimum stopband loss from 54dB to 50dB.

3.4 Converting Active-RC Circuits To Current Conveyor Equivalent Circuits

As previously mentioned, an important attribute of a current conveyor is its ability to *convey* current between two terminals (X and Z) at vastly different impedance levels. In the past this operation of current conveying has largely been associated with unity current gain; however, a nonunity current gain is just as easily implemented. In fact, one could also envision a current conveyor circuit with infinite current gain [31]. In this section we shall demonstrate how many active-RC circuits (specifically those realizing a filtering function) can be converted to a current-conveyor-based circuit with exactly the same component sensitivities. This development is important because it enables one to convert the present repertoire and knowledge of active-RC circuits directly into the current domain without having to "re-invent the wheel" in order to search for low-sensitivity current-mode circuits. The basis for converting between voltage-mode and current-mode circuits is the adjoint network concept. This we discuss next.

3.4.1 Adjoint Networks

A network is considered reciprocal when the same input-output transfer function results as the excitation and the response are interchanged -- see

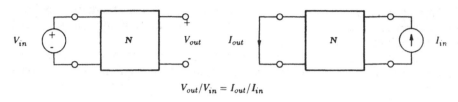

$$V_{out}/V_{in} = I_{out}/I_{in}$$

Figure 3.22: Network N is considered to be reciprocal.

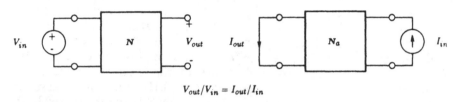

$$V_{out}/V_{in} = I_{out}/I_{in}$$

Figure 3.23: Networks N and N_a are interreciprocal to one another. N and N_a are not necessarily the same network.

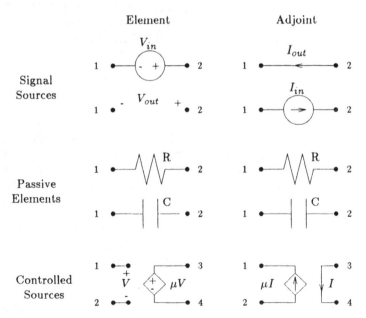

Figure 3.24: Some electrical elements and their corresponding adjoint elements.

Figure 3.22. Although not all networks possess this reciprocal behavior, given any network N, a corresponding network N_a referred to as the *adjoint network*, can be created such that when the excitation and response of network N are interchanged, and network N is replaced by network N_a , the input-output transfer function remains the same. Networks N and N_a are said to be inter-reciprocal to one another [32]. We depict this situation in Figure 3.23 where we illustrate how networks N and N_a have the same input-output transfer function. Reciprocal networks are, by definition, inter-reciprocal with themselves.

Adjoint circuits play an important part in the computation of network sensitivities in many sophisticated circuit simulators, such as SPICE. Basically, component sensitivities of network N are simply the product of a specific current or voltage in network N and the corresponding variable in network N_a [33]. An important attribute of inter-reciprocal networks is that they have identical component sensitivities.

Constructing the adjoint of a given network is a simple task and forms the basis for generating our current conveyor circuits. Basically, the adjoint of a given network is found by replacing each element in that network by another according to the list given in Figure 3.24 [34, 35]. As is apparent from Figure 3.24, the input voltage source is converted to a short circuit and the current through it now becomes the output response variable. Conversely, the port at which the output voltage of the original network is taken is now excited by a current source. By the inter-reciprocal property $V_{out} / V_{in} = I_{out} / I_{in}$. Passive elements R and C in the adjoint network are the same as those in the original network. Lastly, a voltage amplifier with infinite input impedance and zero output impedance transforms into a current amplifier with zero input impedance and infinite output impedance. Although current amplifiers can be realized in various ways, grounding the Y terminal of a current conveyor realizes a high-quality current amplifier. The relationships listed in Figure 3.24 thereby provide the connection between well-known active-RC circuits and current conveyor-based circuits. In the following three sub-sections we shall outline several novel applications of this theory.

3.4.2 Single Current Conveyor Biquads

Circuits that utilize a single voltage gain amplifier and an RC network to realize a second-order voltage transfer function offer a low-cost and low-power approach to the design of active-RC filters. These circuits have been extensively studied [36] and it is well-known which structures have low sensitivity properties. Filter circuits constructed with current conveyors can directly benefit from this wealth of knowledge.

For example, in Figure 3.25(a) we display the well-known lowpass Sallen-Key positive feedback circuit. Its corresponding adjoint network is then simply the network shown in Figure 3.25(b) with the current amplifier replaced by a grounded Y terminal current conveyor. (Note that positive-

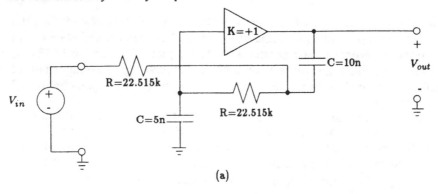

(a)

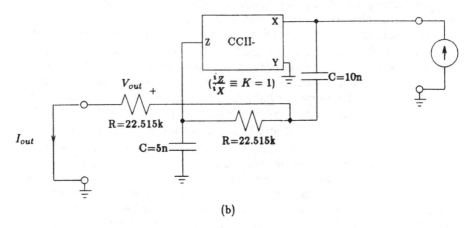

(b)

Figure 3.25: **(a) Voltage amplifier-based SAB (b) Current conveyor-based SAB.**

gain voltage amplifiers transform into negative CCII's). The input-output transfer function T(s) for either one of these circuits is:

$$T(s) = \frac{V_{out}}{V_{in}} = \frac{I_{out}}{I_{in}} = \frac{4KQ^2/R^2C^2}{s^2 + 2/RC[2Q(1-K)+1]s + 4Q^2/R^2C^2}$$

Clearly, the sensitivity of the transfer function of either circuit to any circuit element is the same.

To demonstrate the feasibility of this approach we constructed both circuits of Figure 3.25 and compared their results to the expected behaviour. Specifically, a 2nd-order low-pass Butterworth filter function having a 3dB passband edge at 1kHz was realized by both circuits. In the case of the current-mode circuit (circuit in Figure 3.25(b)) because of the lack of suitable measuring equipment we had to improvise in the way we obtained our current domain transfer function measurements. Instead, we made use of the virtual ground that exists at the input of the circuit shown in Figure 3.25(b) and injected current into this input node via a voltage source and a series resistance. The output current was simply derived from the voltage across the resistor connected to the output port. Except for a possible scale factor correction the voltage transfer function of this voltage driven circuit should be equivalent to the current transfer function I_{out}/I_{in} for the circuit shown in Figure 3.25(b). The measured results are shown in Figure 3.26 together with the ideal response. It is clearly evident that both sets of measurements agree quite closely with the expected results.

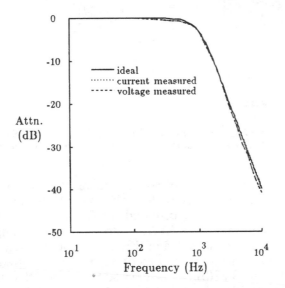

Figure 3.26: Ideal and measured frequency response of voltage and current amplifier-based SAB circuits.

3.4.3 High-Order Filter Realizations

Active-RC SAB circuits can be cascaded or coupled into a feedback loop because of the zero output impedance of the SAB circuit. Likewise, current conveyor-based circuits can also be connected in tandem, or in a coupled structure, but for the opposite reason; the zero input impedance of each single current conveyor biquad. For example, in Figure 3.27(a) we display a 4th-order elliptic high-pass filter circuit constructed with two twin-tee single-amplifier biquad circuits followed by a buffered output attenuator network. This circuit is designed to realize a transfer function having a

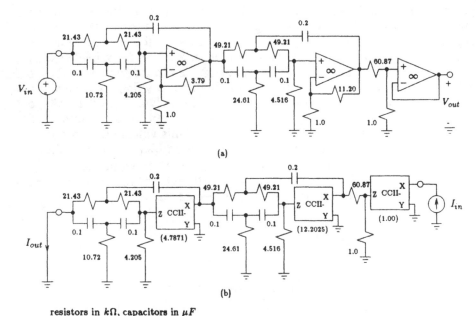

(a)

(b)

resistors in $k\Omega$, capacitors in μF

Figure 3.27: (a) A 4th-order HP elliptic filter circuit as a cascade of op-amp based SABs [37]. (b) a corresponding current conveyor equivalent HP filter circuit.

unity-gain high-pass passband ripple of 3dB beginning at 1000 rad/sec and a stopband between 0 and 500 rad/sec with a minimum attenuation of 40dB. The design details for this circuit can be found in ref. 37.

Converting the circuit of Figure 3.27(a) into its corresponding adjoint network and replacing the current amplifiers with appropriate current conveyor circuits (with nonunity current gain) we obtain the current domain circuit shown in Figure 3.27(b). It is interesting to note that the high-frequency performance of this circuit is far superior to that of the its voltage-mode counterpart shown in Figure 3.27(a). To illustrate this we performed a rather simple circuit simulation assuming that the current conveyors, specifically CCIIs, are realized with identical op-amps having an ω_t of 1M rad/sec as in the active-RC circuit. In addition it is reasonable to assume for low current gains (< 20) that the bandwidth of the current mirrors used in the current conveyors are much larger than the op-amp bandwidth. Accepting these assumptions we can model the current conveyor with the circuit shown in Figure 3.28. The results of the simulations, together with the ideal results, are displayed in Figure 3.29. These results clearly confirm the expectations noted above.

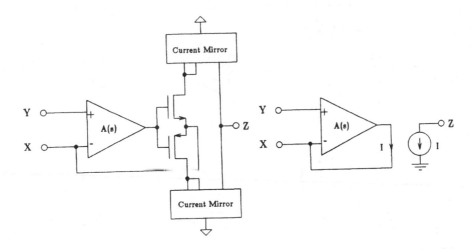

Figure 3.28: A simplified circuit model of a CCII+ taking into account the finite gain A (s) of the op-amp. It is reasonable to assume under moderate current gains that the current mirrors have a much larger bandwidth than the frequency compensated op-amp and therefore its behaviour can be reasonably approximated by an ideal current mirror.

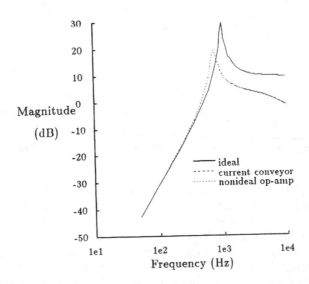

Figure 3.29: The high-frequency response of an op-amp SAB HP filter and that of its corresponding current-conveyor equivalent circuit as compared to the ideal behaviour. Note that the magnitude response of the current conveyor circuit is identical to the ideal response.

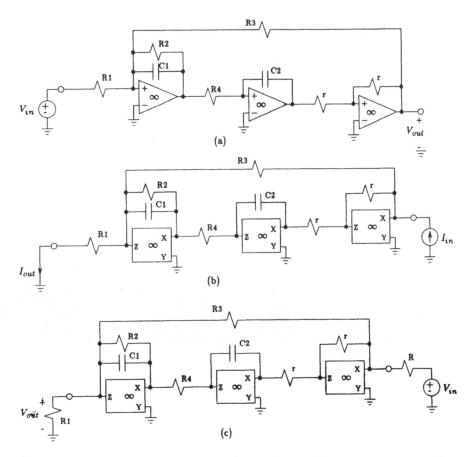

Figure 3.30: (a) Tow-Thomas 3 amplifier biquad circuit (b) corresponding adjoint circuit realized current operational amplifiers (c) same circuit as in part (b) except voltage driven.

3.4.4 Current-Mode Circuits Using Current Operational Amplifiers

In Chapter 4 of this book Toumazou and Lidgey describe an ingenious current-mode circuit that behaves in much the same way as a voltage operational amplifier; however, the input-output signal variables are currents not voltages. Superior large-signal-handling capabilities and wider bandwidths are expected from these circuits than is presently obtainable from a conventional op-amp. They refer to this device as the current operational amplifier. In the context of our work, a current operational amplifier is a special case of a high gain current amplifier with a virtual short circuit appearing at the input terminals. Alternatively, this can also be viewed as a special case of a current conveyor having its Y terminal grounded and possessing large current gain between its X and Z terminals

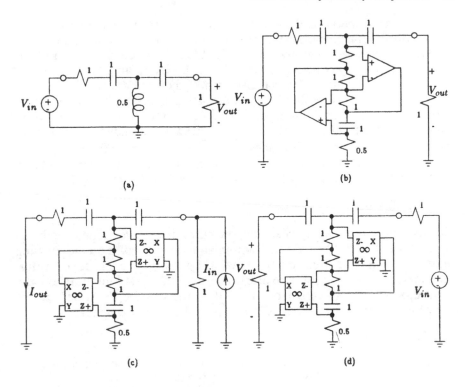

Figure 3.31: **(a) 3rd-order LC ladder (b) simulating the inductor of the LC ladder using an op-amp-based GIC (c) current-based adjoint network of part (b) (d) Thevinin equivalent of part (c).**

Recognizing this, we can apply the sensitivity-preserving adjoint network principle described above to "any" voltage-mode op-amp circuit and convert it directly into the current domain. To perform this we make use of the VCVS - CCCS adjoint relationship listed in Figure 3.24 with μ tending towards infinity.

As an example of this, in Figure 3.30(a) we display the well-known Tow-Thomas 3-amplifier biquad circuit [38]. The adjoint of this circuit is then simply that shown in Figure 3.30(b) with high-current-gain current conveyors realizing the current operational amplifiers. It is interesting to note that further circuit manipulations can be performed on this circuit to create a circuit that has a voltage-to-voltage transfer function instead of the present current-to-current transfer function. Consider that the input current source can be replaced by a feed-in resistor driven by a voltage source connected to the virtual ground appearing at the input of the current conveyor. Conversely, instead of taking the output current as the output response we can consider the voltage appearing across the resistor carrying the output current (R_1) as the output voltage. We depict this new voltage driven current-conveyor circuit in part (c) of Figure 3.30.

The current operational amplifier can also be utilized in the active simulation of inductor elements in LC ladder filter networks. In Figure 3.31(a) we illustrate a high-pass 3rd-order LC ladder network and in Figure 3.31(b) we have replaced the single inductor by its op-amp GIC equivalent circuit. Next, in Figure 3.31(c) we display the adjoint of the circuit displayed in Figure 3.31(b). Note that here we are using a current conveyor with a fully-differential current output. To realize the inverting output an additional current mirror is all that is required to be added to the current operational amplifier circuit proposed in [31]. Converting the input-output portion of this circuit from a current-to-current transfer ratio to a voltage-to-voltage ratio is a simple matter and is displayed in Figure 3.31(d).

3.5 Conclusions

Although more than twenty years old, the current conveyor appears to be an "idea whose time has come". This optimistic view is based on the growing interest in analog current-mode signal processing, the improved fully-integrated conveyor circuit realizations, and the vast literature on the application of current conveyors as well as the method described in this paper for converting voltage-mode circuits to the current domain.

3.6 References

[1]   K. C. Smith and A. S. Sedra, ``The current conveyor--a new circuit building block," Proc. IEEE, Vol. 56, pp. 1368- 1369, Aug. 1968.

[2]   A. S. Sedra and K. C. Smith, ``A second-generation current conveyor and its applications," IEEE Transactions on Circuit Theory, Vol. CT-17, pp. 132-134, Feb. 1970.

[3]   A. S. Sedra, `` A New Approach to Active Network Synthesis," Ph.D. Thesis, University of Toronto, 1969.

[4]   B. Wilson, ``Constant bandwidth voltage amplification using current conveyors," International Journal On Electronics, Vol. 65, No. 5, pp. 983-988, Nov. 1988.

[5]   J. H. Huijsing and J. De Korte, ``Monolithic nullor - a universal active network element," IEEE Journal of Solid State Circuits, Vol. SC-12, pp. 59-64, Feb. 1977.

[6]   B. Wilson, ``Universal conveyor instrumentation amplifier," Electronics Letters, Vol. 25, No. 7, pp. 470-471, March 1989.

[7]   C. Toumazou and F.J. Lidgey, ``Novel current-mode instrumentation amplifier," Electronics Letters, Vol. 25, No. 3, pp. 228-230, Feb. 1989.

[8]   L. T. Bruton, RC-Active Circuits, Prentice-Hall, Englewood Cliffs, New Jersey, 1980.

[9]   K. C. Smith and A. S. Sedra, ``A new simple wide-band current-measuring device," IEEE Transactions on Instrumentation and Measurement, Vol. IM-18, pp. 125-128, June 1969.

[10] B. L. Brennan, T. R. Viswanathan and J. V. Hanson, ``The CMOS Negative Impedance Converter,'' IEEE Journal of Solid-State Circuits, Vol. 23, No. 5, pp. 1273-1275, Oct. 1988.

[11] A. Fabre and P. Rochegude, ``Ultra low-distortion current-conversion technique,'' Electronics Letters, Vol. 20, pp. 674- 676, Aug. 1984.

[12] G. C. Temes and W. H. Ki, ``Fast CMOS current amplifier and buffer stage,'' Electronics Letters, Vol. 23, pp. 696-697, June 1987.

[13] Paul R. Gray and Robert G. Meyer, Analysis and Design of Analog Integrated Circuits, (2nd Ed.), Wiley, New York, 1984.

[14] K. C. Smith and A. S. Sedra, ``Realization of the Chua family of new non-linear network elements using the current conveyor,'' IEEE Transactions on Circuit Theory, Vol. CT-17, pp. 137- 139, Feb. 1970.

[15] K. Pal and R. Singh, ``Inductorless current conveyor allpass filter using grounded capacitors,'' Electronics Letters, Vol. 18, pp. 47, Jan. 1982.

[16] U. Kumar and S. K. Shukla, ``Recent developments in current conveyors and their applications'', Microelectronics Journal, Vol. 16, pp. 47-52, Jan. 1985.

[17] F.W. Stephenson and J. Dunning-Davies, ``Simplified design procedures for a third-order system using current conveyors,'' Electronics Letters, Vol. 17, No. 7, pp. 21-22, March 1979.

[18] R. Senani, ``Novel higher-order active filter design using current conveyors,'' Electronics Letters, Vol. 21, No. 22, pp.1055-1056, Oct. 1985.

[19] C.P. Chong and K.C. Smith, ``Biquadratic filter sections employing a single current conveyor,'' Electronics Letters, Vol. 22, No. 22, pp.1162-1164, Oct. 1986.

[20] P. Aronhime, ``Transfer function synthesis using a current conveyor,'' IEEE Transactions on Circuits and Systems, CAS-21, pp.312-313, March 1974.

[21] G. W. Roberts and A. S. Sedra, ``All-current-mode frequency selective circuits,'' Electronics Letters, Vol. 25, No. 12, pp. 759-761, June 1989.

[22] G. G. A. Black, R. T. Friedmann, and A. S. Sedra, ``Gyrator implementation with integrable current conveyors,'' IEEE Journal of Solid State Circuits, Vol. SC-6, pp. 395-399, Dec. 1971.

[23] B. Wilson, ``High performance current conveyor implementation,'' Electronics Letters, Vol. 20, pp. 990-991, Nov. 1984.

[24] F. J. Lidgey and C. Toumazou, ``Accurate current follower,'' Electron. and Wireless World, Vol. 91, No. 1590, pp. 17-19, April, 1985.

[25] C. Toumazou and F. J. Lidgey, ``Floating-impedance converters using current conveyors,'' Electronics Letters, Vol. 21, No. 15, pp. 640-642, July 1985.

[26] J. L. Huertas, ``Circuit implementation of current conveyor,'' Electronics Letters, Vol. 16, pp. 225-227, March 1980.

[27] M. Nishio, H. Sato, and T. Suzuki, ``A gyrator constructed by CCII with variable current transfer ratio," IEEE International Symposium on Circuits and Systems Proceedings, pp. 93-96, May 1985.

[28] F. Gohh, CMOS Current Conveyors, M.A.Sc. Thesis, University of Toronto, 1988.

[29] J. W. Haslett, M. K. N. Rao, and L. T. Bruton, ``High-frequency active filter design using monolithic nullors," IEEE Journal of Solid State Circuits, Vol. SC-15, pp. 955-962, Dec. 1980.

[30] B. Wilson, F. J. Lidgey and C. Toumazou, ``Current mode signal processing circuits," IEEE International Symposium on Circuits and Systems Proceedings, pp. 2665-2668, June 1989.

[31] C. Toumazou, F. J. Lidgey and M. Yang, ``A novel class AB current amplifier, design and application," Proceedings of the European Conference on Circuit Theory and Design, Brighton, United Kingdom, pp. 315-318, Sept. 1989.

[32] J. L. Bordewijk, ``Inter-reciprocity applied to electrical networks," Appl. Sci. Res., B6, pp. 1-74, 1956.

[33] S. W. Director, and R. A. Rohrer, ``The generalized adjoint network and network sensitivities," IEEE Transactions on Circuits and Systems, CT-16, pp. 318-323, Aug. 1969.

[34] G. C. Temes and J. W. LaPatra , Introduction to Circuit Synthesis and Design, McGraw-Hill, 1977.

[35] P. Penfield; Jr., R. Spence, and S. Duinker, Tellegen's Theorem And Electrical Networks, MIT Press, Cambridge, Massachusetts, and London, England, 1970.

[36] A.S. Sedra, M.A. Ghorab, and K. Martin, ``Optimum configurations for single-amplifier biquadratic filters," IEEE Transactions on Circuits and Systems, CAS-27, pp. 1155-1163, Dec. 1980.

[37] E. J. Kennedy, Operational Amplifier Circuits Theory and Applications, Holt, Rinehart and Winston, Inc. New York, 1988.

[38] A.S. Sedra and P.O. Brackett, Filter Theory and Design: Active and Passive, Matrix Publishers, Inc., Portland, OR, 1978.

Chapter 4

Universal Current-Mode Analogue Amplifiers

Chris Toumazou and John Lidgey

4.1 Introduction

Over the years the semiconductor industry has endeavoured to provide circuit designers with a cheap, high quality, versatile analogue building-block and many amplifier designs have been manufactured with the sole aim of producing a controlled voltage output from a voltage input. The most popular of these networks is the high gain voltage-mode operational amplifier (VOA). It has, thus, become customary for electrical engineers to think of analogue signal processing in terms of voltage variables and this tendency has resulted in numerous voltage signal processing circuits. For example, the operational amplifier is conveniently configured into controlled voltage source amplifiers, such as the voltage controlled voltage source (VCVS) and current controlled voltage source (CCVS), but not so easily configured into controlled current source amplifiers such as the voltage controlled current source (VCCS) and current controlled current source (CCCS).

There is often a demand in analogue signal processing for amplifier circuits that possess well-defined current signal processing properties. Furthermore, current amplifier based circuits can offer certain high performance properties, such as speed, bandwidth and accuracy, which make them preferable to voltage amplifier designs. In view of this there is an obvious need for high quality current amplifier based analogue networks, which can provide comparable, if not superior, performance to traditional voltage amplifier designs. This potential has been recognised by many circuit designers, and a major area of research over the past two decades has been the design and application of current-mode techniques.

The contents of this Chapter have been arranged into two major parts. The first part concentrates upon techniques which extend the capabilities of the VOA to achieve current-mode performance. Several current-converter circuit designs have been reported which employ the VOA, attempting to provide it with a well-defined current output facility. One of the most successful techniques is VOA supply current sensing, where current-mirrors are used to sense the output current of the VOA via its power supply rails. The authors have developed a number of novel circuit applications based upon VOA supply current sensing and a review of this work will be presented.

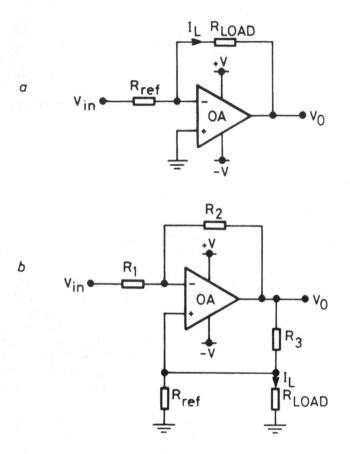

Figure 4.1 Current-Converters using the VOA.(a) Feedback current source [1] (b) Howland [2] current source.

The elegant concept of the current-conveyor, introduced by Smith & Sedra in 1968 (see Chapter 3) is essentially a voltage/current-mode hybrid circuit. The versatility and high performance features of the current-conveyor render it an extremely powerful analogue building block. This is confirmed by the wealth of literature and the well established 'Current-conveyor theory and practice', the title of Chapter 3. Also, in the first part of this Chapter, we show how high performance properties of current-conveyors and current-converters realised using the VOA supply current sensing technique are obtained.

Though it is somewhat ironic, we also show that circuits which use VOAs to achieve current-mode performance can often, in turn, deliver an improved performance when cofigured as voltage amplifier circuits.

Whilst the extended VOA approach to current-mode performance does work well, it is inherently limited by the voltage processing architecture. It is interesting to note that the conventional VOA architecture has remained essentially unchanged, mainly due to technology restrictions, over the last two decades, despite clear evidence that high performance can be obtained from current-mode designs (see Chapter 2 and Chapter 12). Advances in complementary bipolar technology have been catalitic in the development of integrated current-conveyors (Chapter 15) and a new breed of commercially available op-amp, referred to as a transimpedance or current-feedback op-amp (see Chapter 16). Like the current-conveyor, the transimpedance amplifier is a voltage/current-mode hybrid circuit and similarly it is the current-mode Section of the circuit that is responsible for its high performance features.

With recent advances in technology, we are now presented with the opportunity of developing a realizable true current-mode op-amp. The second part of this Chapter deals with the design and development of such a differential input/output current-mode op-amp. We show that such a building block has greater versatility than the VOA and all 4 main amplifier topologies can be accurately configured in closed-loop, including a closed-loop current-converters and current-conveyors.

4.2 Review of Traditional Current Output Circuits Using Voltage Operational Amplifier Designs

The most commonly used amplifier having a controlled bipolar current output and frequently encountered in standard designs [1, 2], is to place the load in the feedback loop of a voltage operational-amplifier as shown in Figure 4.1(a). Although the voltage gain of the amplifier varies with R_{LOAD}, the current in the feedback loop remains fixed, assuming a fixed V_{in} and R_{ref}. However, one of the limitations of this approach is that the current driven load is not ground referred and a desired feature of most current amplifier designs is the ability to feed a well-defined current into a grounded load. Despite this drawback, the circuit does have some interesting features. The input impedance is extremely low, as nearly all the input current is drawn through the load by the action of negative feedback. The input current only differs from the load current by the current flowing into the VOA, which is extremely small due to the high open-loop voltage gain and input impedance of the VOA.

The circuit shown in Figure 4.1(b) and developed by Howland [1], solves this ground referenced problem. The circuit will act as a current source of, $I_L = -V_{in}/R_{ref}$, for the condition that $R_3/R_{ref} = R_2/R_1$. If the ratio of these resistances are equal the circuit will function with a theoretically infinite output resistance, determined by the combined positive and negative feedback action of the VOA. However, a major drawback of this combined feedback approach is that very small departures from ideal balance conditions either dramatically reduces the output resistance or results in instability because the output resistance becomes negative.

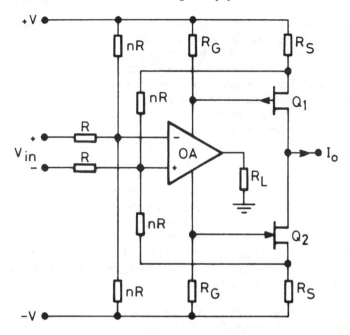

Figure 4.2   Graeme's [3] controlled current source

Despite attempts to improve on the performance of the Howland circuit, the critical resistor matching constraints and potential instability remains a fundamental problem. The Howland design is typical of many similar topologies, some employing several op-amps, which have the same problems of requiring closely matched resistor networks, a combination of positive and negative feedback resulting in potential instability and low bandwidth. A new generation of amplifier circuits, which extend the VOA to achieve a current output capability, is described in the next Section.

4.3 Operational Amplifier Supply Current Sensing: Review and Developments

The supply current sensing technique makes use of the fact that the sum of the currents in the supply leads of the VOA equals the output current, provided that no other connections exist with ground that carry substantial signal currents.   In most VOAs this condition is met.   Current-mirrors are employed to sense the split signal currents flowing through the VOA's supply rails and to recombine them at a single high impedance bipolar output. Using this technique a number of novel current converter topologies have been realised.

The technique of supply current sensing was reported by Graeme [3] in 1973. He showed that a precise voltage-controlled-current source (VCCS) could be achieved by using a pair of complementary field-effect transistors to sense the current flowing in the supply rails of a conventional VOA, as shown in Figure 4.2. The circuit consists of opposing FET current sources Q1 and Q2, controlled by the high gain feedback around the VOA. The difference in FET currents produces the output current, and this difference current is controlled by comparing the input voltage V_{in} to the feedback voltage provided by the current sensing source resistors, R_s. The circuit's output current is controlled by the input voltage to within the accuracies of the resistors selected and within the gain-bandwidth and power supply rejection limitations of the VOA used. In general, the circuit provides good results, such as high precision and wide bandwidth and avoids the output impedance uncertainty of traditional designs. However, as a consequence of using two independent feedback connections, one for each output polarity, the circuit performance is again sensitive to resistive mismatch. Supply-current sensing was not an entirely new mode of VOA operation as it had been employed previously by Garza [4] as a means of producing greater power gain from a low-power VOA, but the full versatility of the technique was not recognised at that time.

In 1975, Hart & Barker [5] suggested an alternative method, in their realization of a class-B voltage to current converter. The circuit similarly employed the VOA as its main gain block, together with a set of complementary current-mirrors used to sense the phase split output current flowing through the collector leads of a class-B output stage as shown in Figure 4.3(a). The class-B cascode transistor connection comprising Q1 to Q4 provides the circuit with high common-emitter current gain together with a high breakdown voltage and high output resistance. Positive and negative current-mirror circuits, denoted by P and N respectively, comprise essentially the improved 4-transistor Wilson, G.R. [6] current-mirror, together with an output connected FET to improve the circuits overall output impedance and maximum output voltage swing. Improved Wilson current-mirrors were used for their high output impedance and accurate current transfer performance, as discussed by Hart and Barker [7].

Although Hart and Barker's [5] class-B scheme catered for much higher load voltage excursion than did Graeme's proposal [3], and avoided the undesirability of resistor matching, the class-B mode in which the circuit operates resulted in considerable cross-over distortion. This problem was identified by Rao and Haslett [8], who showed that much higher frequency performance together with improved output current drive could be obtained if the output circuit were operated in class-AB. In a further paper [9] they related the class-AB voltage following action to that obtained by the classical class AB pushpull output stage of a conventional VOA and showed how the output signal current could now be sensed via the VOA's supply leads using a current-mirror arrangement. This brought together Graeme's [3] original proposal of current sensing VOA supply leads and Hart and Barker's [5] use of complementary current-mirrors as external current sensing elements. The net result was the universal class-AB VOA structure shown in Figure 4.3(b) which could be configured to provide a number of accurate current

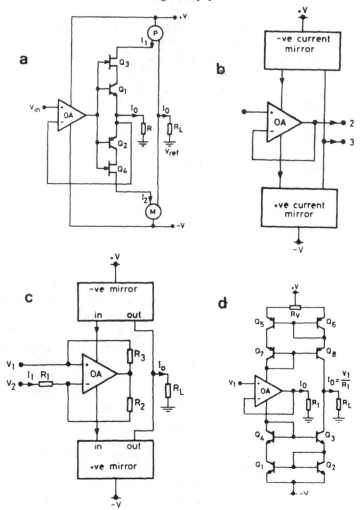

Figure 4.3 Current converter techniques using VOA supply current sensing (a) Hart and Barkers [5] class B voltage to current converter (b) Haslett and Rao's [9] class AB voltage to current converter (c) Hart and Barker' s [10] universal converter network (d) Hart and Barker' s [10] practical converter using modified Wilson current- mirrors.

converter topologies. In fact, as will be shown later in Section 4.5 this circuit realises the very versatile CCII current conveyor.

A similar modified VOA structure had also been reported by Hart and Barker [10] as a universal operational amplifier converter. In this design an arrangement of feedback resistors within the VOA circuit could be selected to configure the network into any of the four main amplifier topologies. A block diagram of the circuit is shown in Figure 4.3(c) and a voltage to

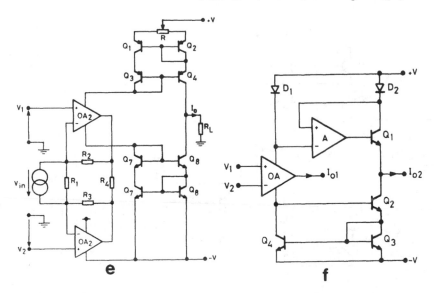

**Figure 4.3 (cont)   (e) Nedungadi' s [11] differential current-converter.
(f) Huijsing and Veelenturf' s [12] Operational Mirrored Amplifier (OMA).**

current converter using current-mirrors in the supply leads is shown in
Figure 4.3(d).

Nedungadi [11] proposed the more versatile current converter structure
shown in Figure 4.3(e).   The circuit operation was similar in principle to
that of the universal converter system of Figure 4.3(c) and could similarly be
configured into all four main amplifier converter topologies by the
appropriate resistor selection. However, the addition of OA2, at the input of
the circuit provides differential input properties and therefore the ability to
configure differential voltage to current (DVCC) and differential current to
current (DCCC) converter functions.

The poor high frequency performance of these schemes, due to the lateral
PNP transistors available at that time, was avoided by Huijsing and
Veelenturf [12]. They replaced the negative current-mirror by an elegant
negative current-mirror source developed by Barker and Hart [13], which
employed NPN transistors and a local amplifier. This total NPN converter
structure was termed an operational mirrored amplifer (OMA) and is shown
in Figure 4.3(f).   As a four-port general purpose analogue building block it
can be configured to provide all the previously described controlled current
converters, in addition to the closed-loop controlled voltage converter
circuits possible with the VOA section of the circuit. However, as a discrete
design the high frequency performance was now limited by amplifier, A, of
the negative current-mirror circuit, but a monolithic realisation was
implemented by Huijsing and Veelenturf [12], providing a much higher
frequency performance.

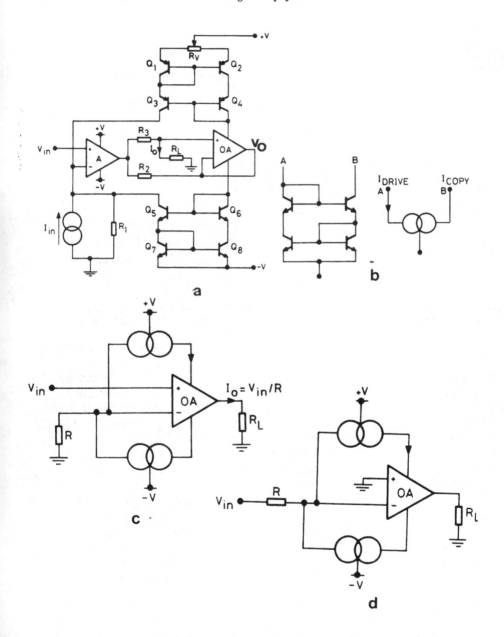

Figure 4.4 Low distortion current-converters (a) Nudungadi's [13] class AB high current converter.(b) Wilson' s current mirror symbol [14] (c) Wilson's [14] voltage to current converter using series feedback (d) Wilson's [14] voltage to current converter using shunt feedback.

A further problem encountered with these class-AB current converter designs, is their current output limitations imposed by either the VOA or transistors used to construct the current-mirror circuits. Nedungadi [13] shows that using a similar converter structure to Figure 4.3(a) together with a high current output, class-AB operational amplifier, current outputs greater than 100 mA could be obtained while still maintaining high conversion accuracy and high efficiency. This high current, converter technique is shown in Figure 4.4(a). The essential difference between Figure 4.4(a) and previous schemes is that the current-mirrors, instead of being outputs, are fed back to the input of the circuit, while the converter output is taken from the operational-amplifier. This current-mirror feedback arrangement ensures that only the input current is passed through the current converter section of the circuit, irrespective of the output current magnitude. For example, with a current gain A_i of -100, and an output current I_0 of 100 mA, the converter need only supply 1 mA, which is simply and accurately achieved using standard VOAs and small signal transistor arrays. Results of Nedungadi's [13] work also show that with the current-mirrors connected in this feedback arrangement an improvement in the converter's output signal distortion is possible. Further work in this area was reported by Wilson, B. [14]. Wilson described the main cause of quite significant output signal distortion, in converters of the type shown in Figure 4.3(a) to Figure 4.3(f) being due to the current-mirrors connected in open-loop at the circuit output, where collector-voltage modulation effects are significant. Wilson [14] showed that by adopting a similar approach to Nedungadi [13] and connecting the current-mirrors in a feedback arrangement, this source of voltage modulation could be removed. Shunt and series feedback voltage-to-current converter schemes were described, and shown to provide an improvement in second harmonic distortion, (SHD), of up to 28 dB at 1 kHz over all previous converter schemes. During this work a current-mirror symbol was also introduced, to be used for convenience when representing the current-mirror circuits. This current-mirror symbol, together with circuits for the low-distortion series and shunt feedback converters are shown in Figure 4.4(b) and Figure 4.4(c) and Figure 4.4(d) respectively. Wilson went on to show how previous current amplifier designs, based upon the open-loop mirror approach could also be improved to the low-distortion type by simply connecting the mirrors in shunt feedback. However, by connecting the current-mirrors in this feedback arrangement is purely a means of referring the main distortion component of the converter to the input side of the circuit, the current transfer accuracy of the converter is still limited by the current transfer performance of the current-mirror circuits. Furthermore, this feedback arrangement results in much poorer frequency performance than with the 'open-loop' converter structure.

This new trend of using current-mirrors to sense the VOAs output current and provide well-defined bipolar output properties proved far superior to traditional feedback techniques and by the late 1970's to early 1980's the advancement in this area had been well documented.

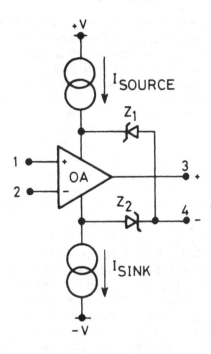

Figure 4.5 Nordholt' s [15] **universal operational amplifier with floating supplies.**

The review of work so far has essentially considered techniques to provide the conventional VOA with a single controlled bipolar current output. In 1982, Nordholt [15] adopted a slightly different approach and described how the supply rails of the conventional VOA could be configured to provide a floating output. This was achieved by feeding the d.c. power to the VOA from a balanced current source and sink and using two series zener diodes to fix the supply voltages. The net result was a universal amplifier structure effectively biased with floating d.c. supplies as shown in Figure 4.5, which could provide differential output as well as differential input properties. Numerous current-converter circuit topologies were then easily realisable using this adapted VOA. Furthermore, with this approach current conversion accuracy no longer relied upon the current transfer properties of current-mirrors.

4.3.1   *Power Supply Current Distribution*

So far we have shown that VOA supply current sensing technique is a very attractive means of effectively accessing the VOAs output current. As a result a number of elegant current converter topologies have been realised. It is thus important to understand how the output current of the VOA distributes itself amongst the power rails of the VOA and to understand whether the

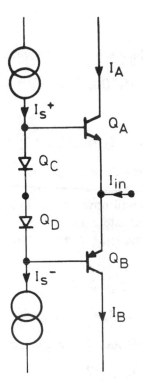

Figure 4. 6 Typical Class AB push-pull output stage.($I_S^+ = I_S^- = I_S$)

distribution is linear or non-linear. In this Section we will give a full analysis of the current sharing action of the VOAs output current between the collector/drains of its class AB output stage.

A typical diagram of the basic push-pull output stage of an operational-amplifier is shown in Figure 4.6. Current I_S is the d.c. bias current provided from a constant current source, which supplies the bias for diode connected transistors Q_C and Q_D, and hence biasses output transistors Q_A and Q_B in the forward active region. Input current I_{in} is in effect the operational-amplifier's feedback current, if the VOA is connected as a current follower or conveyor.

The circuit of Figure 4.6 is a good approximation to a translinear loop comprising transistors Q_A to Q_D. The translinear circuit principle introduced by Gilbert [16] and reviewed in Chapter 2 relates the base-emitter voltages of N transistor junctions in a closed-loop. Assuming each transistor in Figure 4.6 to be at the same temperature, have equal emitter areas, and operate under identical conditions, then by application of the translinear circuit principle around the loop Q_A to Q_B, it can be shown that the balance of currents is given by

$$I_s^2 = I_A.I_B \qquad (4.1)$$

Application of an input current $I_{in} > 0$ it can be shown that

$$I_B = I_{in}/2 + I_s [(I_{in}/2I_s)2 + 1]^{1/2} \qquad (4.2)$$

and

$$I_A = -I_{in}/2 + I_s [(I_{in}/2I_s)2 + 1]^{1/2} \qquad (4.3)$$

If $|I_{in}| << I_s$, then these equations reduce to $I_A = I_s - I_{in}/2$ and $I_B = I_s + I_{in}/2$ and the circuits operating mode can be regarded as class-A only. If, however, $|I_{in}| >> I_s$, the circuit will now operate in a class-B and therefore $I_A => 0$, and $I_B => I_{in}$. It is of interest to examine and quantify the non-linear distribution of I_{in} in I_A and I_B, between these two extremes. I_A and I_B can be rewritten as

$$I_A = I_s - \delta I_{in} \qquad (4.4)$$

$$I_B = I_s + (1 - \delta) I_{in} \qquad (4.5)$$

where δ is the input current distribution factor given by

$$\delta = 1/2 - I_s/I_{in} [((I_{in}/2I_s)2 + 1)^{1/2} - 1] \qquad (4.6)$$

Graphs of measured and theoretical current distribution factor δ versus input current I_{in} for two operational-amplifiers, are shown in Figure 4.7. For an LM741 operational-amplifier (curve a) I_s was measured to be approximately 0.254 mA and for the LF441 operational-amplifier (curve b) I_s was 0.036 mA. From the graphs it can be seen that good correlation for both operational-amplifiers exists between theoretical values of δ and those obtained experimentally.

The graphs indicate the expected wider distribution of input current between the operational-amplifiers supply leads in the case of the larger supply current LM741 operational-amplifier. For example for a value of $I_{in} = 1$ mA, in the case of the LM741 operational-amplifier approximately 30% of the input current is drawn by the positive supply rail and so the remaining 70% is drawn by the negative supply rail. However, for the LF441 operational-amplifier only about 5% of the input current is drawn by the positive supply rail, the remaining 95% being drawn by the negative supply rail. In the converter schemes where the outputs are added the δ factor is not so important. However, in applications where the current-mirrors sense just one of the power rails such as the precision rectifier circuits to be described in Section 4.6.4, then it is important that δ is

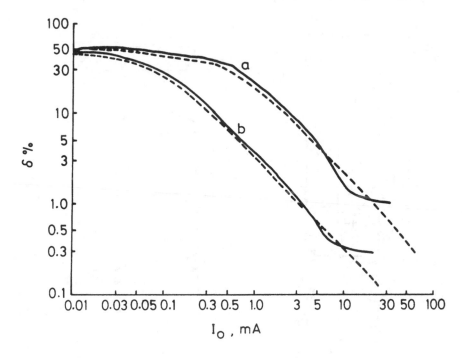

Figure 4.7 Graphs of supply current distribution factor $\delta\%$ versus VOA output current. Curve (a) LM741, Curve (b) LF441

————— δ experimental   ------------ δ theoretical

minimised. This can be achieved by using very low supply current VOAs, or for manufacturers to allow access to the VOAs output collectors (drains), since this is the correct sensing node.

4.4 Current-Followers

A number of proposals to develop a universal operational amplifier structure, based upon the standard VOA, and which could be configured to provide well defined current converter properties have been considered. This work can be categorised into two groups. One group considers techniques that make two equal in-phase output currents available by current-mirroring the VOA's supply currents with respect to the positive and negative supply voltages, as described above, and summing them to a single ended output (supply current sensing). The other group provides the VOA with two equal but opposite output currents, with characteristics similar to that of a "Nullor" (floating VOA).

In order to assess the optimum current converter performance of both the floating VOA structure of Figure 4.5 and supply current-sensing technique of Figure 4.3(b) both structures were configured into unity gain current-controlled current sources, alternatively referred to as current-followers [17,

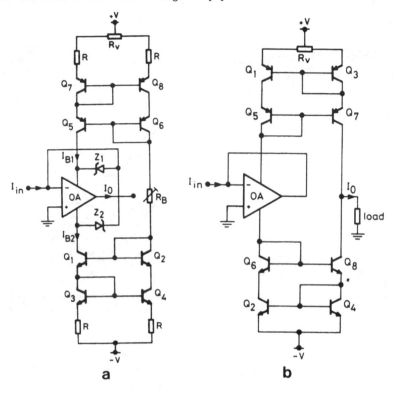

Figure 4.8   High performance Current-Followers.
(a) Floating VOA   (b) VOA supply current sensing.

18], as shown in Figure 4.8(a) and Figure 4.8(b) respectively.   The current-follower is a circuit with extremely low (ideally zero) input impedance, extremely high (ideally infinite) output impedance. The net performance when used with a signal source, is to produce a current drive to a load equal in value to the short circuit current obtainable from the input signal source. The two current-followers shown in Figure 4.8 were constructed using 741 VOAs and CA3096 transistor arrays for the current-mirror circuits.   The circuit of Figure 4.8(a) yields excellent DC current transfer definition with errors between input and output currents less than 1ppm for loads greater than 100kΩ. However it has been shown [17] that the small signal current transfer performance of the follower is given by approximately

$$i_O/i_S = 1/(1 + jf(K_1 + 1)/GB) \qquad (4.7)$$

where GB is the unity voltage gain bandwidth product of the VOA, K_1 is a bandwidth scaling constant $(R_O + R_L)/R_S$, R_O is the open-loop output resistance of the VOA, R_S is the driving source impedance of the amplifier and R_L the load impedance. From equation (4.7) it is clear that the bandwidth is very load dependent as shown experimentally in [17] and high frequency performance is lost for any appreciable load value.

Although the gain definition of the current follower circuit of Figure 4.8(b) is not as good as that for the follower of Figure 4.8(a), current-gain now being equal to λ of the current-mirrors, the advantage of this circuit is the potentially high bandwidth and slew-rate due to the load isolation from input to output that this configuration offers. Also, because the VOA is connected as a voltage follower with a grounded non-inverting input terminal, the output node of the VOA is held at virtual ground, providing a very low input impedance and high slew-rate capability, since the VOA has no appreciable signal swing at its output .

It can be shown [18] that the small signal current transfer performance of the current follower of Figure 4.8(b) is given by approximately

$$io/is = \lambda \, (j\varpi)[\, (1+jf/GB \,)/(1+ jf.K2/GB)] \qquad (4.8)$$

where $\lambda(j\varpi)$ is the frequency dependent current transfer of the current-mirror and the bandwidth scaling constant denoted as K_2 is given by $(1 + R_0/R_S)$. Assuming that the bandwidth is determined by the VOA and not the mirrors then the -3dB bandwidth of this follower is approximately GB/K_2. This relationship demonstrates that the bandwidth of the follower is essentially independent of the load impedance used. This is a major improvement in bandwidth over the previous current-follower designs where the bandwidth is load dependent. To illustrate the effects of R_S upon the followers performance, graph Figure 4.9 shows both theoretical and measured results of the -3dB bandwidth of the follower obtained at different values of R_S. For each value of R_S, R_L was varied from 10 Ω to 10 kΩ, and there was no visible change in bandwidth. This confirms the predicted follower bandwidth insensitivity to R_L. Obviously, there will be a maximum value of R_L before the bandwidth of the follower becomes dominated by the finite output impedance of the current-mirrors.

The relationship in equation (4.8) is the fundamental basis for which all the designs based upon VOA supply current sensing techniques result in high bandwidth capability. It is the elegant impedance transforming properties resulting from the application of a VOA in a current follower from which the important relationship in (4.8) is obtained. It is also apparent from equation (4.8) that if the current follower were driven from an ideal current source such that R_S would be ideally infinite, K_2 would be unity and now the bandwidth of the follower would be totally independent of the VOA! We will show later in Section 4.6.1 that this virtually constant high bandwidth current follower can be used in a universal amplifier structure which yields constant bandwidth independent of transfer function.

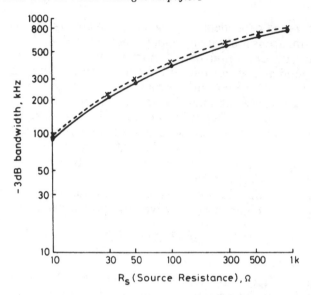

Figure 4.9   Graph of bandwidth versus source resistance (Rs) for VOA supply current sensing current follower of Figure 4.8(b). VOA type 741. Parameters: Ro = 75 Ω , GB (measured) = 800 kHz

4.5   Current-Conveyors

The current-conveyor (CC) introduced by Smith and Sedra [19, 20] in 1968 (described in greater detail in Chapter 3) has proven to be an extremely versatile analogue building block, as it facilitates a voltage tracking input in addition to its current-converter properties.   Numerous analogue circuit functions, some readily apparent and others more unusual, which are not so easily or accurately realisable using standard VOAs can be accurately realised using current-conveyors.   Fairly precise up to date reviews concerning these applications and their authors have been made available by Kumar [21] and later by Kumar and Shukla [22]. Unfortunately, most of the work presented on current-conveyors [21, 22] has been theoretical, and applications in network synthesis. Very little attention has been devoted to a practical high performance realisation of this evidently promising analogue building block until recently. This Section and chapters 3 and 15 describe high performance current-conveyor realisations [23, 24].

A current-conveyor (CC) is a grounded three port network defined by the hybrid matrix

$$\begin{bmatrix} i_Y \\ v_X \\ i_Z \end{bmatrix} = \begin{bmatrix} 0 & a & 0 \\ 1 & 0 & 0 \\ 0 & \pm 1 & 0 \end{bmatrix} \begin{bmatrix} v_Y \\ i_X \\ v_Z \end{bmatrix}$$

(4.9)

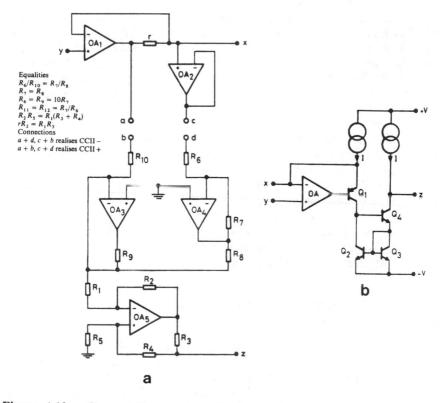

Figure 4.10   Current-Conveyor realisations using VOAs .
(a) Bakhitiar and Aronhimes' s [26] current-conveyor
(b) Pookaiyaudom and Srisarakham' s [27] single ended CCII+
current-conveyor.

where x and y are the input terminals and z the output terminal. In the first generation current-conveyor, CCI, $a = 1$ in the above equation. Thus, an impedance connected at terminal x is also reflected at y. In the second generation current-conveyor, CCII, $a = 0$ in the matrix , the terminal y effectively draws zero current. We will concentrate on the more versatile CCII type current-conveyors in this paper. The $\pm$ sign on h_{32} in the matrix denotes positive or negative current-conveyors respectively. By convention positive is taken to mean i_x and i_y both flow either towards or away from the conveyor, whilst negative means i_x and i_y flow in opposite directions.

In 1971, Black et al. [25] were the first to present an operational-amplifier based CCII current-conveyor, and also looked at the feasibility of integrating current-conveyors to provide standard IC building blocks. The implementation employed a μA749 operational-amplifier connected in the voltage following mode together with additional transistors configured as a current-mirror to sense the output current flowing through the "uncommitted" collector lead of the operational-amplifier; this uncommitted output collector being a special feature of the μA749. Suggestions were made that manufacturers should fabricate the necessary transistors on the same chip as the μA749 providing a single IC package which could by simple pin interconnection be configured into CCII+ and CCII- current-conveyors. This was the first reported attempt at using VOA supply current sensing techniques to realise current-conveyors. Unfortunately, the μA749 operational-amplifier has a class A, PNP current source output stage restricting the useful range of output current operation.

In 1978 Bakhtiar and Aronhime [26] presented a realisation of a CCII using conventional class-AB output operational amplifiers and resistors. The circuit is illustrated in Figure 4.10(a), and can be configured into either a CCII+ or a CCII-. However, the scheme is typical of traditional resistive feedback type current output amplifiers described earlier in Section 4.2. In addition to the excessive number of operational amplifiers employed in this realisation, the circuit demands tightly matched resistors to satisfy the equality constraints, has low bandwidth and uncertainty of output impedance, all common drawbacks of traditional VOA based current amplifier techniques.

4.5.1 High Performance Current-Conveyors Based on Op-Amp Supply Current Sensing

In 1979 Pookaiyaudom and Srisarakham [27] proposed a CCII+ current-conveyor realisation (Figure 4.10(b)), using a single operational amplifier and current-mirror to sense the VOA's output current. The circuit is very similar to that previously described by Black et al. [25], is limited to Class A output operation and requires precise matching between the two current sources, I.

This use of current-mirrors to sense the VOA's output current is very much related to the current-converter techniques described in the earlier Section. Unfortunately, over the last decade the tendency has been to treat current-converter research and current-conveyor research in isolation, whereas in fact they are very much related. As advances in current-converter realisations using standard VOAs began to flourish, particularly since the introduction of the VOA supply current-sensing techniques, little attention had been devoted to employing similar techniques to realise high performance current-conveyors.

Even the more recent designs of current-conveyors by Heurtas [28] and Senani [29] in 1980, are based upon traditional resistive feedback current amplifier techniques and exhibit similar drawbacks. In fact, it is interesting to note that the circuits of Figure 4.3 to 4.5 all have the basic structure of a current conveyor! Various extentions to the basic conveyor principle can be visualised from the circuits of Figure 4.3. For example the circuit of Figure 4.3(e) can be thought of as a differential current conveyor where the differential X terminals of the conveyor are formed by the inverting inputs

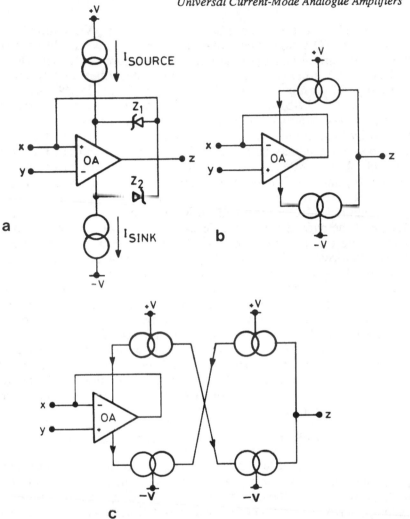

Figure 4.11   High performance Current-Conveyors. (a) CCII - based upon floating VOA structure, Isource=Isink, V(z1)= V(z2) (b) CCII + based upon VOA supply current sensing.   (c) CCII - based upon VOA supply current sensing and using cross-coupled mirrors.

of the two input VOAs and the differential Y terminals of the conveyor by the noninverting inputs.

Fortunately, it is possible to extend the capabilities of the previously described current-followers to high performance current-conveyors. As a two port network, the current-follower has similar matrix characteristics to the current-conveyor but with the y terminal in equation (4.9) earthed. By accessing this normally earthed terminal, the current-follower is readily configured into the more versatile current-conveyor. Two current-followers have been described, one based upon a floating VOA structure and the other using VOA supply current-sensing techniques. Both topologies configured as current-conveyors are shown in Figure 4.11. The floating VOA based current-conveyor of Figure 4.11(a) describes a CCII- and the VOA supply rail sensing scheme of Figure 4.11(b) a CCII+ [30]. In fact the circuit of Figure 4.11(b) or adaptations of the circuit had been reported earlier by several authors but not formally identified as a current conveyor until 1984 by Wilson [30]. If opposite to the above polarity conveyors are required both circuits may be preceded by an inverting current-follower to reverse the phase. Alternatively, in the case of the current-conveyor of Figure 4.11(b), the circuit may be structurally modified by the addition of a second pair of current-mirrors cross coupled to produce phase inversion as shown in Figure 4.11(c). The circuit now describes a CCII-. The principal advantage of the current-conveyors described here over previous conveyor realisations are that they provide class AB bipolar output operation hence higher current drive capability, and they use no resistors, hence avoiding the resistor matching and equality constraints common to earlier circuits. Furthermore, as first shown in [18] and Section 4.4 of this work, the current-follower section of the conveyor has a high, almost constant, bandwidth and this together with the wide bandwidth obtainable from the voltage follower section results in this particular current conveyor realisation exhibiting excellent high speed performance. The structure of the conveyor is simply a VOA with output current-mirrors and is ideally suited for monolithic integration. High precision is obtained using the floating VOA structure, while wide bandwidth and high slewrate obtained using the VOA supply current sensing technique.

4.6 Practical Applications of Supply Current Sensing

The authors have worked extensively on the more practical approach of adapting conventional VOAs to operate as current-mode devices. Primarily working towards the development of controlled current amplifiers, interestingly a new generation of voltage based circuits have been shown to out perform their traditional voltage counterparts in terms of speed, bandwidth and accuracy, and review of this work will now be presented.

Most research in the area of current-conveyor applications has been directed towards using a minimal number of conveyors to realise a particular circuit function. Although it is apparently an advantage to keep the component count down, generally this means that more feedback loops are required which results in poorer performance. The authors approach in applying CCIIs has been to attempt to maintain high overall performance and not necessarily to reduce the number of CCIIs employed. This approach also offers distinct design flexibility.

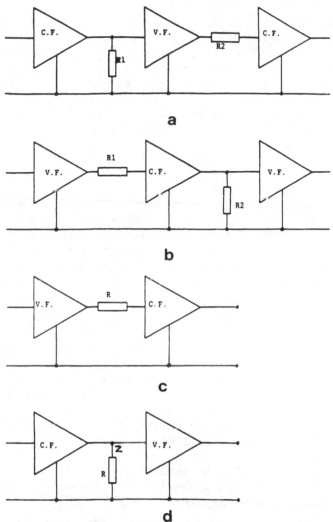

Figure 4.12   **Constant bandwidth follower based amplifier structure.**
(a) Current Amplifier $A_i = R_1/R_2$ (b) Voltage Amplifier $A_v = R_2/R_1$
(c) Transconductance Amplifier $G_T = 1/R$ (d) Transresistance Amplifier
$R_T = R$

4.6.1 Constant Bandwidth Follower Based Voltage Amplifier

In [18], using a similar structure of amplifier to that described by Allen and
Terry [31], the authors demonstrate that a universal amplifier exploiting
voltage and current-follower properties results in a stable, wide bandwidth
amplifier that is not constrained by the usual gain-bandwidth product of the
particular op-amps used. Using two current-followers and two voltage
followers it is feasible to develop a useful, general purpose follower based

operational amplifier network. Such a network can be configured into any of the four main amplifier topologies with the maximum addition of two resistors acting as voltage to current converters. Figure 4.12 illustrates how each amplifier type may be constructed from the followers and an expression for each of their transfer functions is included in the Figure. A feature of this structure of amplifier is the lack of total output to input feedback. Stabilising negative feedback is employed within each follower block so that no additional feedback is necessary when the external transfer function defining elements are added. As a result any phase lag from input through to output is irrelevant in terms of determining the stability of each amplifier system. The bandwidth of each voltage follower Section would be close to the gain bandwidth product of the VOA.

As an example consider the voltage amplifier design shown in Figure 4.12(b). The input into the first voltage follower provides a high input impedance and transfers the input voltage across resistor R_1 converting it into a current drive. The second stage is a current-follower which conveys the current V_{in}/R_1 to a high impedance current source driving R_2. As the third stage is a voltage follower with high input impedance the voltage across R_2 is transferred to the output giving; $V_{out} = (R_2/R_1) V_{in}$ and therefore a voltage gain, $A_v = R_2/R_1$. Using the same approach it is relatively simple to verify the remaining three basic amplifiers shown in Figure 4.12.

It may at first seem a disadvantage that three basic building blocks are required to produce a voltage amplifier when a single operational-amplifier will do. However the complexity is justified by the high performance possible with voltage and current-follower designs. Furthermore, due to the flexibility of this design, transconductance and current amplifiers are easily realisable as shown in the Figure, avoiding the complex multiple pair resistor matching requirements that typify conventional methods that were discussed earlier in Section 4.2.

It is interesting to note that the frequency performance of the current-follower section is determined by the driving source impedance R_s from equation (4.8), which in this case is R_1. Thus the larger value of R_1 the higher the frequency performance. R_1 can be chosen to maximise the bandwidth of the amplifier and the voltage gain can be set independently with R_2. High gain and high bandwidth can thus be set simultaneously. However, there is a limitation on the size of R_2 and hence voltage gain and bandwidth of the amplifier due to the limited output impedance of the current-mirror circuits. Experimental results using conventional VOAs have indicated improvements of more than 50 times the gain bandwidth capability of the individual VOAs used in the system. This high gain bandwidth capability is a feature of most of the current-conveyor applications which employ the supply current sensing VOA architecture in their designs. In fact the circuit of Figure 4.12(b) can be reduced to that of a single current-conveyor with an output voltage follower. The normally earthed non-inverting terminal of the

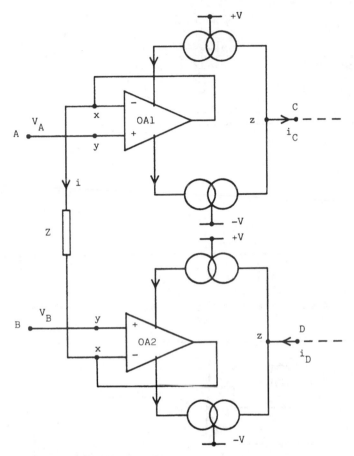

Figure 4. 13   Differential input/output transadmittance cell.

current follower forms the voltage input terminal (Y-input) of the conveyor (as discussed in Section (4.3)), hence eliminating the first voltage follower in the circuit.

It is interesting to note that the transresistance follower-based amplifier of Figure 4.12(d) has the same basic structure as the recent commercially available transimpedance (current feedback) op-amps [32,33]. Assuming that the current-follower is constructed with a current-conveyor with the y terminal grounded, then this point may be accessed to form the non-inverting terminal of the transimpedance op-amp. In the case of the transimpedance op-amp the amplifier's transimpedance is maximised by omitting resistor R, yielding an open-loop high gain amplifier.

The transimpedance op-amp is configured in closed-loop with a conventional resistive network as the so-called z terminal is not externally accessed. Interestingly both approaches to generating well-defined closed-loop gain give wide constant bandwidth and high slew-rate of this amplifier

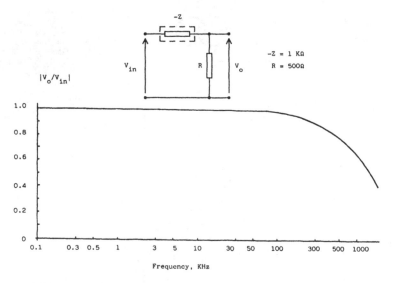

Figure 4.14   Potential divider frequency response (-Z=1k Ω , R=500 Ω)

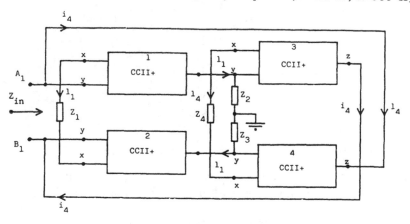

Figure 4.15   Floating GIC circuit using current-conveyors

structure. In light of this comparison one may view the other three circuits of Figure 4.12 as potential internal op-amp structures. For example the current amplifier of Figure 4.12(a) with R_1 open-circuit and R_2 short circuit and the y terminal of the input current-follower/conveyor accessed gives an open-loop current-mode op-amp topology.

In view of these applications the follower-based universal amplifier promises to be a useful general purpose single chip device. The chip could comprise two purpose built voltage followers and two current-followers of the type described in Figure 4.12 or three current-conveyors as these can be configured into either voltage-followers or current-followers simply by connecting the appropriate terminal to ground.

4.6.2   Current-mode Impedance Converters

Using two or more of these CCII+ current conveyors, floating-impedance convertors can be realised [34]. In this Section a floating-negative-impedance convertor and a floating-generalised-impedance convertor (GIC) are described. The floating-impedance convertors are based on the differential input/output transadmittance cell shown in Figure 4.13. OA_1 and OA_2 are conventional operational amplifiers and both act as high-input-impedance voltage-followers to the two voltage inputs V_A and V_B. The potential across the impedance Z is $V_A - V_B$, and this defines the current in the feedback path around each operational amplifier. This feedback current is sensed by the current-mirrors in the power supply leads to OA_1 and OA_2, providing the differential output currents shown on the diagram. If the circuit of Figure 4.13 is modified by connecting the high-output-impedance nodes C and D to A and B respectively, then the circuit behaves as a floating negative impedance -Z. As a test, a negative-resistance potential 'divider' was constructed using 741-type operational amplifiers together with current-mirrors constructed from CA3096 transistor arrays. The standard high-performance four-BJT current-mirror was used in all the circuits described here. The voltage gain against frequency response of the 'divider' is shown in Figure 4.14. The DC voltage transfer was set at -1, and it can be seen that the negative impedance remained constant at $-1k\Omega$ up to a frequency of 100 kHz, which is principally determined by the unity-voltage-gain bandwidth of OA_1 and OA_2.

A positive floating GIC using current-conveyors is shown in Figure 4.15 and an implementation using VOA supply current sensing is shown in Figure 4.16. The circuit comprises two of the basic cells of Figure 4.13, the output current drives from the first cell being converted to voltage inputs for the second cell through impedances Z_2 and Z_3. The current feedback paths from the output of the second cell to the input terminals of the GIC are crosscoupled in order to achieve a net positive floating impedance between nodes A_1 and B_1. A negative GIC is readily cofigured by reversing this polarity.

The principal currents i_1 and i_4 shown in Figure 4.16 are drawn assuming that $V_{A1} - V_{B1} > 0$, that the current-mirrors provide unity 'reflection coefficients' and that the operational amplifiers are ideal. By inspection,

$$i_4 Z_4 = i_1 (Z_2 + Z_3) \tag{4.10}$$

and

$$i_1 Z_1 = V_{A1} - V_{B1} \tag{4.11}$$

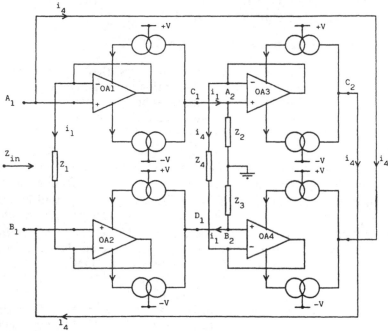

Figure 4.16   Practical Implementation of GIC usin VOA supply current sensing.

then

$$Z_{IN} = (V_{A1} - V_{B1})/i_4 = \frac{Z_1 Z_4}{(Z_2 + Z_3)}$$

(4.12)

To test the validity of the design a floating inductance of 5 mH was realised using two equal-value capacitors of 10 nF for impedances Z_2 and Z_3 and two 1 kΩ resistors for Z_1 and Z_4. The active components were the same type as those used in the negative-impedance convertor. A second-order LCR bandpass filter was constructed using this simulated inductance. The performance was illustrated by a step-response test in reference [34] for the classical three cases of underdamped, critically damped and overdamped responses by changing R.

Measurements showed that the simulated inductance of 5 mH held for frequencies up to 100 kHz, this frequency being determined by the unity-voltage-gain bandwidth of the operational amplifiers used.   From measurements the equivalent series resistance of the inductor was found to be less than 1 Ω. Many other circuit functions , some qiute unusual such as the positive frequency dependent negative resistor can be accurately realised fron this GIC topology. However, in common with all circuits using active

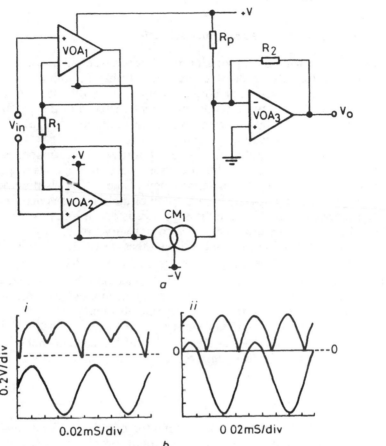

Figure 4.17 **Current-mode precision rectifier**
(a) Circuit diagram.
(b) Output waveforms comparing conventional (i) and current-mode (ii)
precision rectifier performance.

components, it is essential to limit the signal magnitude and frequency to
within the operating constraints of the particular devices used.

4.6.3 Universal current-conveyor Active Filter

In reference 35 current-conveyors have been used to develop a universal
active filter which features low active and passive sensitivites together with
high bandwidth capability. In the design independent control of all filter
parameters can be set with single grounded resistors. A further feature is the
use of grounded capacitors, which are desirable for both fabrication and high
frequency performance. In fact an extremely powerful and elegant
transformation of conventional active RC filters into current conveyor
equivalents is described in Chapter 3 Section 3.4.

4.6.4 Current-Mode Precision Rectifier

In reference 36 a wide-band precision rectifier is described in which an input differential current-conveyor is reconfigured to have a unipolar output, as shown in Figure 4.17(a). The conventional 'fast' two diode, two inverting VOA, precision full-wave rectifier was constructed and the performance compared with that obtained from the circuit of Figure 4.17(a). Identical components were employed in both circuits.

Figure 4.17(b) illustrates clearly the improved performance at a test frequency of 10 kHz. Using this technique very low distortion precision rectification up to the unity gain frequency of the VOA is now possible. Since the rectifier is unipolar, the accuracy of the rectifier is determined by the output current share between the supply rails of the VOA as discussed in Section 4.3.1. It is therefore desirable to use very low supply bias current VOAs in this application. It can be shown that for high performance $I_s \ll V_{in}/R_1 < I_{0(max)}$ where $I_{0(max)}$ is the maximum VOA output current. Clearly when the input signal falls to zero the inequality is not satisfied. Best performance can be obtained for a given range of input signal voltages, with appropriate choice of R so that the VOA operates fairly close to the current limit of $I_{0(max)}$ at the maximum input voltage level. In Chapter 15 of this book an attractive integrated circuit version of the current-mode precision rectifier is described based upon a recent commercially available (PA630) integrated current conveyor!

4.6.5 Current-mode Precision Peak Detector

Conventional diode precision peak detector circuits suffer from similar frequency restrictions to the conventional precision rectifier, again principally because of the VOA switching between the hold and sample mode. VOA supply current sensing can therefore be applied here to provide high accuracy, wide bandwidth precision peak detection. The basic design of a precision positive peak detector is shown in Figure 4.18. Again, low supply bias VOAs should be used for optimum performance. The circuit of Figure 4.18 is essentially that of a positive half wave rectifier. For $V_{in} > 0$, signal current $I_x = (V_{in} - V_0)/R$ will charge capacitor C, until $V_{in} = V_0$ and $I_x = 0$. If V_{in} is reduced, I_x is steered in the opposite direction and capacitor C continues to hold the peak value of V_{in}. Optimum aquisition performance is obtained by setting resistor R equal to zero.

Furthermore, resistor Rp can be adjusted to ensure that the bias current flowing through CM2 supplies all the required bias currents for the circuit. This will therefore reduce the "droop" rate due to capacitor leakage. However, some output droop will be caused by capacitor discharge through the finite output impedance of the current-mirrors. Variations of peak detector circuits may be formed by simple modifications to the circuit of

Figure 4.18. For example, by connecting the negatiive supply rail of OA1 to the input of current-mirror CM1, an absolute value peak detector is created.

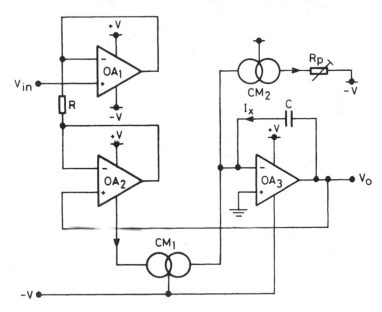

Figure 4.18   Precision positive peak detector.

4.6.6   *Current-Mode Instrumentation Amplifier*

Supply current sensing is used in the following application to achieve a high CMRR voltage instrumentation amplifier [37], the circuit essentially being an evolution of the precision full-wave rectifier of Figure 4.17(a). Conventional resistive feedback differential amplifiers, including the standard 3 op-amp instrumentation amplifier, require precisely matched resistors to achieve high CMRR. A feature of this design is its simplicity and that high CMRR performance over a wider bandwidth than conventional designs is obtained, without the need for precisely matched resistors. A schematic of the instrumentation amplifier is shown in Figure 4.19(a).

Op-amps, OA1 and OA2, both act as high input impedance voltage followers for the non-inverting and inverting inputs, V_2 and V_1 respectively. The potential across R_1 is $V_2 - V_1$ and this defines the current in the feedback path of OA1. This current is sensed by the current-mirrors CM1 and CM2. The outputs of the two current-mirrors are recombined at the current sensing node of the output stage comprising OA3 and R_2.

Assuming ideal performance of the voltage follower and current-mirrors, the output voltage at OA3 is given by

$$V_0 = \lambda \, [R_2/R_1](V_2 - V_1) \tag{4.13}$$

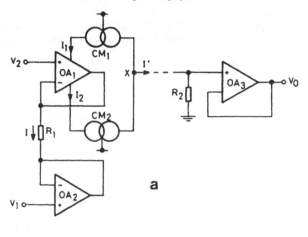

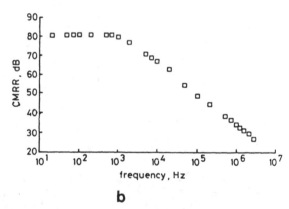

Figure 4.19 Current-mode instrumentation amplifier.
(a) Circuit diagram (b) CMRR frequency performance.

where λ is the current transfer ratio (close to unity) of the current-mirrors.

Clearly from equation (4.13) the common-mode gain is ideally zero. Although ideal op-amps have been assumed, it is worth noting that the analysis shows an 'infinite' CMRR is obtained without any resistor matching constraints, unlike conventional instrumentation amplifier circuits. The potential difference across R_1 is zero in the presence of a common-mode input due to the action of the two voltage followers. This results in no supply current changes in either OA1 or OA2 and hence the output voltage is zero. It should be noted that any mismatch in the λ of the CM1 and CM2 has no effect at all on the common-mode gain, merely causing some distortion in the output signal. The primary contributor to a non-zero common-mode gain is mismatch in the gain-bandwidth products of the two input op-amps, OA1 and OA2, confirmed by small-signal analysis of the circuit which yields the following expressions for differential-mode gain, A_{vd}, and common-mode gain, A_{vc}.

$$A_{vd} = \frac{R_2 \left(1 + jfk \dfrac{GB_1 + GB_2}{2 GB_1 GB_2} \right)}{R_1 \left(1 + \dfrac{jfk}{GB_1} \right) \left(1 + \dfrac{jfk}{GB_2} \right)}$$

(4.14)

$$A_{vc} = \frac{R_2 \left(jfk \dfrac{GB_1 - GB_2}{GB_1 GB_2} \right)}{R_1 \left(1 + \dfrac{jfk}{GB_1} \right) \left(1 + \dfrac{jfk}{GB_2} \right)}$$

(4.15)

where $k = (1 + 2R_o/R_1)$ and R_o is the open-loop output resistance of the op-amp and GB1, GB2 are the gain bandwidth products of OA1 and OA2 respectively. The expression for Avc shows that the non-zero common-mode gain arises from gain-bandwidth product mismatches between OA1 and OA2. It should be noted that the conventional 3 VOA instrumentation amplifier with perfect resistor matching approaches this same result!

As can be seen from the expression for A_{vd} the DC gain of the amplifier is determined by the ratio R_2/R_1. It is apparent that since neither resistor is in the feedback loop of the op-amp s, then differential gain can be varied by R_2 without affecting the bandwidth of the circuit, and as with previous circuits the gain-bandwidth product is generally constant.

To illustrate the performance obtainable a 4MHz AD711 VOA together with four transistor current-mirror circuits from CA3096 transistor arrays were used to construct the instrumentation amplifier. Figure 4.19(b) shows a plot of CMRR against frequency for unity differential mode gain. The DC value of 80dB is due to the CMRR limitations of the VOAs used, and the roll off at high frequencies is due to the mismatch in gain-bandwidth product of the input VOAs. Obviously this performance is not optimum and an integrated circuit version would yield a high CMRR over a wider bandwidth.

4.6.7 *Hardware Reduction*

As an analogue building block, many current conveyors may be employed in a single application [19]. Figure 4.20 shows two hardware reduction schemes. In Figure 4.20(a) the number of current-mirrors has reduced from some 4N to only 4 if output current summation is required. Furthermore, single package matched operational amplifier can be employed since a

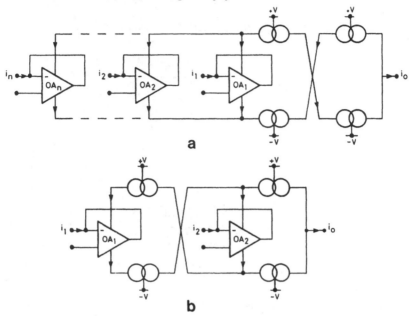

Figure 4.20 Hardware reduction schemes. (a) current addition,
$i_o = i_1 + i_2 + i_3 + ... i_n$ **(b) current subtraction, $i_o = i_1 - i_2$**

common supply is used in this technique. If current subtraction is required then the scheme shown in Figure 4.20(b) similarly may be used to reduce the number of current-mirrors. Since many applications of current conveyors require summing and subtracting of current variables, then application of such hardware reduction schemes could result in a significant reduction in circuit complexity. Furthermore the accumulation of N current-mirror output offsets and current-mirror matching constraints between CCII stages is avoided by using the supply rails of the VOA as summing junctions.

4.7 Seven Terminal Op-Amp

Rather than the five-terminals of a conventional VOA, it would be very useful if a seven-terminal VOA were available, the additional two terminals being the collectors (drains) of the output push-pull pair [36]. This would then allow direct output current sensing, rather than unnecessarily sensing the whole of the VOAs supply current. If such a VOA were available then the dynamic range, precision and noise performance of the supply current sensing technique would significantly improve. The circuit diagram of a typical CMOS operational amplifier with uncommitted output drains is shown in Figure 4.21. The required seven-terminal VOA is created by simply taking the output drains to two external pins. Such a modification can be easily carried out by manufacturers of VOAs and the versatility of the VOA extended to allow true output current sensing as discussed in Section (4.3.1).

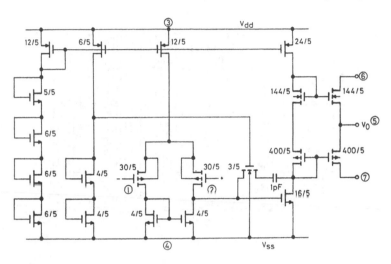

Figure 4.21 Seven terminal CMOS VOA.

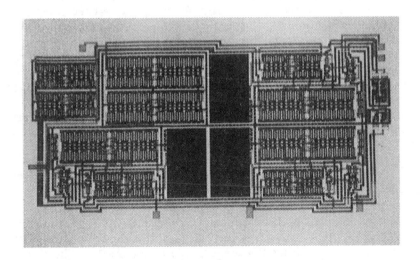

Figure 4. 22 Layout plot of three matched seven terminal VOAs (Figure 4.21) and current-mirror circuits integrated on a single chip using a 2 micron N-well CMOS process.

The VOA shown in Figure 4.21, together with on-chip CMOS current-mirrors has recently been implemented as an integrated circuit on a 2 micron N-well CMOS process. This integrated circuit comprises three matched 7-terminal VOAs and four high performance current-mirrors, allowing configuration into any of the supply current sensing circuits referred to in

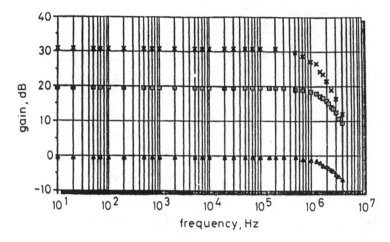

Figure 4.23   Voltage gain frequency performance of follower based amplifier of Figure 4.12 (a) using the integrated circuit of Figure 4.22

this Chapter. To ensure good matching all the VOAs are closely placed together and share the same bias circuit. A layout plot of the integrated circuit is shown in Figure 4.22. The voltage gain of the VOA was measured to be 63 dB and the unity gain bandwidth product measured to be about 2 MHz. Figure 4.23 shows the measured gain frequency response curve of the seven-terminal VOA connected as a current follower (Figure 4.8 b) and used in the universal follower based voltage amplifier of Figure 8. The voltage followers were connected from the remaining two VOAs on the chip. The results demonstrate that the gain can be varied from 0 dB to 30 dB with no significant change in bandwidth, as expected from equation (4.8). Using the integrated 7-terminal VOA, the authors have also constructed the precision full wave rectifier of Figure 4.17 and instrumentation amplifier of Figure 4.19(a), and preliminary results confirm the anticipated performance improvements. For example, distortion in the rectifier was measured to be less than 60 dB up to the unity gain bandwidth of the VOA.   The instrumentation amplifier maintained a high CMRR of 70 dB up to its unity gain bandwidth.   These performance figures although not optimum are clearly superior to conventional voltage-mode counterparts. The CMOS implementation of the current-conveyor shown in Figure 4.21 is clearly attractive and a similar implementation has also be described in Chapter 3.

4.8 Current-Mode Feedback Amplifiers

4.8.1 Translinear Class AB Current Amplifier

In Section 4.6.1 we have demonstrated a universal follower based amplifier. It was shown (Section 4.4) that the op-amp supply current sensing technique resulted in current-follower performance almost independent of the bandwidth and slew-rate performance of the operational amplifier employed.

However, if one considers the voltage amplifier of Figure 4.12(b) for example, the output resistance of the current follower circuit is $R_2//R_0$, R_0 being the output resistance of the current-mirror. In practice, since the output resistance is close to R_2, an output buffer is necessary to provide low output resistance and as a result the potentially high slew-rate feature and high bandwidth of the current-follower (equation 4.8) is lost with this design.

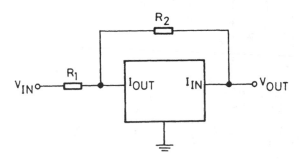

Figure 4.24 Alternative closed-loop voltage amplifier connection.

Figure 4.24 shows an alternative feedback structure to achieve voltage gain [38] this now being given by,

$$V_o/V_{in} = -(\lambda/(\lambda + 1)).R_2/R_1 \qquad (4.16)$$

with output resistance,

$$R_{out} = R_2//(R_0/(\lambda + 1)) \qquad (4.17)$$

and the output resistance of the voltage amplifier is still high. A voltage follower would be required at the output but again this destroys the potentially high bandwidth and slew-rate inherent in the current-followers. However, this structure of voltage amplifier is particularly desirable if $\lambda \gg 1$, resulting in lower output resistance and better gain definition without the need of an output buffer.

In this section we describe a new current-mode analogue building-block which has similar properties to those of the current-follower Figure 4.8(b) but introduces current gain to satisfy the requirement of $\lambda \gg 1$ in equations (4.16) and (4.17)

The new current-mode amplifier architecture was reported [38], and is shown in Figure 4.25. OA1 is imbedded within the translinear cell (T1 to T4) (Chapter 2). The d.c. current-source and sink ensures that all four BJTs are actively biassed with current Io, providing the desirable class AB input/output performance. The voltage following action of the translinear cell (T1 to T4) provides negative feedback around OA1 ensuring that the

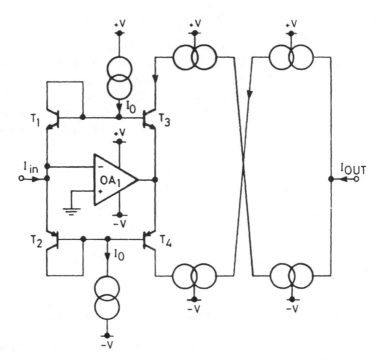

Figure 4.25   Translinear class AB current amplifier .

output of OA1 is a virtual ground and the input impedance of the amplifier is low.   The current gain of the circuit $A_i = \beta_3 = \beta_4 = \beta$ is provided by transistors T3 and T4. The four current-mirrors (CM1 to CM4) are needed to translate the phase-split output currents from T3 and T4 to the output node and provide the correct phase relationship.   The total current gain is approximately $\lambda^2\beta$ and the output impedance is again determined by the R_o of the current-mirrors. Unlike the current follower of Figure 4.8(b) which exhibits unity current gain of accuracy determined by the current-mirror λs, the new current-mode circuit exhibits an 'open-loop' current gain of approximately β. The circuit is readily configured as a closed-loop voltage amplifier shown in Figure 4.24 with a voltage gain of $-R_2/R_1$.

The circuit exhibits some extremely interesting features. The closed-loop output impedance is low and so an output buffer is not necessary as in the previous follower-based voltage amplifier designs, and as a result the potentially wide bandwidth and high slew rate of this design is maintained. Also since the current-mirrors are imbedded within the negative feedback loop, the closed-loop performance is independent of the current-mirror λs. All the desirable performance features of the amplifier are enhanced with greater current gain within the translinear T1 - T4 cell.

Prototype experimental results using standard VOAs with transistor arrays, shown in [38], confirm the expected high slew-rate performance together with the bandwidth once again being virtually independent of closed-loop gain setting. Also it is interesting to recognise that the circuit is essentially a current conveyor with current gain β, the non-inverting input being the Y terminal of the conveyor. In addition this circuit has the potential of creating a closed-loop current-conveyor when feedback is applied rendering the performance essentially independent of the current-mirror imperfections. This new amplifier architecture has potential for high speed continuous time and sampled-data analogue signal processing circuit applications.

4.8.2   Current-feedback Op-Amps

Since the introduction of the first commercially available monolithic VOA (Fairchild's μA709 produced in 1965) there have been steady improvements in the performance. The two most noteable developments resulted from the introduction of active rather than resistive loads, giving greater voltage gain per stage and the introduction of FETs which provide performance enhancements including reduced power consumption, lower input bias current and higher input impedance. Despite these evolutionary advances the internal architecture of the VOA has remained remarkably unaltered. However, it is interesting to see the emergence over the past 2-3 years, of an entirely new architecture VOA, now available from several of the specialist analogue semiconductor manufacturers.

These VOAs are generally referred to as current-feedback or transimpedance devices and are all very similar in structure, to that shown in Figure 4.26. The design relies heavily on the availability of a high quality complementary BJT process in which the NPN devices are well matched in performance to the PNPs. One of the major problems with the standard emitter-coupled, or long-tail pair, input stage of the conventional VOA is that the large signal differential input transconductance saturates at a relatively low input voltage level. This effect results in an output slew-rate limitation being determined by the ratio of the long-tail current to input capacitance of the second voltage gain stage, typically giving a slew-rate performance of the order of 0.2 to 20V/μs. In the new architecture op-amps the long-tail pair has been abandoned for a complementary common-emitter and common-base stage and the quoted slew-rate of these amplifiers is some two orders of magnitude higher.

The purpose of including this discussion on current-feedback op-amps in this paper is not to present a detailed analysis as this is available elsewhere [32, 33], but to reflect on the more fundamental features of the new architecture of these devices and its relationship to the past work in this Chapter on the exploitation of current-mode amplifier designs. These elegant new VOA architecture are effectively IC versions of the supply current sensing VOA topology that has been the main theme of this paper,

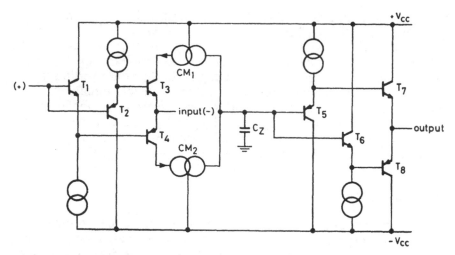

Figure 4. 26   "Current Feedback" transimpedance Operational Amplifier.

and as expected high slew-rate and constant bandwidth is obtained. Examination of the circuit of Figure 4.26, the non-inverting input corresponds to the Y terminal of a current-conveyor, the inverting terminal to the X input of the current conveyor and the Z node, although not accessible in these devices, is the output mode of the current conveyor. The basic structure therefore can be seen to be a current conveyor followed by a voltage-follower, giving a low output resistance for voltage output applications. The new op-amp architecture is a partial step towards a true current-mode op-amp, but it has been designed in such a way that it can be used in an almost identical feedback application to that of the conventional VOA.

4.8.3 Operational Floating Conveyor (OFC)

One of the basic limitations of the conventional VOA is its single ended output generally making accurate current sampling at the output a problem. Recently, due to advances in complementary bipolar technology, current-mode analogue circuits such as the recent current-feedback op-amp [32,33], and the integrated current-conveyor [23] are becoming commercially available, as discussed earlier. It is the current-mode sections of the circuits which achieve the high performance yet current can still not be accurately sampled at the output to achieve well defined closed loop current-amplifier topologies. In this Section a new four port general purpose analogue building block to be termed an operational floating conveyor (OFC) is described; operational because it achieves high gain and floating since it has a differential output. The OFC has very similar transmission properties to the current-feedback op-amp and current-conveyor but with an additional output which allows accurate output current sampling.

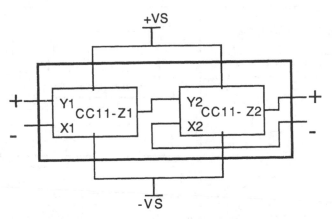

Figure 4.27 Operational Floating Conveyor (OFC).

The operational floating conveyor is described by the 4-port network shown in Figure 4.27. The OFC comprises a series connection of two 2nd generation current conveyors of the CCII- type. The first conveyor may be a CCII+ , but as we shall see later the conveyor type is dependent upon the OFC implementation. The port relationships of the current-conveyor have been defined in equation (4.9). By connecting two CCII- conveyors in the serial connection shown in Figure 4.27 and representing X1 and Y1 of the first CCII as inverting and non-inverting input terminals respectively and X2 and Z2 of the second conveyor as inverting and non-inverting output terminals respectively, results in a very elegant hybrid of amplifier stages between different input and output ports. For example, high open-loop transimpedance gain may be achieved between X2 and X1, current-gain between Z2 and X1, voltage gain between X2 and Y1 and transadmittance gain between Z2 and Y1 assuming a load termination at X2 and Z2 in each case. This transmission of multiple transfer functions allows all of the above four main amplifier types to be accurately configured in closed loop. In particular closed-loop current output circuits are readily connected since output current sampling is accurately achieved via the current following action between outputs X2 and Z2.

A range of closed-loop circuits using the OFC are shown in Figure 4.28 to illustrate the versatility of the OFC in practice. A circuit of particular interest is the closed loop current-conveyor of Figure 4.28(d), since to date most reported high performance current-conveyors have been open-loop output realisations [22] particularly those based upon the op amp supply current sensing technique. Circuit applications even the more unusual, which employ current-conveyors [22], current-feedback op-amps [33] or conventional voltage mode op-amps can be accurately configured using the OFC.

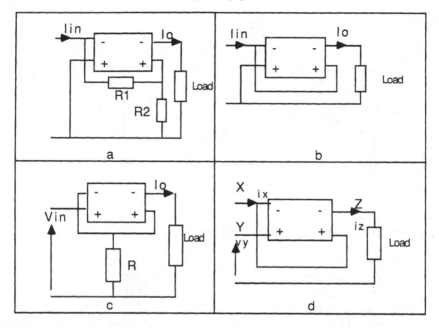

Figure 4.28 Closed loop controlled current output circuits applications.
(a) Current Amplifier (Io/Iin = 1+R1/R2) (b) Current Follower (Io/Iin=1)
(c) Transconductance Amplifier (Io/Vin = -1/R) (d)   Current Conveyor,
type CC11- (see Equation 4.9)

4.8.3.1 OFC Implementation

As shown in Figure 4.27, the OFC can be implemented quite simply using
two current-conveyors. Figure 4.29(a) shows a simple (OFC) prototype
implementation based upon the conveyor realisations of Figure 4.11.
Internal frequency compensation may be necessary and this can be provided
by a single grounded capacitor at node z1, y2. To confirm the closed-loop
operation of the OFC, the circuit of Figure 4.27(a) was constructed using 741
VOAs together with CA3096 mixed bipolar transistor arrays for the current-
mirror circuits. The current-mirrors were the 4-transistor Wilson type.
The circuit was then configured into the closed-loop current-amplifier of
Figure 4.28(a) and its measured closed-loop gain frequency response is
shown in Figure 4.29(b). A gain accuracy to within 0.1% was achieved for
current gain settings from 1 to 20. The bandwidth is not exactly constant as
would be expected with this type of architecture, basically because the design
has not really been optimised for high frequency performance. The
bandwidth of the amplifier varied from about 100kHz for unity current gain
to about 20kHz when the gain was increased to 20. For the tests an internal
compensation capacitor of 100 pF was required which was connected from
the output of the first CCII- to ground. The OFC was also connected into the

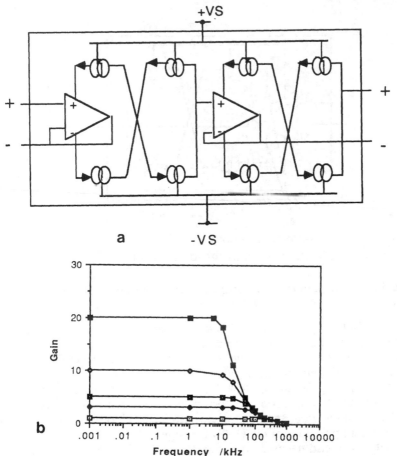

Figure 4.29 Low performance OFC circuit implementation (a) Circuit using operational amplifier supply current sensing.(b) Experimental frequency response curves of current-amplifier of Figure 4.28 (a)

other circuits of Figure 4.28 confirming in each case the expected closed loop parameters.

The circuit of Figure 4.29(a) uses relatively low performance components to confirm the OFC idea. An alternative, high performance integrated circuit version of the OFC can be readily derived from the current-feedback op-amp architecture typically shown in Figure 4.26. As described in Section 4.8.2 the input stage of the current-feedback op-amp has the exact port properties of CCII+ and an integrated current-conveyor based upon this type of architecture is described in Chapter 15. The output stage of the current-feedback op-amp is a voltage buffer and so its output may be regarded as output node X2 of the OFC. Thus to implement the full OFC an additional pair of cross-coupled mirrors (as in Figure 4.29(a)) should be connected to

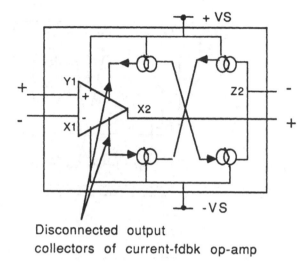

Disconnected output
collectors of current-fdbk op-amp

Figure 4.30 High performance Integrated OFC circuit implementation using a single current-feedback op-amp.

the voltage buffers power supply 'collector leads' T7 and T8 in Figure 4.26 in order to sense the op-amps output current and create current output node Z2 as required. The circuit shown in Figure 4.30 illustrates this proposal. Alternatively, the current-mirrors may be directly connected to the power rails of the op-amp (as in Figure 4.29(a)) to sense the entire supply current but this will lead to non-optimum performance as discussed in Section 4.3.1. Note that since the input stage of the current-feedback op-amp is a CCII+ the X and Z output polarities of the OFC (see Figure 4.27) are reversed in Figure 4.30, and this would necessitate a reversal of the feedback polarities in the applications of Figure 4.28. Simulation results employing the current feedback op-amp architecture of Figure 4.26, set up as the OFC of Figure 4.30, and using typical complementary bipolar transistor parameters indicate constant bandwidth current amplification for a wide range of gain settings. In conclusion a new versatile analogue building block called an OFC has been described which allows the simple formulation of accurate closed loop current-converters and amplifiers. Its realisation exploits the technique of op-amp supply current sensing which can be conveniently applied to current-feedback op-amps. Application of the OFC to all the current converter designs described in this Chapter is new work which is presently being investigated and reported here for the first time.

4.8.4 *Towards a true Current-Mode Op-Amp*

Whilst it is commendable to see the new developments such as current feedback op-amps and current-conveyors, the performance improvements are obtained from the current-mode section of the design and until a fully differential input/output current-mode op-amp is developed, the full benefits of the current-mode approach will not be realised. In this section we briefly describe some work aimed at the development of a fully differential input/output current-mode opamp.

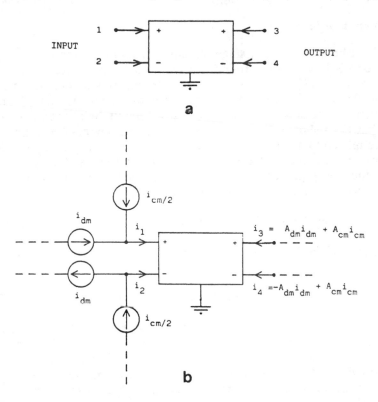

Figure 4.31 Network Characteristics of current-mode op-amp.
(a) Network symbol (b) Network response.

4.8.4.1 Network Characteristics

The current-mode op-amp is a grounded four port device ideally with zero differential input impedance, infinite differential current gain and zero common-mode current gain. A network symbol for such an amplifier is shown in Figure 4.31(a), and the network response summarised in Figure 4.31(b). i_{dm} and i_{cm} are the differential and common-mode input current respectively and A_{idm} and A_{icm} are the differential and common-mode gains of the amplifier respectively. Clearly the output is differential only if A_{icm} is zero and then

$$i_3 = -i_4 = A_{idm}.(i_1-i_2)/2 = A_{idm}.i_{dm} \qquad (4.18)$$

As in a voltage amplifier to achieve high common-mode rejection the designer should attempt to minimise the common-mode gain. It is seen from equation (4.18) that in fact the current differencing gain $(i_3/(i_1-i_2))$ is in fact equal to half the differential gain A_{idm}

4.8.4.2 Differential Input Stage

We first describe a basic input stage suitable for a current-mode op-amp. The input cell is based on the four transistor translinear current cell [39] shown in Figure 4.32(a). The basic circuit comprises a four transistor 'translinear' loop T1 T2 T3 and T4 together with diodes D1 and D2 and current source Io' and sink Io biassing the transistors in their forward active region. It has been shown that for perfectly matched devices the collector currents are related by the equation I1.I3 = I2.I4 = Io2. again by simple application of the powerful translinear circuit principle (Chapter 2). Assuming that the input current is small compared to Io and is purely differential, that is Iin (x) =- Iin (y), then if Iin > 0 ,

$$I2 = I3 = Io + Iin/2 \text{ and } I1 = I4 = Io - Iin/2 \qquad (4.19)$$

Alternatively, if the input current is common, that is Iin(x) = Iin (y) and Iin > 0 then

$$I1 = I2 = Io - Iin/2 \text{ and } I3 = I4 = Io + Iin/2 \qquad (4.20)$$

The above relationships exploit the translinear nature of the four transistor cell for small input signal levels. A further feature of this cell is that a potential connected at node (Z) is reflected to nodes (X) and (Y) via an emitter following action. The combined current differencing and voltage following capability of the cell form the features of a versatile analogue building block.

For example the input stage of the current feedback op-amp (see Figure 4.24) is essentially a single ended version of the cell, with node (Y) being non-inverting, voltage input terminal, node (x) the inverting, current input terminal and the current difference between I2 and I4 (obtained via a set of current-mirrors) forming the current output of the first gain stage. This also realises a CCII + current-conveyor. Attempts have been made in [39] to cancel the common-mode input currents but this has led to complex current-mirror arrangements requiring fairly stringent matching for common-mode cancellation. In the following section the cell is modified to provide a high CMRR by the introduction of a novel common-mode current feedback arrangement [40].

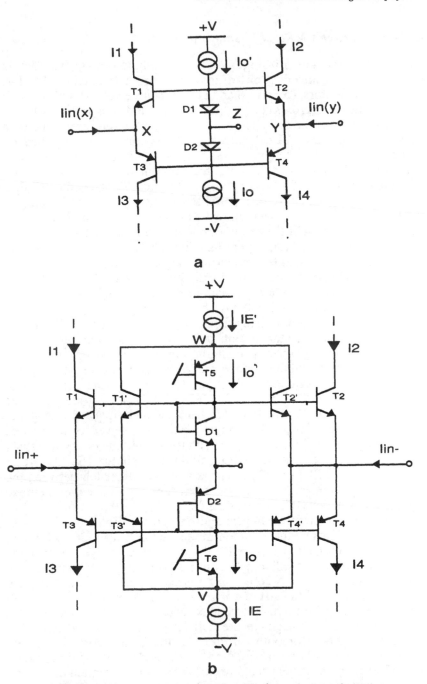

a

b

Figure 4.32 Translinear current input cell of current-mode op-amp.
(a) Basic differential current input cell (b) High CMRR differential current input cell

4.8.4.3.Modified High CMRR differential current cell

The modified cell is shown in Figure 4.32(b). Common-mode feedback is implemented by the introduction of additional input transistors T1'-T4', common-base transistors T5 and T6 and current source IE' and sink IE. It is assumed that IE' is approximately equal to IE and that IE is 3 Io for appropriate biassing of the cell. The circuit operation will be explained with reference to the top half of the cell since the cell is symmetrical. Application of a small signal differential input current to the cell results in an incremental increase in I2 and a similar incremental reduction in I1 (assuming Iin > 0) , according to eqn (4.19). If T1' is matched to T1 and T2' matched to T2 then the collector of T1' and T2', will experience the same incremental change in collector currents. Since the two collectors are connected at node (W) then the differential current will flow in a loop around T1' and T2' not affecting the differential output current operation through T1 and T2. However, application of a common-mode input current, again assuming Iin > 0, results in an incremental current reduction in the collectors of T1, T2, T1' T2' according to eqn (2.20). Assuming that current source IE' has a very high output impedance then Io will increase by the average of the incremental common mode input current causing the base emitter voltages of T1 and T2 to rise and hence their collector currents to increase. Thus, the input voltage is kept almost stationary when a common mode input current is applied, providing the cell with a high common mode rejection capability. A similar operation occurs for a common mode input, Iin < 0. An alternative way of viewing the circuit is to think of T1', T2' and T5 as a folded cascode connection. Since node (W) is a low impedance the high frequency capability of the cell is is not sacrificed. The circuit may also be viewed as the current equivalent to a long tail differential pair with the magnitude of CMRR again very dependent upon the quality of a 'tail' current source IE. A further advantage of this modified arrangement is that any mismatch between the current source and sink values of IE are taken care of by the common mode feedback circuitry which DC stabilizes the cell.

A prototype test circuit comprising the cell of Figure 4.32(b) was constructed to demonstrate the common-mode feedback idea. In addition to the basic circuit, four simple current-mirrors were employed to provide recombination of the phase split output currents I1, I3 and I2, I4 to two single ground reference outputs. Also current source and current sink IE were realized by three parallel transistors for correct biasing of the cell. The cell was constructed using MPQ3904 matched NPN and MPQ3906 matched PNP transistor arrays. Bias current to the cell (IE) was set at 500 µA. DC transfer characteristics for both differential and common-mode current inputs to the cell are shown in Figure 4.33(a) and 4.33(b) respectively. Figure 4.33(a) confirms the expected linear, differential operation of the cell over quite a wide input dynamic range. Figure 4.33(b) clearly shows the expected low common-mode output current which is virtually constant over

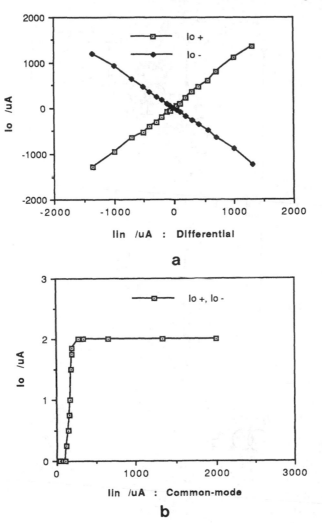

a

b

**Figure 4.33 Experimental results for current cell of Figure 4.32 (b)
(a) Differential-mode DC transfer characteristic (b) Common-mode DC
transfer characteristic.**

a similar input current range. The common mode gain varies from about -
50dB for very low input current levels to about - 65dB for input current
levels of about 2mA. These results are quite significant when considering
that without the common-mode feedback the common mode gain of the cell
would be 0 dB. The differential mode input resistance of the cell was
measured to be around 250 ohms and the common-mode input resistance of
the cell of the order of 100 Kohm. The prototype cell is by no means
optimum. NPN and PNP devices were separately packaged increasing the

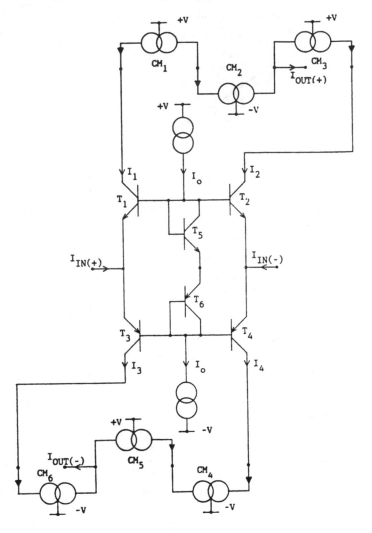

Figure 4.34 Unity gain current-controlled current source.

common-mode gain due to mismatch, and secondly the current source and sink have fairly low output resistance and should be cascoded to improve common mode performance. Furthermore, modifications to the basic input cell which already achieve some common mode rejection capability (replacing diode connected transistors D1 and D2 in Figure 4.32(a) with source followers as in Figure 4.26) could result in greatly improved performance. The cell is presently being optimised for subsequent integrated circuit realization with simulation results using a typical complementary bipolar process indicating common-mode gains of the order of -100dB. This new cell is also ideally suited as the differential input stage of a current-conveyor and the operational floating conveyor described in Section (4.8.3) and will find wide application in numerous other current-mode analogue signal processing circuits.

4.8.4.4 *Current-mode op-amp architecture*

A basic architecture for a current mode op-amp is shown in Figure 4.34. Figure 4.34 shows the basic differential input cell of Figure 4.32(a) embedded within six current-mirrors CM1 to CM6. This complementary current-mirror arrangement provides the circuit with a balanced high impedance differential output, with additional cancelling of common mode components. This may be readily confirmed by referring back to equations (4.19) through (4.20). The circuit of Figure 4.34 is an evolutionary step in the design of the full differential-mode op-amp. It has low impedance differential input and high impedance differential output and good input common-mode current rejection properties. However, the differential-mode current gain is virtually unity. In order to complete the development of a full current-mode op-amp it is necessary to provide high A_{idm}.

Figure 4.35 shows the basic architecture of Figure 4.34 with well-defined current gain provided by two NPN Gilbert current gain cells (see Chapter 2) T7 to T10 and T11 to T14. A simpler symmetrical design using one PNP and one NPN gain cell was initially conceived, but due to the availability of only relatively poor performance lateral PNP transistors at the time, the circuit was modified to that shown. The differential current input cell in fact forms an ideal drive to the Gilbert gain cell with the differentail current gain of the amplifier given by simply

$$Ai = (1+I_E/2I_o)$$ (4.21)

and can be made adjustable by varying the current sink value of I_E. I_E also provides the amplifier with a maximum output current limit. Positive d.c. reference voltages Vx and Vy are provided by two seperately biassed resistor/diode chains.

Accurate open loop gain settings are only achieved if the minimum beta of the gain cell transistors is a lot higher than the open-loop gain setting. For high open-loop gains Darlington transistor pairs may be employed within the gain cells. The circuit of Figure 4.35 was constructed using XRB101 (NPN) and XRB102 (PNP) transistor arrays. The current-mirrors used in the prototype were simple two transistor mirrors. Current Io was set at 100 μA. IE was varied from 1mA to 3mA to give respective differential gain settings from 5 to 15. Experimental d.c. transfer curves for the amplifier are shown in Figure 4.36 indicating the various operating regions of the amplifier, in particular the very linear operation for small input signals ($I_{IN}< Io/4$). Further experimental measurements of the amplifier with enhanced current gains using Darlington transistors can be found in reference [41]. High

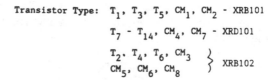

Transistor Type: T_1, T_3, T_5, CM_1, CM_2 - XRB101

T_7 - T_{14}, CM_4, CM_7 - XRD101

T_2, T_4, T_6, CM_3 $\left.\begin{array}{c}\\\\\end{array}\right\}$ XRB102
CM_5, CM_6, CM_8

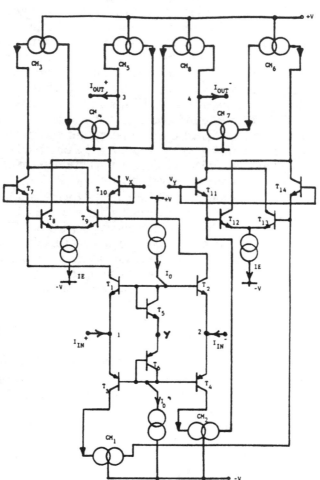

Figure 4.35 Current-mode op-amp employing Gilbert current gain cells.

CMRR can be obtained from the amplifier by connecting the common-mode feedback circuitry of Figure 4. 32(b) to the input stage.

The realisation of a current-mode op-amp brings about a number of exciting circuit possibilities. For example closed loop current conveyors current amplifiers and high performance current-feedback voltage

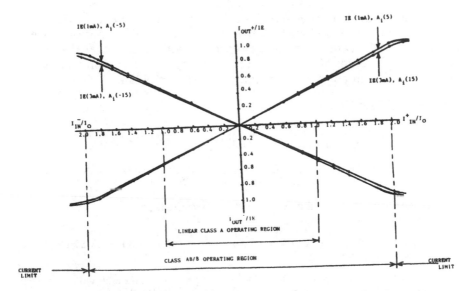

Figure 4.36 Normalised transfer characteristics of Figure 4.35 for open-loop gain settings Ai = (5-15).

amplifiers can be accurately realised with such an op-amp. An interesting concluding thought is that if node Y of the amplifier of Figure 4.35 were also made an input terminal such that the amplifier is a three input, two output device an extremely versatile general purpose analogue building block is created!

4.10 Conclusions

There is clearly a need to incorporate more circuits having current-processing properties into present day analogue circuit design. Many of the current amplifier designs described in this Chapter have the advantage that they use commercially available components and can be tailored to provide the required performance for a specific application. General performance limitations are those typical to voltage amplifier based integrated circuit structures and the high performance expected of current activated circuits is not fully realised for this reason. The semiconductor industry is recognising the current-mode potential and we have recently seen a move towards a new generation of VOAs, known as current-feedback amplifiers which have very similar properties to many of the architectures described in this Chapter. However, in the absence of a high quality, fully integrated current-mode operational amplifier, the development of circuit techniques which extend the output capability of the conventional VOA to obtain a controlled current output property is clearly necessary.

Results of this Chapter and in more recent review papers [42, 43] clearly indicate that the past two decades of research in this area give strong motivation for the integrated circuit development of current amplifier based circuits to be used in situations where conventional VOA performance is limited. Admittedly this somewhat radical step will mean that engineers will need to think in terms of current as well as voltage but only then will the full potential of the current-mode analogue amplifiers be obtained.

4.11 REFERENCES

[1]   TOBEY, G.E., GRAEME, J.G. and HUELSMAN, L.P.: 'Operational amplifiers, design and applications', McGraw-Hill, 1971, Ch. 6, pp. 225-227.

[2]   MEIKSIN, Z.H. and THACKRAY, P.C.: 'Electronic Design with off-the-shelf Integrated Circuits', Parker, 1980, Ch. 5, pp. 195-199.

[3]   GRAEME, J.: 'Applications of operational amplifiers', McGraw-Hill, 1973, Chp. 3, pp. 93-94.

[4]   GARZA, P.P.: 'Getting power gain out of the 741-type op amp', Electron. Int., Feb. 1973, p. 99.

[5]   HART, B.L. and BARKER, R.W.J.: 'A precision bilateral voltage-current converter', IEEE J. Solid-State Circuits, 1975, SC-10, pp. 501-503.

[6]   WILSON, G.R.: 'A Monolithic junction FET-npn operational amplifier', IEEE J. Solid-State Circuits, 1968, SC-3, pp. 341-348.

[7]   HART, B.L. and BARKER, R.W.J.: 'DC matching errors in the Wilson current source', Electron. Lett., 1976, 12, pp. 389-390.

[8].   RAO, M.K.N. and HASLETT, J.W.: 'Class AB bipolar voltage-current converter', Electron. Lett., 1978, 14, pp. 762-764.

[9]   HASLETT, J.W. and RAO, M.K.N.: 'A high quality controlled current source', IEEE Trans., 1979, IM28, pp. 132-140.

[10]   HART, B.L. and BARKER, R.W.J.: 'Universal operational-amplifier converter technique using supply current sensing', Electron. Lett., 1979, 15, pp.496-497.

[11]   NEGUNGADI, A.: 'A dual differential bilateral current converter', Proc. IEEE, 1980, Vol. 68, pp. 932-934.

[12]   HUIJSING, J.H. and VEELENTURF, C.J.: 'Monolithic Class AB operational mirrored amplifier', Electron. Lett., 1981, 17, pp. 119-120.

[13]   NEDUNGADI, A.: 'High current Class AB converter technique', Electron. Lett., 1980, 16, pp. 418-419.

[14]   WILSON, B.: 'A low-distortion bipolar feedback current amplifier technique', Proc. IEEE, 1981, 69, pp. 1514-1515.

[15]   NORDHOLT, E.H.: 'Extending Op Amp capabilities by using a current-source power supply', IEEE Trans., 1982, CAS-29, No. 6, pp. 411-414.

[16]   GILBERT, B.: 'Translinear circuits: A proposed classification', Electron. Lett., 1975, 19, pp. 14-16.

[17]   LIDGEY, F.J.: 'Current Followers', Electronics and Wireless World, Feb. 1984, pp. 40-43.

[18]   LIDGEY, F.J. and TOUMAZOU, C.: 'An accurate current-follower and Universal follower based amplifiers', Electron. and Wireless World, April 1985, Vol. 91, No. 1590, pp. 17-19.

[19]   SMITH, K.C. and SEDRA, A.: 'The current conveyor: a new circuit building block', Proc. IEEE, 1968, 56, pp. 1368-1369.

[20]   SEDRA, A. and SMITH, K.C.: 'A second generation current conveyor and its applications;, IEEE Trans., 1970, CT-17, pp. 132-134.

[21]   KUMAR, U.: 'Current conveyors: a review of the state of the art', IEEE Circuits and Syst. Mag., 1981, 3, pp. 10-14.

[22]   KUMAR, U. and SHUKLA, S.K.: 'Recent developments in current conveyors and their applications', Microelectron. J., 1985, 16, pp. 47-52.

[23]   WADSWORTH, D.C.: 'An accurate current-conveyor topology and monolithic implementation', IEE Proc G, 'Special Issue' on currnt-mode analogue circuit design, to be published April 1990.

[24]   SEDRA, A.S., ROBERTS, C.W. and GOHK, F.: 'The current-conveyor: History, Progress and New Results', IEE Proc G, 'Special Issue' on current-mode analogue circuit design, to be   published April 1990.

[25]   BLACK, G.A., FRIEDMANN, R.G. and SEDRA, A.S.: 'Gyrator implementation with integrable current-conveyor', IEEE J. Solid-State Circuits, 1971, SC-6, pp. 396-399.

[26]   BAKHTIAR, M.S. and ARONHIME, P.: 'A current-conveyor realisation using operational amplifiers', Int. J. Electron., 1978, 45, pp. 283-288.

[27]   POOKAIYAUDOM, S. and SRISARAKHAM, W.: 'Realisation of stable current-controlled frequency-dependent positive resistance', Proc. IEEE, 1979, 67, pp. 1660-1662.

[28]   HEURTAS, J.L.: 'Circuit implementation of current conveyor', Electron. Lett., 1980, 16, pp. 225-226.

[29]   SENANI, R.: 'Novel circuit implementation of current conveyors using an OA and an OTA', Electron. Lett., 1980, 16, pp. 2-3.

[30]   WILSON, B.: 'High-performance current conveyor implementation', Electron. Lett., 1984, 20, pp. 990-991.

[31]   ALLEN, P. and TERRY, M.B.: 'Use of current amplifiers for high performance voltage applications', IEEE J. Solid-State Circuits, 1980, Vol. SC-15, pp. 155-162.

[32]   BOWERS, D.F.: 'A precision dual 'current-feedback' operational amplifier', IEEE, 1988 Bipolar Circuits and Technology Meeting, pp. 68-70.

[33]   Dual high speed current-feedback operational amplifier, Precision Monolithics Inc data Sheet for OP-260, application sheet no 12/88 Rev .B, PMI data book 1988

[34]   TOUMAZOU, C. and LIDGEY, F.J.: 'Floating impedance converters using current conveyors', Electron. Lett., 1985, 21, pp. 640-642.

[35]   TOUMAZOU, C. and LIDGEY, F.J.: 'Universal active filter using current conveyors', Electron. Lett., 1986, 22, pp. 662-664.

[36]   TOUMAZOU, C. and LIDGEY, F.J.: 'Wide-band precision rectification', IEE Proc. G, Electron. Circuits and Syst., 1988, Vol. 134, No. 1, pp. 7-14.

[37]   TOUMAZOU, C. and LIDGEY, F.J.: 'Novel current-mode instrumentation amplifier', Electron. Lett., Issue 3, 1989.

[38]   TOUMAZOU, C. and LIDGEY, F.J.: Translinear class-AB current amplifier', Electronics Letters,1989, Vol 25, No 13, pp 873-874.

[39]   TOUMAZOU, C. and LIDGEY, F.J.: 'Novel bipolar differential input/output current controlled current source', Electron Lett., 1985, 21 pp 199-200.

[40]   TOUMAZOU, C., RYAN, P. and PAYNE, A.: High CMRR fully differential current cell, submitted for publication to Electronics Letters.

[41]   TOUMAZOU, C., 'Universal Current-Mode Analogue Amplifiers'   PhD Dissertation, Oxford Polytechnic September 1986

[42]   TOUMAZOU, C., LIDGEY, F.J. and CHEUNG, P.K.: 'Current-mode analogue signal processing circuits', A review of recent developments 1989, Proc IEEE ISCAS, Portland, 1989, pp. 1572-1575.

[43]   TOUMAZOU, C., LIDGEY, F.J. and MAKRIS, C.A.: ' Extending voltage-mode amplifiers to current-mode performance' IEE Proc. G, Electron. Circuits and Syst, vol 137, no 2, pp 116 - 130, April 1990.

Chapter 5

High Frequency CMOS Transconductors

Scott T. Dupuie and Mohammed Ismail

5.1 Introduction

Current mode signal processing circuits have recently demonstrated many advantages over their voltage mode counterparts including increased bandwidth, higher dynamic range, and better suitability for operation in reduced supply environments (e.g. 3.3V). In addition, current mode processing often leads to simpler circuitry and lower power consumption. Traditionally, however, most analog signal processing has been accomplished using voltage as the signal variable. In order to maintain compatibility with voltage processing circuits, it is often necessary to convert the input and output signals of a current mode signal processor (CMSP) to voltages. Figure 5.1 shows a block diagram of a CMSP with the necessary interface circuits.

Either the transconductor or the transresistor (or both) may be eliminated in those applications where the appropriate variable is in current form. When present, however, the transconductor is a crucial part of the design since it may limit the linearity, frequency response, and noise performance attainable from the system. Therefore, any transconductor intended for use as the "front end" of a CMSP must meet the following criteria:

1. high linearity for large input signals,
2. low noise,
3. no dominant internal poles,
4. large transconductance,
5. low quiescent power dissipation (i.e. high efficiency).

Tunability may also be an important requirement in applications requiring a precise value of g_m (e.g. filters), independent of process and temperature variations.

The primary goal of this Chapter is to highlight the important aspects of transconductor design in the framework of current mode signal processing systems. Emphasis will be placed on large signal performance, with secondary consideration given to frequency response and noise. The Chapter begins with an introduction to the mathematical techniques needed to characterize transconductor performance. Using these techniques, the large

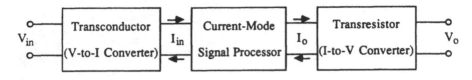

Figure 5.1: Current mode signal processing system

signal and small signal properties of the differential pair are analyzed. It will be shown that while offering excellent high frequency performance and low noise, its large signal characteristics are nonlinear. Therefore, the dynamic range and efficiency of the differential pair are limited.

Several techniques for improving the linearity of MOS transconductance elements have recently been proposed. Most of these schemes can be broadly classified into four categories: adaptive biasing, class-AB, source degeneration, and current differencing [1]. Some transconductors combine two or more of these techniques to achieve linearization. However, only those transconductors which exploit the intrinsic properties (i.e. without feedback) of MOS transistors will be considered here. The remainder of the Chapter is devoted to discussing each one of these techniques in detail. SPICE simulations (Level 2 and Level 3 models are used throughout) and experimental results are provided wherever possible to verify the conclusions reached by theoretical analysis.

5.2 Differential Pair Transconductors

The simplest and most widely used transconductor is the source-coupled differential pair. In addition to its obvious simplicity, the differential pair offers a true differential input and can readily achieve both positive and negative transconductance values. With a slight increase in complexity to implement common-mode feedback, this enables the implementation of a fully-balanced architecture, thus improving the dynamic range, PSRR, and CMRR. Furthermore, the inherent symmetry of the differential amplifier tends to reduce offsets and drift. While offering excellent high frequency performance (fundamentally limited by transmission line effects in the gate) and low noise, its large signal characteristics are nonlinear. As a result, it will be shown that both the dynamic range and the efficiency of the differential pair are limited.

---

[1]Current differencing is the process of subtracting two nonlinear currents with equal nonlinear terms but unequal linear terms, resulting in a linear I/O dependence.

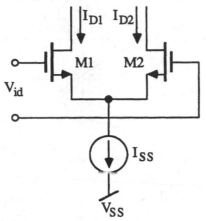

Figure 5.2: Differential pair transconductor

5.2.1   Nonlinearity

The basic source-coupled differential pair is shown in Figure 5.2. A large signal analysis of the circuit will illustrate its signal handling limitations. Using the simplified square-law relationship for a MOSFET (described in Appendix B) in the saturation region and assuming M1 and M2 are perfectly matched, the output current is given by [1]

$$I_o = I_{D1} - I_{D2} = \begin{cases} \sqrt{2I_{ss}K}\, V_{id} \sqrt{1 - \dfrac{K}{2I_{ss}} V_{id}^2} & |V_{id}| \le \sqrt{\dfrac{I_{ss}}{K}} \\[3mm] I_{ss}\, \mathrm{sgn}(V_{id}) & |V_{id}| \ge \sqrt{\dfrac{I_{ss}}{K}} \end{cases} \qquad (5.1)$$

Clearly, the input stage is linear only over a limited range of differential input voltage. The nonlinearity, which is a function of I_{ss} (or the gate bias voltage [V_{GS}-V_T], depending on your point of view), causes two problems. First, notice that as the signal level is increased the transfer function becomes more nonlinear. Therefore, large input signals will result in harmonic distortion and spurious signals being generated due to intermodulation. Second, since the transconductance of the input stage equals the slope of the I_o vs. V_{id} characteristic curve, g_m decreases as the signal level increases. This makes the transconductance a function of differential input signal level.

The most common method of specifying large signal performance is by measuring total nonlinearity. Nonlinearity is defined here as the percent deviation from the ideal value of $g_m V_{id}$ (where g_m is the small signal transconductance) in the output current. It is worth mentioning that this is a more rigorous test than is often used in the literature, where nonlinearity is

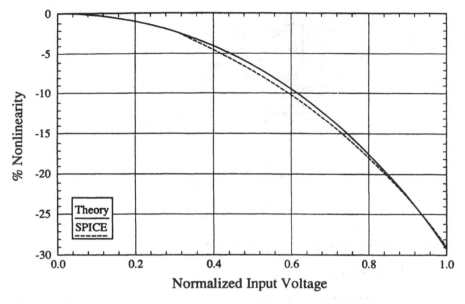

Figure 5.3: Nonlinearity of differential pair versus normalized input voltage (W/L=50/6, I_{SS}=200µA)

often measured with respect to the full scale output current. To maintain less than 1% nonlinearity, (5.1) shows that V_{id} is restricted to the range

$$-.20 \sqrt{\frac{I_{ss}}{K}} \le V_{id} \le .20 \sqrt{\frac{I_{ss}}{K}} . \tag{5.2}$$

Notice that the linear dynamic range is only a small fraction (20%) of the total dynamic range, resulting in poor efficiency [2]. The large signal handling capability can be improved by increasing the bias current, increasing the channel length, or decreasing the channel width. Therefore, tradeoffs must be made when selecting the bias currents and device sizes. For example, increasing I_{SS} to improve the linearity results in larger power dissipation. More importantly, however, increasing I_{SS} or decreasing the W/L (i.e. increasing V_{GS}) will degrade performance due to mobility reduction and reduce the negative common-mode input range. These factors ultimately limit the degree to which large signal handling capability can be improved. Figure 5.3 shows a plot of the percent nonlinearity in I_o versus x, where $x=V_{id}/\sqrt{I_{SS}/K}$ is the normalized input voltage.

---

[2]Efficiency is defined here as the maximum linear output current divided by the total bias current.

The percent deviation in the transconductance from the small signal value of g_m is often used as a measure of linearity as well. The relationship between g_m and V_{id} can be developed by taking the derivative of (5.1) with respect to V_{id}, yielding

$$g_m = \frac{\sqrt{2 I_{ss} K} \left[1 - \dfrac{K V_{id}^2}{I_{ss}} \right]}{\sqrt{1 - \dfrac{K V_{id}^2}{2 I_{ss}}}}. \tag{5.3}$$

As expected, (5.3) reveals that $g_m = \sqrt{2 I_{ss} K}$ for $V_{id} = 0$ and falls to zero as V_{id} approaches $\sqrt{I_{ss}/K}$. Also note that tuning of the differential pair transconductor is accomplished by varying the bias current I_{ss}. Since the large signal handling capability is proportional to $\sqrt{I_{ss}}$, the input signal swing degrades with decreasing tail current. The problem is particularly severe if a wide tuning range is desired.

5.2.2 Harmonic Distortion

Although nonlinearity is the most commonly used method of characterizing the large signal performance of a transconductor, it does not reveal any direct information about the harmonics or intermodulation products generated. Harmonic generation is a function not only of the total nonlinearity, but also of the type of nonlinearity (monotonic, even order, etc.) and the mode of excitation. Since the output current of the differential pair is a continuous monotonic function of the input voltage, the amount of harmonic distortion can be determined by expanding (5.1) into a MaClaurin Series (see Appendix B), resulting in

$$I_o = \sqrt{2 I_{ss} K} \; V_{id} + 0 - \frac{1}{2\sqrt{2}} \frac{K^{3/2}}{\sqrt{I_{ss}}} V_{id}^3 + 0 - \dots \tag{5.4}$$

Two things are worth noting about this result. First, even order products are eliminated due to the differential topology. This significantly reduces the distortion compared to a common-source amplifier since the dominant even-order nonlinearity is eliminated. Second, the coefficients of distortion products above 3rd order decrease rapidly. Therefore, higher order terms can be neglected with little loss in accuracy. Based on (A.5), the total harmonic distortion (THD) is given by

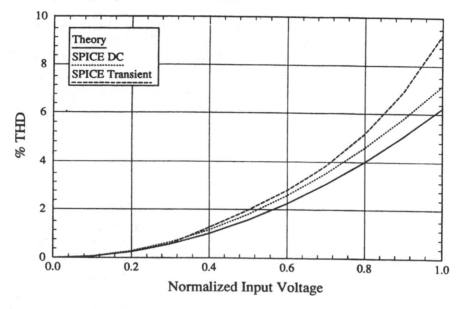

Figure 5.4: THD of differential pair versus normalized input voltage (W/L=50/6, $I_{SS}=200\mu A$)

$$THD \approx HD_3 = \frac{K}{16I_{SS}} V_{id}^2 = \frac{1}{16} x^2 .$$

(5.5)

To maintain less than 1% THD, V_{id} is restricted to the range

$$-.40 \sqrt{\frac{I_{SS}}{K}} \leq V_{id} \leq .40 \sqrt{\frac{I_{SS}}{K}} .$$

(5.6)

Notice that the maximum input signal for 1% THD is twice that for 1% nonlinearity. Therefore, the common practice of specifying nonlinearity rather than THD underestimates the large signal handling capability of the differential pair. The nonlinearity and transconductance as a function of V_{id} can be determined from the Maclaurin series as well and it can also be used to verify (5.2). It can easily be shown that the nonlinearity for any input voltage is four times the THD if the distortion is predominantly third order. Therefore, this technique is sufficient to provide a complete description of the large signal performance.

Two different simulation methods were used to verify the results of (5.5). In the first technique, SPICE is used to generate a DC transfer curve of the circuit. The results are then applied to a least squares polynomial curve fit program which approximates the transfer characteristic in the form of (5.4). The second technique employs the more traditional approach of performing a transient analysis of the circuit with a sinusoidal input followed by a

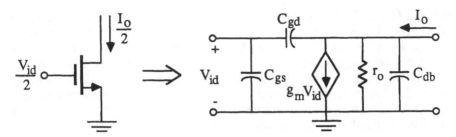

Figure 5.5: Differential-mode half circuit of differential pair

Fourier decomposition of the output signal. The advantage of the first technique is that only one computationally simple analysis has to be performed in order to predict the distortion at any input level. However, the transient analysis method is more accurate, particularly when the nonlinearity is not well behaved (e.g. nonmonotonic). Figure 5.4 compares the theoretical results with those of the two analysis techniques (see Appendix B for the model parameters).

It should be pointed out that neither technique is sufficient to accurately predict distortion levels to a fraction of a percent. Distortion levels this low depend heavily on minute details of device characteristics and may be dominated by device mismatch.

5.2.3 Frequency Response

As stated previously, the differential pair is capable of excellent high frequency performance. In the applications to be considered here, the load impedance seen by the transconductor is equal to the input impedance of the CMSP (ideally zero). Therefore, only the short circuit transconductance frequency response will be considered. Of course, a finite load impedance will cause a pole to exist at the output of the transconductor, but it will be dominated by the input characteristics of the CMSP. Under these circumstances, the frequency response of the differential pair is limited by the magnitude of the source resistance and ultimately by transmission line effects in the gate.

Figure 5.5 shows the differential-mode half circuit of the differential pair and its small signal equivalent. Straightforward analysis shows that the short circuit transconductance is given by

$$\frac{I_o(s)}{V_{id}(s)} = \frac{-g_m\left(1 - s\frac{C_{gd}}{g_m}\right)}{s\,R_S\left(C_{gs} + C_{gd}\right) + 1} \tag{5.7}$$

where R_S is the output resistance of the source. In addition to the pole caused by R_S, a right-half-plane (RHP) zero is formed due to the feedforward path through C_{gd}. A RHP zero introduces excess phase lag at high frequencies just like LHP pole does. This can be the limiting factor in applications requiring a precise phase response (e.g. filters). Notice that (5.7) predicts the pole will disappear when R_S equals zero. For values of R_S comparable to the channel resistance, however, the lumped element small signal model is no longer valid. In this case a more detailed analysis of the distributed channel resistance and gate capacitance reveals an infinite number of poles exist, with an effective nondominant pole approximated by [2]

$$p_{eff} \approx \left[\sum_{i=2}^{\infty} \frac{1}{p_i} \right]^{-1} = 2.5\omega_T$$

(5.8)

where $\omega_T = g_m / C_{gs}$. For modern processes ω_T ranges from several hundred MHz to several GHz. Of course, an effective pole equal to p_{eff} is never achieved in practice due to the polysilicon gate resistance. However, excellent high frequency performance is still obtained due to the fact that the differential pair has no internal nodes to generate parasitic poles.

5.2.4   Noise

The dominant sources of noise in MOS transistor circuits include thermal channel noise, generation-recombination noise, and induced gate noise. Generation-recombination mechanisms give rise to a 1/f noise spectrum and therefore are the dominant source of noise at low frequencies (typically below 10 to 100 KHz). Induced gate current noise arises from the capacitive coupling of the gate to the channel resulting in a $(f/f_T)^2$ (where $f_T = 2\pi\omega_T$) noise spectrum. This noise source becomes significant only at very high frequencies or when R_S is large. At intermediate frequencies, however, thermal channel noise dominates resulting in a constant noise spectrum over several decades of frequency. Since we are primarily concerned with high frequency operation (but well below f_T), only thermal channel noise will be considered.

Thermal channel noise can be modeled by a noise current generator connected from drain to source as shown in Figure 5.6(a). The spectral density of i_d^2 for an MOS transistor in the saturation region is approximately given by [3]

$$\frac{i_d^2}{\Delta f} \approx \frac{2}{3} 4kT \left(g_m + g_{mb} \right) = \frac{2}{3} 4kTg_m \left(1 + \chi \right)$$

(5.9)

where k is Boltzman's constant, T is absolute temperature, and $\chi = g_{mb}/g_m$. Applying this model to the signal processing system of Figure 5.1 results in the equivalent noise circuit (for differential signals) of Figure 5.6(b), where

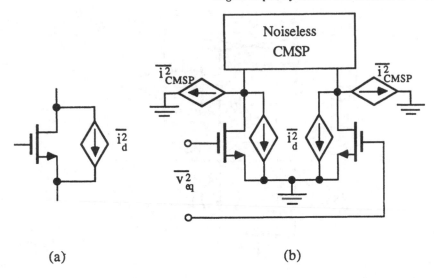

(a)   (b)

Figure 5.6: Equivalent noise circuits of (a) a MOS transistor (b) the entire CMSP

i_{CMSP}^2 is the equivalent input noise current of the CMSP. Since the bias current in the CMSP will generally increase with the bias current of the transconductor, i_{CMSP}^2 can usually be written in the form $N_1 i_d^2$ The constant N_1 depends on the topology and W/L ratio of the devices in the CMSP. Under these circumstances, the equivalent input referred voltage noise of Figure 5.6(b) is given by

$$\frac{v_{eq}^2}{\Delta f} = \left[\frac{\frac{8}{3}kTg_m(1+\chi)}{g_m^2} + \frac{i_{CMSP}^2}{g_m^2} \right] = \frac{16}{3}\frac{kT}{g_m}(1+\chi)\left[1+N_1\right].$$

(5.10)

The first term is due to the transconductor itself while the second term is caused by the equivalent input noise current of the CMSP. This method clearly illustrates the reasoning behind the usual practice of maximizing the transconductance of the input stage to minimize noise. Maximizing g_m requires a large I_{SS} and/or a large W/L ratio. Recall from Section 5.2.1, however, that large signal handling capability requires a large I_{SS} and a small W/L ratio. Therefore, the dynamic range (DR) [3] is limited and can be related to the design parameters by

---

[3]Dynamic range is defined here as the ratio of the maximum input signal for 1% THD, $V_{id(MAX)}$, divided by the equivalent input noise voltage $\sqrt{v_{eq}^2}$.

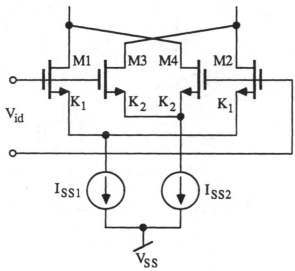

Figure 5.7: Cross-coupled differential pairs

$$DR = \frac{V_{id(MAX)}}{\sqrt{v_{eq}^2}} \propto \frac{I_{SS}^{3/4}}{K^{1/4}}.$$

$$(5.11)$$

As expected, the dynamic range can be improved by increasing I_{SS} since this simultaneously maximizes linearity and minimizes noise. It can also be improved to a lesser extent by decreasing the W/L ratio because linearity is a stronger function of W/L than noise.

5.2.5   Cross-Coupled Differential Pairs

A substantial increase in linearity can be obtained by simply cross-coupling two differential pairs [2] as shown in Figure 5.7. By properly scaling the ratio of W/L's and bias currents, approximate cancellation of the remaining odd order nonlinearities can be achieved. Since anytime more devices are added to a circuit the noise almost invariably increases, the noise performance of the transconductor will also be investigated. It will be shown that the increase in linearity is substantially greater than the increase in noise so that a net increase in dynamic range can be achieved.

The mechanism of nonlinearity cancellation via current differencing is best observed by examining (5.4). Notice that the linear term is proportional to $\sqrt{2I_{SS}K}$ whereas the nonlinear term is proportional to $K^{3/2}/\sqrt{I_{SS}}$. Therefore, nonlinearity cancellation is accomplished by scaling the W/L ratios and tail currents of the differential pairs according to

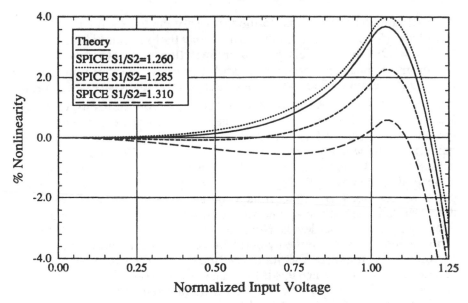

Figure 5.8: Nonlinearity of cross-coupled differential pairs versus normalized input voltage (W/L$_1$=50/6, I$_{SS1}$=200µA, I$_{SS2}$=100µA)

$$\left[\frac{(W/L)_1}{(W/L)_2}\right]^{3/2} = \left[\frac{I_{SS1}}{I_{SS2}}\right]^{1/2}$$

(5.12)

providing $(W/L)_1 \neq (W/L)_2$ and $I_{SS1} \neq I_{SS2}$. This makes the coefficient of the nonlinear term identical for both pairs but the coefficient of the linear term different. The same result can be obtained directly from (5.1) by using the approximation $\sqrt{1-x^2} \approx 1-x^2/2$. When the output currents are subtracted, the nonlinear terms cancel resulting in a linear transconductance given by

$$g_m = g_{m1} - g_{m2} = \sqrt{2I_{SS1}K_1} - \sqrt{2I_{SS2}K_2}.$$

(5.13)

Several design choices must be made at this juncture. For example, increasing the bias current and associated W/L ratios in accordance with (5.12) improves efficiency and increases g_m. Unfortunately, this degrades the accuracy of the nonlinearity cancellation due to secondary effects. In addition, the scaling of device aspect ratios can be performed using a constant length or constant width approach. The constant length approach will generally result in better nonlinearity cancellation since any short channel effects will be the same for both devices. However, it will be shown shortly that constant width scaling can lead to better high frequency performance.

SPICE simulations were performed to verify the theory for $I_{SS1}=2I_{SS2}$. Based on (5.12), perfect nonlinearity cancellation should occur when $(W/L)_1=1.26(W/L)_2$. This particular operating point was chosen as a compromise between efficiency and accuracy in the nonlinearity cancellation. Figure 5.8 shows a plot of the percent nonlinearity in I_o versus x, where $x=V_{id}/\sqrt{I_{SS2}/K_2}$ is the normalized input voltage. Constant length scaling was used to maximize the linearity performance. These results demonstrate a marked improvement in linearity over a simple differential pair, with the.worst case reduced from 30% to 4%. The remaining nonlinearity is due to the fact that higher order terms are being neglected.

Several different optimizations can be performed to improve the performance over a specified voltage range. Two such optimizations are demonstrated in Figure 5.8. The first optimization attempts to maximize linearity for small input levels. Note that extremely low levels of distortion can be achieved over a limited voltage range. The second optimization was performed to minimize the total nonlinearity for normalized voltages between zero and one, since this is the range in which both pairs are active. In all cases, the THD remains below 1% for normalized voltages in excess of one. This is possible because M1 and M2 remain active even after M3 and M4 have saturated.

As one might expect, the frequency response of cross-coupled differential pairs is essentially identical to a single differential pair, as described in Section 5.2.3, with one notable exception. If the transconductor is driven with fully balanced signals and the load provided by the CMSP is symmetrical, the RHP zero caused by the feedforward paths through C_{gdi} is given by

$$z = \frac{g_m}{C_{gd1} - C_{gd2}}$$

(5.14)

as opposed to $z=g_m/C_{gd}$. If the devices are scaled so that all four transistors have the same width (resulting in the same gate overlap capacitance), the zero is effectively eliminated. Qualitatively, this occurs because the displacement current through C_{gs1} is canceled by an equal magnitude but opposite phase displacement current through C_{gs2}. The cancellation is exactly valid only if the pole at the input circuit of both differential pairs is the same ($C_{gs1}=C_{gs2}$ or $R_S=0$). A partial cancellation is still obtained, however, even if this is not the case.

As a final note, the noise performance of the CMSP with a cross-coupled differential pair transconductor will be considered. Using the techniques outlined in Section 5.2.4, the equivalent input referred voltage noise is given by

$$\frac{v_{eq}^2}{\Delta f} = \frac{16}{3} \frac{kT}{g_m} (1 + \chi) \left[1 + N_2 + \frac{2g_{m2}}{g_m} \right]$$

(5.15)

where $g_m = g_{m1} - g_{m2}$. Note that (5.15) has the same form as (5.10) except for the additional term $2g_{m2}/g_m$. In most cases, this term is insignificant compared to N_2. Of greater importance, however, is the fact that $N_2 > N_1$ because of the increased level of bias current necessary to attain a specific g_m (due to the g_m subtraction). Therefore, optimum noise performance is obtained by minimizing g_{m2} in accordance with (5.12), consistent with the desired degree of nonlinearity cancellation. In many cases the increase in noise is smaller than the increase in large signal handling capability, resulting in a net improvement in dynamic range.

5.3   Adaptively Biased Transconductors

Circuits which employ one or more signal dependent bias currents (as opposed to DC bias currents) are said to be *adaptively biased*. The concept of adaptive biasing was first described in [4], where it was used to improve the slew rate of micropower amplifiers. In that application, linearity was not an issue since it was simply desired to increase the maximum current which could be delivered to a load. It wasn't long, however, before several authors employed the technique to linearize MOS transconductors and multipliers [5-7]. In this section we will examine the basic principles behind adaptively biased linear transconductors and compare two specific design examples. It will be shown that excellent linearity can be achieved using adaptive biasing without increasing the noise. The main penalty is in high frequency performance, which is typically not as good as a differential pair.

5.3.1   Basic Principles

The concept of adaptive biasing can be best understood by referring back to the characteristic equation of the differential pair given in (5.1). Let the constant bias current I_{SS} be replaced by a voltage dependent bias current of the form

$$I_{SS} = I_{DC} + k' V_{id}^2$$

(5.16)

where I_{DC} is a constant current (equal to the original I_{SS}). Substituting (5.16) into

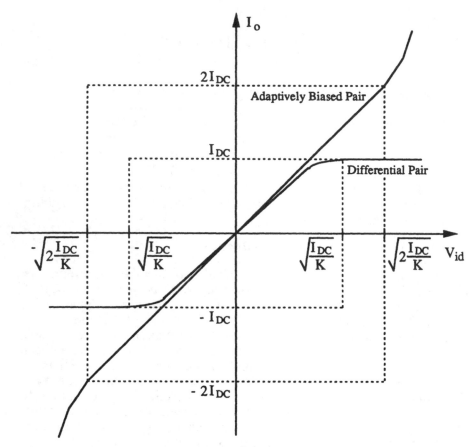

Figure 5.9: Transfer characteristics of differential pair with and without adaptive biasing

$$I_o = I_{D1} - I_{D2} = \begin{cases} \sqrt{2I_{DC}K}\; V_{id} \sqrt{1 - \dfrac{K - 2k'}{2I_{DC}} V_{id}^2} & |V_{id}| \le \sqrt{\dfrac{2I_{DC}}{K}} \\[3ex] I_{DC} + k'V_{id}^2 & |V_{id}| \ge \sqrt{\dfrac{2I_{DC}}{K}} \end{cases}$$

(5.17)

which results in a linear transconductance $g_m = \sqrt{2I_{DC}K}$ when $k' = K/2$. Figure 5.9 compares the transfer characteristic of an adaptively biased differential pair to a statically (DC) biased differential pair. Notice that not only is the transconductor perfectly linear, but the range over which both transistors are active is increased by a factor of $\sqrt{2}$. Therefore, the large signal handling capability for 1% THD is increased by a factor of approximately $\sqrt{2}/.4$, or 11dB. Similarly, the efficiency is improved from 40% to 200% since the maximum linear output current is twice the DC bias current. Also note that the adaptively biased transconductor does not show any clipping outside its linear range. Instead, the output current varies parabolically with respect to the input voltage in this region due to the square law bias current source.

Although (5.17) predicts that perfect nonlinearity cancellation can be achieved, some residual nonlinearity will remain due to second order effects not predicted using the simplified square law model. The dominant second order effects for long channel transistors operating in strong inversion include mobility reduction, channel length modulation, and the body effect. In many cases, the body effect can be largely eliminated by placing M1 and M2 in their own well. Furthermore, mobility reduction is usually a much stronger effect than channel length modulation due to the low input impedance of the CMSP. Therefore, mobility reduction is the dominant source of systematic error.

Including the effects of mobility reduction as described in Appendix B, (5.17) becomes

$$I_o = \sqrt{2I_{DC}K}\, V_{id} \sqrt{1 - \frac{K-2k'}{2I_{DC}}V_{id}^2} - \theta V_{id}[I_{DC} - (K - k')V_{id}^2] \qquad (5.18)$$

where θ is the mobility reduction parameter (.01 to .25 V^{-1}). Expanding (5.18) into a Maclaurin series results in

$$I_o = \left[\sqrt{2I_{DC}K} - \theta I_{DC}\right]V_{id} - \left[\frac{\sqrt{2}}{4}\sqrt{\frac{K}{I_{DC}}}(K - 2k') - \theta(K - k')\right]V_{id}^3. \qquad (5.19)$$

Note that a small positive third order distortion component will remain when $k'=K/2$. Therefore, mobility reduction causes a slight decrease in the small signal transconductance and introduces only odd order harmonic terms (since the symmetry of the circuit is unchanged). The new optimum value for k' can be derived by setting the coefficient of V_{id}^3 to zero, resulting in

$$k'_{opt} = K\left(\frac{\theta - \frac{1}{2\sqrt{2}}\sqrt{\frac{K}{I_{DC}}}}{\theta - \frac{1}{\sqrt{2}}\sqrt{\frac{K}{I_{DC}}}}\right). \qquad (5.20)$$

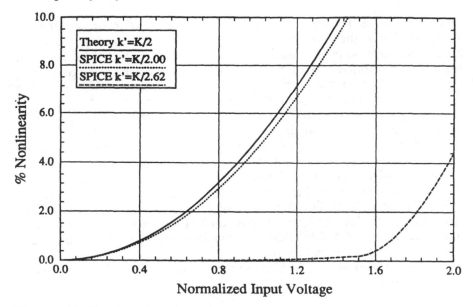

Figure 5.10: Nonlinearity of adaptively biased differential pair versus normalized input voltage (W/L=10/6, I_{DC}=300μA)

This assumes, of course, that a perfect quadratic bias current can be generated according to (5.16). Nonideal effects in the generation of the signal dependent bias current will be discussed later.

The optimum value of k' as predicted by (5.20) is often considerably less than K/2. To demonstrate this, an adaptively biased differential pair was simulated in SPICE (Level 3) using typical 2μm model parameters (see Appendix B). Based on these parameters, (5.20) predicted a new optimum value of k'=K/2.62 (where K is calculated from the zero bias mobility). Figure 5.10 compares the nonlinearity performance for both values of k'. The input voltage has been normalized to $x = V_{id}/\sqrt{I_{DC}/K}$ for direct comparison with the differential pair analysis. Ideally the nonlinearity should be zero for $x \leq \sqrt{2}$. Note that a significant improvement in performance is obtained by including a first order estimate of mobility reduction. Clearly, the simple square law model used in the original analysis is inadequate to predict distortion levels below a few percent.

The above analysis assumes that all the devices are perfectly matched according to the designated ratios. Device mismatch can also be a significant source of distortion. In particular, threshold voltage mismatch and K factor mismatch lead to random errors which can be reduced through proper design and layout. Although the effects of device mismatch will not be derived, it can be shown that K factor mismatch gives rise primarily to second order harmonics since it destroys the symmetry of the circuit. Threshold voltage mismatch causes a DC offset term but does not create added distortion. We may conclude, therefore, that the second order distortion coefficient is due

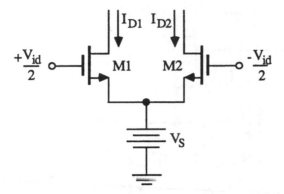

Figure 5.11: A two transistor squaring circuit

primarily to K factor mismatch and the third order coefficient is dominated by mobility reduction.

5.3.2 The Two-Transistor Squaring Circuit

One way to generate a signal dependent bias current which is proportional to the square of the input signal is to directly exploit the square law properties of MOS transistors in saturation. Consider the circuit depicted in Figure 5.11 [6], which is identical to a differential pair except that the common source node is voltage biased rather than current biased. Assuming the transistors are perfectly matched and the input signal is fully-balanced (around a common-mode value V_{ic}), the sum of the drain currents is given by

$$I_{D1} + I_{D2} = 2K \left(V_{CM} - V_T\right)^2 + \frac{K}{2} V_{id}^2 \tag{5.21}$$

where

$$V_{CM} = V_{ic} - V_S \tag{5.22}$$

provided both transistors remain active (i.e. $|V_{id}| \leq 2(V_{CM} - V_T)$). Note that (5.21) has the same form as (5.16), with

$$I_{DC} = 2K \left(V_{CM} - V_T\right)^2 \tag{5.23}$$

$$k' = \frac{K}{2}.$$

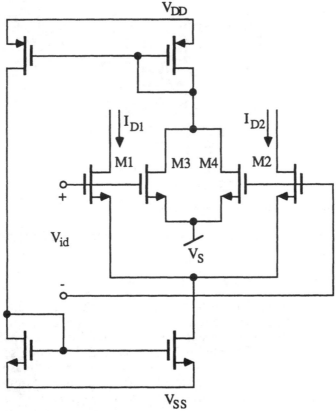

Figure 5.12: Adaptively biased differential pair using two transistor squaring circuit

Therefore, the two transistor squaring circuit is ideally suited to generate the necessary input dependent bias current. A straightforward implementation of this technique is shown in Figure 5.12. Neglecting mobility reduction for the moment, (5.23) predicts that optimum nonlinearity cancellation will occur when M1-M4 are matched. An alternative implementation which uses level shifters instead of a voltage source to bias the squaring circuit is given in [6]. Some form of level shifting is necessary in order to ensure that the floating differential pair (i.e. M1 and M2) has an adequate common-mode input range since all four transistors have the same V_{CM}. In either case, the class of input signals which can be processed is limited due to the requirement of fully-balanced signals.

Notice in Figure 5.12 that all of the noise generated by the squaring circuit is injected at the common source connection of the differential pair. Since this noise appears as a common-mode current, it is significantly reduced by the high common-mode rejection ratio of the differential pair. As a result, the increase in large signal handling capability is directly transformed into an equivalent increase in dynamic range. Unfortunately, the frequency response

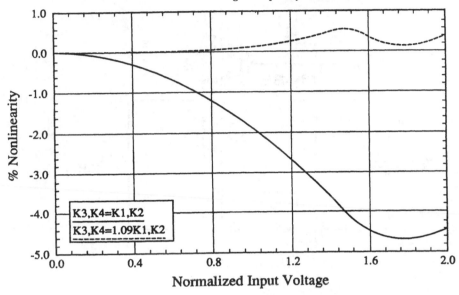

Figure 5.13: Nonlinearity of adaptively biased differential pair using two transistor squaring circuit versus normalized input voltage (W/L=10/6, I_{DC}=300μA)

is poor because the output from the squaring circuit has to propagate around the feedback loop before it is applied to the differential pair. As a result, the linearity improvement is best at low frequencies, where the time delay between application of a signal and generation of the adaptive bias current is negligible.

Due to the effects of mobility reduction, the squaring circuit will not produce a perfect quadratic transfer function. Including the effects of mobility reduction, (5.21) is modified to

$$I_{D1} + I_{D2} \approx 2K\left(V_{CM} - V_T\right)^2\left[1 - \theta\left(V_{CM} - V_T\right)\right] + \frac{K}{2}V_{id}^2\left[1 - 3\theta\left(V_{CM} - V_T\right)\right]$$

$$(5.24)$$

resulting in

$$I_{DC} = 2K\left(V_{CM} - V_T\right)^2\left[1 - \theta\left(V_{CM} - V_T\right)\right] \quad (5.25)$$

$$k' = \frac{K}{2}\left[1 - 3\theta\left(V_{CM} - V_T\right)\right].$$

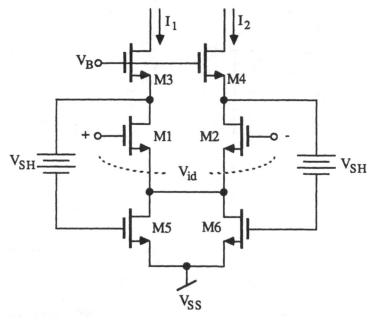

Figure 5.14: Improved adaptively biased differential pair using two transistor squaring circuit

Notice that mobility reduction causes a small decrease in both the DC bias current and the coefficient of the quadratic term. More importantly, however, for a constant V_{CM} the output current is still a square function of V_{id}. Therefore, a first order cancellation of mobility reduction effects is feasible by selecting a new optimum value of k'. The new value could be obtained by substituting (5.24) into the characteristic equation of a differential pair (also including the effects of mobility reduction) for I_{SS} and proceeding as before. Since the equations quickly become unwieldy, however, a more empirical solution appears attractive.

Figure 5.13 shows the nonlinearity generated by the circuit of Figure 5.12 for k'=K/2 as predicted by SPICE (Level 3). Comparing Figure 5.10 (solid curve) to Figure 5.13 (solid curve) shows that mobility reduction in the squaring circuit actually compensates for the same effect in the differential pair. The nonlinearity, which is now in the opposite direction, is approximately 40 percent of the nonlinearity when an ideal quadratic bias current was assumed. This compensating effect may result in adequate performance in many applications without the need for further optimization. In critical applications, even better performance can be obtained by increasing k' (i.e W/L of M3 and M4) by roughly 40 percent of the magnitude it was decreased in (5.20), as shown in Figure 5.13. Once again, a significant improvement in performance is obtained by considering mobility reduction effects. The accuracy of the empirical approach used above may vary depending on the bias point selected. However, it should yield a good starting point for computer optimization.

Although good linearity can be obtained with the straightforward implementation shown in Figure 5.12, the circuit has several shortcomings. As mentioned before, the input signal must be fully-balanced in order for the squaring circuit to function properly. Furthermore, the common-mode input level (V_{CM}) is simultaneously a bias voltage since it determines the magnitude of I_{DC}. Finally, a low impedance voltage source or a voltage level shifting circuit must be implemented to maintain adequate common-mode range. Such limitations place severe restrictions on the class of input signals which can be applied to the transconductor, thus reducing its versatility. These deficiencies can be largely overcome with the circuit shown in Figure 5.14 [7]. This circuit eliminates the need for fully-balanced signals while offering signal independent biasing and a floating input.

Assuming all six transistors are matched and have their bulks connected to their sources, the operation can be described as follows. Since M1 and M3 are matched and conduct equal amounts of current, their gate-to-source voltages are equal (ignoring the effect of drain voltage on V_T). Similarly, the gate-to-source voltages of M2 and M4 are equal. Therefore, the voltage difference at the sources of M3 and M4 is equal to the differential input voltage. Notice, however, that these source voltages are fully-balanced even if the input signal is single-ended due to the common-mode rejection of the differential pair. This balanced version of V_{id} is copied to the gate voltages of M5 and M6 by the level shifters V_{SH}, resulting in

$$V_{GS5} = V_{CM} - \frac{V_{id}}{2}$$

(5.26)

where

$$V_{CM} = \frac{V_B - V_{SH} - V_{SS}}{2}.$$

(5.27)

By applying (5.21) it can easily be shown that the sum of the drain currents in M5 and M6 contains the necessary quadratic dependence for nonlinearity cancellation. Unlike the previous implementation, however, the DC operating current is determined by V_B and V_{SH}, independent of the common-mode input level. In addition, fully-balanced signals are no longer required. The level shifters needed in this circuit may be implemented using simple source followers.

As was the case in Figure 5.12, the noise generated by the components in the squaring circuit is virtually eliminated by the common-mode rejection of M1 and M2. The noise generated by the cascode transistors M3 and M4 is negligible due to the large impedance seen looking down from their sources (resulting in a low effective g_m). The high frequency performance should be somewhat improved, however, since the feedback signal does not have to propagate through several current mirrors. This is particularly true if M3 and M4 have their bulks connected to V_{SS}, thus reducing the parasitic capacitances at their sources. Unfortunately, this would significantly degrade linearity due to the body effect.

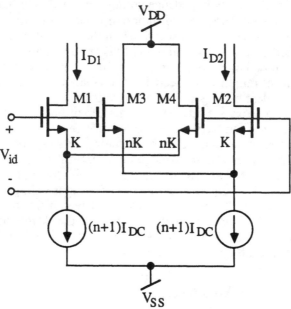

Figure 5.15: Cross-coupled quad cell

5.3.3   *The Cross-Coupled Quad Cell*

Another popular circuit for the generation of a square law bias current is the cross-coupled quad cell shown in Figure 5.15 [5]. The circuit consists of two unbalanced differential pairs, each with one transistor n times larger than the other. The output is taken as the sum of the drain currents in the two smaller transistors. Assuming all devices are operating in the saturation region, these drain currents are

$$I_{D1} = I_{DC}\left[1 + \gamma x^2 + \frac{\alpha}{2} x \sqrt{1 - \beta x^2}\right]$$

$$I_{D2} = I_{DC}\left[1 + \gamma x^2 - \frac{\alpha}{2} x \sqrt{1 - \beta x^2}\right]$$

(5.28)

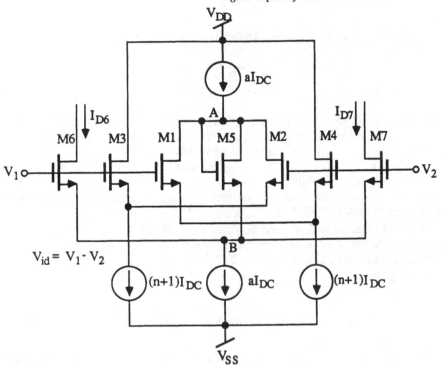

Figure 5.16: Adaptively biased differential pair using cross-coupled quad cell

where $x = V_{id}/\sqrt{I_{DC}/K}$ is the normalized input voltage and the remaining constants are defined as

$$\alpha = \frac{4n}{n+1}$$

$$\beta = \frac{n}{(n+1)^2}$$

$$\gamma = \frac{n(n-1)}{(n+1)^2} \qquad (5.29)$$

When the two drain currents are summed, the square root terms disappear resulting in

$$I_{D1} + I_{D2} = 2I_{DC}\left[1 + \gamma x^2\right]$$

$$= 2I_{DC} + 2K \frac{n(n-1)}{(n+1)^2} V_{id}^2 \qquad (5.30)$$

provided $|x| \leq \sqrt{(n+1)/n}$. Note that (5.30) has the same form as (5.16), where

$$k' = 2K \frac{n(n-1)}{(n+1)^2}$$

$$(5.31)$$

Choosing $k'=K/2$ for optimum nonlinearity cancellation results in n=2.155.

Figure 5.16 shows a complete circuit for a linear transconductor based on the cross-coupled quad cell. Transistors M1-M4 form the cross-coupled quad cell while M6 and M7 constitute the differential pair. Notice that this circuit has a unique way of delivering the input dependent bias current to the differential pair. Rather than use current mirrors, an additional device M5 and a current sink aI_{DC} are used to level shift the summed drain currents of M1 and M2 from node A to node B. Assuming all the devices are matched, it can be shown that a value of $a>4n/(n+1)$ is necessary to keep M5 conducting over the desired input range. For n=2.155, the transfer characteristic of this circuit is given by

$$I_o = I_{D6} - I_{D7} = 2\sqrt{KI_{DC}} \, V_{id} \qquad |x| \leq 1.21.$$

$$(5.32)$$

The transconductance is $\sqrt{2}$ times the value of previous circuits because the DC bias current is twice as much (see (5.30)). Note that the linear range is limited by the squaring circuit and not the differential pair. Beyond this range the behavior of I_o depends on the value of a selected. Larger values of a typically result in better nonlinearity cancellation at the cost of increased power dissipation. Normally a is chosen to be four since this makes the quiescent gate-to-source voltages of M1-M7 identical.

The circuit of Figure 5.16 was simulated at the optimum value of n for a=4 [5]. The percent nonlinearity in I_o versus the normalized input voltage is shown in Figure 5.17. The other curves show the effect of small variations in n corresponding to typical manufacturing tolerances. Notice that the nonlinearity is extremely small throughout the entire usable range. This is due to the fact that the effects of mobility reduction are self canceling as was the case for the two transistor squaring circuit. Although the math is too involved to examine here, it can be shown that the optimum value of n changes little when mobility reduction is considered [8].

The circuit of Figure 5.16 overcomes many of the limitations of the previous implementations. Fully-balanced signals are not required since it is inherently a floating input system. Therefore, no complicated level shifting or voltage biasing is necessary. In addition, the frequency response is often better since no current mirrors, level shifters, or cascode devices are necessary to generate the square law bias current. Noise performance is not sacrificed either since the noise produced by the cross-coupled quad circuit is

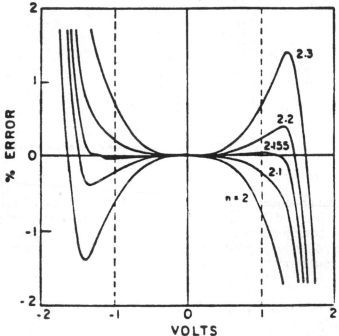

Figure 5.17: Nonlinearity of adaptively biased differential pair using cross-coupled quad cell versus normalized input voltage [5] (© 1984 IEEE)

still injected as a common-mode signal. Therefore, the noise performance is identical to a differential pair with the same bias current.

5.4 Class-AB Transconductors

Transconductors in which the maximum output current is greater than the quiescent bias current (i.e. $\eta > 100\%$) are generally operating in the class-AB mode. Adaptively biased transconductors also have this property. The difference is that adaptively biased circuits achieve this efficiency by employing dynamic current sources, whereas class-AB circuits obtain the extra current directly from the supply. As a result, class-AB transconductors typically exploit the square law characteristics of an MOS transistor in the saturation region to achieve linearization [9]. In this section, we will examine the basic principles behind class-AB transconductors and compare several different designs. We will show that although excellent linearity and efficiency can be achieved, class-AB circuits often (but not always) have poor common-mode rejection and hence are unsuitable for many applications. Where applicable, however, they offer low noise and the capacity for excellent high frequency performance.

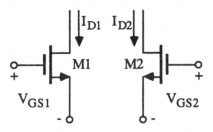

Figure 5.18: Two transistor circuit used to demonstrate class-AB principle

5.4.1   Basic Principles

The fundamental principles of class-AB transconductors based on the square law characteristics of MOS transistors can be understood by examining the two transistor configuration shown in Figure 5.18 [7]. Assuming both transistors are perfectly matched and operating in the saturation region, the differential output current is given by

$$I_{diff} = I_{D1} - I_{D2} = K\left(V_{GS1} + V_{GS1} - 2V_T\right)\left(V_{GS1} - V_{GS1}\right). \qquad (5.33)$$

Eq. (5.33) reveals that a linear transconductance can be achieved by ensuring that the sum of the gate-to-source voltages is a constant [4]. With the sum constant, if V_{id} is equal to V_{GS1}-V_{GS2} then (5.33) reduces to

$$I_{diff} = 2K\left(V_{CM} - V_T\right)V_{id}. \qquad (5.34)$$

where $V_{CM} = (V_{GS1}+V_{GS2})/2$ is the common-mode input level. Note that the transconductance $g_m = 2K(V_{CM} - V_T)$ is perfectly linear and may be varied electronically by adjusting the common-mode input level. The mechanism of nonlinearity cancellation is graphically illustrated in Figure 5.19.

An useful by-product of this technique is that the sum of the drain currents displays a quadratic relationship to the differential input voltage. Under the same conditions as above, the sum of the drain currents is given by

$$I_{sum} = I_{D1} + I_{D2} = 2K\left(V_{CM} - V_T\right)^2 + \frac{K}{2}V_{id}^2. \qquad (5.35)$$

---

[4]More precisely, that $V_{GS1}+V_{GS2}-2V_T$ is a constant. It will be shown later that the body effect on V_T is either insignificant or can be reduced through proper design.

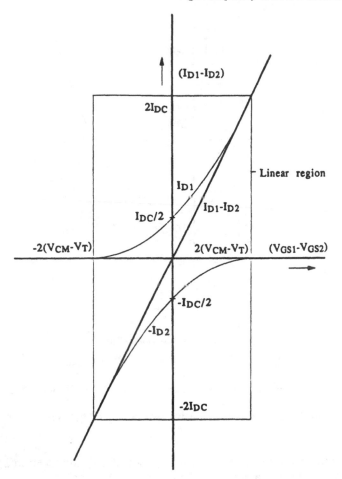

Figure 5.19: Graphical illustration of linear class-AB operation

The DC term can be eliminated by subtracting off the common-mode current if necessary. It is interesting to note that the reverse of the above condition is also true. If a sum current having the form given in (5.35) is forced upon the two transistors, the sum of the gate-to-source voltages will be constant and therefore a linear transconductance is obtained. This was the fundamental principle used to linearize the differential pair in Section 5.3. Therefore, adaptive biasing can be completely described in terms of the square law properties of MOS transistors. This viewpoint offers more insight into the operation of the adaptively biased differential pair than was previously obtained. The mechanism of generating a square law current is graphically illustrated in Figure 5.20.

From the above analysis, the fundamental principles of class-AB operation can be defined. Under the condition of a constant sum of gate-to-source

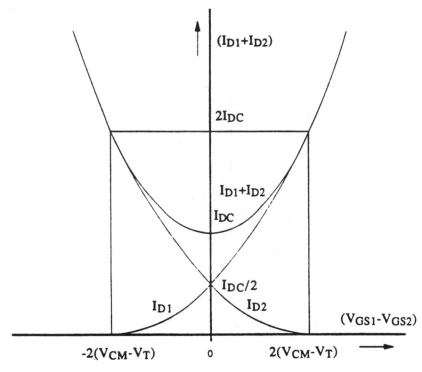

Figure 5.20: Graphical illustration of square law class-AB operation

voltages, two matched MOS transistors operating in the saturation region display

1. a linear relationship between the difference of the gate-to-source voltages and the difference of the drain currents, and
2. a quadratic relationship between the difference of the gate-to-source voltages and the sum of the drain currents

These conclusions are valid provided both transistors remain active, which requires

$$|V_{id}| \leq 2\left(V_{CM} - V_T\right) = \sqrt{\frac{2I_{DC}}{K}} \, .$$

(5.36)

In this region, the output currents vary between

$$- 2I_{DC} \leq I_{diff}, \, I_{sum} \leq 2I_{DC}$$

(5.37)

where

$$I_{DC} = 2K \left(V_{CM} - V_T\right)^2 \tag{5.38}$$

is the total DC bias current. Note that these limits are identical to those of the adaptively biased differential pair. Therefore, many of the conclusions reached for the adaptively biased differential pair are valid here as well. In particular, the large signal handling capability is approximately a factor of $\sqrt{2}/.4$ (11dB) better than a classical differential pair. Also, since the maximum linear output current is twice the DC bias current, the efficiency of class-AB transconductors based on the above principle can reach 200%.

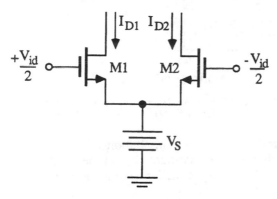

Figure 5.21: Voltage biased source-coupled pair

5.4.2 *Voltage Biased Source-Coupled Pairs*

The first class-AB transconductor we will consider is the voltage biased source-coupled pair shown in Figure 5.21 [7,10]. The circuit is identical to the classical differential pair except that the high impedance tail current source has been replaced by a low impedance voltage source. This circuit was first analyzed in Section 5.3.2, where it was shown that the sum of the drain currents is proportional to the square of the input signal. In order for the sum of the gate-to-source voltages to be a constant, however, the circuit must be driven with fully balanced signals. In this case, the equations derived above are directly applicable, resulting in a perfectly linear transconductance given by

$$g_m = 2K \left(V_{CM} - V_T\right) \tag{5.39}$$

where $V_{CM} = V_{ic} - V_S - V_T$. Novel techniques for implementing an adjustable voltage bias source can be found in the literature [11].

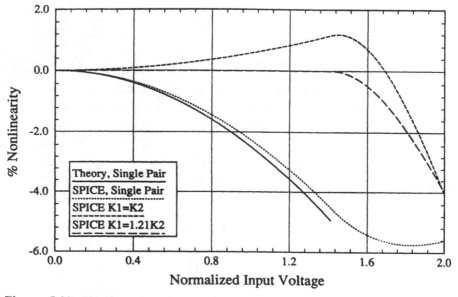

Figure 5.22: Nonlinearity of voltage biased source-coupled pairs versus normalized input voltage (W/L=10/6, V_{CM1}-V_T=2.5V, V_{CM2}-V_T=1.5V)

Due to the theoretically perfect cancellation of nonlinearities, mobility reduction once again plays an important role in the limits of performance. Including the effects of mobility reduction (see Appendix B), the output (i.e. difference) current becomes

$$I_o \approx 2K\left(V_{CM}-V_T\right)\left[1-2.5\theta\left(V_{CM}-V_T\right)\right]V_{id}-\frac{\theta K}{4}\left[1-2\theta\left(V_{CM}-V_T\right)\right]V_{id}^3.$$

$$(5.40)$$

As in our previous results, mobility reduction causes a slight decrease in the small signal transconductance and introduces only odd order distortion. From (A.5), the dominant third order distortion is approximately

$$HD_3=\frac{1}{32}\frac{\theta}{V_{CM}-V_T}V_{id}^2=\frac{\theta}{16}\left(V_{CM}-V_T\right)x^2$$

$$(5.41)$$

where $x=V_{id}/\sqrt{I_{DC}/K}$ is the normalized input voltage. As one might expect, mobility reduction becomes more of a problem as the gate overdrive voltage (V_{CM}-V_T) increases. Figure 5.22 shows the theoretical and simulated (SPICE Level 3) percent nonlinearity in I_o versus x for a typical bias of V_{CM}-V_T=2V and θ=.05V^{-1}. Since the third order distortion component is dominant, the total nonlinearity is four times the THD. Therefore, the THD is only 1.25% for a maximum input voltage of $x=\sqrt{2}$. For a discussion of distortion in the sum of the drain currents, see Section 5.3.2.

It is interesting to note that the differential-mode half circuit of the voltage biased source-coupled pair is identical to that of the classical differential pair shown in Figure 5.5. Therefore, the high frequency performance of the two circuits are identical for small fully-balanced signals. The noise performances are identical as well, since noise injected at the common source node is rejected due to symmetry. As a result, the discussion of frequency response and noise of the differential pair in Section 5.2 is directly applicable here as well. Similarly, the conclusions regarding device mismatch in Section 5.3.1 are also applicable. In particular, K factor mismatch gives rise primarily to second order harmonics, while threshold voltage mismatch causes a DC offset but no added distortion.

A significant improvement in linearity can be obtained by cross-coupling two transconductors as in Section 5.2.5. By properly scaling the W/L ratios in proportion to the different bias voltages (V_{CMi}-V_T), approximate cancellation of the nonlinearity due to mobility reduction can be achieved. The optimum ratio of W/L's can be determined by making the coefficient of the nonlinear term identical for both pairs. From (5.40), this occurs when

$$\frac{K_1}{K_2} = \frac{1 - 2\theta \left(V_{CM2} - V_T \right)}{1 - 2\theta \left(V_{CM1} - V_T \right)} \tag{5.42}$$

providing $V_{CM1} \neq V_{CM2}$. When the output currents are subtracted, the nonlinear terms cancel resulting in a linear transconductance given by

$$g_m = 2K_1 \left(V_{CM1} - V_T \right) - 2K_2 \left(V_{CM2} - V_T \right). \tag{5.43}$$

Notice that the sensitivity of the transconductance to changes in the threshold voltages is significantly reduced by the subtraction. This minimizes shifts in g_m due to processing variations and reduces noise coupling from the substrate.

SPICE simulations (Level 3) were performed to verify the theory for V_{CM1}-V_T=2.5V and V_{CM2}-V_T=1.5V. Based on (5.42), perfect nonlinearity cancellation should occur when K_1=1.13K_2. Due to the body effect on K_i, the optimum ratio of W/L's will be slightly different (1.21 in this case). Since the source voltages of both pairs are constant, however, the body effect does not degrade linearity. Figure 5.22 compares the linearity performance of voltage biased source-coupled pairs with and without cross-coupling. The input voltage in the cross-coupled case has been normalized to $x=V_{id}/\sqrt{2}(V_{CM2}-V_T)$ since this pair limits the linear range. A significant improvement in linearity over a single voltage biased pair is achieved even without optimizing the ratio of W/L's. In fact, essentially perfect nonlinearity cancellation is predicted when the ratio of W/L's is optimized. Of course, such performance is contingent upon the accuracy of the SPICE models and could never be achieved in practice due to device mismatch and process variations.

As was the case for the classical differential pairs, cross-coupling increases the noise of the CMSP due to the increased level of bias current necessary to attain a specific g_m (due to g_m subtraction). Therefore, optimum noise performance is obtained by minimizing $V_{CM2}-V_T$ in accordance with (5.42). Unfortunately, this also reduces the large signal handling capability since nonlinearity cancellation only occurs while both pairs are active (i.e. $|V_{id}| \le 2(V_{CM2}-V_T)$). Since the linearity is quite good without cross-coupling, the increase in noise would probably reduce the overall dynamic range in most situations.

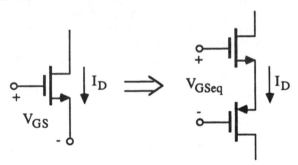

Figure 5.23: Replacing single transistor by CMOS double pair

5.4.3 CMOS Double Pair

The class-AB mode of operation has been restricted to single channel MOS thus far due to the requirement of matched transistors. This limitation can be overcome by replacing each transistor with the CMOS double pair shown in Figure 5.23 [12]. Using the simple quadratic MOS model (see Appendix B), the gate-to-source voltage of each transistor is given by

$$V_{GSi} = \sqrt{\frac{I_D}{K_i}} + V_{Ti}$$

(5.44)

where V_{Ti} is the bulk dependent threshold voltage (see (B.5)). The equivalent gate-to-source voltage $V_{GSeq}=V_{GSN}+V_{GSP}$ of the CMOS double pair is then given by

$$V_{GSeq} = \left(\frac{1}{\sqrt{K_N}} + \frac{1}{\sqrt{K_P}}\right)\sqrt{I_D} + V_{TN} + |V_{TP}|$$

(5.45)

$$= \sqrt{\frac{I_D}{K_{eq}}} + V_{Teq}$$

where

$$V_{Teq} = V_{TN} + |V_{TP}| \tag{5.46}$$

$$K_{eq} = \frac{K_N K_P}{\left(\sqrt{K_N} + \sqrt{K_P}\right)^2}.$$

Therefore, a pair of opposite polarity MOS transistors acts as a single transistor with an equivalent threshold voltage and transconductance parameter given by (5.46).

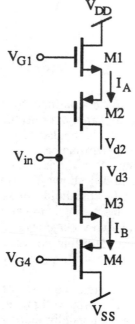

Figure 5.24: CMOS double pair transconductor

Based on this conclusion, several variations of the class-AB scheme can be implemented with CMOS double pairs [10]. One specific implementation which has received considerable attention is shown in Figure 5.24 [13]. This configuration is popular due to its simplicity and the fact that only one single-ended input voltage is required. Assuming like channel devices are perfectly matched and applying (5.45), the differential output current equals

$$I_0 = I_A - I_B = -2K_{eq}\left(V_{G1} - V_{G4} - \Sigma V_T\right)V_{in} - K_{eq}\left(V_{G1} - V_{G4} - \Sigma V_T\right)\Delta V_T$$

(5.47)

where

$$\Sigma V_T = V_{TN1} + V_{TN3} + \left|V_{TP2}\right| + \left|V_{TP4}\right|$$

(5.48)

$$\Delta V_T = \left(V_{TN3} - V_{TN1}\right) + \left(\left|V_{TP4}\right| + \left|V_{TP2}\right|\right) + \left(V_{G1} - V_{G4}\right)$$

provided all of the devices remain in the saturation region, which requires

$$V_{G4} + V_{TN3} + \left|V_{TP4}\right| \leq V_{in} \leq V_{G1} - V_{TN1} - \left|V_{TP2}\right|$$

(5.49)

and

$$V_{d2} - V_{TN3} \leq V_{in} \leq V_{d3} + V_{TN3}.$$

(5.50)

The output difference current can be obtained by employing current mirrors in the traditional manner or by simply connecting the drains of M2 and M3 together. The latter method avoids the use of current mirrors which tend to reduce the high frequency performance of the circuit. This is a distinct advantage over single channel methods which do not have this capability. In this case, however, (5.50) shows that the limits on linear operation are severely restricted by the threshold voltages of M2 and M3. The linear range can be extended by increasing V_{TP2} and V_{TN3}, which may be achieved artificially by connecting two voltage level shifters between V_{in} and the gates of M2 and M3, respectively. Simple source followers can be used to implement the level shifters since strict symmetry between these two voltages is not required.

Notice in (5.47) that the output current consists of the desired input dependent term and a DC bias term. If a twin-well process is available, each transistor can be placed in its own well, thus eliminating the body effect on V_T. Therefore, selecting $V_{G1} = -V_{G4}$ will minimize the offset term. In general, however, perfect symmetry between V_{G1} and V_{G4} is not needed for nonlinearity cancellation. Under the assumption of a constant ΣV_T, the transconductance is perfectly linear and is given by

$$g_m = 2K_{eq}\left(V_{G1} - V_{G4} - \Sigma V_T\right).$$

(5.51)

Tuning can be accomplished by varying the gate bias voltages V_{G1} and V_{G4}. For simplicity, V_{G1} is usually used for transconductance control while V_{G4} provides offset nulling. One distinct advantage of the CMOS double pair is that the transconductance is independent of the supply voltages. Therefore, no low impedance tunable power supply is required and the PSRR is much improved.

If a twin-well process is not available, the major source of nonlinearity in (5.51) is the body effect, which makes $\sum V_T$ a function of V_{in}. Several techniques for reducing or eliminated the body effect have been proposed [13]. For example, a first order cancellation of the body effect can be obtained by proper selection of the gate bias voltages V_{G1} and V_{G4}. Unfortunately, the necessary voltages are often impractical and necessitate complicated biasing schemes. A more viable technique which can achieve the same degree of nonlinearity cancellation is to ratio the dimensions of n-channel and p-channel devices. Although the device dimensions for best linearity depend on the body effect parameters (γ_P and γ_N) and bias conditions, the optimum K_P/K_N ratio is typically around .5 to 1. Given that $\mu_N \approx 3\mu_P$, this criterion is satisfied when $(W/L)_P \approx 1.5$ to $3(W/L)_N$.

Due to the addition of extra devices to facilitate the dual channel mode of operation, the frequency response of the CMOS double pair will not be as good as the voltage biased source-coupled pair. A small signal analysis of the circuit reveals that the optimum frequency response occurs when $W_P L_P \approx W_N L_N$. Combined with the criteria above for body effect cancellation, the optimum device dimensions for best linearity and frequency response are

$$\frac{W_P}{W_N} \approx \sqrt{3} \qquad \frac{L_N}{L_P} \approx \sqrt{3} \tag{5.52}$$

assuming $(K_P/K_N)_{opt}=1$. Keep in mind that the optimum dimensions are highly process dependent. However, both the nonlinearity minima and the frequency response maxima tend to be noncritical. Therefore, these results should provide a good starting point for computer optimization. More details on the body effect and frequency response can be found in [13].

5.4.4 Cross-Coupled Double Quad

The class-AB implementations considered thus far all suffer from limitations which limit their versatility. For example, the voltage biased source-coupled pair requires fully-balanced signals for nonlinearity cancellation. Furthermore, the common-mode input level must be precisely controlled since it determines the transconductance and linear range. To this end, a low impedance voltage source or a voltage level shifting circuit must be

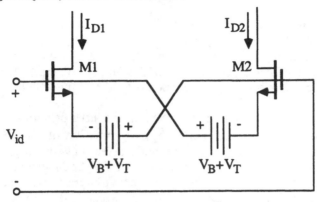

Figure 5.25: Linear MOS transconductor circuit principle

implemented. Also, since the voltage biased source-coupled pair has no inherent common-mode rejection, all of the common-mode rejection must be obtained in later stages (i.e. CMSP). As a result, the CMRR is typically much worse than a classical differential pair. On the other hand, the CMOS double pair transconductor has only one single-ended input and can realize only negative transconductance values. The common-mode input level must also be controlled to prevent large offsets and common-mode output currents.

An alternative implementation of the class-AB circuit principle which overcomes many of these deficiencies is shown in Figure 5.25 [5,12]. Recall from Section 5.4.1 that the sum of the gate-to-source voltages of M1 and M2 must be constant for nonlinearity cancellation to occur. Applying KVL around the input loop results in

$$V_{GS1} + V_{GS2} = 2(V_B + V_T) \tag{5.53a}$$

and

$$V_{GS1} - V_{GS2} = 2V_{id}. \tag{5.53b}$$

Since no assumptions were made concerning the nature of the input signal V_{id}, (5.53) is valid regardless of whether or not the input signal is fully balanced. Furthermore, the voltage sources used to maintain a constant sum of gate-to-source voltages are referenced to the input signal rather than the power supplies. Therefore, current source biasing can be used to provide floating inputs and restore inherent common-mode rejection. As a result, the common-mode input level no longer affects the transconductance or linear range (provided all devices remain saturated). With these properties in mind, substituting (5.53) into (5.33) yields

$$I_o = I_{D1} - I_{D2} = (4KV_B)V_{id}. \tag{5.54}$$

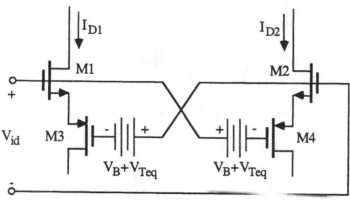

Figure 5.26: Replacing input transistors by CMOS double pairs

The transconductance $g_m=4KV_B$ is perfectly linear and can be tuned by changing the bias voltage V_B.

The circuit principle shown in Figure 5.25 requires two floating DC voltage sources. The magnitude of these sources must remain constant regardless of the current flowing through them. In addition, the voltage sources must not draw any current from the input terminals in order to maintain the high input resistance. The first attempt at implementing this scheme was described in [5]. The circuit is identical to the cross-coupled quad described in Section 5.3.3. Referring to Figure 5.15, the floating voltage sources are implemented with two additional n-channel transistors, M3 and M4. If these devices were biased with a constant current, the requirements described above would be satisfied. Note, however, that the currents in M3 and M4 are a function of the input signal level. Therefore, to obtain reasonable voltage sources with low impedance, these devices must be wide and biased with currents that are large compared to the signal currents through them. This requires a large value of n and results in poor efficiency.

A much better solution can be realized by replacing the single transistors (M1 and M2) in Figure 5.25 with CMOS double pairs [12], as shown in Figure 5.26. Since the CMOS double pair acts like a single transistor, the circuit behaviour is unchanged except that K is replaced by K_{eq} and V_T is replaced by V_{Teq}. Notice, however, that the drain currents in M1 and M2 no longer flow through the floating voltage sources. This simplifies the design of these elements considerably. The required floating voltage sources can now be realized by diode connected CMOS double pairs biased with a DC current, as shown in Figure 5.27. From (5.44), the V_B part of the bias voltage is given by

$$V_B = \sqrt{\frac{I_B}{K_{eq}}} .$$

(5.55)

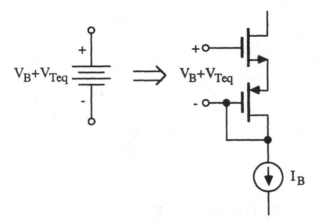

Figure 5.27: Double pair implementation of a floating voltage source

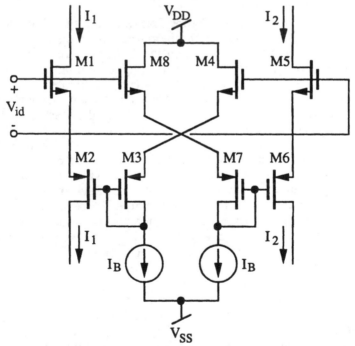

Figure 5.28: Cross-coupled double quad transconductor

Introducing this bias network into the circuit of Figure 5.26 results in the final linear transconductor shown in Figure 5.28. The transfer characteristic can be obtained by substituting (5.55) into (5.54), yielding

$$I_o = I_1 - I_2 = 4\sqrt{I_B K_{eq}}\, V_{in} \qquad |V_{in}| \leq \sqrt{\frac{I_B}{K_{eq}}}.$$

(5.56)

The transconductance is perfectly linear and can be tuned with the bias current I_B. As with other class-AB transconductors, the linear range is considerably better than the differential pair (since $K_{eq} < K_N, K_P$). However, since both the total quiescent current and the maximum linear output current are $4I_B$, the maximum efficiency is only 100% (as opposed to 200%). The efficiency can easily be extended beyond 100% if necessary by decreasing the W/L ratio of the inner quad transistors with respect to the outer quad. Reasonable increases in efficiency should be possible without a significant degradation in the linearity. The output current can also be obtained from either the top or the bottom. This is a distinct feature of the circuit which enhances its versatility.

As was the case for previous transconductors, mobility reduction and device mismatch will cause deviations from the ideal theory. The increased complexity of the circuit as compared to previous implementations makes a direct analysis of mobility reduction effects tedious. To simplify the calculations, the mobility reduction can be modeled by a series source resistance [7,12]. Modifying (5.44) to account for the effects of mobility reduction yields

$$V_{GS} - V_T \approx \sqrt{\frac{I_D}{K}} + \frac{\theta}{2K}I_D.$$

(5.57)

The term $(\theta/2K)I_D$ can be interpreted as an ohmic voltage drop across a resistance $R_S = \theta/2K$ in series with the source. For a CMOS double pair, the value of R_S for the n-channel and p-channel devices are added to yield an equivalent source resistance given by

$$R_{Seq} = \frac{\theta_N}{2K_N} + \frac{\theta_P}{2K_P}.$$

(5.58)

Using this model, the output current of the cross-coupled double quad can be approximated by

$$I_o \approx 4\sqrt{I_B K_{eq}}\, V_{id} - 2K_{eq}^2 R_{eq} V_{id}^3$$

(5.59)

As expected, the primary effect of mobility reduction is to introduce odd order distortion. From (A.5), the dominant third order harmonic distortion is

$$HD_3 \approx \frac{K_{eq}R_{eq}}{8V_B} V_{id}^2 = \frac{K_{eq}R_{eq}V_B}{8} x^2$$

(5.60)

where $x = V_{id}/\sqrt{I_B/K_{eq}} = V_{id}/V_B$ is the normalized input voltage. This technique could also be used to simplify the calculation of mobility reduction effects in other circuits as well.

Another source of distortion in the circuit of Figure 5.28 is the body effect, which makes the threshold voltages a function if V_{id}. Fortunately, the common-mode component of V_{id} does not cause distortion since it changes all the source potentials by the same amount. The differential-mode component, however, causes the bias voltages of the inner quad and the threshold voltages of the outer quad transistors to change in a nonlinear way. Since this component drives the gates of each double pair in a fully-balanced manner, choosing equal transconductance for the n-channel and p-channel devices will make the common source node of each double pair a virtual ground. As a result, the body effect can be minimized by making $K_N = K_P$, as was the case for the previous double pair implementation. Unlike before, however, optimum nonlinearity cancellation occurs when the wells are connected to their respective supply voltages, not to the sources. As in previous circuits, an analysis of device mismatch effects reveals that K factor mismatch causes second-order distortion. In addition, both K factor mismatch and threshold voltage mismatch introduce DC offset terms.

The cross-coupled double quad transconductor was recently fabricated in an n-well, 5μm CMOS process [12]. All transistors have a W/L of 100μm/10μm and are biased at $I_B = 100$μA, resulting in $K_{eq} = 87.5$μA/V and $R_{eq} = 491\Omega$. The n-wells in the layout are connected to the respective p-channel sources, in contrast to the optimization strategy described above. The measured THD as a function of input signal amplitude is shown in Figure 5.29, where $V_{id(MAX)} = 2.4V_{pp}$. Note that the linearity remains quite low (below .2%) for most of the input range but shows a more complex behavior than predicted by (5.59). This is probably due to compensation by other nonlinearities (including the body effect) which were not included in the analysis. A spectrum analysis of the output signal near the minimum points of distortion indicates that a first order cancellation of the third order component is occurring. The remaining distortion is predominantly fifth order.

It is interesting to note that the cross-coupled double quad has the same noise performance as a simple differential pair biased at the same current. Referring to Figure 5.28, it can be seen that the circuit actually consists of two independent transconductors consisting of M1-M4 and M5-M8. Each of these subcircuits contains four transistors (as opposed to two for the

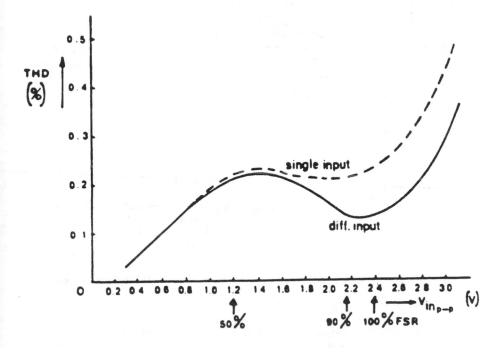

Figure 5.29: Measured THD of cross-coupled double quad versus input voltage [12] (© 1987 IEEE)

differential pair) and therefore generates twice as much noise power but the same amount of signal power as a differential pair biased at the same current. Combining the output current from the two transconductors results in four times as much noise power (assuming uncorrelated noise sources). Since the output current is available in complementary form, it also results in four times as much signal power. Therefore, at a total bias current of $2I_B$, the signal-to-noise ratio is approximately identical to a differential pair biased at I_B. As a final note, although an extensive frequency response analysis is not available, measurements and simulations indicate that a reasonably good bandwidth is obtainable

5.5 Triode Region Transconductors

All of the transconductance elements considered so far have employed MOS transistors operating in the saturation region. Resistive elements can also be used as transconductors, provided certain topological requirements are satified. Several transconductors employing MOS transistors in the nonsaturation (i.e. triode) region have recently been described [17-21]. They possess certain advantages over the circuits considered previously including simplicity, excellent tunability, and zero quiescent power dissipation. In this section, we will examine the basic principles behind linear triode region transconductors. For illustrative purposes, the two most popular techniques of canceling MOS nonlinearities will be described. It will be shown that only two MOS transistors are required to cancel the dominant even order nonlinearity of the MOSFET. A four transistor version of the same principle will also be described which cancels odd order nonlinearities as well and achieves a much better high frequency performance.

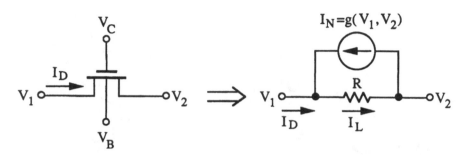

Figure 5.30: Large signal model of a MOS transistor in the triode region

5.5.1   *Basic Principles*

Consider an NMOS transistor illustrated in Figure 5.30 along with its large-signal model. Assuming the device is operating in strong inversion, the drain current is accurately given by (B.1). To simplify the analysis of triode region circuits, the 3/2 power terms in (B.1) can be expanded in Taylor series with respect to V_1 and V_2. Then I_D can be written in the general form

$$I_D = K\left[a_1(V_1 - V_2) + a_2(V_1 - V_2)^2 + a_3(V_1 - V_2)^3 + ...\right] \tag{5.61}$$

where K is given in (B.1). The coefficients a_i are independent of V_1 and V_2 but are functions of the gate and substrate potentials (V_C and V_B) and the process parameters. By applying (A.1), it can be shown that a_i are given by

$$a_1 = 2(V_C - V_T) \tag{5.62}$$

$$a_2 = -\left[1 + \frac{1}{2}\gamma(\phi_B - V_B)^{-1/2}\right]$$

$$a_i = \gamma A(i)(\phi_B - V_B)^{-(2i-3)/2} \qquad \text{for} \quad i \geq 3$$

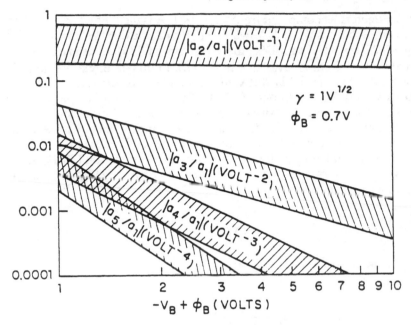

Figure 5.31: Typical values for coefficients of the nonlinear terms [17] (© 1983 IEEE)

where $A(3)=-1/12$, $A(4)=+1/32$, $A(5)=-1/64$ ..., etc., and V_T is the bulk dependent threshold voltage of the transistor given in (B.5).

The inverse of Ka_1 in (5.61) is the small signal resistance R of the transistor and is given by

$$R = \frac{1}{Ka_1} = \frac{1}{\mu C_{ox}\dfrac{W}{L}(V_C - V_T)} = \left(\frac{L}{W}\right)R_S$$

(5.63)

where R_S is a ''sheet resistance''. The aspect ratio L/W is used to set the nominal value of R. R can also be tuned by changing V_C (henceforth called the control voltage). The second and higher order terms in (5.61) are nonlinear. If these terms are canceled, the MOS transistor behaves like a linear resistor provided the device is operating in the triode region, which requires

$$V_1 \le V_C - V_T$$

(5.64)

An indication of how the nonlinear higher order terms in (5.61) affect the transistor characteristics is given by the relative magnitude of the coefficients a_i compared to a_1. Figure 5.31 shows the ratios a_2/a_1, a_3/a_1, etc., for common process parameters as computed from (B.1) [17]. A typical practical situation is illustrated by taking $V_2=0V$, $V_1=1V$, $V_C-V_T=2V$, $\mu Cox(W/L)=10\mu A/V^2$ and $V_B=-5V$ (for $\pm5V$ supplies, the substrate of NMOS transistors are connected to the minimum available potential). In this case, the first term in the right-hand side of (5.61) is $20\mu A$, the second term is $-6\mu A$, the third term is $3x10^{-2}\mu A$, and the fourth one is $-2x10^{-3}$ μA, ..., etc. It is clear that the dominant deviation from linearity comes from the second order term. In the following subsections we will discuss simple transconductance elements which can cancel even order terms or both even and odd order terms.

In order to further simplify the analysis, the current equation (5.61) can be divided into linear terms and nonlinear terms. Referring to Figure 5.30, the drain current can be expressed as

$$I_D = I_L - I_N \tag{5.65}$$

where I_L is the linear term and is given by

$$I_L = \frac{1}{R}(V_1 - V_2). \tag{5.66}$$

Similarly, I_N is a nonlinear function of V_1 and V_2 and can be expressed as

$$I_N = g(V_1) - g(V_2) \tag{5.67}$$

where

$$g(V) = g_e(V) - g_o(V). \tag{5.68}$$

The functions $g_e(V)$ and $g_o(V)$ are even and odd, respectively, so that

$$g_e(V) = g_e(-V) \qquad \text{and} \qquad g_o(V) = -g_o(-V). \tag{5.69}$$

Rewriting I_N in terms of g_e and g_o results in

$$I_N = \left[g_e(V_1) - g_e(V_2) \right] + \left[g_o(V_1) - g_o(V_2) \right]. \tag{5.70}$$

The odd term $[g_o(V_1)-g_o(V_2)]$ is very small compared to the linear term I_L in (5.65) (0.1% of it or less). Depending on V_1 and V_2, the term $[g_e(V_1)-g_e(V_2)]$ can be large and therefore its effect must be eliminated. Two methods of nonlinearity cancellation are described next. We start with a four-transistor transconductance which achieves complete cancellation of nonlinearities [18].

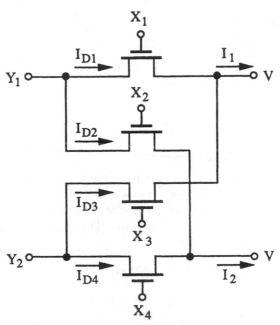

Figure 5.32: MOS four transistor triode region transconductor

5.5.2 Four Transistor Transconductor

A high linearity MOS transconductor can be constructed using four identical MOSFETs, as shown in Figure 5.32. The devices may be enhancement or depletion types, n-channel or p-channel. Assuming that each drain current is expressed according to (5.65), the nonlinear component I_{N1} of the current I_1 can be written as

$$I_{N1} = I_{ND1} + I_{ND3} \tag{5.71}$$
$$= \left[g(Y_1) - g(V)\right] + \left[g(Y_2) - g(V)\right]$$

where $g(Y)$ and $g(V)$ are nonlinear functions in Y and V, respectively, and are independent of the gate voltage X. It is clear that I_{N1} is exactly equal to the nonlinear component I_{N2} of the current I_2. Therefore, complete cancellation of these nonlinear terms will occur in the current difference $I_1 - I_2$. The linear component of I_1 is given by

$$I_{L1} = I_{LD1} + I_{LD3} \tag{5.72}$$
$$= \mu C_{OX} \frac{W}{L} \left[(X_1 - V_T)(Y_1 - V) + (X_3 - V_T)(Y_2 - V)\right]$$

whereas that of I_2 is given by

$$I_{L2} = I_{LD2} + I_{LD4} \tag{5.73}$$
$$= \mu C_{OX} \frac{W}{L} \left[(X_2 - V_T)(Y_1 - V) + (X_4 - V_T)(Y_2 - V) \right].$$

Taking the current difference $I_1 - I_2 = I_{L1} - I_{L2}$ yields

$$I_1 - I_2 = \mu C_{OX} \frac{W}{L} \left[(X_1 - X_2) Y_1 + (X_3 - X_4) Y_2 + (X_2 + X_4 - X_1 - X_3) V \right]. \tag{5.74}$$

The nodes V are usually connected to the input terminals of an approximately ideal op-amp. The voltage V is a nonlinear function of the signals Y_1 and Y_2 [19]. Therefore, for a truly linear current difference $I_1 - I_2$, the V term in (5.74) must be forced to zero. This results in the general nonlinearity cancellation condition that

$$X_1 + X_3 = X_2 + X_4 \tag{5.75}$$

providing all of the MOS devices are operating in the triode region. This condition is satisfied when

$$Y_i \leq \min (X_j - V_T) \tag{5.76}$$

where j=1,2 for i=1 and j=3,4 for i=2.

A special case of the general transconductance property described above occurs when Y_1 and Y_2 are the AC input signals to the transconductor. In this case, the X_i signals are used as control DC voltages with $X_1 = X_4$ and $X_2 = X_3$. This satisfies (5.75) and results in an equivalent small signal transconductance given by

$$G_{eql} = \frac{i_1 - i_2}{y_1 - y_2} = \mu C_{OX} \frac{W}{L} (X_1 - X_2) \tag{5.77}$$

where

$$y_1, y_2 \leq \min \left[X_1 - V_T, X_2 - V_T \right] \tag{5.78}$$

to ensure linear operation for all MOS devices.

Another possible configuration is obtained by interchanging the input and bias voltages [18]. That is, x_1 ($=x_4$) and x_2 ($=x_3$) are input AC signals, whereas Y_1 and Y_2 are used as bias (i.e. control) voltages. The equivalent transconductance is then given by

$$G_{eq2} = \frac{i_1 - i_2}{x_1 - x_2} = \mu C_{OX} \frac{W}{L} (Y_1 - Y_2)$$

(5.79)

where

$$x_1, x_2 \le \max \left[Y_1 + V_T, Y_2 + V_T \right]$$

5.80)

to ensure linear operation for all MOS devices.

It is important to point out that AC inputs to the transconductor, whether x or y signals, need not be fully balanced. Therefore, the circuit can be used to develop single ended MOSFET-C continuous-time filters using regular single ended output op-amps [19], thus avoiding the complexity of fully balanced op-amps. Fully balanced structures, however, eliminate the dominant second order nonlinearity generated as a result of the 'mobility degradation' phenomenon [20] which was neglected in the analysis. Finally, note that Y_i and X_j can simultaneously have AC input signals applied as long as (5.75) is satisfied and all transistors operate in the triode region. In this case, the resulting equivalent transconductance will be time varying. This has recently been used in the development of continuous-time four quadrant multipliers [21]. Assuming balanced signals for convenience (i.e. $x=x_1=-x_2$ and $y=y_1=-y_2$), then using (5.78) or (5.80) the condition for linear region operation is

$$x + y \le V_T.$$

(5.81)

This condition is easy to satisfy if depletion devices are used since they have a negative threshold voltage and can accept AC signals at their gates which vary around a zero average. If enhancement devices are used, the DC level of the input signals must be shifted for proper operation.

5.5.3 Two Transistor Transconductor

A simpler MOS transconductor that uses only two MOS transistors can be derived from the general topology of Figure 5.32 by eliminating the cross-coupled devices. In this case $I_1 = I_{D1}$ and $I_2 = I_{D4}$. Hence, the nonlinear component of I_1 is given by

$$I_{N1} = g(Y_1) - g(V).$$

(5.82)

Using (5.68) - (5.70), (5.82) can be written as

$$I_{N1} = g_e(Y_1) + g_o(Y_1) - g(V).$$

(5.83)

Similarly, the nonlinear component of I_2 is given by

$$I_{N2} = g_e(Y_2) + g_o(Y_2) - g(V).$$

(5.84)

Examining (5.83) and (5.84) reveals that the dominant even nonlinearity in the current difference can be canceled by using fully balanced signals (i.e. $Y_2 = -Y_1$). This yields

$$I_{N1} - I_{N2} = 2g_o(Y_1).$$

(5.85)

The linear component of $I_1 - I_2$ is given by

$$I_1 - I_2 = \mu C_{OX} \frac{W}{L} \left[Y_1 (X_1 + X_4 - 2V_T) - (X_4 - X_1) V \right].$$

(5.86)

The nonlinear term V in (5.86) can be canceled by making $X_4 = X_1$. By ignoring the odd nonlinearities given by (5.85) we can define a small signal linear transconductance given by

$$G_{eq3} = \frac{i_1 - i_2}{y_1} = 2\mu C_{OX} \frac{W}{L} (X_1 - V_T)$$

(5.87)

The transconductance as such turns out to be dependent on V_T with a limited signal handling capability compared to the four transistor transconductor.

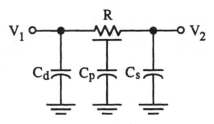

Figure 5.33: Small signal high frequency model of a MOS transistor in the triode region

5.5.4   *Frequency Response*

Figure 5.33 shows the high frequency small signal model of a MOS transistor including the parasitic capacitances, where the small signal resistance, R, is given by (5.63). The extrinsic parasitic capacitances C_d and C_s have a limited effect on the frequency response of the transconductors discussed earlier. This is primarily because in most cases C_d and C_s are either voltage driven or connected to an input terminal of an operational amplifier (virtual short).

Therefore, their effect is not significant if the op-amp approximates ideal behavior. The effect of the distributed capacitance C_p, however, can be more significant and ultimately limits the useful frequency range.

The distributed effects of a MOS transistor can be accurately modeled by a uniform RC transmission line. If the device is operating deep in the triode region, the first order y parameters are given by [22,23]

$$y_{11} = y_{22} \approx \frac{\frac{s\tau}{2} + 1}{R\left(\frac{s\tau}{6} + 1\right)} \tag{5.88}$$

$$y_{12} = y_{21} \approx \frac{-1}{R\left(\frac{s\tau}{6} + 1\right)}.$$

Using this simplified model, the frequency response of both the two transistor and the four transistor transconductors are readily determined.

For the two transistor transconductor with fully balanced inputs (i.e. $y_2 = -y_1$) and identical gate voltages $X_1 = X_4$, both transistors will have the same small signal y parameters. In this case, the small signal currents i_1 and i_2 are

$$i_1 = -[y_{22}]V - [y_{21}]y_1 \tag{5.89}$$

$$i_2 = -[y_{22}]V + [y_{21}]y_1.$$

Subtracting the two currents, it is readily shown that the actual transconductance is given by

$$G_{eq3(actual)} = G_{eq3} \frac{1}{\frac{s\tau}{6} + 1}. \tag{5.90}$$

Note that the transconductor has a parasitic pole at $s = -6/\tau$ which limits its usefulness in high frequency applications.

The four transistor transconductor displays significantly better high frequency performance. Referring to Figure 5.33, let $X_1 = X_4 = V_{C1}$ and $X_2 = X_3 = V_{C2}$ and consider the special case where Y_1 and Y_2 are the AC input signals. The small signal currents i_1 and i_2 are

$$i_1 = -[y_{22} + y'_{22}]V - [y_{21}]y_1 - [y'_{21}]y_2 \tag{5.91}$$

$$i_2 = -[y_{22} + y'_{22}]V - [y_{21}]y_2 - [y'_{21}]y_1$$

where y_{22}, y_{21} and y'_{22}, y'_{21} are defined for the gate voltages V_{C1} and V_{C2}, respectively. Once again, subtracting the two currents results in

$$G_{eq1(actual)} = G_{eq1} \frac{1 + s\dfrac{\tau_1 + \tau_2}{6}}{1 + s\dfrac{\tau_1 + \tau_2}{6} + s^2\dfrac{\tau_1\tau_2}{36}} \tag{5.92}$$

where $\tau_1 = R_1 C_p$ and $\tau_2 = R_2 C_p$. R_1 and R_2 are given by (5.63) for $V_C = V_{C1}$ and $V_C = V_{C2}$, respectively. The actual transconductance expression given by (5.92) has equal s terms in the numerator and denominator. Therefore, the four transistor structure originally developed for nonlinearity cancellation is also self compensating with respect to intrinsic parasitic capacitances. This allows successful signal processing at relatively high frequencies when compared to the two transistor transconductor. To clarify this point, let's consider the s^2 term in (5.92) at a frequency $\omega = \omega_0$, where ω_0 is the pole frequency of a second order transconductance-C filter implemented using the transconductor of Figure 5.32. For $\omega_0 = G_{eq1}/C$ (C is the integrating capacitance), the s^2 term in (5.92) is given by

$$s^2\frac{\tau_1\tau_2}{36} = -\frac{C_p^2}{36C^2}\left[\frac{(V_{C1} - V_{C2})^2}{(V_{C1} - V_T)(V_{C2} - V_T)}\right]. \tag{5.93}$$

For practical values of V_{C1} and V_{C2} where distortion levels are acceptable, the quantity in brackets in (5.93) is much less than one. Therefore, the s^2 term can be neglected if $C_p \ll C$, resulting in almost complete cancellation of parasitic effects.

5.6   Closing Remarks

As pointed out in the introduction, transconductance elements form a vital interface between CMSP's and the voltage dominated world. The transconductor should inhibit the performance of the CMSP system as little as possible. Therefore, transconductors intended for use in these systems must meet rigid specifications with respect to linearity, noise, and high frequency capability. Several transconductance architectures which can meet these requirements were presented including adaptive biasing, class-AB, and current differencing. However, no attempt was made to provide a complete review of every transconductance architecture or circuit. Many other techniques exist for implementing high performance transconductors including source degeneration [24,25] and the inverse function method [26,27]. These approaches may provide a viable alternative in some applications.

5.7 Acknowledgements

The authors wish to thank Dr. Steven Bibyk for many hours of fruitful discussions, Rich Kaul for proofreading the chapter, and Doug Yarrington for assisting with the document preparation.

This work is supported in part by NSF Grant No. MIP. 8896244, The Semiconductor Research Corporation Contract No. 89-DJ-066, and by a grant from AT&T Technologies.

5.8 References

[1]   P.E. Allen and D.R. Holberg, *CMOS Analog Circuit Design*, Holt, Rinehart, and Winston, New York, 1987.

[2]   H. Khorramabadi and P.R. Gray, "High frequency CMOS continuous-time filters," *IEEE J. Solid-State Circuits*, vol. SC-19, pp. 939-948, Dec. 1984.

[3]   G. Nicollini, D. Pancini, and S. Pernici, "Simulation-oriented noise model for MOS devices," *IEEE J. Solid-State Circuits*, vol. SC-22, pp. 1209-1212, Dec. 1987.

[4]   M.G. Degrauwe, J. Rijmenants, E.A. Vittoz, and H.J. De Man, "Adaptive biasing CMOS amplifiers," *IEEE J. Solid-State Circuits*, vol. SC-17, pp. 522-528, June 1982.

[5]   A. Nedungadi and T.R. Viswanathan, "Design of linear CMOS transconductance elements," *IEEE Trans. Circuits Syst.*, vol. CAS-31, pp. 891-894, Oct. 1984.

[6]   J.N. Babanezhad and G.C. Temes, "A 20-V four-quadrant CMOS analog multiplier," *IEEE J. Solid-State Circuits*, vol. SC-20, pp. 1158-1168, Dec. 1985.

[7]   K. Bult, "Analog CMOS square-law circuits," Ph.D. dissertation, Univ. of Twente, Enschede, The Netherlands, Jan. 1988.

[8]   A.P Nedungadi, "Effects of nonidealities on the performance of a CMOS transconductance linearization scheme," *Proc. Int. Symp. Circuits Syst. (ISCAS)*, 1987, pp. 758-761.

[9]   K. Bult and H. Wallinga, "A class of analog CMOS circuits based on the square-law characteristics of an MOS transistor in saturation," *IEEE J. Solid-State Circuits*, vol. SC-22, pp. 357-365, June 1987.

[10] F.J. Fernandez and R. Schaumann, "Techniques for the design of linear CMOS transconductance elements for video-frequency applications," *Proc. 28th Midwest Symp. Circuits Syst.*, 1985, pp. 499-502.

[11] K.D. Peterson and R.L. Geiger, "A fully balanced CMOS OTA for high frequency monolithic filters," *Proc. 29th Midwest Symp. Circuits Syst.*, 1987, pp. 208-211.

[12] E. Seevinck and R.F. Wassenaar, "A versatile CMOS linear transconductor/square-law function circuit," *IEEE J. Solid-State Circuits*, vol. SC-22, pp. 366-377, June 1987.

[13] C.S. Park and R. Schaumann, "A high-frequency CMOS linear transconductance element," *IEEE Trans. Circuits Syst.*, vol. CAS-33, pp. 1132-1138, Nov. 1986.

[14] E. Fong and R. Zeman, "Analysis of harmonic distortion in single-channel MOS integrated circuits," *IEEE J. Solid-State Circuits*, vol. SC-17, pp. 83-86, Feb. 1982.

[15] P. Antognetti and G. Massobrio, *Semiconductor Device Modeling with SPICE*, McGraw-Hill, New York, 1988.

[16] Y.P. Tsividis, *Operation and Modeling of The MOS Transistor*, McGraw-Hill, New York, 1987.

[17] M. Banu and Y. Tsividis, "Fully-integrated active-RC filters in MOS technology," *IEEE J. Solid-State Circuits*, vol. 18, pp. 644-651, Dec. 1983.

[18] M. Ismail, "Four-transistor continuous-time MOS transconductor," *Electronics Letters*, vol. 23, pp. 1099-1100, Sept. 1987.

[19] M. Ismail, S. Smith, and R. Beal, "A new MOSFET-C universal filter structure for VLSI," *IEEE J. Solid-State Circuits*, vol. 23, pp. 183-194, Feb. 1988.

[20] P.J. Ryan and D.G. Haigh, "Novel fully differential MOS transconductor for integrated continuous-time filters," *Electronics Letters*, vol. 23, pp. 742-743, July 1987.

[21] N.I. Khachab and M. Ismail, "MOS multiplier/divider cell for analog VLSI," *Electronics Letters*, vol. 25, pp. 1550-1552, Nov. 1989.

[22] J.M. Khoury and Y. Tsividis, "Analysis and compensation of high frequency effects in integrated MOSFET-C continuous-time filters," *IEEE Trans. Circuits Syst.*, vol. CAS-34, pp. 862-875, Aug. 1987.

[23] S. Smith, M. Ismail, and L. Kesting, "A new active-RC leap-frog structure for high-frequency MOSFET-C integrated filters," *Proc. 31st Midwest Symp. Circuits and Syst.*, pp. 689-692, Aug. 1988.

[24] Y. Tsividis, Z. Czarnul, and S.C. Fang, "MOS transconductors and integrators with high linearity," *Electronics Letters*, vol. 22, pp. 245-246, Feb. 1986.

[25] F. Krummenacher and N. Joehl, "A 4-MHz CMOS continuous-time filter with on-chip automatic tuning," *IEEE J. Solid-State Circuits*, vol. 23, pp. 750-758, June 1988.

[26] R.R. Torrence, T.R. Viswanathan, and J.V. Hanson, "CMOS voltage to current transducers," *IEEE Trans. Circuits Syst.*, vol. CAS-32, pp. 1097-1104, Nov. 1985.

[27] E. Klumperink, E. Zwan, and E. Seevinck, "CMOS variable transconductance circuit with constant bandwidth," *Electronics Letters*, vol. 25, pp. 675-676, May 1989.

Appendix 5A

Power Series Representation of Nonlinear Functions

One way to determine the amount of harmonic distortion which will be generated by a nonlinear circuit is to represent its transfer function in the form of a power series [14,15]. This technique is valid at low frequencies, where reactive effects can be ignored, provided the circuit has no discontinuities or gross nonlinearities. Under these circumstances, the Maclaurin series[5] expansion of a nonlinear transconductor (chosen for convenience) with $I_o = f(V_{id})$ is given by

$$I_o(V_{id}) = I_o(0) + \frac{\delta I_o(0)}{\delta V_{id}} V_{id} + \frac{\delta^2 I_o(0)}{\delta V_{id}^2} \frac{V_{id}^2}{2!} + \frac{\delta^3 I_o(0)}{\delta V_{id}^3} \frac{V_{id}^3}{3!} + \dots \tag{A.1}$$

which is of the general form

$$I_o = a_0 + a_1 V_{id} + a_2 V_{id}^2 + a_3 V_{id}^3 + \dots \tag{A.2}$$

$$= \sum_{n=0}^{\infty} a_n V_{id}^n$$

The a_0 term represents a DC offset due to device mismatch, the $a_1 V_{id}$ term is the desired output, and all higher-order terms represent undesired nonlinearities which cause harmonic and intermodulation distortion. To determine the percent harmonic distortion, assume V_{id} is sinusoidal with the form

$$V_{id} = V_p \cos\omega t . \tag{A.3}$$

Substitution of (A.3) into (A.2) reveals

$$I_o = a_0 + a_1 V_p \cos\omega t + a_2 V_p^2 \cos^2\omega t + a_3 V_p^3 \cos^3\omega t + \dots \tag{A.4}$$

$$= a_0 + a_1 V_p \cos\omega t + \frac{a_2 V_p^2}{2}(1 + \cos 2\omega t) + \frac{a_3 V_p^3}{4}(3\cos\omega t + \cos 3\omega t) + \dots$$

Notice that the $a_2 V_{id}^2$ term results in output component at DC and a component at twice the fundamental frequency. Therefore, the DC output of the transconductor (nominally zero) will shift and a harmonic component at twice the fundamental will be generated. In a similar fashion, the $a_3 V_{id}^3$

---

[5]Taylor series around the operating point $V_{id}=0$.

terms results in a component which adds to the fundamental and generates a harmonic at three times the fundamental frequency. This process can be expanded to the nth harmonic, with each a_n term contributing to the nth harmonic, and to lesser extent, other harmonic terms.

Harmonic distortion (HD_n) in the output current is defined as the amplitude ratio of the nth harmonic (at frequency $n\omega$) to the fundamental. For small distortion, higher order terms may be neglected, resulting in

$$HD_2 \approx \frac{1}{2}\frac{a_2}{a_1}V_p \tag{A.5}$$

$$HD_3 \approx \frac{1}{4}\frac{a_3}{a_1}V_p^2$$

Similarly, total harmonic distortion (THD) is defined as the RMS sum of the n harmonic distortion terms

$$THD = \sqrt{\sum_{n=2}^{\infty} HD_n^2}\ . \tag{A.6}$$

Therefore, the harmonic distortion generated by any transfer characteristic expressed in a Maclaurin Series can be readily determined. Intermodulation distortion (IMD) can be predicted in the same way by applying two sinusoids simultaneously at different frequencies and repeating the above analysis. In addition, the transconductance as a function of V_{id} can easily be obtained by taking the derivative of (A.2) with respect to V_{id}. Therefore, this technique is sufficient to provide a complete description of the large signal performance.

In some cases it is more convenient to derive the reverse transfer function of a circuit (e.g. $V_{id}=f(I_o)$). The Maclaurin series expansion of a nonlinear transconductance expressed in this form is

$$V_{id} = c_0 + c_1 I_o + c_2 I_o^2 + c_3 I_o^3 + \dots \tag{A.7}$$
$$= \sum_{n=0}^{\infty} c_n I_o^n$$

To determine the amount of distortion generated by the circuit (in the forward direction), the coefficients a_n can be expressed in terms of the coefficients c_n. The first several terms are [7]

$$c_1 = \frac{1}{a_1} \tag{A.8}$$

$$c_2 = -\frac{a_2}{3a_1}$$

$$c_3 = -\frac{a_1 a_3 - a_2^2}{5a_1}$$

where a_0 and c_0 have been ignored since they are just DC bias terms. The harmonic distortion can now be calculated as before by solving for a_n.

It should be pointed out that the Maclaurin series approach is only valid at frequencies well below the 3db point of the transconductor where reactive effects (i.e. linear or voltage dependent capacitances) can be neglected. At higher frequencies, the coefficients a_n are no longer constant but become functions of frequency. Under these circumstances, or when extremely low values of distortion are predicted by the Maclaurin series method, a more complicated analysis using the Volterra series must be employed. We will only consider the case where reactive effects can be ignored.

Appendix 5B   Device Modeling

Although extremely complex models for the MOS transistor are available, the purpose of hand calculations is to highlight basic principles and aid in the design process. Therefore, the models used throughout this Chapter contain the minimum amount of complexity to predict the fundamental performance characteristics of a given circuit. SPICE simulations (Level 2 and Level 3 models are used throughout) and experimental results are provided wherever possible to verify, and demonstrate the limitations of, the first order theory.

B.1   Nonsaturation Model

Consider an N-channel MOS transistor operating in the nonsaturation region with terminal voltages defined (with respect to ground) as shown in Figure B.1(a). For a long channel device operating in strong inversion[6], the drain current is accurately modeled [16] by

---

[6]Whether a device is effectively operating in the long channel regime depends on many factors including channel length, drain-to-source voltage, and process variables. For our purposes, "long channel" will refer to devices with $L \geq 5\mu m$ and $V_{DS} \leq 5V$.

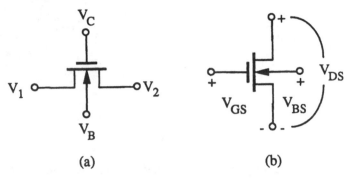

(a)   (b)

Figure B.1: MOS transistor terminal voltage designations

$$I_D = 2K\{(V_C - V_B - V_{FB} - \phi_B)(V_1 - V_2) - \frac{1}{2}[(V_1 - V_B)^2 - (V_2 - V_B)^2] -$$

$$\frac{2}{3}\gamma[(V_1 - V_B + \phi_B)^{3/2} - (V_2 - V_B - \phi_B)^{3/2}]\} \qquad (B.1)$$

with

$$K = \backslash F(\mu C_{OX}, 2) \backslash F(W \qquad (B.2)$$

$$\gamma = \frac{1}{C_{OX}} \sqrt{2qN_A\varepsilon_s}$$

and the remaining constants defined as

$W, L =$		channel width and length, respectively
μ	$=$	effective carrier mobility
V_{FB}	$=$	flat band voltage
C_{OX}	$=$	gate oxide capacitance per unit area
ϕ_B	$=$	approximate surface potential in strong inversion for zero backgate bias
N_A	$=$	substrate doping concentration
ε_s	$=$	silicon dielectric constant
q	$=$	electron charge

The same equation holds for P-channel devices as well by redefining the polarity of some voltages. This form of the drain current equation is convenient when dealing with triode mode transconductors where the current is bidirectional (drain and source terminals alternate). If the drain current is unidirectional, then the terminal voltages can be redefined as in Figure B.1(b). Rewriting (B.1) in terms of these variables results in the more familiar equation

$$I_D = 2K\{(V_{GS}-V_{FB} - \phi_B)V_{DS} - \frac{1}{2}V_{DS}^2 - \frac{2}{3}\gamma\,[(V_{DS}-V_{SB}+\phi_B)^{3/2} - (V_{SB} + \phi_B)^{3/2}]\}$$

$$(B.3)$$

Equations (B.1) and (B.3) provide good accuracy but are too complicated for use in hand calculations of most circuits. The complexity arises due to the 3/2 power terms, which are caused by the nonlinear change in depletion charge along the channel. Fortunately, this variation is only weakly nonlinear so that good results can be obtained with a linear approximation. Under these circumstances, the drain current is given by

$$I_D = 2K\left[(V_{GS} - V_T)V_{DS} - \frac{1}{2}(1 + \delta)V_{DS}^2\right] \qquad V_{DS} < V_{DS(sat)} \qquad (B.4)$$

where

$$V_T = V_{FB} + \phi_B + \gamma\sqrt{\phi_B + V_{SB}} \qquad (B.5)$$

$$= V_{TO} + \gamma(\sqrt{\phi_B+V_{SB}}-\sqrt{\phi_B})$$

$$\delta \approx \frac{\gamma}{2\sqrt{1+\phi_B+V_{SB}}}$$

$$V_{DS(sat)} = \frac{V_{GS}-V_T}{1+\delta}$$

Similar expressions for $V_{DS(sat)}$ can be derived for (B.1) and (B.3), although the results are more complicated. An upper limit on $V_{DS(sat)}$ is often used by assuming $\delta \approx 0$. Note that when γ is small or V_{SB} is large, this approximation is valid and (B.4) reduces to the familiar square law equation used in most hand calculations.

B.2 Saturation Model

An approximate expression for the drain current of an MOS transistor in the saturation region can be found by substituting $V_{DS(sat)}$ into (B.4), resulting in

$$I_D = \frac{K}{1 + \delta}(V_{GS} - V_T)^2 \qquad V_{DS} > V_{DS(sat)} \qquad (B.6)$$

where $V_{DS(sat)}$ is defined in (B.5). Alternatively, the $1+\delta$ term can be accommodated by simply redefining K as

$$K = \frac{\mu C_{OX}}{2(1 + \delta)} \frac{W}{L}$$
(B.7)

This definition of K will be assumed when the square law saturation model is being used. Since δ is a constant, however, it will not affect the basic conclusions derived from nonlinearity computations.

Due to the simplicity of these equations, they do not take into account many second order effects including mobility reduction, velocity saturation, and channel length modulation. The later two anomalies are primarily short channel effects and therefore won't be considered here. Mobility reduction, however, can cause a significant departure from the behavior predicted above for large V_{GS}. The effect of mobility reduction can be modeled to a first order by

$$\mu = \backslash F(\mu_o, 1 + \theta(V_{GS} - V_T)$$
(B.8)

where

μ_o	=	zero field mobility
θ	=	$1/T_{OX}E_{CR}$
T_{OX}	=	gate oxide thickness
E_{CR}	=	critical field normal to the channel

Typical values for θ range between .01 to .25 V^{-1}. Therefore, a one volt change in V_{GS} can result in up to a 20% decrease in the transconductance. Mobility reduction will only be considered when the simple square law models predict perfect nonlinearity cancellation.

The model cards used for many of the simulations in this Chapter were derived from a typical 2µm process and are shown below.

```
.MODEL NMOS1 NMOS (LEVEL=2 LD=0.25U TOX=400E-10 NSUB=1.85E+16 VTO=.80
+UO=650 UEXP=0.13 UCRIT=7.0E+4 DELTA=1.5 VMAX=5.0E+4 XJ=0.2U NEFF=3.0
+RSH=34 CGDO=2.25E-10 CGSO=2.25E-10 CJ=2.41E-04 MJ=0.65 CJSW=4.70E-10
+MJSW=0.3 PB=0.7)

.MODEL PMOS1 PMOS (LEVEL=2 LD=0.25U TOX=400E-10 NSUB=6.0E+15 VTO=-.80
+UO=245 UEXP=0.35 UCRIT=9.0E+4 DELTA=1.0 VMAX=3.0E+4 XJ=0.1U NEFF=1.5
+RSH=121 CGDO=2.15E-10 CGSO=2.15E-10 CJ=2.88E-04 MJ=0.5 CJSW=4.00E-10
+MJSW=0.3 PB=0.7)
```

The important DC parameters for N-channel devices are $\mu C_{OX} = 56\mu A/V$, $V_{TO} = 0.8$ V, $\gamma = 0.9$ $V^{1/2}$, and $\phi_B = 0.6$ V. Although SPICE (Level 2) uses a different model for mobility reduction than (B.8), the equivalent θ is approximately equal to 0.05 V^{-1}. The value of δ can be calculated from (B.5). Level 3 models with parameters similar to those above are used in some cases.

Chapter 6

Bipolar Current Mirrors

Barrie Gilbert

6.1 Introduction - The Ideal Current Mirror

Few readers of this book will need to be introduced to the concept of the current mirror; it has become a familiar icon of modern analog design. Accordingly, we will deal only briefly with the well-established foundations of the subject, already adequately presented in many excellent texts[1], and concentrate instead on developments of the basic current-mirror forms, having properties suited to special, though not uncommon, applications. Some of these developments have been included only to illustrate the wide variety of possibilities, and may not have any immediate practical value; this is true of the high-ratio forms presented in Section 6.5.2.

In principle, the basic function could have been realized before the advent of the bipolar junction transistor (BJT), but the unique properties of this device opened the door to efficient and eminently practical forms of mirrors, now to be found in a large proportion of analog ICs (and in many digital ones, too). The design approach used here is based strongly on the translinear view of the BJT. In many cases, this will mean the invocation of strict-TL forms of which the classical current mirror is the simplest possible example. Just as probable, however, will be a strong dependence on the basic translinearity of the BJT, that is, the predictable behavior of many special-purpose current mirrors will frequently hinge on the unique relationship between collector current, I_C, and base-emitter voltage, V_{BE}; they are "TN" circuits, although not always "TL" (see Chapter 2).

Current mirrors find endless uses not only in biasing applications of low to moderate accuracy, where their high output impedance makes them valuable as good approximations to ideal current sources. More complex mirrors provide special capabilities, such as high accuracy over many decades of current, exceptionally high output resistance, very low or high transfer ratios, and so on. Depending on how broad is one's definition of a "current mirror", the term can also embrace various types of useful nonlinear behavior and temperature shaping; we will certainly want to examine these important aspects of current-mirror design. Mirrors are also employed as broadband signal conveyors (as, for example, in current-mode amplifiers, described in other chapters of this book). We shall therefore spend some time examining the noise performance of common mirrors. A thorough discussion of dynamic behavior (AC gain and phase, and large-signal

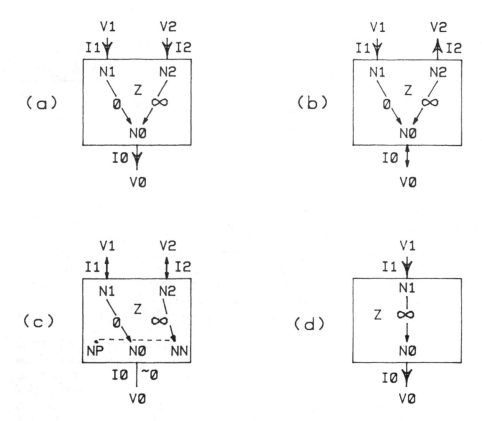

Figure 6.1 **Generalized forms of (a) current mirror (b) current reflector (c) current conveyor and (d) current source.**

transient response) is not included here, since it is a large topic in itself, and dependent on numerous details which lie beyond the scope of this brief survey of useful topologies. However, where appropriate, some aspects of dynamic behavior will be mentioned.

The simplest current mirror is a three-terminal device, Figure 6.1a, having an input node, N1, capable of accepting a current, I_1, of only one polarity, an output node, N2, into which a replication, I_2, of the input current flows in the same direction, and a common node, N0, in which the sum of the input and output currents flow. For now, we will assume the basic mirror to be built with NPN transistors; obviously, the polarity of bias voltages and direction of current flow will be reversed for PNP types. A practically-useful mirror is characterized by three desiderata, listed roughly in order of importance:

1) The current I_2 in the output branch should be essentially independent of the voltage V_2 on node N2, which may be biased at any potential from a

few hundred millivolts above the common node to many volts above it; that is, the incremental output resistance[1], r_o (more generally, impedance, z_o) should be relatively high, ideally infinite.

2) Both the large-signal mirror ratio, $M = I_2/I_1$, and the small-signal ratio $G = DI_2/DI_1$, should be essentially independent of the magnitude of the currents over many decades; that is, the ideal mirror is a *linear* element (so $G \int M$). In signal-path applications, the gain magnitude and phase response should be benign functions of frequency, ideally, completely independent of it.

3) The DC voltage, V_1, at the input node N1 should be small (say, within a few hundred millivolts of the common node) and it, and any AC voltage generated at this node, should be essentially independent of the input current, I_1; that is, the input resistance, r_i, (more generally, impedance, z_i) should be relatively low. In a formal context, it might be convenient to define an ideal mirror as one having an input impedance which approaches zero, but in practice this is rarely as important as achieving a near-zero output admittance (item 1).

We shall see that it is possible to design mirrors in which one or more of these characteristics can be highly refined. Frequently, such optimization is at the expense of other properties. For example, in directing attention to improving output resistance, the input resistance often has to increase. The particular way in which a mirror is optimized depends, of course, on the application. In many bias applications, for example, the emphasis will be on maintaining the highest possible output impedance (both the conductance and capacitance may need to be minimized): the input characteristics are less important here. On the other hand, in signal-path applications, it will often be found that quite high output conductances can be tolerated, while the input impedance (the reactive part often containing both capacitive and inductive components) must be minimized.

6.1.1 Mirrors, Reflectors, Conveyors, Sources

Before proceeding, it will be useful to compare the general properties of the current mirror with other cells having similar properties. The term "current reflector", for example, is sometimes applied to a three terminal network (Figure 6.1b) which satisfies all of the basic criteria for a mirror, but in which the direction of the output current is reversed[2]. However, this

---

[1]Lower-case variables are used here to denote small-signal parameters.

[2]The terms "mirror" and "reflector" invite comparison with their optical counterparts. In conventional ray tracing, the direction of the ray is shown as reversing at the reflection surface.

terminology is not standardized. As signal-path cells, the mirror is an inverting stage while the reflector is non-inverting; a folded cascode can be viewed as a reflector. Current conveyors, dealt with at length in Chapter 3, can be viewed as a special kind of "double-mirror" capable of both sinking and sourcing currents at input and output. Figure 6.1c depicts an ideal current conveyor. Nodes N1 and N2 can now accept and deliver currents of either polarity. The voltage at N1 closely follows that at N0; little (ideally zero) current flows in N0. In practice, two extra terminals, NP and NN are required to provide biasing and a source of positive and negative current to the output. Many of the techniques for improving the performance of current mirrors to be described in this chapter can be applied to current conveyors.

Finally, Figure 6.1d shows a floating current source, which requires only two terminals. Since there is no external control of the current magnitude, sources of this kind are generally only used in DC biasing applications[3]. While it is perfectly possible to design two-terminal current-source circuits, it is customary to utilize three-terminal controlled sources because of the simplicity (particularly, as is often the case, when multiple sources are required) and flexibility they afford. Current mirrors are widely used in such applications, although where the emphasis is on current *generation* rather than *replication* even simpler circuits often suffice. Some transducers generate current-mode signals directly (such as photodiodes), but most signals, and all accurate fixed references, are in the form of voltages, requiring voltage-to-current (V/I) conversion. This is a large topic, and there have been some interesting developments, particularly in the field of wideband V/I conversion, in recent years. However, it is beyond the scope of the present discussion, which for the most part assumes that we are dealing with variables (bias levels and signals) which are already in the form of currents. Section 6.5.3 will show some simple techniques for V/I conversion.

6.2 One-transistor Mirrors

In the simplest possible scenario, a single BJT can be used as a mirror: node N1 (of Figure 6.1a) is the base, N2 the collector and N0 the emitter. Of course, the practical objection to this proposal is that the mirror ratio, M, is much higher than generally needed and poorly-controlled, being just the common-emitter current-gain, β, and not very linear. While these objections are all true, it is nevertheless useful to begin here, because we will discover that many of the properties of more familiar mirrors can be predicted from the behavior of the single transistor circuit.

Figure 6.2a shows the biasing for an NPN device. We wish to determine

---

[3]It is worth mentioning, however, that two-terminal current-source circuits can be devised in which the current is a function of some environmental factor, such as temperature.

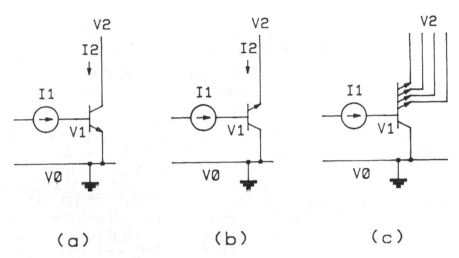

Figure 6.2 Single NPN transistor used as a rudimentary mirror; (a) usual forward active mode (b) inverse-mode (c) multiple outputs from inverse-mode transistor.

the input and output resistances. It will be assumed that β is independent of current over the range of interest. If β were also independent of V_2, the output I_2 would always be simply βI_1 and the output resistance r_o would be infinite. In fact, β increases with V_2 due to base-width modulation, and in the customary BJT model would be exactly doubled[4] when the collector bias voltage, V_{CB} (roughly, V_2), is equal to the forward Early voltage, V_{AF}.

It is important to note that for an ideal BJT operating at a fixed base current (I_1 in this case) V_{BE} does not vary *at all* with the collector voltage V_2, surprising as that may seem. It follows that if we were to replace the fixed current drive I_1 with a fixed *voltage* drive precisely equal to $V_{BE(Ib=I1)}$ the output would vary in exactly the same way with V_2, and a constant current $I_B=I_1$ would flow in the base. To model this, the collector current can be written

$$I_C = I_2 = (1+V_{CB}/V_{AF})I_{2(VCB=0)} \qquad (6.1)$$

Thus, the output resistance, near $V_{CB} \approx 0$, is unaffected by whether the base is current- or voltage-driven, and is simply

$$r_o = V_{AF}/I_2 \qquad (6.1a)$$

For a transistor having some fixed but general value of V_{BE}, and having any beta (even *infinite*), we will find that Eq. 6.1 still applies, when the

---

[4]Indeed, one can usefully define VAF as that value of VCB which exactly halves the base width, thus doubling the collector current for a given base charge.

expression

$$I_C = I_2 = (1+V_{CB}/V_{AF}) I_S(T) \, exp(V_{BE}/V_T) \qquad (6.1b)$$

is more appropriate[5]. Here, $I_S(T)$ is the "saturation current" for the device, typically 10^{-16}A, and $V_T = kT/q$ is the thermal voltage, approximately 26mV at T=300K.

To trace the improvement in the design of current sources as we progress, it will be useful to assign bench-mark values to some of the parameters of a hypothetical "standard BJT" and the operating conditions. Thus, let V_{AF} be 100V, β_0 (the value of β at $V_{CB} = 0$ and also at zero frequency) be 100, the mirror's output current, I_2, be 1mA and the temperature, T, be 300K. Then, this primitive mirror has a calculated output resistance of 100kΩ. At low values of V_2, r_0 may fall due to quasi-saturation effects. When V_2 is less than V_{BE} (that is, the transistor enters saturation) the output resistance falls suddenly; thus, the lower limit on V_2 is about 200mV above the common node. At high values of V_2, r_0 may again diminish, due to the onset of avalanche multiplication in the collector. The absolute upper limit on V_2 is constrained by the BV_{CEO} of the transistor. These effects and limitations are found, to a greater or lesser extent, in fully-developed current mirrors.

As well as being poorly defined, the mirror ratio M is a function of V_2. This raises an aspect of current mirror behavior which may be overlooked: not only does the output node exhibit a finite resistance, but the *small-signal gain*, G, varies with V_2, being in this case β_0 at $V_2 = V_{BE}$ and, for example, $2\beta_0$ at $V_2 = V_{BE}+V_{AF}$. Beta is also a function of current and temperature.

The small-signal input resistance r_i is $\Delta V_1/\Delta I_1$, which in this case is the incremental emitter resistance $r_e = \Delta V_{BE}/\Delta I_2$ multiplied by β_0, that is, $\beta_0 V_T/I_2$. Stated in terms of the input current, $r_i = V_T/I_1$, independent of beta. For I_1=10μA, r_i is about 2.6kΩ. The incremental input resistance is often of less significance than the absolute variation in input voltage over the current range: following classical junction behavior, this varies by roughly 18mV for a 2:1 change, or 60mV for a decade change, in I_2.

Uncontrolled mirrors of this sort are rarely used, although there are special applications in monolithic design where the beta-dependence of output current can be of value in compensating for the base current of other transistors in the circuit. An indirect example is the input bias-current cancellation scheme for an op-amp, described by Laude[2].

In some cases, an inverted transistor can be used as a single-transistor mirror (Figure 6.2b). This approaches the behavior of a more typical current mirror, having a much lower, and more predictable, value of M, which is now the inverse beta, typically ranging from less than 1 to rarely more than 10, depending on the fabrication process and device geometry.

---

[5]For simplicity, we will uniformly assume that exp(VBE/VT - 1) and exp(VBE/VT) are essentially equal for most practical values of VBE (see Chapter 2).

Within any given process, inverse beta has a much narrower range of values, typically varying by less than ±20% for all production lots under given operating conditions, and often as close as ±3% for all devices within a circuit, thus allowing fairly predictable design. V_2 must not exceed the open-base emitter-collector breakdown voltage, BV_{ECO}, typically 6V. Many modern circuits operate from a single 5V supply, so this need not be a serious shortcoming. An example of the use of an inverted transistor as a near-unity-gain mirror is found (as an incidental feature of the output-stage design) in a voltage-to-frequency converter described by Gilbert [3].

The output resistance r_o is now V_{AR}/I_2, where V_{AR} is the inverse Early voltage, typical values being 2 to 10V. Since this is very much lower than V_{AF}, the output resistance is correspondingly lower than for normal-mode operation. However, the output capacitance of this rudimentary mirror is very low, being just C_{JE}, whereas the normally-operated device has an output capacitance of $C_{JC}+C_{JS}$. It is usually possible to realize a factor of at least ten in the reduction of the output capacitance. This may commend the use of inverted transistors as current sources in biasing applications of high-frequency circuits. For example, in biasing a differential pair, the capacitance at the common emitter node may be troublesome in creating a nuisance pole in the open-loop transfer function, sometimes called the "tail-pole" [4]. In such cases, the inexact value of the bias current may be more tolerable than the high capacitance of a standard current source. One further advantage of the inverted-mode operation is that it is an easy matter to create multiple current outputs from a single multi-emitter transistor (Figure 6.2c). This consumes very little chip area, because only a single collector region is needed and each additional emitter may require only 100 or 200 square microns including its surround. When carefully designed, the matching between these currents can be surprisingly good - typically within ±1%. The effective value of the mirror ratio depends on the number of emitters and the particular geometry; the interested reader should consult the classical literature on I²L design, which uses NPN transistors in the inverse mode [5, 5a].

6.2.1 *Single-Transistor Mirrors Using Lateral PNPs*

Many junction-isolated IC processes are limited by the unavailability of a high-performance isolated-collector PNP transistor, and recourse to the lateral PNP is necessary for mirrors which source current from the positive supply. Later, we will discuss accurate mirrors using these devices, but at this point we will mention briefly some special aspects of lateral PNPs that affects their use as current sources.

The lateral PNP is often treated as a very poor relative in monolithic design texts. True, it has poor frequency response, begins to suffer from high-level injection effects at much lower currents than an NPN transistor, and exhibits other non-ideal aspects. Nevertheless, when used creatively, this device often provides very adequate performance in biasing applications, and can even have advantages over fully-isolated vertical PNP transistors in some cases. Its structure can be found in most standard texts on monolithic design, but certain important peculiarities of its behavior are rarely discussed. For example, it is often said that the beta of a lateral PNP is quite low. In fact, in a well-controlled fabrication process, the *low-current* gain can be as high as that of the NPN transistor; values of 200-300 at $I_C = 1\mu A$ are not uncommon. However, due to the low doping concentration in the epi-layer, which forms the base of a lateral PNP, its beta is a very strong function of current, since the injected carriers modulate the effective base doping and thereby reduce emitter efficiency. This is simply the nature of high-level injection, the effects of which begin to be felt at about 1mA for a typical minimum-geometry NPN, but at current of perhaps one-hundredth of this for a lateral PNP. It can be shown that *in high-level injection* its large-signal beta, $\beta(I_C) = I_C/I_B$, is more appropriately specified, not as a constant, but as a function of the collector current and a parameter which will be called the "beta scaling current", I_{BS}, such that

$$\beta(I_C) = I_{BS}/I_C \qquad (6.2)$$

For a typical structure, I_{BS} is about 10mA, and is surprisingly stable over production lots, being traceable to the doping profile, device geometry, and other well-controlled parameters. I_{BS} is only a weak function of temperature. Thus, the beta would have a reliable value of one at $I_C=10mA$, ten at $I_C=1mA$ and so on. Collector currents of devices having their emitter-base junctions in parallel, and driven by a single source of I_B, or driven independently, will exhibit good matching, even though the V_{BE} is not "classical", increasing by much more than 60mV/decade in the high-injection region, and also having a considerable excess component due to ohmic resistance. In low current operation, the beta matching of lateral PNPs is little different than for an NPN (possibly a little better).

Many interesting consequences of operation at high-level injection follow. First, the base current is

$$I_B = I_C^2/I_{BS} \qquad (6.3)$$

that is, a *parabolic* function of I_C. Alternatively, the collector current is seen to be a square-root function of I_B, that is,

$$I_C = \sqrt{I_B I_{BS}}. \qquad (6.4)$$

At lower injection levels, the base current becomes asymptotic to its maximum (low-current) value, β_{max}, and at very low currents it shows the customary decline again, as space-charge recombination begins to dominate the base current. We can empirically formulate the base current as

$$I_B = I_C \sqrt{\{ (I_C/I_{BS})^2 + (1/\beta_{max})^2 \}} \qquad (6.5)$$

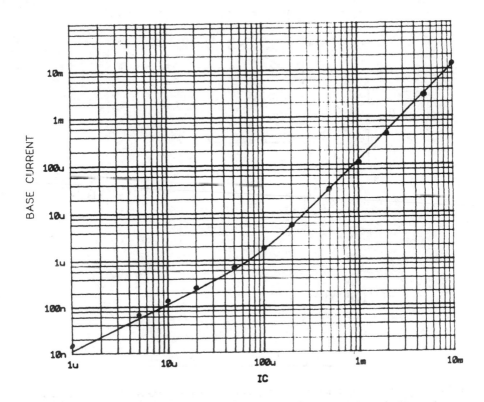

Figure 6.3 Measured base current of lateral PNPs compared to the empirical fit given in Eq. 6.5

Figure 6.3 shows the good agreement between this suggested formulation and measurements on a lateral PNP from a production monolithic process. In this case, I_{BS} was 8.1mA and β_{max} was 90. Similar close conformance has been found for a variety of different processes.

While it would be unwise to depend too heavily on this generally-reliable high-level behavior to implement some desired nonlinear function, it does suggest that a single lateral PNP may be valuable as a single-transistor mirror in biasing applications. An example is shown in Figure 6.4, where the "slewing" node of a high-speed op-amp must be driven by a fairly large current, say, 2mA. Using a conventional current mirror, using high-beta transistors and a mirror ratio of, say, 2, would require a driving current of 1mA. By relying on the nonlinear beta of a lateral PNP, with I_{BS} = 10mA, we can calculate (using Eq. 6.4) that I_B needs to be only 0.4mA.

There are other advantages to using single lateral PNPs in this way. Obviously, the chip area is less than that required by a two-transistor mirror; note in this connection that in multiple-output mirrors the common-base node corresponds to a single isolated epi region, another factor leading to

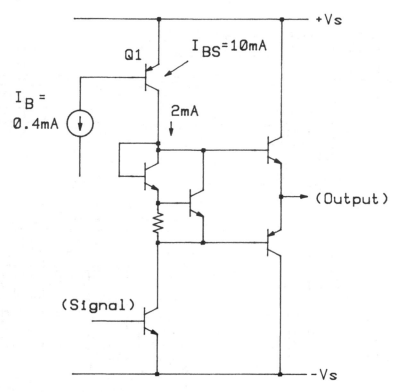

Figure 6.4 A simple open-base PNP current mirror (Q1) used to bias an op-amp slewing node with improved efficiency; remaining components are for illustrative purposes only.

reduced chip area. The output capacitance is only C_{JC} (the epi-substrate junction now only affects the input capacitance) and since the mirror gain is usually low, Miller multiplication of this capacitance is only moderate. A more subtle advantage is that the output resistance is higher than might be expected from low-current measurements of V_{AF}. This is because of another consequence of high-level injection, namely, the progressively-increasing charge concentration in the base at higher collector currents reduces the rate at which the depletion layer moves into the base, that is, the Early voltage of a lateral PNP increases at high values of I_C as the reduction of base width with collector bias decreases. In fact, it can be shown that the output conductance, g_o, is proportional to the square-root of collector current; for a typical device, its value can be pragmatically stated as $5\sqrt{I_C}$ microsiemens, when I_C is in milliamps. Figure 6.5 shows the close agreement with experimental results for g_o versus I_C, measured on seven samples.

Where a very low output capacitance is of dominant importance, a lateral PNP can also be operated in an inverse mode, with the output being taken from the emitter - usually a small circular diffusion of 5 to 20μm in diameter

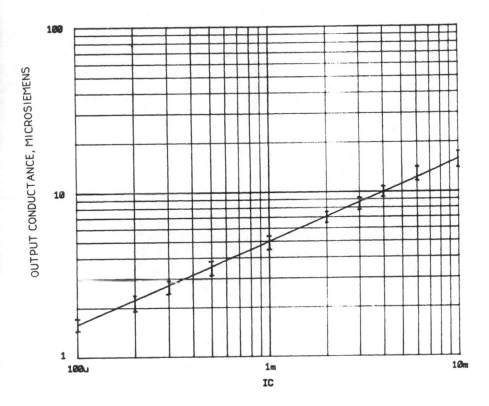

Figure 6.5 Measured output conductance of lateral PNP's compared to empirical fit of $5Ic^{0.5}$ microsiemens (Ic in mA)

- with the collector ring diffusion treated as the emitter common terminal. The inverse beta is lower, partly because more than half of the current injected from this ring flows outward to the surrounding isolation wall. However, the base scaling current is higher, due to the larger area of the effective emitter, so useful amounts of output can be generated before the lowered beta renders the device inefficient. Since the breakdown voltage of the emitter and collector are the same (that is, the BV_{CBO} of the corresponding NPN) the bias restrictions which arose with the inverted-mode operation of the NPN do not apply here. Note, however, that lateral PNP transistors can exhibit breakdown due to punch-through if the emitter-collector spacing is insufficient for the bias voltage; this occurs when the depletion layers penetrate into the base region far enough to reduce the effective base width to zero.

   Current sources and mirrors based on lateral PNPs can take advantage of another possibility afforded by the structure, namely, the use of a split collector diffusion. Figure 6.6a shows a typical geometry, in which the two segments are of equal size; unequal segments can be used, and of course more

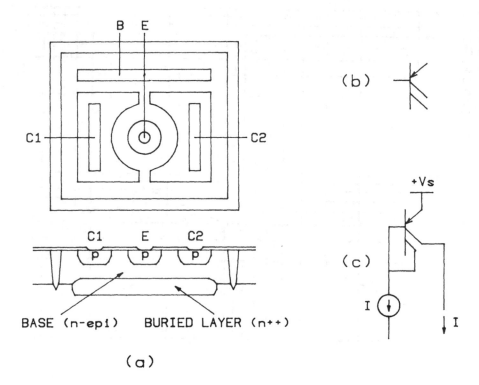

Figure 6.6 Split-collector lateral PNP: (a) plan view; (b) symbol; (c) connections to realize a unity-gain current mirror.

than two segments are possible. The unique nature of this device makes it valuable in numerous biasing applications. The current matching between halves of a two-collector lateral PNP is very good, routinely better than ±1%, and using special layout precautions, employing cross-connected pairs of split-collector transistors, sources matched to better than ±0.1% (at equal V_{CB}) have been maintained in production. This is valuable when using the device to provide a pair of active loads for an amplifier input stage, for example. This matching does not depend on the V_{BE} matching of two separate transistors, as is the case when using normally-biased NPN transistors as current generators. One way of visualizing this is to imagine that the single 'dot' of the emitter is like a light bulb positioned at the centre of the collector ring - emitting holes rather than photons - which illuminates all points on the ring with equal 'flux'. Since the emitter and collector diffusions are defined on the same mask, the radial symmetry is excellent. Even more intriguing is the almost-zero sensitivity of this current match to thermal- or stress-gradients of the chip; these can severely degrade the matching of collector currents in a pair of NPNs, via their effect on V_{BE},

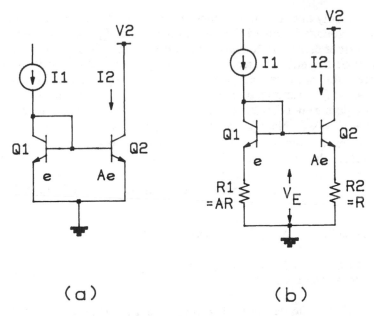

Figure 6.7 The classical NPN current mirror (a), with the addition of emitter degeneration (b).

which results in a current ratio error of 8% for a roughly 2mV ΔV_{BE} induced by a 1°C temperature difference between devices.

A split-collector lateral PNP transistor (two 180° collector segments) is occasionally used to realize a near-unity-gain mirror, by connecting one of the segments back to the base (Figure 6.6c). This is particularly effective at low currents, where the beta is high. Obviously, this idea can be extended to provide a ratio of 2 or 0.5 (using three 120° segments), or 3 or 0.33 (using four 90° segments). It is difficult to use more segments, limited by the minimum feature dimension and required spacing between these segments. Non-integral splits can be used, with reduced ratio accuracy.

6.3 Two-transistor Mirrors

The basic NPN current mirror (Figure 6.7a) is the simple theme from which endless variations and transformations can be spun, many of fundamental and pervasive value, others of which are mere curiosities awaiting exploitation. Although much has been written about this circuit in the many fine standard texts, there remain certain points of view that may not have been expressed or sufficiently stressed. It happens, for example, to be the simplest translinear circuit of practical interest, and we will begin here with the translinear view of this mirror.

Recall from Chapter 2 that in any loop containing an equal number of voltages across clockwise- and counterclockwise-biased PN junctions, the product of the current-densities in the two directions must be equal. Here, the loop contains only one junction voltage in each direction, and these are one and the same voltage in this trivial case, namely, the V_{BE} of both transistors. Since the current densities are equal, the ratio of the collector currents must be identical to the emitter area ratio, A. In practice, this would only be exactly true if this ratio were an integer and realized by exact replication of a basic geometry, for example, by making Q2 from four basic units of Q1, and if both transistors were operated at identical values of V_{CB}, which demands in this case that $V_2 = V_{BE}$. Note that the finite ohmic resistances do not affect the accuracy of the currents, since they can be assumed to scale in proportion to A and therefore the excess V_{BE} which they contribute is identical for each device. Nor is the finite, temperature-dependent beta, including its nonlinearity with current, a source of error in the ratio of the *collector currents*.

Errors in the actual mirror ratio, M, arise because

1   V_2 is, in general, not equal to V_{BE}, and the finite output resistance will cause I_2 to increase with V_2.

2   Some of I_1 is lost to the bases of both Q1 and Q2, so that $I_{C1} < I_1$.

3   Unless strict integer replication of geometry is used, there will be errors in the effective value of A and the ratio of junction resistances.

4   Even then, small random variations in emitter area and other sources of V_{BE} mismatch (including thermal gradients and strain) will occur.

6.3.1   *Effect of Practical Device Imperfections*

Errors in delineating the emitter area ratio A will result in essentially identical errors in the actual mirror ratio M. This is sometimes explained in terms of "V_{BE} mismatch", but this expression is helpful only as a reminder that a very small voltage difference between emitter nodes can have a large effect on accuracy. Indeed, there really can be no V_{BE} mismatch in this circuit, since in the current mirror both devices *operate at identical V_{BE}*, having their emitter-base junctions in parallel. However, we can translate an uncertainty in area ΔA into an equivalent V_{BE} error for the case (not true in the current mirror) where the transistors are operated at *identical collector currents*. This is well-known to be

$$\Delta V_{BE} \;=\; V_T \, ln \, (1 + \Delta A) \qquad\qquad (6.6)$$

which simplifies to

$$\Delta V_{BE} \;\approx\; V_T \, \Delta A \qquad\qquad (6.6a)$$

for $\Delta A \ll 1$, amounting to $\pm 260\mu V$ at $T = 300K$ for a $\pm 1\%$ area uncertainty. V_{BE} is also a function of emitter depth and other local variations in the device; these, and any further variabilities in V_{BE}, can be converted to a modified area ratio A', which may have a value greater or less than A, by rearranging Eq. 6.6 to first calculate ΔA then

$$A' \approx A \ exp \ (\Delta V_{BE} / V_T) \tag{6.7}$$

The error due to finite beta can be quickly determined by noting that

$$I_1 \ = \ I_{C1} + I_{B1} + I_{B2} \ = \ I_{C1} + I_{C1}/\beta_1 + I_{C2}/\beta_2 \tag{6.8}$$

and, if $V_2 = V_{BE}$ and good scaling disciplines have been observed, we can assume that $\beta_1 = \beta_2 = \beta_o$, so that

$$I_1 \ = \ I_{C1} + I_{C1}/\beta_o + A I_{C1}/\beta_o$$

while

$$I_2 \ = \ A I_{C1}$$

The actual mirror ratio, I_2/I_1 is therefore

$$M \ = \ \frac{A}{1 + \ (1 + A)/\beta_o} \tag{6.9}$$

For a classical mirror having $A = 1$ and our "standard" condition of $\beta_o = 100$, the error in M is -2%. Note that for very large values of A the above expression correctly shows the current ratio is asymptotic to β_o.

Finite Early voltage causes the output current to increase with the collector bias as given in Eq. 6.1. Combined with the loss due to finite beta, the complete expression is

$$I_2 \ = \ \frac{A(1 + V_2/V_{AF})}{1 + \ (1 + A)/\beta_o} \ I_1 \tag{6.10}$$

Note that this expression includes the voltage-dependence of beta, since I_2 increases at the same rate as β_2, so the base current of Q_2 is constant with V_2. The output resistance remains V_{AF}/I_2, independent of β or A. The small-signal input resistance r_i is now essentially just the r_e of Q_1, that is, V_T/I_1, and is also independent of β, A, and V_{AF}. Note, however, that this is exactly the same as for the rudimentary one-transistor mirror.

6.3.2 *Use of Emitter Resistors to Improve Accuracy*

We will return to further developments of the translinear view of the current mirror later, but first examine the effect of adding resistors in the emitter branches (Figure 6.7b). This is usually referred to as emitter degeneration, because the naturally-high gm of the devices is lowered in this way.

To understand how this works, first note that the V_{BE} of Q_2 is a function of both its collector current I_C and collector voltage V_{CB}. Starting with Eq. 6.1b

$$V_{BE} = V_{BE0} - V_T \ln (1 + V_{CB}/V_{AF}) \qquad (6.1c)$$

Here, V_{BE0} the value of V_{BE} for some I_C and temperature, and at $V_{CB} = 0$, given by

$$V_{BE0} = V_T \ln (I_C/I_S(T)) \qquad (6.1d)$$

So, if the mirror is to have a high output resistance, it must somehow be desensitized to the variation in the V_{BE} of Q2 caused by output voltage variations. This is the function of emitter degeneration.

With reference to Figure 6.7b, it can be seen that the larger the "degeneration voltage", V_E, that is allowed to develop across resistors R_1 and R_2, the less any variation of V_{BE} affects the output current I_2. When V_E is very large, Q2 might be considered to be operating essentially as a common-base stage, current-driven in its emitter, but there is a subtle and important difference. We can readily calculate the output resistance for this limiting case by considering just the variation of β_2 with V_2. For now, assume that the resistor ratio R_1/R_2 is made equal to the emitter area ratio A, as indicated in the figure. This would normally be the case; the consequences of altering it are discussed later. Then, we can immediately write

$$V_E = I_2(1+1/\beta_2)R = (I_1-I_2/\beta_2)AR$$

from which it follows that

$$M = \frac{I_2}{I_1} = \frac{A}{1 + (1 + A)/\beta_2} \qquad (6.11)$$

Now, if β_2 were fixed, this is essentially the same result as given already in Eq. 6.9, but in fact

$$\beta_2 = \beta_0(1+V_{CB}/V_{AF}) \qquad (6.12)$$

Inserting this into Eq. 6.11 and finding the derivative dI_2/dV_{CB} yields the output conductance, from which the general output resistance can be shown to be

$$r_0' = \frac{V_{AF}}{A(A+1)I_1} \frac{\{ \beta_0(1 + V_{CB}/V_{AF}) + 1 + A \}^2}{\beta_0} \qquad (6.13a)$$

This is not very insightful, so let us consider some special cases. First, for the classical mirror with A = 1 and $\beta_0 \gg 1$, operating at with V_2 much less than

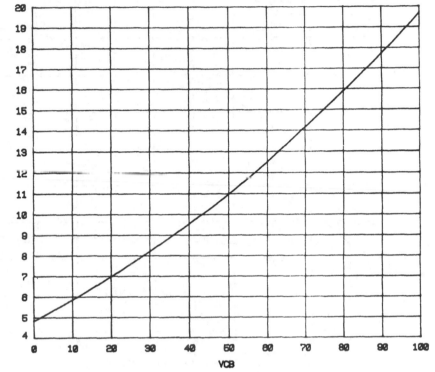

Figure 6.8 Output resistance of a highly-degenerated mirror operating at $I_C = 1mA$ varies from about 5MΩ at $V_{CB} = 0$ to 20MΩ at $V_{CB} = V_{AF}$.

V_{AF}, Eq. 6.13a simplifies to approximately

$$r_o' = \frac{V_{AF}\beta_o}{2I_2} \qquad (6.13b)$$

with I_1 replaced by I_2, appropriate for the conditions stated.

For example, using our benchmark device having $\beta_o = 100$ and $V_{AF} = 100V$, r_o' at 1mA is 5MΩ, 50 times the value for the basic mirror. The reason it is not beta times as high, as might be expected, is that a reduction in the base current of Q2 due to an increase in V_2 simultaneously reduces *two* sources of error: the first is in the loss of current conveyed from the emitter branch of Q2 to its collector (that is, alpha increases with V_2); the second is in the reduction of the base current lost from the input current I_1. If the base node were grounded and the emitter of Q2 driven by an ideal current source of value I_2, then r_o' would be simply $V_{AF}\beta_o/I_2$, the value sometimes given for the asymptotic output resistance of the highly-degenerated mirror. Errors in analysis can usually be traced to the use of small-signal modeling procedures,

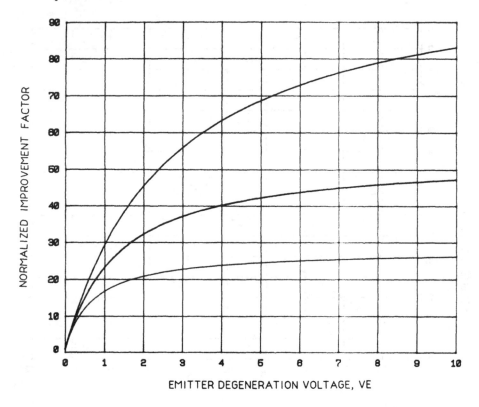

Figure 6.9 Improvement in output resistance versus degeneration voltage V_E, normalized to case where $V_E = 0$, for $\beta = 50, 100$ and 200.

which do not accurately reflect alteration in the operating point of all devices.

Eq. 6.13a shows that the output resistance is also a function of the collector bias voltage, which is not often recognized; for the case where $A = 1$ and $\beta_o \gg 1$, it simplifies to

$$r_o' = \frac{V_{AF}\beta_o}{2I_2} (1+V_{CB}/V_{AF})^2 \qquad (6.13c)$$

Thus, at $V_{CB} = V_{AF}$ we would expect r_o' to be four times higher than at $V_{CB} = 0$, and that a curve of I_2 versus V_2 should be nonlinear. This is exactly what happens for our semi-ideal transistor, having $\beta = 100$, $V_{AF} = 100$, operating at $I_C = 1mA$ and using a very large value for R. Figure 6.8 shows a simulation result; the incremental r_o' which varies from about $5M\Omega$ to $20M\Omega$ between $V_{CB} = 0$ and $V_{CB} = V_{AF}$, as predicted by Eq. 6.13c. Nonlinearity in r_o' is rarely of concern; at high mirror ratios, and for *lower* values of V_E, the nonlinearity is reduced.

When the emitter resistors are of more practical value, the output resistance does not reach these asymptotic values. Figure 6.9 plots the ratio r_o'/r_o, that is, the improvement in output resistance compared to the basic mirror (Figure 6.7a), versus degeneration voltages of zero to 10V, with β_o set to 50, 100 and 200; note that the asymptotic value at $\beta_o = 100$ is 50, as previously found. These curves were calculated for $V_{AF} = 100V$.

It is sometimes desirable to achieve a mirror ratio other than unity entirely by the use of unequal emitter resistors, that is, keeping $A = 1$ but making $R_2 \neq R_1$. This situation could arise, for example, when a current gain of, say, five is needed, while the collector capacitance of the output device has to be minimal. The use of equal transistor sizes operating at unequal currents will inevitably result in a ΔV_{BE} between the emitter voltages. In the special case where I_1 is PTAT, this will track the PTAT voltages across the resistors, and the *ratio* will remain stable over temperature, for all combinations of emitter areas and resistor ratios. More commonly, I_1 will be temperature-stable and this ΔV_{BE} will now add (when $M < 1$) or subtract (when $M > 1$) from the *stable* voltage $V_E = I_1 R_1$ across emitter resistor R_1 (Figure 6.7b), resulting in a temperature dependence in the voltage across R_2, and hence in the actual mirror ratio. If M is not greatly in error (because it is close to one, or because V_E is large, or because the resistor ratio has been adjusted to first-order correct the error), it can be shown that the residual drift has a fractional magnitude $(V_T/V_E \; ln \; M)/300$ at $T = 300K$. For example, if the adjusted M were 5 and a V_E of 500mV were used, the drift would be $((26mV/500mV) \; ln \; 5)/300 = 0.0028$ or 280 ppm/°C.

The input resistance of a mirror using emitter degeneration is approximately $V_T/I_1 + R_1$. This may be troublesome; if nothing else, the voltage at the common base node N1 erodes the output range, which now cannot go below $I_1 R_1 + V_{BE}$. In signal-path applications, the pole formed by R_1 and the C_{JS} of Q1 plus other parasitic capacitive components, will affect the HF response. Note that emitter resistors do not reduce the output deficit caused by finite beta. Beta compensation can be included by adding a resistor R_B of the correct value in the base of Q1, as shown in Figure 6.10. It can be shown that for V_{CB} close to zero (that is, neglecting base-width modulation errors) the necessary value is

$$R_B = \frac{\beta + 1}{\beta - A} (1 + A)(r_e + R_1) \tag{6.14}$$

For example, using $A = 2$, $I_1 = 1mA$ (so $r_e = 26\Omega$) and $R_1 = 500\Omega$ (so introducing about 500mV of degeneration) R_B should be 1.578kΩ if the factor involving beta is ignored, or 1.626kΩ if it is included and beta is assumed to be 100. Clearly, the method becomes unpredictable when A is comparable with β, and can never be precise since β_1 and β_2 are not in general equal; nevertheless, the technique is of considerable practical utility.

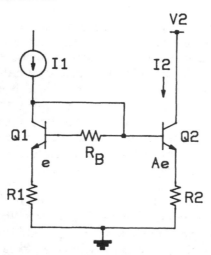

Figure 6.10 **The inclusion of a resistor in the base of Q_1 helps to compensate for the finite beta of Q_2.**

6.3.3 Noise and Drift Considerations

Current mirrors generate noise. This arises from two sources: shot noise in the junctions and Johnson noise in the resistances (extrinsic and diffusion). At low frequencies, flicker noise will also be present.

Consider first the ideal non-degenerated mirror, that is, one using semi ideal transistors (having high beta and Early voltage, and no ohmic resistances), with a general emitter ratio A. It is easy to show that the flatband noise spectral density (NSD) in the output due to shot noise is

$$s_I = \surd\{2qI_1A(A+1)\} \tag{6.15a}$$

where q is the electronic charge, 1.6×10^{-19}C. This can be pragmatically stated as

$$s_I = \surd I_1 \ \surd\{A(A+1)\} \ \times \ 17.9 \text{pA}/\surd\text{Hz} \tag{6.15b}$$

when I_1 is expressed in mA. Thus, an ideal unity-gain current mirror operating at 1mA generates a NSD of 25.3pA/√Hz. As a practical aside, it is important in verifying these expressions by simulation experiments using SPICE to note that collector currents must first be converted to voltages in order to use the NOISE analysis mode, and that this should be performed using the "H" element (current-controlled voltage-source, or ideal transresistance element) controlled by the bias voltage source for the collector, rather than by the use of a load resistor, which will also be modeled as having Johnson noise. Even a 1Ω current-monitoring resistor will generate about 128pV/√Hz, far more than the voltage generated in this same resistor by the 25pA/√Hz of the mirror in our example.

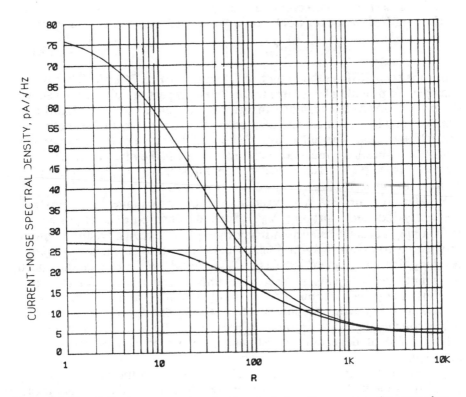

Figure 6.11 Collector current noise-spectral density versus degeneration resistance for a unity-gain mirror operating at 1mA, with $r_{bb'}$ = zero (lower curve) or 100Ω.

The inclusion of Johnson noise arising in the base resistance, $r_{bb'}$ in Q1 and Q2 is most simply carried out by invoking superposition, that is, by first calculating its contribution to the noise spectral density and later calculating the total noise. Base resistance noise can be shown to be

$$s_J = 2qI_1 \sqrt{\{A(A+1)r_{bb'}/kT\}} \qquad (6.16a)$$

where $r_{bb'}$ is the base resistance of Q1, which evaluates to

$$s_J = I_1 \sqrt{\{A(A+1)r_{bb'}\}} \times 4.976pA/\sqrt{Hz} \qquad (6.16b)$$

when I_1 is expressed in mA and $r_{bb'}$ is in ohms. Thus, the contribution to the total NSD due to 100Ω of $r_{bb'}$ in the otherwise-ideal unity-gain mirror at 1mA is about 70.4pA/√Hz, considerably more than the shot noise under these conditions. These expressions assume that $r_{bb'}$ scales with emitter area. Emitter contact resistance $r_{ee'}$ does not increase the overall noise in a properly-scaled mirror, because it reduces the transconductance of Q2 faster than it increases the noise contribution of Q1.

The total NSD of the uncorrelated shot- and Johnson-noise sources sum to

$$s_T = \sqrt{\{s_I^2 + s_J^2\}} \qquad (6.17)$$

Thus, the total NSD at the output of Q2 calculates to 74.8pA/√Hz for our example. To provide a better feel for the likely impact of this noise level, we can express 75pA/√Hz on a bias of 1mA as a dynamic range of 82.5dB in a 1MHz bandwidth. Emitter resistors reduce the effect of both the shot noise and Johnson noise. The full analysis is beyond the scope of this Chapter, but Figure 6.11 shows the reduction possible as the equal emitter resistors R are increased from 1Ω to 10kΩ in the "standard mirror" ($A = 1$, $I_1 = 1$mA) with an $r_{bb'}$ of zero and 100Ω.

Emitter resistors are also used to reduce errors due to strain- and thermally-induced shifts in V_{BE}. Such desensitizing is important in maintaining accurate balance in many monolithic circuits, where a power-dissipating transistor may be located close to a critical mirror[6] , and the resulting thermal gradient induces a ΔV_{BE}, or where inadequate die-attach uniformity leads to chip stresses. The improvement arises for the same reasons that degeneration reduces random noise, and can be readily calculated: it is roughly proportional to $(R+r_e)/r_e$, or $(V_R+V_T)/V_T$, where V_R is the voltage across the emitter resistors. Thus, by including just 240mV (about $9V_T$) of degeneration the sensitivity is reduced by an order of magnitude.

Degeneration can also reduce the absolute error in the mirror ratio, by transferring the matching requirement to resistors, rather than emitter areas. Well-designed resistors can match to ±0.1% or better, while the emitter-area match of small, single NPNs can be as poor as ±1% (corresponding to a ΔV_{BE} of 260µV at 300K). However, there is an interesting trap, here, which is not generally recognized. Imagine a unity-gain current mirror with nominally-identical emitter areas, but mismatched by ΔA. In the absence of emitter resistors, there would be an essentially identical error, ΔM, in the mirror ratio, but this would remain fixed over temperature; that is, there would be no *drift* in M. Likewise, if we inserted emitter resistors of sufficiently large value, the error due to ΔA would be swamped, and any error in the mirror ratio would be caused solely by ΔR. But for moderate values of R the ΔV_{BE} arising across the emitter nodes due to the ΔA will induce a drift in the output current, and at some critical value of R this drift will be maximized. It's easy to show that this will occur when the voltage across the resistors is kT/q, at which point the drift rate for small values of ΔA will be given approximately by

$$\Delta I_C/\Delta T = \Delta A \times 750\text{ppm/}°C \qquad (6.18)$$

---

[6]As well as fixed VBE errors due to a stationary thermal gradient, distortion often arises in a poorly designed IC when the gradient fluctuates with load variations.

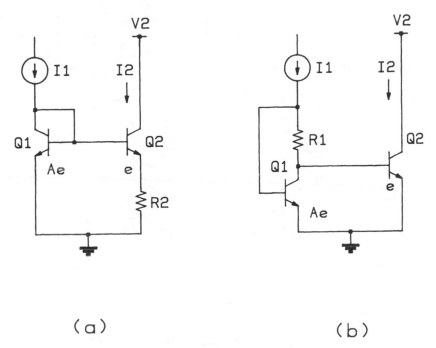

(a)   (b)

Figure 6.12 Two variants of the basic form: (a) the Widlar mirror for generating low currents and (b) the "gm-compensated" mirror.

For example, when $\Delta A=0.1$ (a rather severe 2.5mV V_{BE} mismatch at T=300K) the peak drift in I_C is 75ppm/°C; over the full military temperature range this results in a change of 1.4% in I_C.

6.3.4   Nonlinear Two-Transistor Mirrors

Numerous ways exist to alter the two-transistor mirror to effect various types of useful nonlinear behavior. Such circuits can be used to modify the shape of the transfer characteristic and the shape of the output over temperature. One of the earliest examples of this is the Widlar mirror [6, 6a] in which a resistor is added only to the emitter of Q2. This was intended to be useful in producing the small bias currents often needed in monolithic circuits without requiring the use of a large resistor R_o in the supply branch. Figure 6.12a shows the circuit, with Q1 shown as optionally having a larger area than Q2, since it is expected to be used in a current-lowering mode, aided by R_2. The governing equation is

$$I_2 R_2 \ = \ V_{BE1} - V_{BE2} \ = \ V_T \, ln \, (I_1/AI_2) \qquad (6.19$$

the solution of which is transcendental for I_2 but which can easily be solved

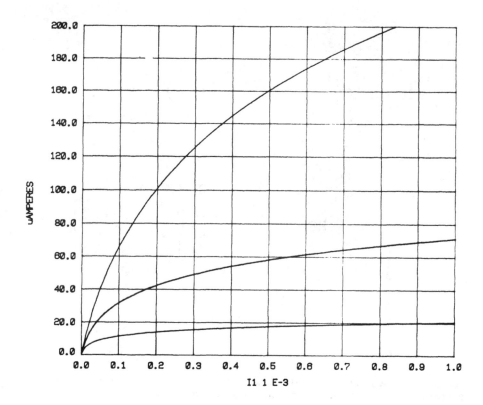

Figure 6.13 Output current versus input current for a Widlar mirror, having equal emitter areas and $R_2 = 200\Omega$ (upper curve), 1kΩ or 5kΩ.

for R_2 given the values of I_2 and I_1, when

$$R_2 = (V_T/I_2)\ ln\ (I_1/AI_2) \tag{6.20a}$$

or, for the needed input I_1 given the values of I_2 and R_2, when

$$I_1 = AI_2\ exp(I_2R_2/V_T) \tag{6.20b}$$

It is sometimes said of this mirror that I_2 is proportional to absolute temperature (PTAT) but clearly this is only approximately true under certain conditions, when the ratio I_1/AI_2 is very large. Figure 6.13 shows how I_2 varies with I_1 for various values of R_2, using the general operating conditions maintained throughout the earlier examples; in this case $A = 1$. Note that the ratio I_2/I_1 is invariant if I_1 is PTAT; this "persistence of PTAT" is a common aspect of the current-mode transfer function of mixed (junction/resistor) circuits.

By altering the topology, placing R_1 in the *collector* of Q_1, as shown in Figure 6.12b, a dramatic change in behavior occurs. This is sometimes called

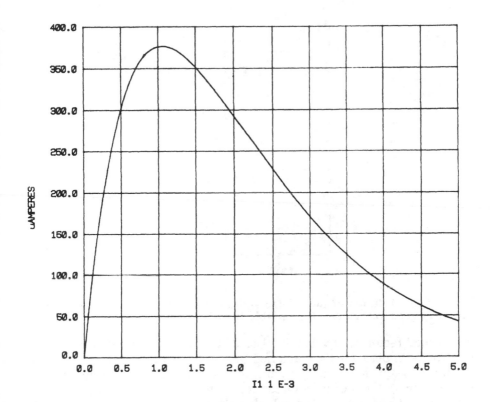

Figure 6.14 Output of the gm-compensated mirror versus input current, for A = 1 and R_1 = 26Ω. Note that peak ratio of e^{-1} occurs at I_1 = 1mA.

the "gm compensated" mirror, because when R_1 is made equal to $1/g_m$, that is, the r_e of Q_1 operating at some specified I_1, the increase in V_{BE1} caused by an increase in I_1 is exactly compensated by the increase in the voltage drop across R_1. Consequently, the voltage delivered to the base-emitter junction of Q_2 is rendered insensitive to I_1 and therefore so is the output I_2.

In fact, the behavior is much more interesting and general than this, and the circuit has many uses. In this case, an analytic expression for I_2 can be found:

$$VBE1 - I1\ R1 = VBE2$$

thus

$$I_2 = AI_1\ exp\ (-I_1 R_1/V_T) \tag{6.21}$$

It is readily shown that I_2 peaks when $I_1 R_1 = V_T$ which is the condition previously noted as resulting in zero sensitivity to I_1, at which point the

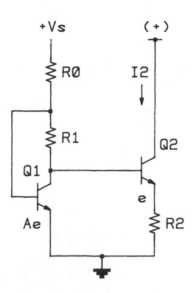

Figure 6.15 **A fusion of the Widlar and gm-compensated mirrors provides relatively good regulation of a variable input current.**

exponential factor has a value of e^{-1} or 0.368. Note that the area factor, A, has here been placed in the emitter of Q2, so that we can optionally restore the mirror ratio, M, to unity by making A = e = 2.72 (roughly 11/4 in integer terms). Of course, in these recent analyses, we have neglected to account for the effects of beta and Early voltage; their inclusion is sometimes tractable, sometimes not, and always obscures the basic concepts. If high accuracy is essential, additional circuit elements invariably are required.

Figure 6.14 shows I_2 versus I_1 from 0 to 5mA, using $R_1 = 26\Omega$ and our "standard" transistors, with A = 1. It will be apparent that in addition to its utility in reducing the effect of input current variations when operated near the peak point, the circuit is useful in generating small currents, operating with voltage drops of much more than V_T across R_1. In this way, much smaller resistor values are needed to produce a given output current than for the circuit of Figure 6.12a. For example, suppose we are using a Widlar mirror as part of a bias scheme to operate from a 5V supply and generate $I_2 =$ 1μA. In order to keep the total resistance within reasonable bounds, we might set I_1 at 100μA, requiring an R_0 of about 43kΩ, and choose to use two minimum-geometry transistors, so A = 1. Then, R_2 must be 59.5kΩ (from Eq. 6.20b). On the other hand, using the circuit of Figure 6.12b we find R_1 can be 100 times smaller - 595Ω - since it has 100μA, rather than 1μA, flowing in it.

Now, in the Widlar mirror I_2 increases slightly with the supply voltage, while using the "gm-compensated" form it decreases. It is therefore likely that a lower sensitivity to the supply can be obtained by combining the two

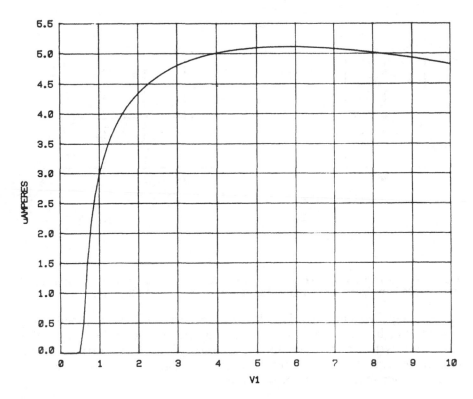

Figure 6.16 **Output of the circuit of Figure 6.15; see text for details.**

circuits, as shown in Figure 6.15. Under certain conditions, this can prove useful. Figure 6.16 shows the result for one such optimization, in which I_2 is designed to be 5µA for a 5V supply; here, $A = 1$, $R_0 = 18k\Omega$, $R_1 = 100\Omega$ and $R_2 = 15k\Omega$; the output is within 4% of its nominal value for supply voltages from 3V to 10V.

6.4 Three-transistor Mirrors

As is apparent, several variations can be found on the two-transistor theme: it is not surprising, therefore, to find that the addition of a third transistor considerably broadens the palette of the designer. Only a few of the numerous possibilities can be included in this brief survey. Most will have as their aim the improvement of accuracy, which most often means the reduction of beta-related errors.

6.4.1 The EF-Augmented Mirror

Figure 6.17a shows a commonly-used current-mirror, in which Q3 operates as an emitter-follower (EF) and provides the base currents of Q1 and Q2, so

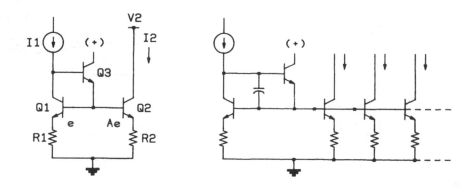

(a)   (b)

Figure 6.17   The EFA mirror in its basic form (a); this circuit is often used to provide multiple outputs (b); capacitor ensures HF stability.

augmenting their current gain. This mirror seems to have escaped being named, but, because we will refer to it several times in the rest of this Chapter, it will here be called the EF-augmented, or EFA, mirror. The use of emitter degeneration resistors and a non-unity area ratio is optional. It is readily shown that, assuming $R_1 = AR_2$ the mirror ratio is now

$$M \; = \; \cfrac{A}{1 + \left\{ \cfrac{1}{\beta_1} + \cfrac{A}{\beta_2} \right\} \cfrac{1}{\beta_3}} \qquad (6.22)$$

so that even for quite large values of A, the beta errors are usually negligible, certainly now much less than other practical errors. For A = 1 and $\beta \gg 1$, this simplifies to

$$M \; = \; 1 - \cfrac{2}{\beta_1 \beta_3} \qquad (6.22a)$$

where it has been assumed that the betas of Q1 and Q2 are equal. When this is not true (for example, when the output collector is very positively biased) the more complete expression must be used.

The output resistance, r_o', using no emitter degeneration is the same as for the basic mirror - V_{AF} divided by the output current - but the maximum r_o' is now $V_{AF}\beta_2/I_2$, twice as high, for large amounts of degeneration.

The high internal current gain introduces the possibility of HF instability, since there are two "beta poles" in the basic circuit: from the base of Q3 to the base of Q1 is the beta pole of Q3; from here back to the base of Q3 is the beta pole of Q1. In fact, a third pole may be generated by the emitter resistance of Q3 and the total capacitance on the base node, which includes the $C\pi$ of Q1 and Q2 (or more devices); this may become complex since the driving-point impedance is somewhat reactive. A complete discussion of the general HF response will not be included here. Suffice it to say that this system is often stable only because the much lower current in Q3 lowers its effective f_T to the point where it creates a dominant pole. Also, there is a small amount of feedforward around this slow device via the C_{JC} of Q1. By adding a further small capacitance across the base-emitter nodes of Q3 its beta is effectively removed from the HF loop. This is shown in Figure 6.17b, which also shows a more general scheme, including emitter degeneration and multiple outputs, now possible because of the higher effective beta.

6.4.2 Use of Cascodes to Raise Output Resistance

Some applications of current mirrors and sources require the highest possible output resistance. In these cases it is sometimes possible to place less emphasis on accuracy of the mirror ratio. A cascode transistor - a common-base stage, inserted between the output node and the mirror proper - can sometimes help. The maximum output resistance of a simple cascode, that is, when driven by a perfect current-source at its emitter, is $V_{AF}\beta/I$, and since the last mirror discussed can already attain that value using large amounts of emitter degeneration, we may need to look for further improvement. However, the use of a cascode allows the input resistance of the main mirror cell to be kept lower than possible using degeneration, and it can operate with a minimum output bias voltage which is lower, for the same output resistance.

Working on the assumption that it is the variation of beta with V_{CB} that limits r_o we might try a Darlington[7] cascode, to lower the base current variation, or an MOS or JFET device, which has essentially zero gate current. Such combinations are becoming very popular with the advent of combined Bipolar-MOS processes. In some cases, however, beta is not the limiting factor: to understand why, we need to look again at Eq. 6.1c, which shows the dependence of V_{BE} on V_{CB}:

$$V_{BE} = V_{BE0} - V_T \, ln \, (1 + V_{CB}/V_{AF}) \qquad (6.1c)$$

For small variations ΔV_{CB} near $V_{CB} = 0$, we can calculate the reverse transfer coefficient of the cascode to be

$$DV_{BE}/DV_{CB} = -V_T/V_{AF} \qquad (6.1e)$$

that is, neglecting the beta-modulation effects of the output resistance, any

---

[7]The term "Darlington" configuration here refers to the use of a pair of transistors with common collectors and the emitter of the first connected to the base of the second so as to result i n a very high effective beta, equal to the product of the individual betas.

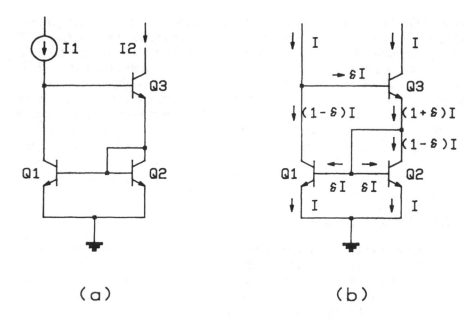

(a)   (b)

Figure 6.18 **The Wilson mirror (a) and a simple way to understand how it achieves good accuracy for finite beta.**

variation in the voltage at the final output will cause the output node of the basic mirror to move and thus modulate the current. In short, it is still important to maximize the output resistance of the main mirror cell if the full benefits of a cascode are to be realized. The variation of V_{GS} with V_{DS} in a FET differs markedly, but the general idea of a reverse transfer ratio in the cascode still applies; in fact, the magnitude of the error is even greater.

6.4.3 The Wilson Mirror and Improvements

Sometime in 1967, George Wilson and the author set each other a mutual challenge as an overnight assignment: find a way of making a better current mirror using only three transistors. George won, and the Wilson mirror has become the first choice in many applications where good beta-immunity and high output resistance are needed.

The standard form (which, like the basic mirror, hardly needs repeating) is shown in Figure 6.18a. The circuit is not a panacea; its low beta-sensitivity, as we shall show, depends on beta-matching (unlike the EFA mirror just discussed) and the trick only works at unity-gain; significant errors can arise due to inequalities in the collector voltages of Q1 and Q2; the voltage at the output cannot approach the common node as any of the other mirrors presented so far; its noise performance is worse than the basic mirror; its AC response is strongly peaked. Nevertheless, it has proven to be a valuable cell, and we will later show some ways in which its utility can be extended.

The circuit is interesting in several ways. First, note that it contains a sub-cell, Q1/Q2, which is just a basic mirror; the input of this sub-cell is essentially the output of the complete mirror, with Q3 viewed as a cascode, and the output of the sub-cell becomes the input of the complete mirror. Thus, the Wilson mirror is a current-feedback circuit, the only mirror so far discussed that relies on feedback to establish its mirror ratio. We can think of its operation in this way: following the application of the input current, I_1, Q3 would like to respond by generating a current beta times as large, but its emitter current is sensed by Q2, and mirrored back to the input, where a current-balance between I_1 and I_{C1} is established, requiring only the base current of Q3 to maintain, this current being in the nature of the error signal in a feedback loop. We might at this point be inclined to suggest that Q3 should therefore be made a Darlington, to reduce this "error" still further. If the sub-cell had no beta errors of its own (essentially true of the EFA mirror) that might be the correct thing to do. In fact, the finite base current of Q3 is needed to ensure that the mirror ratio is close to unity, as we will now see.

Figure 6.18b shows an approximate way of analyzing this circuit. We begin by assuming that the emitter currents of Q1 and Q2 are equal and of value I. This follows from the fact that they are assumed to be identical transistors and are operating with the same V_{BE}. Then, using the convention $\delta = I_B/I_E$, we find the collector current of Q1 to be $(1-\delta)I$, while the combined collector current of Q2 and the base currents of Q1 and Q2 sum to $(1+\delta)I$, which becomes the emitter current of Q3. To a good approximation, the base current of Q3 is δI, so the mirror ratio is

$$M = \frac{I2}{I1} = \frac{(1+\delta)I - \delta I}{(1-\delta)I + \delta I} \equiv 1 \qquad (6.23)$$

whatever the value of δ (which is roughly $1/\beta$). A more careful analysis, but still assuming all betas are equal, yields

$$M = 1 - \frac{2}{\beta^2 + 2\beta + 2} \approx 1 - \frac{2}{\beta^2} \qquad (6.24)$$

This result, given in most texts on the subject, suggests that the error in M due to finite beta is essentially the same as that for the EFA mirror (see Eq. 6.22a). In fact, it obscures an important detail, namely, that when the betas of Q1 and Q2 are *different*, there is a significant reduction in the accuracy promised by Eq. 6.24 and the Wilson mirror can very possibly be an order of magnitude worse that the EFA mirror in this regard. The complete expression is

$$M \approx 1 - \frac{2(\beta_1 - \beta_3) + 2}{\beta_1\beta_3 + 2\beta_1 + 2} \qquad (6.24a)$$

Thus, if we assume $\beta_1 = 100$ and $\beta_3 = 95$ (a plausible 5% beta mismatch) the numerator evaluates to 12 rather than 2, so the error in the mirror ratio is

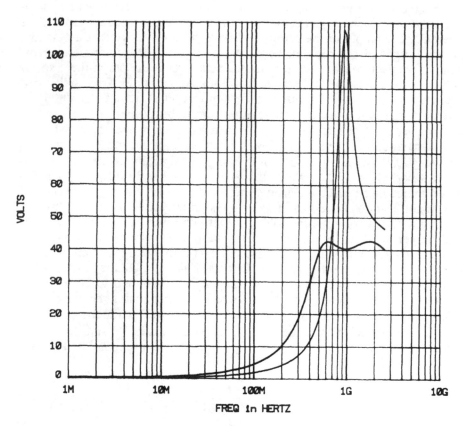

Figure 6.19 Impedance looking into the emitter of Q3, at a bias level of 1mA; see text for explanation of method used to lower the resonance.

six times worse than for the EFA mirror. The actual error in this case is still a fairly harmless 0.124%, but if this were a mirror built from lateral PNP transistors, where the betas might not just be low and mismatched, but strongly different because of high-level injection effects, the errors can be quite large.

It can be shown that the output resistance for the Wilson mirror is the same as for the heavily-degenerated basic mirror, that is

$$r_0' = \frac{V_{AF}\beta_o}{2I_2} \; (1 + V_{CB}/V_{AF})^2 \qquad (6.25)$$

Emitter degeneration does not raise the output resistance of the Wilson mirror, because the output bias V_2 does not materially alter the V_{CB} of Q1; in fact, it is easily shown that the "attenuation factor" is V_{AF}/V_T, typically 4000:1.

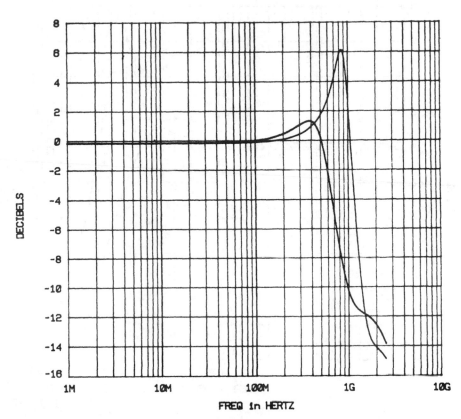

Figure 6.20 Current-mode AC response from the emitter of Q3 (Figure 6.19) to the output, with and without compensation capacitor of 5pF.

However, degeneration is often used for other reasons: it can transfer the burden of ratio-control from the emitter areas to a resistor ratio; it may obviate an error due mismatch in Q1 and Q2 (see below); it lowers the output noise.

Before discussing various improvements which can be made, one other useful property is worth mentioning. The impedance at the emitter of Q3, call it Z3, is quite low. Imagine a signal current of i applied to this node, and a bias current of I_1 applied to the normal mirror input. Q1 is forced to conduct I_1, which requires that the V_{BE} of Q1 must remain at the constant value required by I_1, whatever the value of i, which is therefore absorbed by Q3, and thus appears at the output. It follows that Z3 ought to be nearly zero. Of course, a component δi flows in the base of Q3, thus modifying the collector current of Q1 and perturbing its V_{BE}. We can guess (correctly!) that at low frequencies Z3 approximates to a resistance of $V_T/\beta I_1$, or about

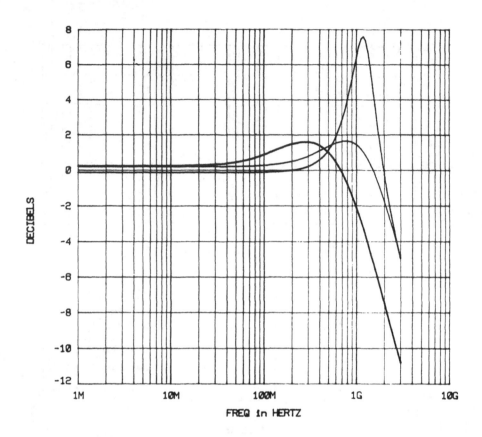

Figure 6.21 **Comparison of AC responses of the three most popular mirrors: the Wilson is the most peaky; the EFA mirror has about half the bandwidth of the basic mirror, but the same 1.8dB of peaking.**

0.26Ω for $I_1 = 1\text{mA}$. At high frequencies, the circuit becomes resonant (at about $f_T/3$) but Z3 remains under 10Ω up to about $f_T/10$. Of course, device capacitances play a more important role at lower currents. Figures 6.19 and 6.20 show typical results for a circuit using equally-sized transistors with an f_T of about 3GHz. The flatness of both the impedance and the AC transfer function, with respect to this non-standard input, can be improved by the addition of a small capacitor across the collector-base junction of Q1.

While on the subject of AC response, we show in Figure 6.21 a comparison of the three most popular current mirrors: the basic mirror, the EFA mirror and the Wilson mirror. In all cases, equal transistor geometries were used, typical of a 3GHz NPN process, operating at a bias current of 1mA and without the use of emitter degeneration. The severe peaking of the Wilson mirror is apparent. In designing mirrors for current-path applications, much can be done to control the gain and phase response. In fact, the first

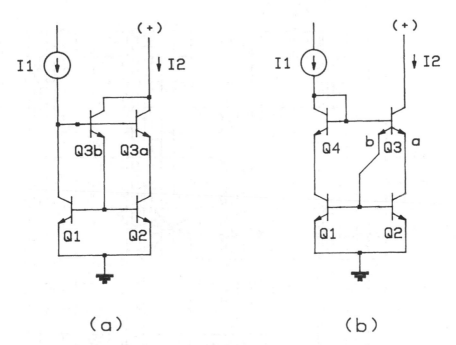

(a)   (b)

Figure 6.22 Improvements to the Wilson mirror: (a) separate path to bases of Q1 and Q2 lowers peaking in AC response; (b) extra diode Q4 lowers static errors due to V_{CB} mismatch between Q1 and Q2.

improvement to the Wilson mirror to be described has a flatter HF response as one of its benefits.

Figure 6.22a shows the circuit, in which an extra transistor Q3b has been added to Q3 to supply the base currents of Q1 and Q2; in practice, this would simply be an additional emitter, as shown in Figure 6.22b. Figure 6.23 shows a useful improvement in the AC response. This is due to the radical alteration of the incremental resistance driving the common-base node. In the basic Wilson mirror, this resistance is the r_e of Q3; in the modified circuit it is increased by the factor $\beta/2$, thus forming a pole with the C_π of Q1 and Q2 at a much lower frequency, resulting in a more stable feedback path.

This modification has a second advantage, only of interest in low-voltage applications, which is that it allows the output node to drop closer to the common node before finally crashing (Figure 6.24). A more important improvement is the addition of an extra diode-connected transistor, Q4, in series with the collector of Q1 (Figure 6.22b). Its purpose is to equalize the V_{CB} of Q1 and Q2 and thereby remove a significant source of error in the mirror ratio. The magnitude of this error is quickly estimated: a ΔV_{CB} of one V_{BE} can be shown to alter the ratio of the sub-cell Q1/Q2 from unity to

$$M' = exp(V_{BE}/V_{AF}) \qquad (6.26)$$

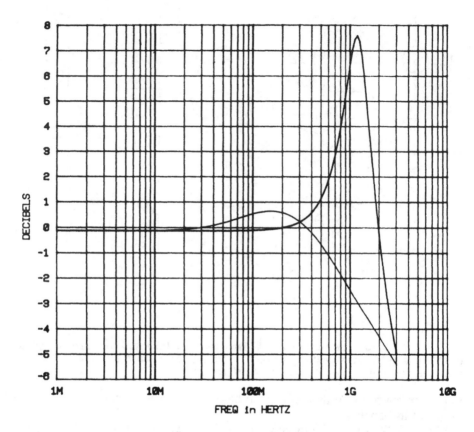

Figure 6.23 **Reduction of HF peaking for the Wilson variant of Figure 6.22a.**

which evaluates to about 1.007 for $V_{AF} = 100V$, and causes the output of the complete mirror to 0.7% too low. An approximate compensation can be provided, when the operating current is constrained to some small range, by replacing Q4 with a resistor.

The sub-cell can be rendered more accurate by the use of cross-connected transistors physically arranged on the chip as a quad [7]. This would be useful when a wide current range is to be handled, or when it is important to keep excess voltage drops to a minimum, but a narrow current range, and where headroom considerations allow, it is usually preferable to use emitter degeneration to improve the ratio accuracy.

6.5 Special-purpose Current Mirrors

It will be apparent that the idea of a "current mirror" has many extensions, and we will end this Chapter with a brief consideration of some of these; we

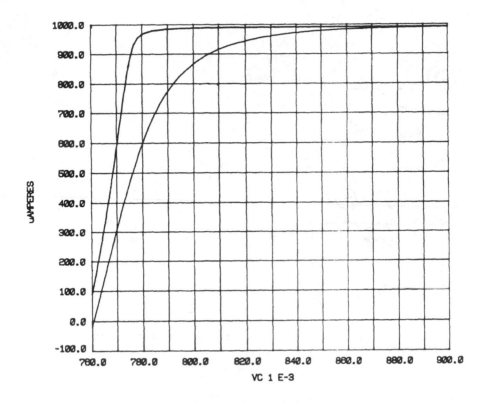

Figure 6.24 Improved low-voltage behavior for the variant of Figure 6.22a.

will touch upon the subject of current-mode amplifiers based on mirrors, which have many possible forms. Contemporary complementary bipolar processes are a great asset in this context; we will show a basic scheme which allows signals to be "ping-ponged" between two very low-voltage supplies, accruing gain in the process. Mirrors have also played a traditionally important role in op-amp design (in converting differential current signals from the input gm stage to single-sided form prior to the integration stage) and as active loads [1]. We will here consider a few generic forms that may be useful in special circumstances.

6.5.1 Variable-Ratio Mirrors

All the mirrors so far discussed have a current ratio which is fixed, either by emitter-area ratios alone, or in combination with resistor ratios. Often it is desirable to be able to control the mirror ratio in some other way, for example, by varying a current or voltage. Clearly, such circuits overlap the domain of multipliers, and many similarities are found; the differences are largely in the way in which we approach the form.

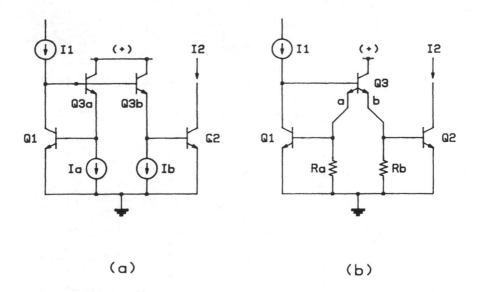

Figure 6.25 Variable ratio forms: (a) current-programmable; (b) resistor-programmable.

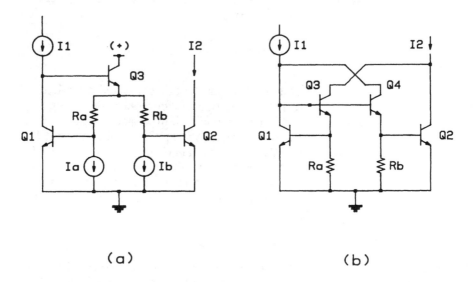

Figure 6.26 Further variable-ratio mirrors. The variant in (b) also exhibits a very high output resistance, through cancellation of the similar dependence of the V_{BE} of Q3 and Q2 on the output bias.

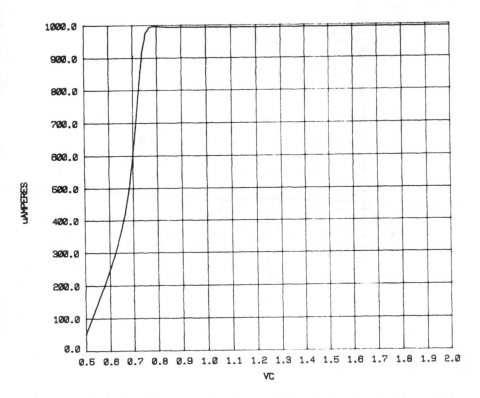

Figure 6.27 Output current of the circuit of Figure 6.26b at low voltages.

Figure 6.25a shows a cell that illustrates this ambiguity: it can be viewed as a simple one-quadrant translinear multiplier, or as a slight variant of the EFA current mirror (it would be exactly that if the two emitters of Q3 were shorted). The use of separate current biases to Q3a and Q3b allow us to control the mirror ratio, which, neglecting errors due to beta, is now

$$M = \frac{Ia}{Ib} \qquad (6.27a)$$

In Figure 6.25b, the use of a single two-emitter transistor for Q3 illustrates a practical simplification: in this case, the current sources have been replaced by resistors Ra and Rb, allowing resistor control of the mirror ratio:

$$M = \frac{V_{BE1}/Ra}{V_{BE2}/Rb} \approx \frac{Rb}{Ra} \qquad (6.27b)$$

In this case, the final ratio is not exact, because the V_{BE} of Q1 and Q2 are, in general, no longer equal.

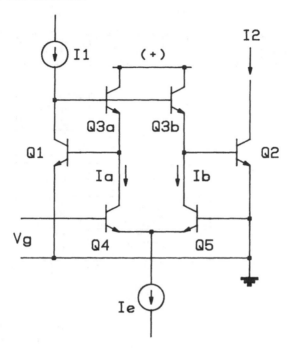

Figure 6.28 **A variable current-gain cell with "linear-decibel" response and a practical range of -40dB to +20dB.**

A more useful variant is shown in Figure 6.26a, where the differential voltage across the bases of Q1 and Q2 can be controlled by the currents Ia and Ib. The mirror ratio in this case is

$$M = exp \ \frac{IaRa - IbRb}{V_T} \tag{6.28}$$

Stable operation over temperature requires that either the resistors or the currents be PTAT.

A further variant is shown in Figure 6.26b. In this case, the controllable ratio is effected in the same way as in Figure 6.25b, except that separate transistors Q3 and Q4 are used and their collectors have been cross-connected, which results in the V_{BE} of Q3 responding to the output bias (V_{C2}) in almost exactly the same way as the V_{BE} of Q2. Since these two devices are facing in opposite directions in the TL loop the output resistance is very much higher than it would be without this feature. Figure 6.27 shows the output characteristic from 0.5V to 2V; the current very rapidly attains its final value, beyond which point the incremental output resistance is much higher than that of the Wilson mirror.

A current-mirror having a voltage-variable ratio is shown in Figure 6.28. This will be recognized as another twist on the basic form shown in Figure

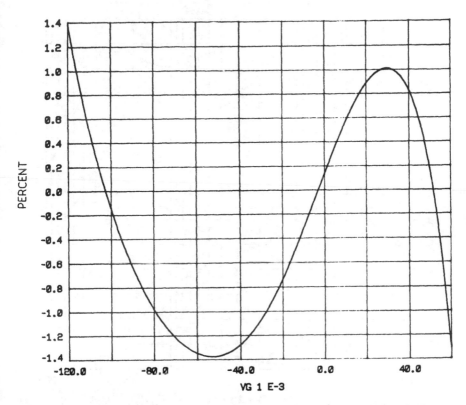

Figure 6.29 Percentage error in the mirror ratio for the circuit of Figure 6.28 is less than ±1.4% from Vg = -120mV (-40dB) to +60mV (+20dB).

6.25a. In this case, the currents Ia and Ib are provided by Q4 and Q5:

$$Ia = \frac{exp(g)}{exp(g) + 1} \, Ie \qquad (6.29a)$$

and

$$Ib = \frac{1}{exp(g) + 1} \, Ie \qquad (6.29a)$$

where $g = Vg/V_T$. It follows from Eq. 27a that

$$M = exp(g) \qquad (6.30)$$

Thus, the decibel gain or loss is a linear function of the control voltage Vg.

Figure 6.29 shows that the gain error in a practical realization of this concept remains within ±1.4% from Vg = -120mV, corresponding to a loss

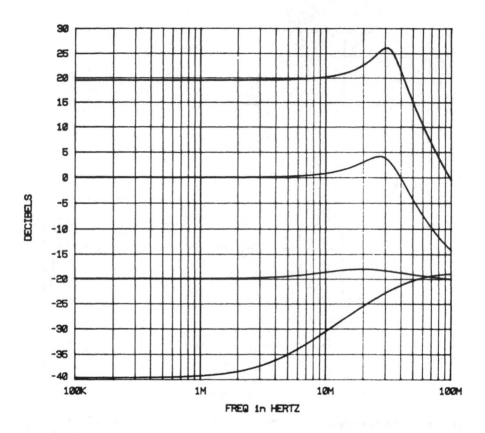

Figure 6.30 **AC response for the circuit of Figure 6.28 at four gain settings.**

of 40dB at T = 300K, up to Vg = +60mV, which provides a gain of 20dB. Some adjustment of emitter areas was used to achieve this result, for which the bias currents were I_1 = 15μA and Ie = 600μA. The AC response for a 3GHz IC process operating under these conditions is shown in Figure 6.30. Some peaking is evident at the higher gains, and signal feedthrough limits HF performance at high attenuations. Nevertheless, this novel circuit may have applications in certain types of gain control, particularly since it can be extended using a complementary bipolar process to handle "two-quadrant" signals.

6.5.2 *Mirrors Having Large Current Ratios*

There is no particular difficulty in achieving high mirror ratios. In many cases, simple emitter area ratios will suffice, supported by current gain if necessary (as in the EFA mirror, which is not limited to the use of a single

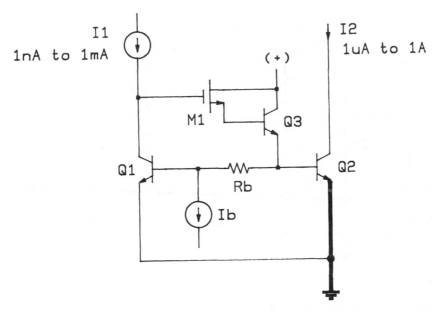

Figure 6.31 A method for augmenting the area ratio (100:1) between Q2 and Q1 to achieve a very high mirror gain at high current.

transistor for current boosting). In practical monolithic designs, emitter area ratios of up to 100:1 are not hard to achieve. Beyond that, one might wish to use some of the ideas presented in the last section: although the focus was on providing variable mirror ratios, it is evident that large ratios can be achieved by introducing fixed voltage offsets between the bases of an EF mirror. Figure 6.31 shows a mirror having a ratio of 1000:1, partly achieved by the use of a large area for the output device, Q2, needed anyway to handle the maximum current of 1A, and partly by the insertion of 59.5mV generated by Ib across the resistor Rb. To maintain an accurate ratio over temperature either Ib or Rb must be PTAT; at the relatively low value required here, Rb could be fabricated from the aluminum metallization (see Chapter 2 for a similar situation), allowing Ib to be fixed with temperature. Many practical ways exist to introduce the needed ΔV_{BE} to good accuracy. Note that the maximum base current of Q1, assuming a beta of 100 and a maximum I_{C1} of 1mA, will cause the peak output to be 1% too high. A combination of MOS and BJT is used to provide the large peak base current of Q2, and also maintain low errors at the smallest input of 1nA. Special attention is directed to the grounding of the emitters of Q1 and Q2; at high currents only a few tens of milliohms of improperly-located aluminum interconnect can severely degrade accuracy.

We have already seen how one can devise mirrors to generate a moderately small current from a larger one, using nonlinear mirrors, but suppose one needed a linear mirror to generate an accurate current only

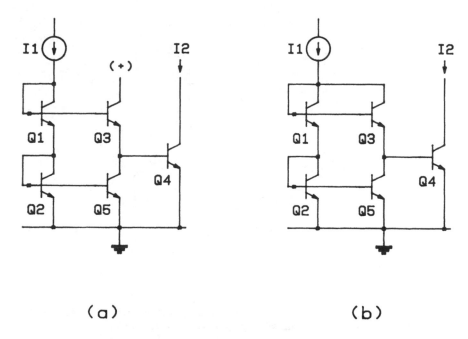

(a)   (b)

Figure 6.32   Translinear mirror cells with capability for very high ratios.

1/1000th as large as its input over a wide range. This is the inverse of the last problem, and although there are few applications for such a function, it is a useful point of departure in showing that the current mirrors we have discussed so far are only the simplest of a more general class of translinear-loop (TL) circuits.

Figure 6.32 shows two more complicated mirrors, differing only in the biasing of the collector of Q3. It is easily determined using the translinear principle that the mirror ratio of the (a) variant is

$$M = \frac{A_3 A_4}{A_1 A_5} \qquad (6.31a)$$

while that of the (b) variant is

$$M = \frac{A_2}{A_2 + A_5} \frac{A_3 A_4}{A_1 A_5} \qquad (6.31b)$$

(The reader might wish to verify that the absence of A_2 in Eq. 6.31a is not an error in analysis; it suggests that in the ideal case Q2 could be omitted. Is this correct?). These composite quotients make it easy to realize very large mirror ratios, and if beta were not finite, it would be as easy to have $M > 1$ as $M < 1$. In practice, the form is limited to operation with $M < 1$. With quite

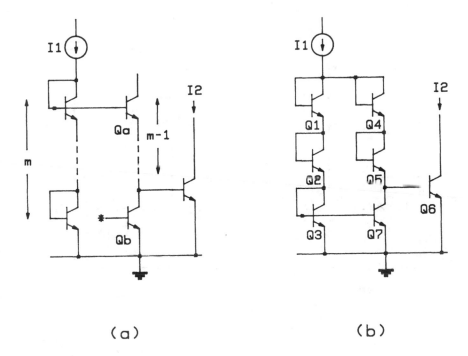

(a)   (b)

Figure 6.33 (a) The "general current mirror" and (b) an example using m=3.

moderate emitter areas, the (b) variant can realize exact attenuation ratios of several thousand.

Figure 6.33a shows the general form of these mirrors. Since the collector of Qa can be connected to (1) the input node, (2) the output node or (3) a supply voltage, and the base of Qb can be connected to (a) the base of the input diode or (b) the base of the output transistor, there are six further variants. Figure 6.33b shows a typical example, where m = 3. Its mirror ratio is

$$M = \frac{A_3^2 A_4 A_5 A_6}{(A_3 + A_7) A_7^2 A_2 A_3} \tag{6.32}$$

An interesting (but probably valueless) outcome of this is that one can generate a maximum attenuation ratio of 187272:1 using only 33 unit emitters. The reader is invited to find the minimum number of unit emitters required to achieve an attenuation ratio of one million, using this or any other form that preserves the ratio over a wide current range.

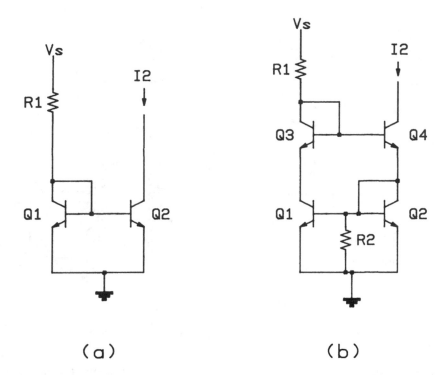

(a)   (b)

Figure 6.34 Supply-energized current-sources: (a) does not provide an output which is proportional to Vs; the circuit shown in (b) does.

6.5.3 Supply-Tracking Current Sources

It is occasionally required to generate a current which is exactly proportional to a voltage. Numerous methods exist to achieve this, and where accuracy is of paramount concern, operational amplifier techniques are often employed. In many cases, however, simpler techniques not only suffice but can operate under conditions not possible using "high-level" solutions.

For example, suppose one wished to generate a current which was to be used as the scaling voltage for a nonlinear function, such as a multiplier or logarithmic amplifier, and the circumstances were such that this must track the supply voltage, V_S, with an accuracy of better than "8-bits" (0.25%) over the temperature range -55°C to +125°C. This situation might arise if V_S were also to be used as the reference for a subsequent A/D converter. Then a simple mirror, resistively biased from the supply, such as shown in Figure 6.34a, would not be appropriate, partly because its current-in/current-out ratio is not very accurate, but more importantly because the voltage across R_1 is not the full supply voltage, but only $V_S - V_{BE}$. If we use an EFA or Wilson

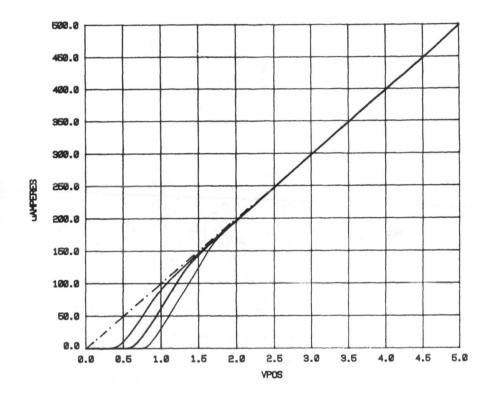

Figure 6.35 Demonstration of the very close proportionality in the output current (Figure 6.34) at -55°C, 25°C and 125°C.

mirror to deal with the basic accuracy issues, the deficit is doubled, that is, the output current is

$$I_2 = (V_S - 2V_{BE})/R_1 \tag{6.33}$$

However, Figure 6.34b shows how this problem can be sidestepped. An extra resistor, R_2, has been added, in which flows a current V_{BE}/R_2, which adds to the output:

$$I_2 = (V_S - 2V_{BE})/R_1 + V_{BE}/R_2 \tag{6.34}$$

Thus, when $R_2 = R_1/2$ the output current is proportional to V_S, although not tracking it all the way down to zero. Note that the"recovered" V_{BE} is exactly equal to the "lost" V_{BE}'s, ensuring accurate compensation. The effectiveness of this solution can be judged from Figure 6.35 which shows the output for a circuit using $R_1 = 10k\Omega$ from $V_S = 0$ to +5V at -55°C, +35°C and +125°C. R_2 was set to 4.9kΩ, slightly lower than theoretically required, to further reduce the already low temperature coefficient, which is now about

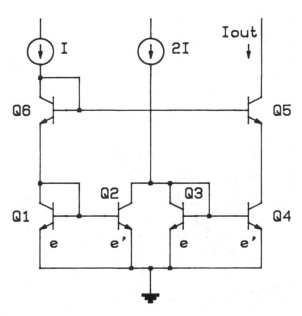

Figure 6.36   A novel circuit capable of extremely high output impedance.

40ppm/°C at $V_S = +2V$, 5ppm/°C at $V_S = +3V$ and only 2ppm/°C at $V_S = +5V$. The output error drops below 2.5μA for supply voltages as low as +2.5V, even at the minimum temperature.

6.5.4   *Mirrors Having Very High or Negative r_o*

Many methods can be devised to increase the output resistance r_o of a current mirror. To remain within the subject of bipolar mirrors, we will exclude those methods depending on the use of MOS cascodes. Two novel techniques are proposed, each having certain merits.

The circuit shown in Figure 6.36 can exhibit remarkably high output resistance, and because the cascode transistor, Q5, can have a small geometry, the output capacitance can be low, too. The two merits of this scheme are

1) The output resistance can be infinite and with practical tolerances held to many hundreds of megohms (positive or negative) at 1mA. Stated more generally, the Thevenin equivalent voltage can be reliably as high as 100 times the βV_{AF} product, typically 1 megavolt.
2) The output can approach the common terminal to within much less than one volt while still exhibiting the maximum value of r_o.

The disadvantages are:

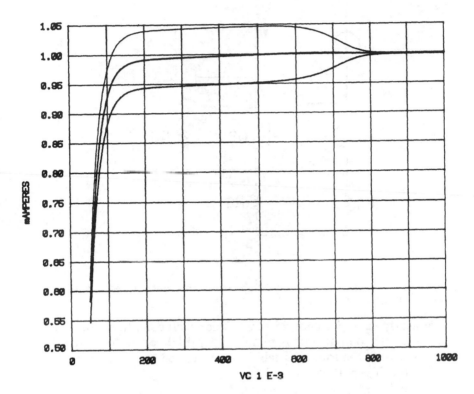

Figure 6.37 Output current versus output bias voltage (Figure 6.36), with a small adjustment to the ratio e'/e. (See text.) Note that the current is quite accurate even down to +100mV of bias.

A) It requires two input currents; in fact, the output is the difference of these currents, but only to full accuracy when they are in the 2:1 ratio shown in the figure.

B) The two inputs are at different potentials; this is easily remedied by the addition of another diode-connected transistor.

C) The technique relies on good matching of beta.

The basic idea is to return the base current of Q5 through the unity-gain mirror Q1/Q2 to the input of the second unity-gain mirror Q3/Q4, and hence add this base current back to the effective input 2I - I. The full analysis will not be given here, but it will be found that the beta errors of the two simple mirrors cancel. Now, in a practical implementation, it may happen that the emitter-area ratios of these mirrors may be slightly mismatched, as shown by the area factors e and e' in the figure. If we properly arrange these four transistors into a cross-connected quad, we can arrange for the ratio e'/e to

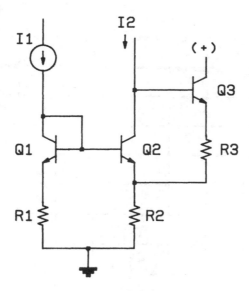

Figure 6.38 A simple circuit providing adjustable output resistance through infinity to negative values.

be very nearly equal for both mirrors. Under such conditions the absolute accuracy of the output and r_o remain very high even for quite large variations in e'/e. The following table was generated by simulation, for I = 1mA, $\beta = 100$, $V_{AF} = 100$:

Ratio e'/e	I_{out} at $V_{C5} = +5V$ (mA)	r_o (MΩ)
1.00	0.9991	130
1.02	0.99995	284
1.04	1.00006	-1330
1.06	0.99944	-196
1.08	0.998	-105

The peak output (1.00014mA) occurred for e'/e = 1.036, which was also the value at which the output resistance passed through infinity. The sensitivity to the beta of Q5 was least at this value; Iout varied from 0.99985mA at $\beta = 50$, peaked at 1.00014mA at $\beta = 80$ and fell to its starting value at $\beta = 200$. Predictably, the temperature sensitivity of the simulated circuit was negligible.

An interesting feature of this circuit is its behavior at low output voltages, V_{C5}. When Q5 saturates, its base current increases rapidly; but the feedback path works to cancel this error, with the result that I_{out} holds up well, even when V_{C5} is only +100mV! Further, it was found that the ratio e'/e had about

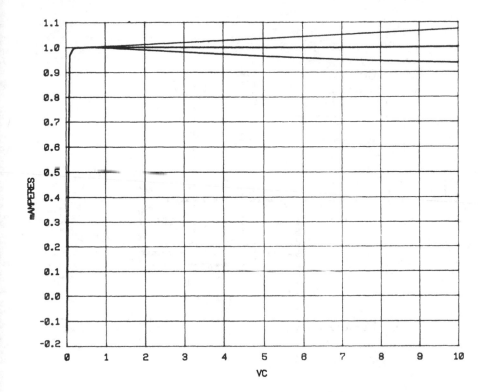

Figure 6.39 Output characteristic (Figure 6.38), showing variable r_0.

the same optimal value for this purpose, too. Figure 6.37 shows I_{out} in the region of V_{C5} from +50mV to +1V, for e'/e = 1.00, 1.04 and 1.08.

Figure 6.38 shows another method for allowing the output resistance to be adjusted through infinity. In this case, a sample of the voltage at the output node is fed back to the emitter of Q2, via Q3 and R_3. (This is an application where an MOS device would be better suited for Q3.) The theoretical value of R_3 depends on *a priori* knowledge of V_{AF}, hence the circuit is not truly designable. Nevertheless, it is of some interest in that it can provide a strongly negative output resistance, if desired. Figure 6.39 shows a simulation in which $R_1 = R_2 = 10\Omega$, and R_3 was either omitted (positive r_0), set to 26kΩ (the theoretical value for infinite r_0 when $V_{AF} = 100V$) or set to half the ideal value.

6.5.5 Voltage-following Mirrors: The GCM

In 1975 the author proposed[7] a type of current mirror in which the input node (N1 of Figure 6.1a) followed the output node, N2. The application of this was in loading the collectors of a differential translinear cell (usually, a

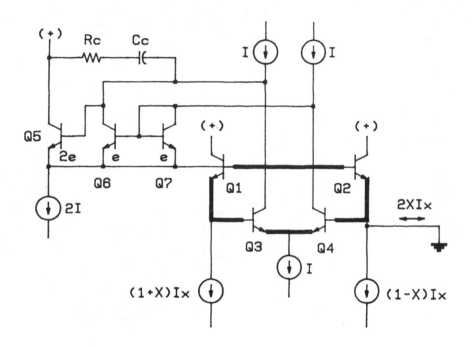

Figure 6.40 A wideband negative current mirror ("GCM").

multiplier) so as to convert the differential current output to a single-sided form, while simultaneously maintaining close equality in the voltage on these collectors, important to achieving very low distortion in certain case. This type of mirror has since been loosely termed the "GCM" (Gilbert current mirror) for want of a better name.

The original version of the circuit used lateral PNPs, but managed to by-pass them at high frequencies. Seevinck suggested an all-NPN method of loading the main translinear cell[8] reproduced here, slightly altered, as Figure 6.40. Transistors Q1, Q2, Q3 and Q4 form a translinear loop, shown by the bold line, in which Q3 and Q4 ideally operate at identical currents, I. Consequently, Q1 and Q2 are forced to operate at equal currents, essentially the "half-input" current $(1+X)Ix$. It follows directly that the current into the load must be this current, replicated in Q2, minus the other "half-input" $(1-X)Ix$, which amounts to just $2XIx$.

The GCM has some nice features: note, for example, that the base current of Q3 is added to the signal $(1+X)Ix$, but a nominally-identical current is later subtracted at the base of Q4.   Also, the collectors of Q1 and Q2 may be

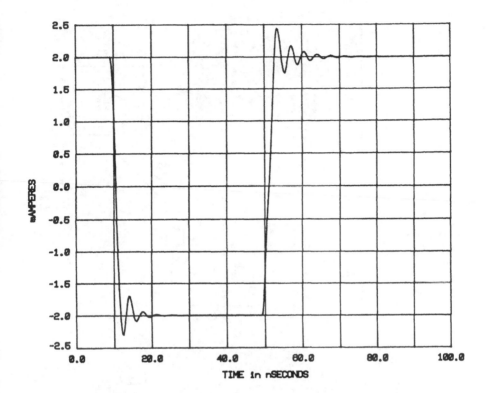

Figure 6.41 **Typical transient response for the GCM.**

taken to the full positive supply (to maximize the f_T of these devices) will little concern for supply-sensitivity, because of the inherent balance of the circuit. HF compensation has been added (Rc and Cc), without which the circuit will be unstable over a wide range of tail currents. In a recent simulation, based on a 3GHz IC process, the following parameters were used: Ix = 1.5mA (to comfortably handle a ±1mA signal modulation, although the GCM operates right up to the limits with very low distortion); I = 250μA; Cc =3.5pF, Rc = 350Ω. The static nonlinearity was less than 0.05%. The input resistance (the effective resistance between the emitters of Q1 and Q2) was <40mΩ up to 100kHz, and less than 1.5Ω at 10MHz. The typical full-scale transient response is shown in Figure 6.41; it shows a rise time of about 2ns. No serious attempt was made to optimize the pulse response. The GCM is worth reconsideration in these days of BiMOS and CB processes.

6.5.6 *Applications of Complementarity to Mirrors*

Apart from the obvious advantage of permitting "reversed-polarity" mirrors, the availability of high-quality PNP transistors which can be freely

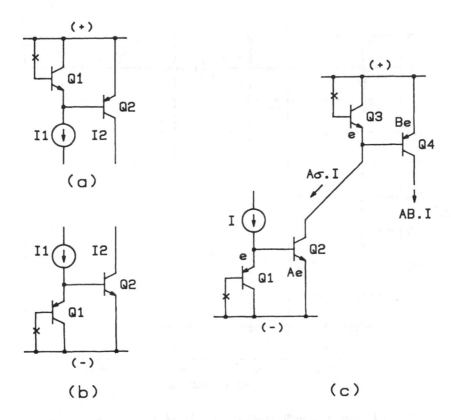

Figure 6.42 **Adaptation of the basic current mirror to complementary design. In (a) and (b) the mirror ratio is uncertain; by alternating stages, the ratio accuracy is restored (see text).**

mixed into a wideband signal path brings many benefits and encourages the invention of novel topologies, or the dusting-off of old ideas that just could not be realized in pre-CB days. We close this chapter, therefore, with a glimpse of just a few of the possibilities, and begin by introducing the concept of *balance* in a CB process, and proposing the use of a quantitative measure of this balance, the "sigma factor".

In monolithic circuit design, a very large dependence is placed on the reliability of device matching. Over the years, we have learned what to expect of V_{BE} matching, or resistor matching, or beta matching. We've also learned, incidentally, to take advantage of certain dependable correlations, such as that between beta and pinch resistance. Matching is an attribute of like devices. Resistors of a given construction match *because* they are all made in precisely the same way. In complementary bipolar design, however, there is clearly another dimension. The beta, or V_{BE}, or Early voltage, or base resistance, or breakdown voltages of a certain PNP geometry are not

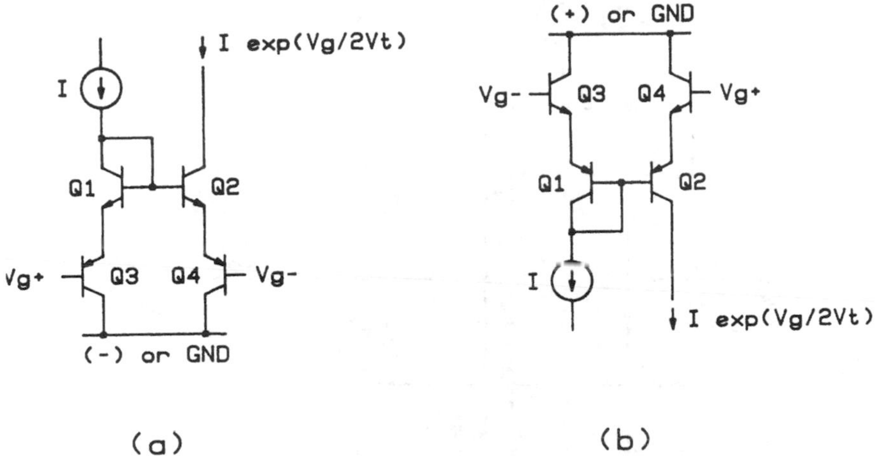

$(+)$ or GND

I exp$(Vg/2Vt)$

I

Q1   Q2

Vg$+$ Q3   Q4   Vg$-$

$(-)$ or GND

(a)

Vg$-$ Q3   Q4   Vg$+$

Q1   Q2

I

I exp$(Vg/2Vt)$

(b)

Figure 6.43   Mirrors having current-gain which can be controlled at a differential "high impedance" port.

expected to *match* the corresponding parameter of the NPN.

However, we welcome a high degree of *balance* in a CB process. For example, if the process designer had pushed for the highest possible speed in the NPN, it would probably all be lost if the speed of the PNP were not within a factor of, at most, two. Certainly, it would not help to have the extra NPN speed in making a fast, fully-complementary wideband signal path. It might also have necessitated some other compromises in the NPN performance. Likewise, it would be frustrating to design with PNPs having 40V breakdown voltages if the NPNs could only operate at 5V.

Obviously, balance in a complementary process is a valuable asset, and we might, in an ideal world, wish for it to be perfect, so that, for example, the base currents of an NPN and PNP exactly cancel at an input node. In the present context, we might wish for V_{BE}s to be very similar, to allow a stronger interplay between devices and the preservation of accuracy in mixed-polarity mirrors. Such is not the case, but we may find that we can depend on certain parameters to reliably *track* in some way, either over all production lots, or at least over one chip length.

Thus, it is proposed that those parameters in a CB process needing this kind of predictability use a factor, σ, specific to the parameter to quantify the otherwise fuzzy notion of balance. The σ-factors for most parameters, such as beta, Early voltage, BV_{CEO}, and so on, will have a some simple ratio, with a definable mean and standard deviation. It is proposed that in each case, σ is

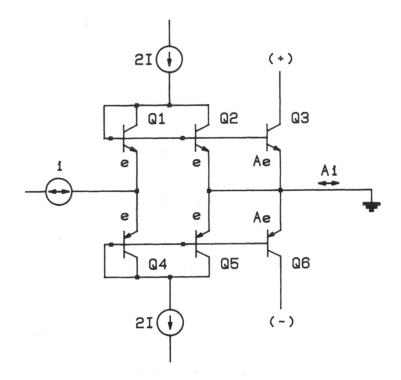

Figure 6.44 **A pair of complementary mirrors placed back-to-back, plus another pair of diodes, makes a novel bidirectional mirror.**

defined as the ratio of the NPN parameter to the PNP parameter. For translinear circuits, we are interested in current densities, and the appropriate metric here is the saturation current, I_S. Accordingly,

$$\sigma = I_{SN}/I_{SP} \tag{6.35}$$

Now, if we try to make a current-mirror out of mixed polarity devices, as shown in Figure 6.42a and 6.42b, we will find that this factor σ enters into the mirror ratio, which is therefore inexact. However, we can cancel this factor by cascading mirrors of opposite polarity, as shown in Figure 6.42c.

A useful attribute of these mixed-polarity mirrors is that they provide an uncommitted base, shown by the "x" in the above figures, which can be optionally used to voltage-control the ratio. Alternatively, we can avoid the σ-factor by using a circuit such as shown in Figure 6.43a and 6.43b. Here, we are back in the familiar territory of "matching" like against like, but, because of the good translinear properties of modern CB-process PNP transistors, we can confidently apply all the theory presented in Chapter 2. These mirrors come with a further bonus: a differential voltage-input port at which the mirror ratio can be controlled. The reader is invited to think about some of

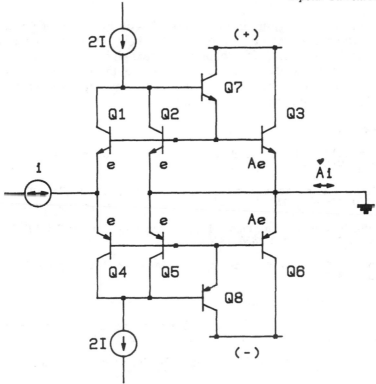

Figure 6.45 **A development of the bidirectional mirror, providing higher gain.**

the many ways in which this feature can be harnessed.

Finally, we propose a current-mirror form (Figure 6.44) which provides bidirectional current transport and a gain defined, in this case, by a fixed emitter area, A. The analysis follows standard TL practice. Figure 6.45 shows a development of this idea, which can be viewed as two EFA mirrors back-to-back, having the benefit of additional internal current gain to improve accuracy.

There is little doubt that the future of current-mode signal-processing will depend more and more on the unique benefits which are provided by the availability of a "translinear PNP" and the increasing use of complementary design techniques.

6.6 Acknowledgements

Paul Brokaw took at look at an early draft of this material and made some helpful observations, including some about my reckless disregard of rigor in connection with dimensional consistency in certain places (for example, Eqs. 6.15b and 6.16b). Normally, I'd agree, but there are times, I feel, when a simple cook-book formula is of value, and in these cases, a result can be quickly obtained by "plugging in the numbers" with little risk of fire and brimstone. Alicia and Monica, who thought their job was done after checking over the manuscript for Chapter 2, started all over again when the present Chapter was finally completed at the eleventh hour. My sincere thanks to them for their diligence in digging for my slips, and to Mark Elbert for his useful comments along the way.

6.7 References

[1] P. R. Gray and R. G. Meyer, "Analysis and Design of Analog Integrated Circuits", New York: Wiley, 1977, 1984.

[2] D. P. Laude, "A Low-Noise High-Precision Operational Amplifier", IEEE JSSC, Vol. SC-16, No. 6, pp. 748 - 750, 1981.

[3] B. Gilbert, "A Versatile Monolithic Voltage-to-Frequency Converter", IEEE JSSC, Vol. SC-11, No. 6, pp. 852 - 864, 1976.

[4] J. E. Solomon, "The Monolithic Op-Amp: A Tutorial Study", IEEE JSSC, Vol. SC-9, pp. 314 - 322, 1974.

[5] J. E. Smith, "Integrated Injection Logic", IEEE Press, 1980.

[5a] H. H. Berger, "The Injection Model - A Structure-Oriented Model for Merged Transistor Logic (MTL)", IEEE JSSC, Vol. SC-9, pp. 218 - 227, 1974.

[6] R. J. Widlar, "Some Circuit Design Techniques for Linear Integrated Circuits", IEEE Trans. of Circuit Theory, Vo. CT-12, pp. 586 - 590, 1965.

[6a] R. J. Widlar, "Design Techniques for Monolithic Operational Amplifiers", IEEE JSSC, Vol. SC-4, pp. 184 - 191, 1969.

[7] B. Gilbert, "Wideband Negative-Current Mirror", Electronics Letters, Vol. 11, No. 6, pp. 126 - 127, 1975.

[8] E. Seevinck, "Analysis and Synthesis of Translinear Integrated Circuits", Elsevier, 1988.

Chapter 7

Dynamic Current Mirrors

Eric A. Vittoz and George Wegmann

7.1 Introduction

The current mirror is a ubiquitous building block in analog integrated circuits. First applied in bipolar technology (see Chapter 6), it is now extensively used in CMOS to duplicate, multiply or divide bias currents or signal currents. Unfortunately, the precision of an MOS mirror is degraded by the large values of threshold offset and 1/f noise of MOS transistors. The classical approach to deal with these limitations will be reviewed in Section 7.3. Section 7.4 will introduce the principle of dynamic current mirrors (or current copiers), which eliminates completely the previous limitations and moves the achievable precision to new limits. Sections 7.5 to 7.9 will discuss these limits, that are essentially due to charge injection by switches, drain conductance and drain-to-gate capacitance, leakage current and sampled noise. Sections 7.10 and 7.11 will show how multiple copies and multiplying or dividing mirrors can be created. Applications are discussed at length in Chapter 14 but some of them will be mentioned in Section 7.12, which also suggests some possible extensions of the dynamic scheme to other functional blocks.

7.2. Summary of the MOS transistor characteristics and model

The symbols and definitions that will be used for n- and p-channel transistors are shown in Figure 7.1. Since the device is symmetrical, source, drain and gate voltages V_S, V_D and V_G are referred to the local substrate which is either the general substrate of the circuit, or a separate well. The substrate will be omitted when it is connected to one rail of the power supply (V^+ for p-channel, V^- for n-channel).

In a first approximation, and for design purposes, the MOS transistor may be fully characterized by just 3 parameters [1]: The gate threshold voltage V_{TO} for zero source voltage, a factor n which represents the reduction of the effect of gate voltage due to fixed charges in the channel and the transfer parameter

$$\beta = \mu C_{ox} W/L \qquad (7.2.1)$$

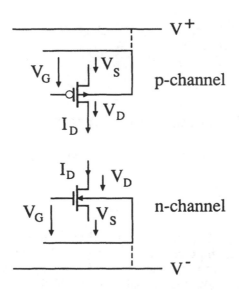

Figure 7.1   Symbols and definitions for MOS transistors

where μ is the mobility of charge carriers and C_{ox} is the gate oxide capacitance per unit area. The channel length L and channel width W of the transistor are defined by the layout. The third parameter may alternately be the specific current

$$I_s = 2n\beta U_T^2 = 2n\mu C_{ox} U_T^2 W/L \qquad 7.2.2)$$

where $U_T = kT/q$. The values of β and I_s of each transistor can be set by varying W/L. The value of the factor n is usually between 1.5 and 2 for small values of V_G and tends to 1 with increasing V_G.

Figure 7.2 illustrates qualitatively the variation of drain current I_D with drain voltage V_D. The source voltage is assumed to be constant and positive. For V_D larger than a saturation value V_{Dsat}, the transistor behaves like a current source I. For $I_D \ll I$, it behaves like a linear conductance g. The detailed variation of drain current with voltages V_S, V_D and V_G depends on the mode of operation of the transistor.

If the saturation current is large ($I \gg I_s$), the transistor operates in strong inversion and can be modelled by the following relations:

$$I_D = \beta (V_D - V_S)\left[V_G - V_{T0} - \frac{n}{2}(V_D + V_S)\right] \quad \text{conduction, for } V_D \leq V_{Dsat}$$
$$(7.2.3)$$

$$I_D = I = \frac{\beta}{2n}(V_G - V_{T0} - n V_S)^2 \qquad \text{saturation, for } V_D \geq V_{Dsat}$$
$$(7.2.4)$$

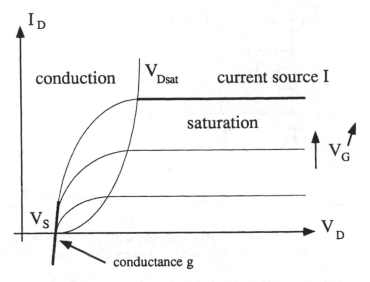

Figure 7.2   Variation of drain current I_D with drain voltage V_D

$$\text{where } V_{Dsat} = \frac{V_G - V_{T0}}{n} = V_S + 2U_T\sqrt{I/I_s} \tag{7.2.5}$$

If the saturation current is small ($I \ll I_s$), the transistor operates in weak inversion and the drain current is given by

$$I_D = I_s e^{\frac{V_G - V_{T0}}{nU_T}} \left[e^{-\frac{V_s}{U_T}} - e^{-\frac{V_D}{U_T}} \right] \tag{7.2.6}$$

which reaches its saturation value when the term in V_D becomes negligible, thus for the minimum possible value of the saturation voltage

$$V_{Dsat} = V_S + 3 \text{ to } 6\ U_T \quad \text{(weak inversion)} \tag{7.2.7}$$

The small signal transconductance in saturation $g_m = dI / dV_G$ is an important design parameter. It is given by

$$g_m = \sqrt{2\beta I/n} = \frac{2I}{n(V_{Dsat} - V_S)} = \beta (V_{Dsat} - V_S)$$

$$\text{in strong inversion} \tag{7.2.8}$$

$$g_m = \frac{I}{nU_T} \qquad \text{in weak inversion} \tag{7.2.9}$$

Furthermore, the residual drain to source conductance g_{ds} in saturation due to the channel shortening effect can be approximated by

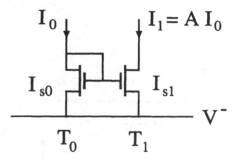

Figure 7.3   Standard n-channel current mirror

$$g_{ds} = \frac{I}{V_E}$$

(7.2.10)

where V_E is the extrapolated or Early voltage, which is proportional to channel length L. The Early voltage also increases slowly with the gate voltage and thus is minimum in weak inversion.

The noise of a transistor in saturation can be referred to its gate and represented by an equivalent input noise resistor

$$R_N = \frac{\gamma}{g_m} + \frac{\rho}{fWL}$$

(7.2.11)

The first term is due to channel noise, with a noise factor γ equal to n/2 in weak inversion and to 2n/3 in strong inversion. The second term is due to the silicon-oxide interface flicker noise, with a factor ρ (in Vm^2/As) that is strongly dependent on the process and slightly increases with the gate voltage.

The mismatch of supposedly identical transistors must be characterized by two independent statistical values: threshold mismatch ΔV_T and $\Delta\beta/\beta = \Delta I_s / I_s$. The RMS value of each of them depends on the process and can be minimized by applying design and layout rules for optimum matching [1].

7.3 Standard MOS mirror design considerations and limitations

Correct operation of the standard mirror shown in Figure 7.3 requires that all transistors are in saturation to produce

$$I_1 = \frac{I_{s1}}{I_{s0}} I_0 = A I_0$$

(7.3.1)

Weak inversion operation provides a minimum saturation voltage V_{Dsat}, but is usually not acceptable for noise, speed and precision considerations.

The output noise current spectrum S_{Nout} (in A^2/Hz) can be calculated from Figure 7.4:

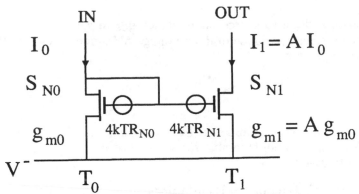

Figure 7.4 Noise sources in a current mirror

$$S_{N1} = A^2 S_{N0} + g_{m1}^2 \cdot 4kT(R_{N0} + R_{N1})$$ (7.3.2)

By introducing (7.2.11) and (7.2.8) with $V_S = 0$:

$$S_{N1} = A^2 S_{N0} + (A^2 + A) \frac{16kT}{3} \frac{I_0}{V_{Dsat}} + \frac{4\rho\, kT}{f} \left[\frac{1}{W_0 L_0} + \frac{1}{W_1 L_1} \right] \left[\frac{2AI_0}{nV_{Dsat}} \right]^2$$ (7.3.3)

The effect of the channel noise (second term) can only be reduced by increasing the saturation voltage. This solution also drastically reduces the effect of interface noise (third term) which can be further lowered by increasing the gate area WL of the transistors.

The speed of the mirror is limited by the time constant τ associated with the gate node capacitance C. Since C and β are both proportional to channel width W, (7.2.8) yields

$$\tau = \frac{C}{g_{m0}} \sim \frac{1}{V_{Dsat}}$$ (7.3.4)

The precision of the current ratio (or of the output current) can never be better than the mismatch of β of the two transistors, but it can be strongly degraded by the threshold mismatch ΔV_T which produces an error in the ratio A

$$\frac{\Delta A}{A} = \frac{g_{m1}}{I_1} \Delta V_T = \frac{2\Delta V_T}{nV_{Dsat}} \qquad \text{in strong inversion} \quad (7.3.5)$$

$$= \frac{\Delta V_T}{nU_T} \qquad \text{(maximum) in weak inversion} \quad (7.3.6)$$

The non-zero drain to source conductance g_{ds} in saturation is another cause of error, since a variation of drain voltage ΔV_D produces a variation of output current given by

$$\frac{\Delta I}{I} = \frac{\Delta V_D}{V_E}$$

(7.3.7)

This variation could be reduced by increasing the channel length to increase V_E, but it is not favorable for speed. The best solution is to use a cascode configuration.

7.4 Principle of dynamic current mirrors

The most important limits of standard mirrors with respect to their use in high-precision analog circuits are the current error due to mismatch and the low-frequency 1/f flicker noise. Both of them can be reduced by increasing the gate area WL and the saturation voltage V_{Dsat}, but this classical approach cannot eliminate flicker noise or reduce the error much below 1%. The performance of a mirror with a low saturation voltage can be largely improved by using the lateral bipolar mode of operation of the MOS transistors [2], but only one type of mirror is possible (source or sink) and the precision can still not be much better than 1%.

The principle of dynamic matching [3], which uses a chopper technique to reject both 1/f noise and offset to high frequencies, has found very useful applications in bipolar technology. It could be applied to CMOS as well, but the residual high frequency current ripple would have to be filtered out for most applications, and multiple mirrors or mirrors with non-unity ratio are difficult to implement.

Unlike bipolar transistors, MOS transistors need no gate current to control their drain current. This property is exploited by dynamic analog techniques to store some analog information on the gate capacitor [4]. One application is the dynamic comparator in which the analog storage capability is exploited to sequentially use the same transistor as the two devices of a differential pair. The very notion of mismatch disappears, since there is only one transistor. The same idea can be applied to build a dynamic current mirror, also called current copier, the principle of which is shown in Figure 7. 5 [5 - 10].

A single transistor T_m is combined with 3 switches S_x, S_y, S_z that are implemented by means of additional transistors, and a capacitor C. In a first phase (phase 0), T_m operates as the input device of a mirror, with its gate and drain connected to the input current source. When equilibrium is reached, capacitor C at the gate is charged to the gate voltage V required to obtain $I_D = I_0$. The value of I_0 is thus stored as a voltage across C. In the second phase (phase 1), T_m operates as the output device of a mirror, with its drain disconnected from the gate and connected to the output node. It sinks an output current I_1 that is controlled by the same gate voltage V and thus is

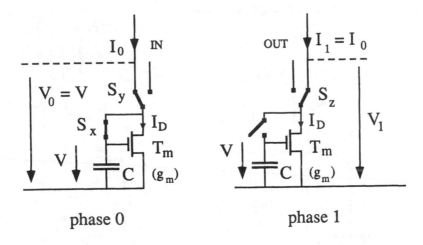

phase 0   phase 1

Figure 7.5 Principle of dynamic current mirror (or current copier)

equal to I_0. This very simple and elegant scheme suffers from various limitations. These limitations will be discussed in the following chapters together with many possible improvements and possible extensions.

7.5 Charge injection

A first important limitation to the precision of the above scheme is due to the realization of the various switches by means of transistors. To close the switch, the switching transistor is made conductive by mobile carriers that are attracted into the channel by the gate voltage. The resulting on-conductance g for $V_D \cong V_S$ is obtained by differentiating (7.2.3) and (7.2.6), which yields

$$g = \beta (V_G - V_{T0} - nV_S) = \sqrt{2n\beta I} \qquad \text{in strong inversion} \quad (7.5.1)$$

$$g = I_S e^{\frac{V_G - V_{T0} - nV_S}{nU_T}} = \frac{I}{U_T} \qquad \text{in weak inversion} \quad (7.5.2)$$

The total charge of mobile carriers in the channel is given in strong inversion by

$$q = WLC_{ox}(V_G - V_{T0} - nV_S) \qquad (7.5.3)$$

and is therefore related to the on-conductance by

$$q = gL^2/\mu \qquad (7.5.4)$$

This relationship can be shown to be valid for weak inversion as well.

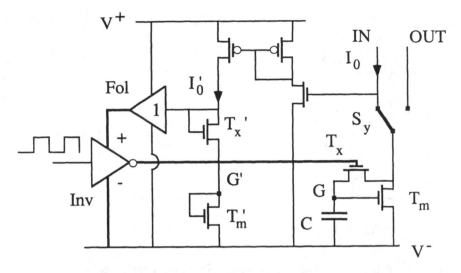

Figure 7.6   Control of the on-conductance of switch S_x

When the switch is opened, these carriers are released from the channel in order to block the transistor and most of them flow to the source or to the drain [11]. Thus, when switch S_x opens, a fraction Δq of q flows onto capacitor C, which causes an error

$$\Delta V = \Delta q/C \tag{7.5.5}$$

in the stored voltage V. This voltage error in turn creates a relative error in output current

$$\frac{\Delta I_1}{I_1} = \frac{g_m}{I_0} \Delta V \tag{7.5.6}$$

ΔV can be decreased by increasing C, with one limit given by the area of the capacitor. It can also be decreased by reducing the total charge q in the channel in order to reduce the fraction Δq that flows onto C. This can be obtained by minimizing the gate area WL and/or by controlling the gate control voltage of the switch in order to adjust its on-conductance g to the minimum required value. Figure 7.6 illustrates a possible implementation of such a control.

A current $I_0' = I_0$ flows through transistor T_m' matched with T_m. Nodes G' and G are therefore at the same potential. Now transistor T_x implementing switch S_x is made conductive by driving its gate by the gate voltage of T_x' (through the voltage follower Fol and the inverter Inv). Thus, if T_x' is matched with T_x, these two transistors have the same saturation current I_0 and the on-conductance g_x of T_x is given by equations (7.5.1) and (7.5.2):

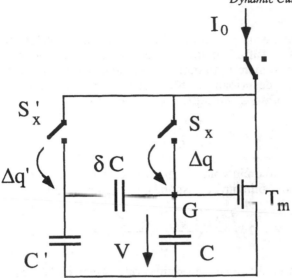

Figure 7.7 Reduction of the effect of charge injection by low sensitivity auxiliary input

$$g = \sqrt{2n\beta I_0} \quad \text{and thus} \quad q = \sqrt{2nC_{ox}I_0WL^3/\mu} \qquad \text{in strong inversion} \quad (7.5.7)$$

$$g = \frac{I_0}{U_T} \qquad \text{and thus} \quad q = \frac{I_0 L^2}{\mu U_T} \quad \text{in weak inversion} \quad (7.5.8)$$

These results must be multiplied by I_{sx} / I_{sx}' if the transistors T_x and T_x' are not identical.

For a given value of the total charge q, Δq can be reduced by employing one of several possible strategies. One of them is based on a charge compensation scheme by means of dummy switches. It requires an adequate design of the switching parameters and a carefully optimized layout [1, 4, 11, 12]. The residual charge Δq is ultimately limited by the degree of matching between the main switch and the dummy switches.

Another possibility which is based on feedback rather than matching is illustrated in Figure 7.7 [8]. A second sample-and-hold formed by capacitor C' and switch S_x' is added, and is connected to the critical node G through a capacitive attenuator $\delta C/C$ (with $\delta \ll 1$). Two steps are needed to disconnect the gate from the drain. The main switch S_x is opened first, dumping a charge Δq onto C. If the system is allowed to settle to equilibrium, this charge is eliminated by the second loop through δC. The auxiliary switch S_x' is then opened as well, which dumps a charge $\Delta q'$ onto C', but the resulting voltage variation at the critical node G is attenuated by the capacitive divider.

This scheme is equivalent to that of offset compensation by a low sensitivity auxiliary input used in the design of operational amplifiers [5].

The value of δ should be as low as possible to attenuate the effect of $\Delta q'$, but large enough to keep a sufficient voltage gain in the auxiliary loop in order to compensate Δq. Thus, there is an optimum value of δ that can be calculated with the approach described in [5]. Assuming the worst case with uncorrelated values of $\Delta q/C$ and $\Delta q'/C'$, this calculation yields:

$$\delta_{opt} = \sqrt{\frac{|\Delta q/C|_{max}}{A_v \, |\Delta q'/C'|_{max}}} \tag{7.5.9}$$

where A_v is the gate to drain voltage gain limited by the output conductance of the transistor. This optimum corresponds to a residual voltage step

$$\Delta V = 2\sqrt{\frac{|\Delta q/C|_{max} \, |\Delta q'/C'|_{max}}{A_v}} \tag{7.5.10}$$

which can be much smaller than the original value given by (7.5.5)

Whatever the strategy used for disconnecting the gate from the drain, the effect of the voltage step ΔV that is produced by this critical operation is weighted by g_m / I_0, which can be reduced by biasing T_m deep into strong inversion. Then, according to (7.2.8), for $V_S = 0$:

$$\frac{\Delta I_1}{I_1} = \frac{2 \, \Delta V}{n V_{Dsat}} = \frac{2 \, \Delta V}{V - V_{T0}} = \frac{\Delta V}{U_T} \sqrt{I_{sm}/I_0} \tag{7.5.11}$$

The lower limit of this error is given by the maximum acceptable value for V_{Dsat}. Further reduction of g_m / I_0 can be obtained by forcing a fraction $\alpha < 1$ of current I_0 to flow outside of transistor T_m, as shown in Figure 7.8 [13]. The dynamic mirror then only copies the complement to 1 of this fraction and the effect of ΔV is reduced to

$$\frac{\Delta I_1}{I_1} = \frac{2 \, (1 - \alpha) \, \Delta V}{n \, V_{Dsat}} \tag{7.5.12}$$

Since the current flowing through T_m cannot be negative, the maximum possible value of α is one. In this optimum situation, no current flows through T_m because no complement to αI_0 is needed. Now the bypassing current αI_0 must obviously be created in some way from I_0 by a combination of p- and n-channel standard mirrors.

Therefore, the value of α cannot be very accurate and will range between 1 and $1 - \Delta\alpha$, with a center value of $1 - \Delta\alpha/2$. The worst case will then correspond to the minimum value of α, for which the effect of ΔV is given by

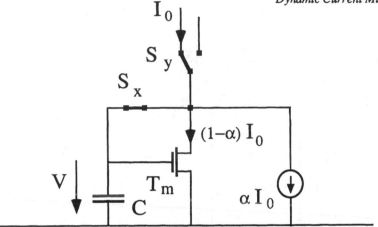

Figure 7.8 Improved mirror for a reduced transconductance of T_m

$$\left|\frac{\Delta I_1}{I_1}\right|_{max} = \frac{2 \Delta \alpha \, |\Delta V|}{n \, V_{Dsat}} \tag{7.5.13}$$

The efficiency of the current bypassing scheme of Figure 7.7 thus depends on the degree of accuracy achievable for αI_0.

The various approaches to reduce the effect of charge injection on the precision of the dynamic mirror may be combined. It must be pointed out that most of them tend to increase the time needed to achieve equilibrium in the circuit and therefore decrease the maximum frequency of switching. Since the input current is sampled, the bandwidth of the mirror is in turn reduced. This tradeoff between speed and precision will be discussed in Section 7.8.

7.6 Cascode configurations

In the basic structure illustrated in Figure 7.5, the drain voltage of T_m must return to the value $V_0=V$ at each storage phase (phase 0). During the copying phase (phase 1) it must jump to a value V_1 imposed by the load of the mirror. Since $V_1 \neq V_0$, this difference in drain voltage during the two phases produces important additional contributions to the inaccuracy of the mirror (Figure 7.9).

The first contribution is due to channel length modulation which is represented by the drain to source conductance g_{ds}:

$$\frac{\Delta I_1}{I_1} = \frac{g_{ds}}{I_1}(V_1 - V_0) = \frac{V_1 - V_0}{V_E} \tag{7.6.1}$$

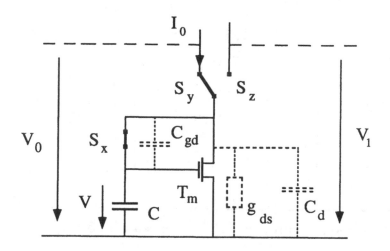

Figure 7.9   Effects of drain voltage variations

Since the Early voltage V_E is of the order of tens of volts, steps of hundreds of millivolts in V_1-V_0 would produce an error larger than 1%.

The second contribution is caused by the drain to gate capacitance C_{gd}. This capacitance transfers a fraction of the drain voltage step to the gate. For $C \gg C_{gd}$ the resulting gate voltage error is

$$\Delta V = \frac{C_{gd}}{C}(V_1 - V_0)$$

$$(7.6.2)$$

This voltage error will be converted into a current error according to (7.5.6).

The third effect is due to the additional charge that must be provided from the output node to change the voltage across C_{gd} and the drain to ground capacitor C_d. This charge creates a transient current (current glitch) which has no importance if the output current is only used after equilibrium is reached. However, if the current is available continuously, as is made possible by using 2 or more cells in alternation, these glitches are not acceptable and produce an error in the average output current given by

$$\Delta I_1 = f_s (V_1 - V_0)(C_{gd} + C_d)$$

$$(7.6.3)$$

where f_s is the frequency of operation of the switches [12]. These considerations show that an accurate implementation must include some means to keep the drain voltage of T_m as constant as possible in spite of the difference V_1-V_0. An operational amplifier could be used for this purpose [8], but a much more compact solution is obtained by adding a common gate transistor T_c in series with the main transistor T_m to build a cascode as shown in Figure 7.10.

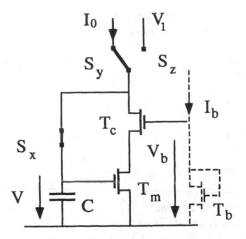

Fig 7.10   Cascode dynamic mirror

The bias voltage V_b of the cascode transistor T_c should be as low as possible to reduce the minimum possible value of the output voltage V_1. The limit is given by the need to keep T_m in saturation. Calculation of this minimum value yields

$$V_b = V + 2nU_T \sqrt{I_0/I_{sc}} = V_{T0} + 2nU_T\left[\sqrt{I_0/I_{sm}} + \sqrt{I_0/I_{sc}}\right] \qquad (7.6.4)$$

and provides

$$V_{1min} = V_{Dsat\,c} = \frac{V_b - V_{T0}}{n} = 2U_T\left[\sqrt{I_0/I_{sm}} + \sqrt{I_0/I_{sc}}\right] \qquad (7.6.5)$$

This value must be lower than $V_0 = V$ in order to maintain the saturation of transistor T_c during the storage phase (phase 0). This condition can be expressed as

$$V_{T0} \geq 2U_T\left[(1 - n)\sqrt{I_0/I_{sm}} + \sqrt{I_0/I_{sc}}\right] \qquad (7.6.6)$$

which is fulfilled for any positive value of the threshold V_{T0} if

$$I_{sm}/I_{sc} \leq (n - 1)^2 \qquad (7.6.7)$$

The condition (7.6.6) can usually be fulfilled, since according to (7.5.11) $I_0 \gg I_{sm}$ is needed to reduce the effect of ΔV (T_m deeply in strong inversion) whereas T_c should be close to (or even in) weak inversion to minimize its effect on V_{1min} (7.6.5). In the case that (7.6.6) cannot be fulfilled, a source follower can be introduced between the storage capacitor C and the gate of T_m [8]. The drawback of this scheme is an increase of the input voltage V_0.

The bias voltage V_b of T_c can be produced by imposing a current I_b through a bias transistor T_b as shown by the dotted line in Figure 7.10. Application of relations (7.2.5) and (7.2.2) yields

$$V_b = V_{T0} + 2nU_T\sqrt{I_b/I_{sb}} \qquad (7.6.8)$$

The required value given by (7.6.4) is thus obtained for

$$I_b/I_{sb} = \left[\sqrt{I_0/I_{sm}} + \sqrt{I_0/I_{sc}}\right]^2 \qquad (7.6.9)$$

which can be precisely determined by controlling I_b / I_0 (by means of standard mirrors) and the relative dimensions of the three transistors [14].

The cascode combination of T_m and T_c is equivalent to a single transistor T_m with values of V_E and C_{gd} respectively increased and decreased by a factor equal to the source to drain voltage gain of T_c. This gain can be as high as several hundred if T_c operates close to weak inversion. Errors (7.6.1) due to output conductance and (7.6.2) due to drain to gate capacitive coupling are thus drastically reduced. However, the output glitches and the corresponding average error given by (7.6.3) remain, because the voltage steps V_1-V_0 are still applied across C_d and C_{gd} of transistor T_c. This simple cascode configuration is therefore only applicable when the output current is used after equilibrium of the copying phase (phase 1) is reached.

For continuous operation, the toggle switch S_y-S_z must be placed between T_c and T_m and two identical cells working in opposite phases must be used, as shown in Figure 7.11. The bias voltage V_b can be obtained as in Figure 7. 10.

The structure may also be self-biased by connecting the gates of T_{c0} and T_{c1} to the input node (drain of T_{c0}). The saturation of T_{m0} and T_{m1} is then only obtained if T_{c0} and T_{c1} are placed in separate wells that are connected to their sources, and if the condition

$$\sqrt{I_{sm}/I_{sc}} \leq 1 - 1/n \qquad (7.6.10)$$

is fulfilled. The minimum value of output voltage is then

$$V_{1min} = 2U_T\left|n\sqrt{I_0/I_{sm}} - (n-1)\sqrt{I_0/I_{sc}}\right| \qquad (7.6.11)$$

The condition (7.6.10) for self-bias can be avoided by connecting switch S_x also between T_c and T_m, as shown in Figure 7.12. The minimum value of V_1 is then given by

$$V_{1min} = V_{T0} + 2U_T\left[n\sqrt{I_0/I_{sm}} + \sqrt{I_0/I_{sc}}\right] \qquad (7.6.12)$$

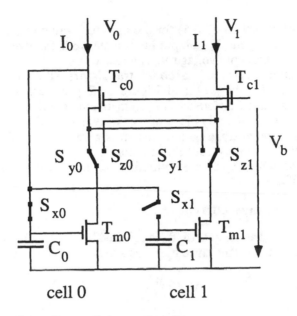

Figure 7.11 Mirror for continuous operation

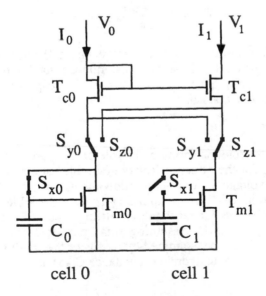

Figure 7.12 Self-biased dynamic continuous mirror

which is much larger than (7.6.5) for the optimum value of V_b. This result does not change if T_{c0} and T_{c1} are put in separate wells that are connected to their sources, only the input voltage V_0 is reduced.

The mismatch between cascode transistors T_{c0} and T_{c1} in the configurations of Figure 7.11 and Figure 7.12 produces some residual glitches due to the small difference ΔV_S in their source voltages [12]. If these transistors operate sufficiently close to weak inversion, $\Delta V_S = \Delta V_T$ and the resulting error in the average output current is given by (7.6.3) with $V_1 - V_0$ replaced by ΔV_T [12]. Care must be taken not to leave the source node of T_{c0} and T_{c1} floating while S_y-S_z are switching, because ΔV_S would increase considerably [12].

7.7. Effect of leakage current

The voltage V stored at the gate of T_m is affected by any leakage current $I_L \ll I_0$ flowing from this node. This current causes a peak to peak variation

$$\Delta V_{pp} = \frac{I_L t_1}{C}$$

$$(7.7.1)$$

where t_1 is the time duration of the copying phase 1, and an average error

$$\overline{\Delta V} = \frac{I_L (t_0 + t_1)}{g_x t_0}$$

$$(7.7.2)$$

where t_0 is the time duration of the storage phase 0 and g_x is the on-conductance of the switch S_x. These relations impose a maximum value for t_1 and minimum values for t_0 and g_x, which must be respected even if the current to be mirrored is constant.

7.8. Settling time

If the current to be mirrored is variable, the role of the storage phase 0 is not only to compensate for the effect of leakage current, but also to update the value of the stored voltage V, in order to allow the variations to be followed by the output current I_1. Correct updating is only possible if the time duration t_0 of this phase is longer than the settling time of the sample-and-hold formed by T_m, S_x and C. Assuming small perturbations, the settling behavior of this circuit can be examined by means of the small signal model of Figure 7.13, where the gate to drain capacitance C_{gd} is neglected

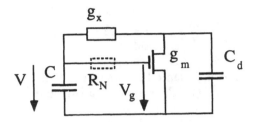

Figure 7.13   **Small signal model for the storage phase**

Opening the loop between capacitor C and the gate, the open loop transfer function is

$$G(s) = \frac{V(s)}{V_d(s)} = - \frac{1}{s\tau_1(1 + s\tau_2)}$$

(7.8.1)

where

$$\tau_1 = \frac{C + C_d}{g_m}$$

(7.8.2)

$$\tau_2 = \frac{C \cdot C_d}{g_x(C + C_d)}$$

(7.8.3)

The two poles of the closed loop circuit (roots of $1 - G(s) = 0$) are

$$s_{1,2} = - \frac{1}{2\tau_2} \pm \sqrt{\frac{1}{4\tau_2^2} - \frac{1}{\tau_1\tau_2}}$$

(7.8.4)

For $4\tau_2 > \tau_1$, the response is a damped oscillation with an envelope time constant $2\tau_2$. For $\tau_1 \gg 4\tau_2$, it settles exponentially with time constant τ_1. The global settling time constant may be reasonably approximated by

$$\tau_s = \tau_1 + 2\tau_2$$

(7.8.5)

which must be 5 to 7 times smaller than t_0 to ensure that equilibrium is reached accurately. Since the switching frequency $f_s < 1/t_0$, this condition puts an upper limit on f_s and C and a lower limit on g_x and g_m. The resulting tradeoff between settling time and charge injection can be expressed by first combining (7.5.4), (7.5.5) and (7.5.6) to yield

$$\frac{\Delta I_1}{I_1} = \frac{\alpha_i L^2}{\mu I_0} \cdot \frac{g_m g_x}{C}$$

$$(7.8.6)$$

where $\alpha_i = \Delta q / q$ and L is the channel length of the switching transistor T_x. Then, introducing (7.8.2) and (7.8.3) with the realistic assumption that $C \gg C_d$ gives

$$\frac{\Delta I_1}{I_1} = \frac{\alpha_i L^2 C_d}{\mu I_0} \cdot \frac{1}{\tau_1 \tau_2}$$

$$(7.8.7)$$

which shows that neither τ_1 nor τ_2 should be too small. Introducing (7.8.5), their optimum values are

$$\tau_1 = 2\tau_2 = \tau_s / 2$$

$$(7.8.8)$$

which results in the minimum effect on output current

$$\frac{\Delta I_1}{I_1} = \frac{8 \alpha_i L^2 C_d}{\mu I_0} \cdot \frac{1}{\tau_s^2}$$

$$(7.8.9)$$

The significance of this result can be evaluated by considering a critical numerical example:

$\alpha_i = 0.5$ (no compensation of charge injection)
$L = 2\mu m$ (conservative process)
$C_d = 0.1 pF$ (conservative process)
$\mu = 700 \text{ cm}^2/Vs$ (standard for n-channel)
$I_0 = 1\mu A$ (very low current)
$\tau_s = 0.1 \text{ }\mu s$ (compatible with $f = 1 MHz$)

which results in $\Delta I_1/I_1 = 2.10^{-3}$, which is a small error.

7.9 Noise

The input signal of the dynamic mirror is sampled at a frequency f_s by the switch S_x during phase 0 and held on the capacitor C during phase 1. Therefore, all spectral components above the Nyquist frequency $f_s / 2$ are folded down to lower frequencies, including those of the input noise.

This undersampling process also operates on the noise generated inside the mirror. Since the bandwidth must be much larger than f_s ($5f_s$ to $10f_s$) in order to reach equilibrium during settling, the switching noise due to undersampling dominates.

Let us first consider the sources of white noise. If $C \gg C_d$, it can be shown that the thermal noise generated in the conductance g_x (Figure 7.13) is negligible with respect to that of the transistor T_m, which is expressed by the term γ / g_m in the equivalent noise resistance given by (7.2.11). The latter corresponds to a noise voltage source of spectral density $4kT\gamma / g_m$ inserted in series with the gate of T_m.

When the switch is closed (phase 0), this source of noise produces a noise voltage V_N across C, the variance of which is

$$V_N^2 = \frac{4\gamma\, kT}{g_m} \Delta f_N \qquad (7.9.1)$$

where

$$\Delta f_N = \int_0^\infty \frac{df}{\left|\, 1 - 1 / G(j\omega)\, \right|^2} \qquad (7.9.2)$$

is the equivalent noise bandwidth. By introducing (7.8.1) with $s = j\omega$, this integral simplifies to

$$\Delta f_N = \frac{1}{4\tau_1} \qquad (7.9.3)$$

which is surprisingly independent of the time constant τ_2 [15]. After introducing this result into (7.9.1) with τ_1 given by (7.8.2) and $C \gg C_d$:

$$V_N^2 = \frac{\gamma\, kT}{C} \qquad (7.9.4)$$

The noise variance is not changed by the sampling process (opening of S_x), which only concentrates the spectrum essentially below $f_s / 2$. According to (7.5.6) and with $\gamma = 2n / 3$ in strong inversion, the noise voltage across C produces a relative output noise variance

$$\left(\frac{I_{N1}}{I_1}\right)^2 = \left(\frac{g_m}{I_0}\right)^2 V_N^2 = \frac{8kT}{3nCV_{Dsat}^2} \qquad (7.9.5)$$

Assuming $C = 1pF$ and $V_{Dsat} = 1V$, this expression gives a relative RMS output current noise value I_{N1} / I_1 of 53ppm at ambiant temperature. This results does not take into consideration the 1/f component of transistor noise.

Without sampling, the spectrum of the total output noise current would be

$$S_{N1cont.} = 4kT \, R_N \, g_m^2 \qquad (7.9.6)$$

or, in strong inversion, with R_N given by (7.2.11) and g_m by (7.2.8):

$$S_{N1cont.}(f) = \frac{16 \, kT \, I_1}{3V_{Dsat}} \left(1 + \frac{f_c}{f} \right) \qquad (7.9.7)$$

where

$$f_c = \frac{3 \, \rho \, g_m}{2n \, WL} \qquad (7.9.8)$$

is the noise corner frequency (frequency for equal values of white and 1/f spectral densities).

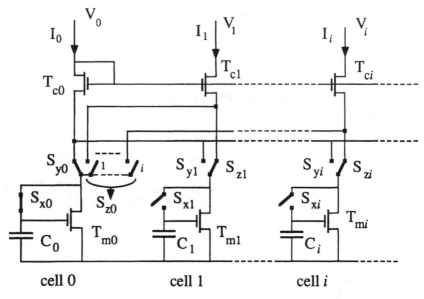

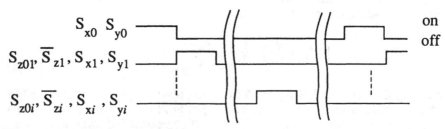

Figure 7.14   Multiple output dynamic current mirror

After sampling, the white noise component is multiplied by the undersampling ratio, but the 1/f noise as such is eliminated by the differentiation properties of the sample-and-hold process [5]. However, undersampling of the high-frequency components of 1/f noise produces an additional component of noise which is approximately constant for f«f_s. This additional contribution has been evaluated in [16] and yields a total noise current spectral density for f « f_s:

$$S_{N1} = \frac{16\,kT\,I_1}{3V_{Dsat}} \left(\frac{1}{2\tau_1 f_s} + \frac{2\,f_c}{f_s}\,(1 + \ln\,(\frac{1}{3\,\pi\,\tau_1\,f_s})) \right)$$

(7.9.9)

The first term in the parenthesis is due to the undersampling of white noise, the second term is due to the undersampling of 1/f noise.

7.10 Multiple and multiplying current mirrors

The configurations of Figures 7.11 and 7.12 can be extended to provide several copies of the input current by simply repeating the output cell (cell 1). As shown in Figure 7.14, n+1 phases are needed to provide n copies. The voltage stored on capacitor C_i of each output cell i is refreshed or updated during one phase per cycle. The cell being refreshed is replaced by cell 0, which requires n different switches S_{z0}.

Obviously, these n continuous output currents may be added to realize a 1 to n multiplying current mirror. Figure 7.15 shows that the set of cascode transistors T_{c1} to T_{cn} can then be shared by all the output cells, which avoids the need to split S_{z0} into several switches and reduces the number of interconnections.

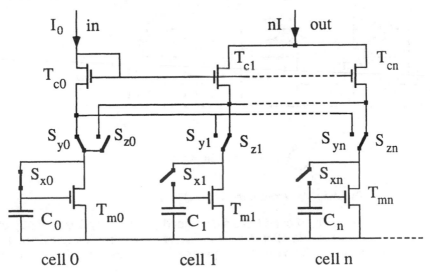

Figure 7.15 One to n multiplying dynamic mirror

7.11. Dividing dynamic current mirrors

Storing the input current in n transistors connected in parallel and copying the current of a single transistor at the output would provide a divide-by-n current divider if all the transistors were exactly identical. However, because of unavoidable mismatch, the input current is not evenly shared among the transistors and the attainable precision is not better than that of a standard mirror.

A calibration scheme must be added to force exactly the same current though each of the parallelled transistors. Figure 7.16 illustrates such a scheme for n=2 [16]. T_m is the main transistor of the mirror which delivers

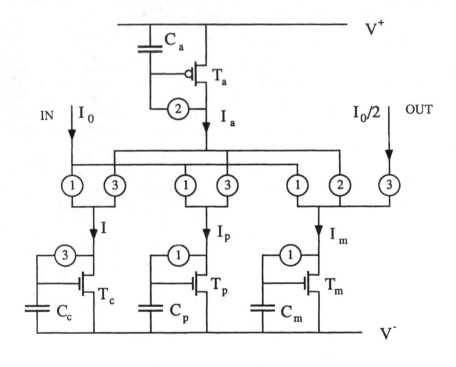

$\textcircled{3}$ = switch on in phase 3

Figure 7.16 Dividing dynamic mirror (n=2)

the output current. The second transistor needed to divide by 2 is made up of the combination of the parallel transistor T_p and the calibration transistor T_c. The circuit operates in 3 phases and requires a few cycles to reach the correct value of output current. For the sake of clarity, each switch is symbolized by a circled number that indicates its phase of conduction.

During phase 1, transistors T_m and T_p share the input current I_0 from which the previously established calibration current I_c is substracted. Thus, for cycle j:

$$I_{mj} = \frac{I_{0j} - I_{cj-1}}{2} (1 + \varepsilon)$$

(7.11.1)

$$I_{pj} = \frac{I_{0j} - I_{cj-1}}{2} (1 - \varepsilon)$$

(7.11.2)

where ε represents the mismatch between T_m and T_p. These currents are then memorized on the capacitors C_m and C_p.

During phase 2, the current I_m is temporarily stored in the p-channel auxiliary current copier T_a. Then, during phase 3, the difference between this stored current $I_a = I_m$ and the current I_p memorized by T_p is stored in copier T_c as the new value of the calibration current I_c:

$$I_{cj} = I_{mj} - I_{pj} = \varepsilon (I_{0j} - I_{cj-1})$$

(7.11.3)

Since I_c must always be positive, T_m and T_p must be designed to ensure $\varepsilon > 0$. The evolution of the calibration current can be obtained by introducing z-transforms into (7.11.3), which yields

$$I_c(z) = \frac{\varepsilon z}{z + \varepsilon} I_0(z)$$

(7.11.4)

Thus, after a step of input current to a value I_0, the calibration current I_c settles to

$$I_{c\infty} = \frac{\varepsilon}{1 + \varepsilon} I_0$$

(7.11.5)

Introducing this equilibrium value of the calibration current into (7.11.1) provides the equilibrium value of the output current I_m:

$$I_{m\infty} = \frac{I_0}{2}$$

(7.11.6)

independent of the mismatch ε. This equilibrium is reached exponentially with a time constant

$$\tau = -\frac{1}{\ln \varepsilon} \text{ cycles} \quad \text{(for example: 0.5 cycle for } \varepsilon = 12\%)$$

(7.11.7)

This principle can be extended to n > 2 by using n-1 separate pairs T_{pi} - T_{ci}. Phase 3 is split into n-1 phases $3i$, each of them being used to update separately one of the calibration transistors T_{ci}. A total of n+1 phases are thus needed and the output current I_0 / n is available during all the phases $3i$.

7.12. Examples of applications and possible extensions

The most immediate application of very accurate curent mirrors is in D/A and A/D converters, which are discussed extensively in Chapter 13. The output current of a D/A converter can be generated by summing the appropriate number of unit currents provided by a multiple current mirror [9, 13]. This approach ensures monotonic conversion but it is limited by the exponential increase in the number of required current copies. The complexity can be largely reduced by using binary weighted currents. These currents can be generated by cascading multiply-by-2 or divide-by-2 dynamic mirrors.

The multiplier approach requires simpler cells, but errors are progressively accumulated from cell to cell and the largest current (most significant bit, MSB) has the worst relative precision. The large resulting gain error can be corrected by adjusting the reference. A more severe limitation is due to noise. The accumulation of noise culminates at the MSB and can by no means be compensated. Moreover, the relative values of noise and error are inversely proportional to the saturation voltage V_{Dsat} of the mirror transistors T_m, as expressed by (7.9.4) and (7.5.11). Keeping a large value of V_{Dsat} at low current levels requires a very low value of the specific current I_s of T_m, as given by (7.2.5), and therefore unrealistically long transistors. Consequently, the first cells of the multiplying cascade generate large values of noise and error, which are propagated to the MSB current.

These limitations can be avoided by using the dividing scheme, which confines the larger error and noise to the least significant bits. This advantage is probably worth the increased complexity of the cells. Analog to digital converters can be based on similar approaches [17, 18].

The high precision of dynamic mirrors can also be exploited in the realization of continuous filters based on C / g_m time constants [19]. Matching between capacitors is usually excellent, as is demonstrated by the high precision obtained in switched-capacitor circuits, whereas the mismatch of transconductances is usually larger by orders of magnitude. As expressed by (7.2.9), the transconductance g_m of a transistor operated in weak inversion only depends on the parameter n and the bias current I. Matching of n is usually good, since this parameter does not depend on device dimensions or on mobility. Thus, biasing the critical transistors by means of dynamic mirrors should provide high-precision continuous filters.

Yet another interesting application can be found in switched current filters, discussed in Chapter 11. These filters emulate switched-capacitor filters by exploiting the storage property of capacitor-transistor combinations. They could be modified to exploit the high precision of dynamic mirrors to achieve very accurate integer weightings. Further applications are discussed in Chapter 14.

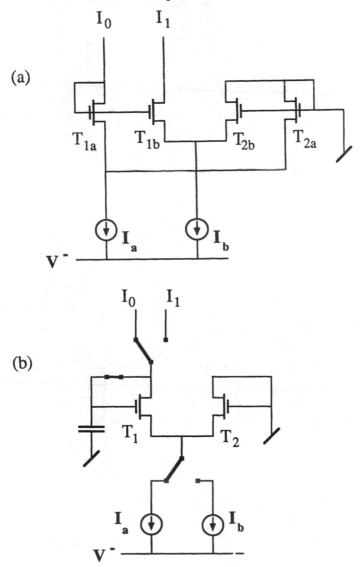

Figure 7.17 Adjustable current mirror

The basic principle used in dynamic current mirrors (or current copiers) can be extended to implement more complicated functions. Some suggestions will be given here but none of them has been experimentally verified.

Figure 7.17 shows an adjustable mirror based on the translinear principle [20] which was presented in Chapter 2. Since this principle exploits their

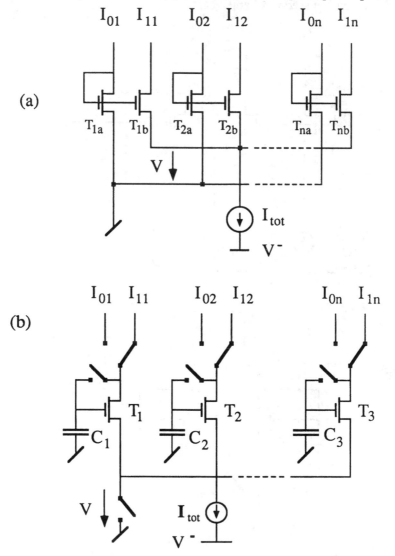

Figure 7.18   Normalization circuit

exponential voltage-to-current characteristics, the MOS transistors must be operated in weak inversion. The variable current mirror of Figure 7.17a then provides

$$\frac{I_1}{I_0} = \frac{I_b}{I_a}$$

(7.12.1)

According to (7.2.6), the threshold mismatch ΔV_T between transistors results, in weak inversion, to a drain current error

$$\frac{\Delta I_D}{I_D} = \frac{\Delta V_T}{nU_1}$$

(7,12.2)

which can reach as much as 30%.

Mismatch can be eliminated by using a single transistor per pair to obtain the dynamic version of the same circuit represented in Figure 7.17b. This circuit could be further improved by introducing the various improvements discussed for the current mirror.

An extension of this circuit provides the normalizer shown in Figure 7.18 [21] together with a potentially much more precise dynamic version. As long as all transistors operate in weak inversion, the common voltage difference V between the sources of all input transistors and those of all output transistors implies that the mirror ratio I_{1i} / I_{0i} is identical for all channels. This ratio is automatically adjusted, for any set of input currents, to maintain the sum of output currents equal to I_{tot}.

Figure 7.19 shows a circuit which exploits weak inversion operation to generate a voltage proportional to U_T and thus to the absolute temperature (PTAT voltage) [15]. Application of (7.2.6) to transistors T_{1a} and T_{1b} yields:

$$V_R = U_T \ln K$$

(7.12.3)

where K is the gain of mirror T_{2a}-T_{2b}. If this mirror is operated deep into strong inversion, the error in V_R is dominated by the threshold mismatch of the pair T_{1a}-T_{1b}. This error is suppressed in the dynamic version. The residual error due to T_{2a}-T_{2b} could be reduced by using a 1-to-K multiplying mirror.

As a last suggestion, Figure 7.20 shows a current conveyor which provides a low-impedance virtual ground input. The dynamic version should reduce the input offset and improve the equality between the input and output currents.

7.13. Conclusion

Based on a simple idea, the dynamic current mirror (or current copier) eliminates the main limitations of standard current mirrors, that are due to offset and 1/f noise. A new problem is created by the charge injected from the switches but the resulting error can be kept very low by adequate design procedures, especially when speed is not a limitation. Except in very special cases, the cascode configuration (or equivalent means) is needed to avoid spoiling the excellent intrinsic precision by errors due to the nonzero output conductance and to the gate-to-drain capacitance.

The basic cell, which provides a single one-to-one discontinuous copy of

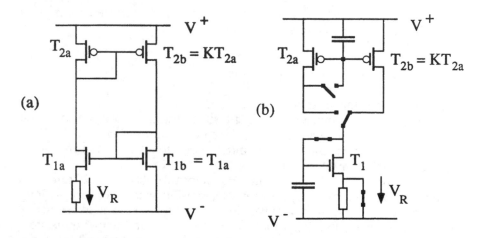

Figure 7.19   PTAT voltage generation.

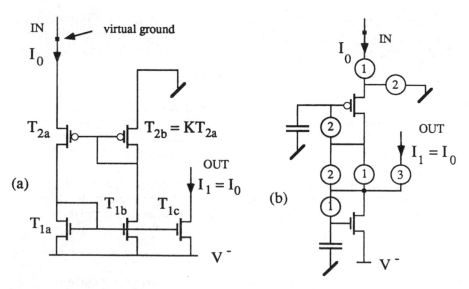

Figure 7.20   Current conveyor.

the input current, can be extended to obtain continuous multiple copies as well as current multiplication or division by integer numbers.

Published applications include a 16-bit DAC [9, 13] and a 10-bit ADC. A precision of 100 ppm has been achieved with continuous mirrors [7, 10, 12] and their application in accurate continuous filters is under investigation.

The principle of dynamic mirrors can probably be extended to a variety of different circuits, to create very precise analog CMOS building blocks.

7.14 References

[1]   E.A.Vittoz, "The design of high performance analog circuits on digital CMOS chips", IEEE Journal of Solid-State Circuits, Vol. SC-20, pp.657-665, June 1989.

[2]   E.A.Vittoz, "MOS transistors operated in the lateral bipolar mode and their applications in CMOS technology", IEEE Journal of Solid-State Circuits, Vol. SC-18, pp.273-279, June 1983.

[3]   R.J.Van de Plassche, "Dynamic element matching for high accuracy monolithic D/A converters, "IEEE Journal of Solid-State Circuits, Vol. SC-11, pp.795-800, Dec. 1976.

[4]   Y.S.Yee *et al*, "A 1mV CMOS comparator", IEEE Journal of Solid-State Circuits, Vol. SC-13, pp.294-297, June 1978 .

[5]   E.A.Vittoz, "Dynamic analog techniques", in *VLSI Circuits for Telecommunications* , edited by Y.P.Tsividis and P.Antognetti, Prentice-Hall, 1985.

[6]   H.Oguey, Private communication, 1978.

[7]   E.A.Vittoz and G.Wegmann, "High-precision current mirrors", final seminar on project PN-13 "CMOS functional blocks" of the Swiss National Research Fundation, Bern, May 1988.

[8]   S.J.Daubert *et al*, "Current copier cell", Electronics Letters, vol.24, pp1560-1562, 8th Dec. 1988.

[9]   W.Groeneveld *et al*, "A self-calibration technique for monolithic high-resolution D/A converters, "ISSCC Dig. of Tech. Papers, pp.22-23, February 1989.

[10]   G.Wegmann and E.A.Vittoz, "Very accurate dynamic current mirrors", Electronics Letters, vol.25, pp. 644-646, 11th May 1989.

[11]   G.Wegmann *et al*, "Charge injection in analog MOS switches", IEEE Journal of Solid-State Circuits, Vol. SC-22, pp.1091-1097, Dec. 1987.

[12]   G.Wegmann and E.A.Vittoz, "Analysis and improvements of highly accurate dynamic current mirrors", Dig. Tech. Papers ESSCIRC '89, Vienna, pp.280-283.

[13]   W.Groeneveld et al , "A self-calibration technique for monolithic high-resolution D/A converters", IEEE Journal of Solid-State Circuits, Vol. SC-24, pp.1517-1522, Dec.1989.

[14]   T.C. Choi *et al*, "High-frequency CMOS switched-capacitor filters for communications application", IEEE Journal of Solid-State Circuits, Vol. SC-18, pp.652-664, Dec. 1983.

[15]   E.A.Vittoz, "Micropower techniques", in *VLSI Circuits for Telecommunications* , edited by Y.P.Tsividis and P.Antognetti, Prentice-Hall, 1985.

[16]   J. Robert *et al* , "Very accurate current divider", Electronics Letters, vol.25, pp. 912-913, 6th July 1989.

[17]   D.G. Nairn and C.A.T. Salama, "Ratio-independent current mode algorithmic analog-to-digital converters", Proc. ISCAS '89, May 1989, Portland, pp.250-253.

[18]   D.G. Nairn and C.A.T. Salama, "Current mode analog-to-digital converters", Proc. ISCAS '89, May 1989, Portland, pp.1588-1591.

[19]   H.Khorramabadi and P.R.Gray, "High-frequency CMOS continuous-time filters", IEEE Journal of Solid-State Circuits, Vol. SC-19, pp.939-948, Dec. 1984.

[20]   B.Gilbert, "Translinear circuits: a proposed classification", Electronics Letters, vol.11, pp. 14-16, 9th January 1975.

[21]   B.Gilbert, "A monolithic 16-channel analog array normalizer", IEEE Journal of Solid-State Circuits, Vol. SC-19, pp.956-963, Dec. 1984.

Chapter 8

Gallium Arsenide Analogue Integrated Circuit Design Techniques

Chris Toumazou and David Haigh

8.1 Introduction

Many high speed analogue circuits have been developed for bipolar technology. These circuits have generally used current as the main processing variable, and some of them are reviewed in Chapters 2, 4, 15 and 16. Work aimed at extending the maximum frequency capability of CMOS technology, described in Chapters 5 and 9, has also tended to employ current processing techniques. In designs for both bipolar and CMOS technology, a number of key current-based components have been found to be particularly valuable. These include the current mirror (Chapters 6 and 7), the high gain (or operational) transconductance amplifier (Chapter 4) and the high precision transconductor, which provides a tunable linear resistive element for integrated circuit filter realisation as described in Chapter 9.

Modern communication systems, both microwave and optical, are beginning to rely heavily on the high frequency circuit performance capabilities provided by III-V semiconductor materials, of which the most mature is Gallium Arsenide (GaAs). The applications of GaAs can be divided into digital and analogue. The analogue applications can be further subdivided into microwave circuits, and sampled data circuits based on switched capacitor circuit techniques [1]. Both the high quality switches and the high gain, fast settling operational amplifiers required in these systems can be realised using the GaAs MEtal Semiconductor FET, or MESFET.

Examination of the designs developed for high speed GaAs analogue sampled data circuits reveals that they too rely extensively on current-processing and utilise a range of high quality fast current-based building blocks. This Chapter presents these building blocks and some of their applications. We include current mirrors (Section 8.3), single-ended to differential converters and differential to single-ended converters (Sections 8.4 and 8.5), a linearisation technique leading to a high quality transconductor (Section 8.6), operational transconductance amplifiers (Section 8.7), and buffer amplifiers (Section 8.8). Section 8.9 covers applications ranging from switched capacitor filters, microwave amplifiers and circuits for optical communication systems.

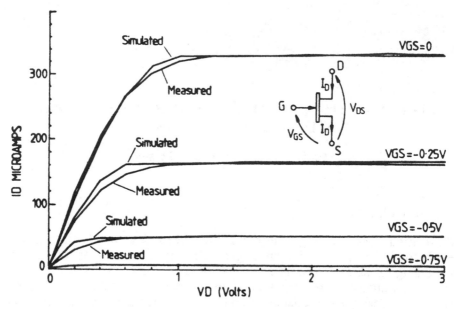

Figure 8.1   Measured and simulated characteristics for typical GaAs MESFET

8.2   Technology, Devices and Simulation

Circuit design techniques are heavily influenced by the fabrication technology involved. High speed GaAs technology is generally regarded as a challenge for circuit designers, principally on account of the lack of a P-channel device, low device gain and, in general, limitation to depletion mode devices. The most generally adopted active device in GaAs technology is the depletion mode N-channel Metal Semiconductor FET (MESFET) which can be produced with a relatively high degree of uniformity and which has a typical gate length in the range 0.5 to 1 micron. Gate widths vary between a minimum of about 10 microns and several hundred microns.

In view of the cost of GaAs technology and the time taken for processing, it is even more important than for other technologies that computer simulation forms the central component of circuit design. In this work, the circuit simulators used are SPICE 2G6 and HSPICE. In the case of SPICE, the GaAs MESFET is represented by the model for a silicon JFET with appropriately chosen parameters. For HSPICE, on the other hand, three levels of specific MESFET model are available. The parameters for these models are extracted from measurements on actual devices.

Measured characteristic curves for a 100 micron gate width MESFET produced by a typical 1 micron (gate length) GaAs process [2] are shown in Figure 8.1. The parameters given in Table 8.1 for a SPICE JFET model were obtained by optimisation (using TECAP) to give a best fit to the measured curves, as indicated by 'simulated' in Figure 8.1. The ideal JFET

Table 8.1 Key parameters for JFET model of GaAs MESFET

Parameter Name	Value	Units
VTO	-1	V
BETA	0.067E-3	mAV-2
LAMBDA (LF)	0.06	V-1
LAMBDA (HF)	0.3	V-1
RD	2920	Ohms
RS	2920	Ohms
CGS	0.39E-15	F
CGD	0.79	F
PB	0.79	V
IS	0.075E-15	A

Table 8.2 HSPICE MESFET parameters

Parameter	Value	Parameter	Value
Level	2	NI	1.45E10
BETA	0.067E-3	XTI	3
LAMBDA	0.3	AF	1
VTO	-1.0	RD	2920
EG	1.16	RS	2920
GAP1	1108	CGD	0.39E-15
GAP2	7.2E-4	CGS	0.39E-15
IS	0.075E-15	PB	0.79
N	1		

model assumes that saturation of Id occurs at Vds = Vgs - VT, whereas actual devices can exhibit 'early saturation' at a lower voltage than this. It can be seen from Figure 8.1 that, for Vgs = 0, the saturation region begins when Vds equals about 1 V, which is the device threshold voltage. Thus these devices do not exhibit the 'early saturation effect' of some GaAs processes [3]. Since the curves in Figure 8.1 are obtained by a semiconductor parameter analyser and are therefore low frequency measurements, the gradient in the saturation region does not reflect the high frequency output conductance (go) which can be many times higher due to dispersion effects [4], as indicated by the LF and HF values for LAMBDA in Table 8.1. As a result working device gain (transconductance/output conductance) figures as low as 20 are typical. Values of Gate-Source and Gate-Drain capacitances are derived by C-V measurements and by optimisation of the model parameters to fit measured s- parameter data. The parameters for the more sophisticated level 2 HSPICE model, which will be used in some sections or this Chapter, are given in Table 8.2. This model is based on a cubic expression for the dependence of drain current on gate source voltage, rather than the simple square-law charcteristic of the classical SPICE FET model.

The modelling of GaAs MESFET's is at a relatively immature stage compared with other technologies. Moreover, correct modelling is vitally important for successful design of linear, high frequency circuits and systems. Consequently, the development and utilisation of improved models forms an important component of on-going work in this field.

8.3   Current Mirrors

8.3.1   Classification of Current Mirrors

There are many different types of current mirror and we begin by considering some definitions and terms commonly used. Current mirrors are firstly categorised as positive or negative. Positive current mirrors have an output in the form of a current sink (quiescant output current positive). Thus they require a power supply which is at a relatively negative potential. Negative current mirrors have a current source output (quiescant output current negative) and require a more positive power supply.

Current mirrors may also be classified according to the orientation of the output signal current. If the output current variation has the same orientation as the input current variation, the mirror is described as non-inverting. If the orientations are opposite, then the mirror is inverting.

Finally, it is possible to combine voltage following and current following (mirroring) in a single component [5]. This component is referred to as a voltage-following current mirror. Such circuits can be regarded as special cases of the 2nd generation current conveyor CCII in which the voltage following input port Y is connected to the current following output port Z - see Chapter 3.

In a technology which provides high quality complementary devices, such as state of the art bipolar (see Chapters 6, 15 and 16) or CMOS (Chapters 5, 6, 7, 11, and 12), positive current mirrors may be realised simply using a pair of N-type (or NPN) devices. Negative current mirrors can be realised using a pair of P-type (or PNP) devices.

In a technology which does not provide complementary devices, the realisation of the negative current mirror has proved difficult. For example, in the past, bipolar technology was unable to provide high quality PNP devices. A circuit realisation which eliminated the need for high quality PNP transistors was proposed by Gilbert [6] but the circuit required identical input and output voltages to operate correctly and so level shifting circuitry was required, leading to increased complexity. An alternative scheme described by Barker and Hart [7] was based on the introduction of a local operational amplifier but this limited high frequency performance.

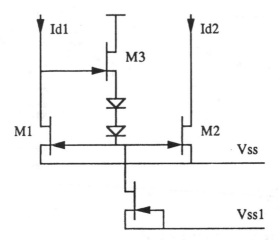

Figure 8.2 Positive GaAs current mirror

For GaAs technology using N-channel MESFET's the negative current mirror is the most useful because there is a requirement for a current mirror to act as a load for N-channel driver devices. The difficulty in realising fast negative current mirrors in NPN bipolar technology may seem discouraging to the development of current mirrors for GaAs technology where devices are limited to N-channel. However, we shall show that in the case of GaAs technology we can in fact utilise the special properties of the depletion mode N-channel MESFET devices to realise high performance negative current mirrors.

8.3.2   Positive Non-inverting Current Mirror

Figure 8.2 shows an existing positive non-inverting current mirror for GaAs technology [8]. Let us approximate GaAs MESFET drain current in the saturation region by a general expression of the form,

$$Id = W \, F(Vgs) \qquad (8.1)$$

where W is gate width, Vgs is the gate-source voltage and F is a non-linear function independant of Vds (The effect of Vds variations will be included later in computer simulations). Referring to Figure 8.2, the subsidiary feedback loop making use of the source follower M3 and diode chain maintain the gate of M1 at an appropriate voltage to absorb the incoming current Id_1.

Thus, the drain current in M1 is equal to the input current and generates a gate-source voltage Vgs_1 according to Equation 8.1,

$$Id_1 = W_1 \, F(Vgs_1) \qquad (8.2)$$

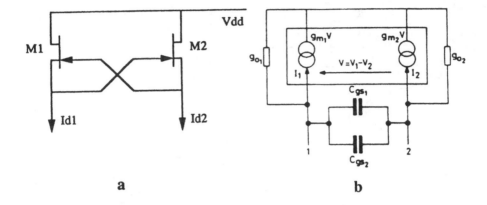

Figure 8.3   Negative GaAs current mirror
a Circuit   b Model

This gate source voltage is applied directly to the gate-source port of M2, yielding an output current

$$Id_2 = W_2 \, F(Vgs_1) = (W_2/W_1) \, Id_1 \qquad (8.3)$$

Equation 8.3 shows that this mirror has a linear current transfer characteristic even for large signals. It has been used successfully as part of a sophisticated commercial GaAs operational amplifier circuit having a gain-bandwidth product of 500 MHz [8]. However, we shall show later that by avoiding the subsidiary feedback loop, even better high frequency performance is possible.

8.3.3   *Negative Inverting Voltage-following Current Mirror*

8.3.3.1   *Basic Analysis*

Figure 8.3a shows a pair of N channel GaAs MESFET's with gates and sources cross-coupled [2,9]. The operation of the circuit may be explained using the small-signal equivalent circuit in Figure 8.3b. Here, the controlled sources of the small-signal models for the two MESFET's are shown in a central block with the (parasitic) passive components outside. Analysis of the central block leads to the transmission matrix description

$$\begin{vmatrix} V_1 \\ I_1 \end{vmatrix} = \begin{vmatrix} 1 & -1/gm_2 \\ 0 & gm_1/gm_2 \end{vmatrix} \begin{vmatrix} V_2 \\ I_2 \end{vmatrix} \qquad (8.4)$$

The term gm_1/gm_2 in the transmission matrix is the current gain of the mirror, showing it has an inverting property when we take into account the transmission matrix sign convention. The unity term in the matrix embodies the voltage following property. The term $-1/gm_2$ is an undesired term and its effect can be evaluated by considering the input impedance of the mirror.

If port 2 is terminated in a load impedance ZL_2, the input impedance at port 1 is given by

$$Zi_1 = (gm_2/gm_1)ZL_2 - 1/gm_1 \qquad (8.5)$$

Similarly, for a load impedance ZL_1 at port 1, the input impedance at port 2 is given by

$$Zi_2 = (gm_1/gm_2)ZL_1 + 1/gm_2 \qquad (8.6)$$

Provided the load impedances ZL_1 and ZL_2 are much greater than $1/gm_1$ and $1/gm_2$, the input impedance at each port is proportional to the load impedance at the other port. These conditions for the proportional impedance property, or impedance transforming property, are generally satisfied in the circumstances where it is wished to utilise the current mirror, as we shall show in the applications of the current mirror to be presented later.

The impedance transforming property of the current mirror is of crucial significance for the design of high speed circuits. Although the input and output nodes of the current mirror are physically seperate, the fact that parasitic capacitance at one node may be equivalently transferred to the other node means that they can be regarded in some respects as a single node. When the current mirror is used in a feedback loop, as is frequently required, this single-node feature minimises the number of poles introduced leading to increased stability margins and optimum dynamic response.

Turning to the large signal behaviour of the current mirror, we now assume a simple non-linear expression for MESFET drain current, in the saturation region, of the form

$$Id = (KW/2) [Vgs - Vt]^2 \qquad (8.7)$$

where K is a transconductance factor and Vt is the threshold voltage. The parameters K, Vt and gate length are assumed to be the same for all MESFET's in a circuit and channel length modulation is omitted at this stage. Assuming that both MESFET's in Figure 8.3a are biassed in the saturation region, the drain currents can be described by

$$Id_1 = Idss_1 (Vgs_1/Vt - 1)^2 \qquad (8.8)$$

and

$$Id_2 = Idss_2 (Vgs_2/Vt - 1)^2 \qquad (8.9)$$

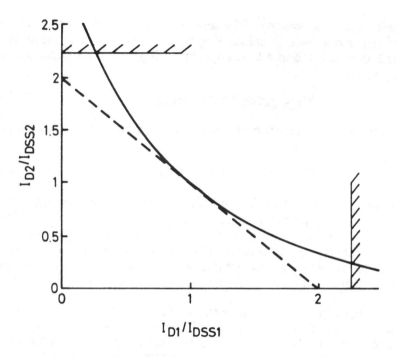

Figure 8.4   Large-signal current transfer characteristic

where Idss = $(KW/2)Vt^2$ is the drain current for zero gate source voltage. The cross-coupled connection imposes the constraint Vgs_1 = - Vgs_2 and therefore we obtain

$$\frac{Id2}{Idss2} = \frac{Id1}{Idss1} + 4\left[1 - \sqrt{\frac{Id1}{Idss1}}\right]$$

(8.10)

Figure 8.4 shows $Id_2/Idss_2$ plotted against $Id_1/Idss_1$. The currents for which the device Schottky diodes become forward biassed (Vgs = 0.5 V) are 2.25 Idss for a typical Vt of -1V, and these limits are shown in Figure 8.4. Although the current transfer characteristic in Figure 8.4 is not linear, it is sufficiently close to the linear curve (shown dashed) for many applications, as we shall show in Section 8.7.

The above analysis of the basic current mirror in Figure 8.3a has ignored the MESFET output conductances go which are very significant for GaAs technology (typical go = gm/20). This parameter will seriously affect the current transfer accuracy and the port impedances of the current mirror. In order to obtain acceptable performance, we introduce cascoding techniques which will now be described.

8.3.3.2 Cascode Current Mirror Using a Seperate Bias Chain

Cascode circuit techniques have an overiding effect on the voltage gains of operational amplifiers using GaAs MESFET's. Without cascoding, the low device gain for GaAs MESFET's (gm/go) of typically 20 leads to amplifier gains of about 10. The introduction of single cascode techniques will lead to gains of the order of 100. Double cascode techniques will provide gains of the order of 1000. For applications in analogue signal processing systems, such as switched capacitor filters [10,11], gains of 1000 are generally required necessitating double cascode techniques. Special circuits which have the property of finite gain insensitivity [12], can tolerate amplifier gains of 40 dB produced by single cascode techniques. Since the current mirror is an important building block for the design of operational amplifiers, as well as other analogue components, we need to be able to realise both single cascode and double cascode current mirrors. Looked at it from an impedance point of view, the small-signal output (drain) conductance of a single MESFET, or basic current mirror as in Figures 8.2 or 8.3a, is of the order of go. For a single cascode MESFET or current mirror, this conductance is reduced to a figure of the order of go^2/gm, and for a double cascode MESFET or current mirror, we obtain the much lower figure of go^3/gm^2.

The voltages on the gates of cascode transistors must be defined in such a way that the current mirror devices are maintained in the saturation region as far as possible. In general, the gate voltage bias can be provided in two ways, namely by using level shifting diodes or by using a 'double level shifting' technique which relies upon the ratioing of device gate widths [2]. The double level shifting approach has the advantage that both the voltage level shift introduced and the voltages required to keep the devices in saturation scale proportionally to the device threshold voltage Vt making circuit performance less sensitive to process variations. In this paper, we shall adopt the double level shifting technique.

In order to understand the double level shifting biassing technique, we begin by considering the following simple arrangement. A GaAs MESFET MA has its gate connected to its source so that its current is equal to its Idss. Its drain is connected to the source of a second MESFET MB such that both devices have the same drain current. Using the expression in Equation 8.7, it may be shown that the Vgs of MESFET B is given by

$$V_{gsB} = V_t [1 - \sqrt{(WA/WB)}]$$

(8.11)

By making the width of MESFET MB greater than that of MA, we obtain a negative Vgs for MB which amounts to a positive level shift between gate and source.

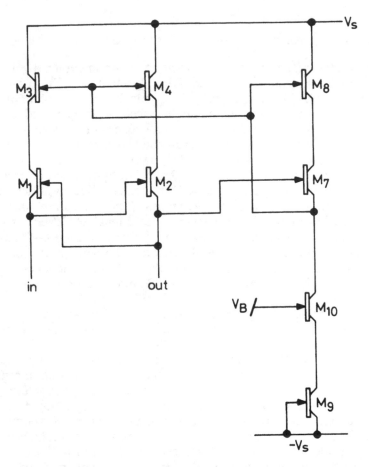

Figure 8.5   Single cascode current mirror using seperate bias chain

In double level shifting, we effectively use a cascade of two such level shifts in order to maintain a MESFET current source in the saturation region. In the absence of 'early saturation', this requires a minimum Vds, and hence a minimum level shift, equal to Vt. Thus each level shift stage requires a Vgs equal to Vt/2, which can be met, from Equation 8.11, by using a minimum device width ratio WB/WA of 4.

A single cascode current mirror using the double level shifting biassing technique is shown in Figure 8.5. M1 and M2 comprise the basic current mirror   and M3 and M4 are the cascode transistors. A seperate bias chain principally comprises source follower M7 fed by a single cascode current source M9 with cascoding MESFET M10. The gate width of M7 is made greater than that of M9, and M3 and M4 have gate widths larger than M1 and M2, using practical gate width ratios of about 5. The combined level shifts produced by M7 and M3/M4 are sufficient to maintain M1 and M2 in the

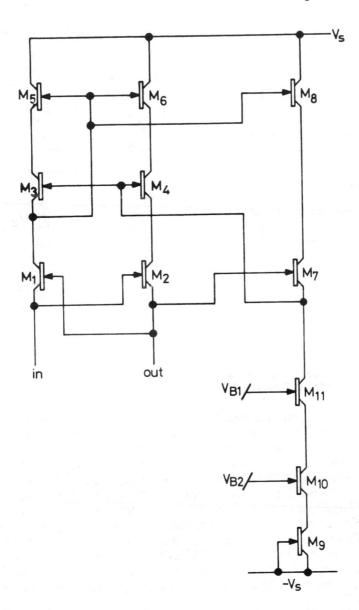

Figure 8.6   Double cascode current mirror using seperate bias chain

saturation region, as required. MESFET M8 bootstraps the drain of M7, improving the source follower performance. Although this double level shifting circuit technique is intended to tolerate devices which do not exhibit early saturation [2], it is still advantageous for devices with early saturation, since it allows device width ratios to be be reduced below the above figures.

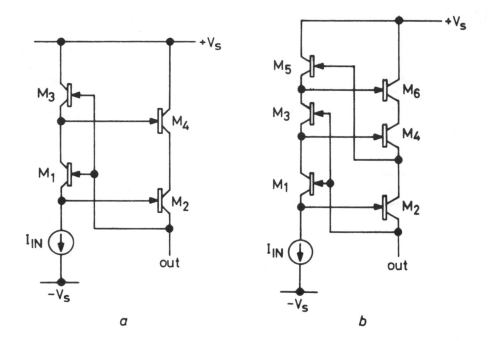

Figure 8.7 Cascode current mirrors
a **Single cascode   b Double cascode**

For the realisation of a double cascode current mirror, another pair of cascode MESFET's must be introduced above M3 and M4 in Figure 8.5. In principle, the gates of these MESFET's could be biassed via a second seperate bias chain fed from the source of M7. However, such a current mirror has an excessive phase shift which leads to relatively large settling times when used to realise amplifiers. This might be attributable to the presence in the circuit of large loops encompassing relatively large numbers of devices. For the double cascode current mirror, therefore, the upper pair of cascode MESFET's are biassed in an alternative way as indicated in Figure 8.6. This more direct method of biassing MESFET's M5 and M6 from the source of M3, which also provides bias for the gate of M8, allows the design of high gain operational amplifiers with very fast settling times, as we shall show in Section 8.7.

8.3.3.3 Economical Cascode Current Mirrors

We have shown that cascode current mirrors can be developed using a seperate bias chain to bias the gates of the cascode MESFET's. The presence of the additional bias chain increases the power consumption of the current mirror and also increases the chip area requirement. In fact, it is possible to

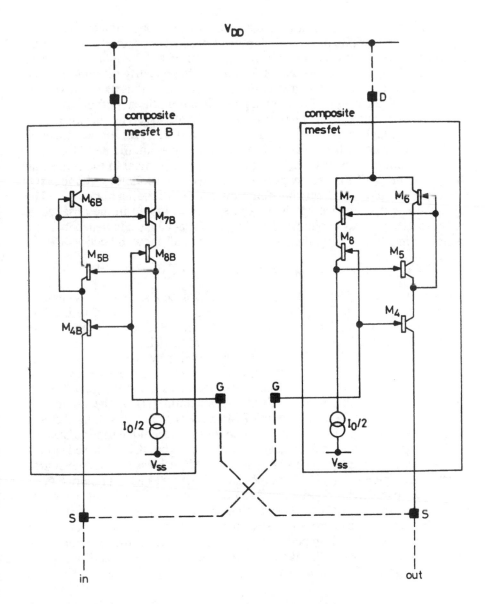

Figure 8.8 Current mirror using composite MESFET's

develop cascode current mirrors without additional bias chains in which the cascode MESFET's themselves provide the required voltages for their gates.

A single cascode current mirror based on this idea is shown in Figure 8.7a [13,14]. MESFET M3 is biassed from M1 using self-bootstrapping [3]. Since the drain-gate voltage of M1 is equal to the negative Vgs for a single

MESFET (M3), it follows from Equation 8.11 that in order to achieve a Vgs equal to Vt, an infinite device width ratio is needed. In practice, we achieve an acceptable approximation to full saturation of M1 by making the gate width of M3 about 10 times that of M1 (assuming no early saturation). The width of M4 is made much greater than that of M2 so that its negative Vgs assists the negative Vgs of M3 in keeping the more critical output MESFET M2 fully saturated. The realisation of a double cascode current mirror using this idea is shown in Figure 8.7b [2,9,13]. Additional cascode MESFET's M5 and M6 are biassed in a similar way as are M3 and M4.

The current mirrors of Figure 8.7 are highly economical in the sense that the cascode MESFET's are performing two functions simultaneously, namely a double level shifting function and a cascoding function. The disadvantage that device M1 is not quite fully saturated in the absence of early saturation leads to increased drain-gate capacitance and hence increase in the settling time of amplifiers realised, as we shall show in Section 8.7.

8.3.3.4 Current Mirror Using Composite MESFET's

The current mirrors described in Section 8.3.3.2 used a seperate bias chain in order to allow every MESFET to be properly maintained in saturation, which leads to optimum performance. In this Section, we present an alternative method of achieving the same aim, using the concept of a composite MESFET [15]. We illustrate the development of a current mirror using composite MESFET's for the double cascode case. Consider MESFET M4 in Figure 8.8. Its gate and source are treated as being the gate (G) and source (S) of a composite MESFET. The drain (D) of the composite MESFET is connected to the drain of M4 via cascode MESFET's M5 and M6, which are biassed using the double level shift technique used to bias the current mirror in Figure 8.6. The gate of M5 in Figure 8.8 is biassed from a seperate bias chain as was M4 in Figure 8.6. M6 is biassed from the source of M5 as in Figure 8.6, and M7 has the same function as M8 in Figure 8.6.

Having developed a double cascode composite MESFET with very low output conductance, the next stage is to combine two such composite devices according to the basic cross-coupled current mirror architecture of Figure 8.3a to form a fast high impedance current mirror, as shown in Figure 8.8. In this current mirror, all devices are properly saturated, facilitating the design of high performance pushpull operational amplifiers, as we shall see in Section 8.7.

8.3.4 Linear negative current mirror

In this Section, we combine two of the negative current mirrors of Figure 8.3a, to form the new current mirror in Figure 8.9. We now assume the general MESFET drain current expression given by Equation 8.1.

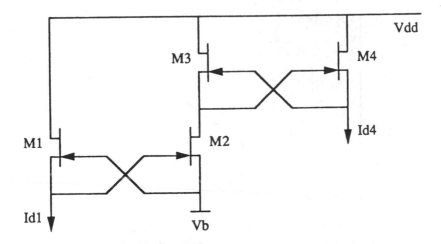

Figure 8.9 Linear negative GaAs current mirror

Referring to Figure 8.9, the drain current in M1 is equal to the input current and generates a gate-source voltage Vgs1 according to Equation 8.1 of

$$Id_1 = W_1 \, F(Vgs_1) \qquad (8.12)$$

This gate source voltage is inverted by the first current mirror and applied to M2. Since the drain current in M3 is equal to that in M2, its gate-source voltage is the same. Finally the gate-source voltage of M3 is inverted and applied to M4. Hence, the output current of the circuit is given by

$$Id_4 = W_4 \, F(Vgs_1) = (W_4/W_1) \, Id_1 \qquad (8.13)$$

Thus, we see that, although this current mirror is formed from two non-linear current mirrors, it has a linear transfer characteristic, even for large signals. Furthermore, the new mirror does not have the voltage following property of the mirror in Figure 8.3a and it is non-inverting.

8.3.5 *Current mirror performance evaluation*

We now compare the performances of the three basic, or non-cascoded, current mirrors in Figures 8.2, 8.3a and 8.9 by computer simulation using HSPICE with the level 2 MESFET parameters of Table 8.2. All devices had gate widths of 16 microns, which is a typical minimum gate width, and bias voltages were chosen to maintain MESFET drain-source voltages at around 3 V.

The DC transfer characteristics are shown in Figure 8.10. Curve 1 illustrates the predicted linearity of the current mirror of Figure 8.2 and curve 2 the non-linearity of the inverting current mirror of Figure 8.3a,

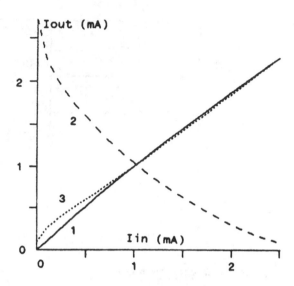

Figure 8.10   Current mirror DC current transfer curves
Curve 1 - Figure 8.2;   Curve 2 - Figure 8.3a; Curve 3 - Figure 8.9

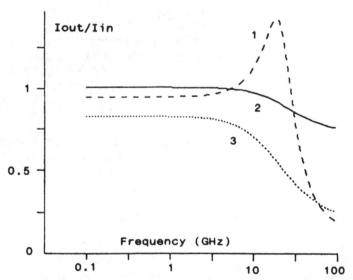

Figure 8.11   Current mirror output current/frequency responses
Curve 1 - Figure 8.2; Curve 2 - Figure 8.3a; Curve 3 - Figure 8.9

which was expected from the theoretical plot of Figure 8.4. Curve 3, for the new current mirror of Figure 8.9, confirms its linear operation.

The simulated frequency responses of the three current mirrors are compared in   Figure 8.11, for short circuit conditions.   These results illustrate the expected tendency towards instability of the mirror in Figure 8.2, which was expected on account of the presence of an internal feedback loop. The new mirror of Figure 8.9 has the excellent stability and bandwidth of the mirror in Figure 8.3a but with slightly lower current gain due to the additional stage. The current mirrors will be used later as versatile, high performance, building blocks.

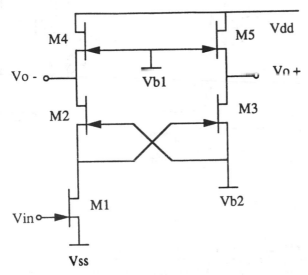

Figure 8.12   Single-ended to differential converter

8.4   Single-ended to Differential Converter

Differential mode circuits have the desirable features of reduced DC offset voltages and improved power supply rejection ratio compared with single-ended circuits. Thus there is a requirement to be able to convert a signal from single-ended mode to differential mode and vice versa.

A circuit which can be used to generate complementary signals from a single unbalanced signal is shown in Figure 8.12. The circuit, in which all devices have the same gate width, makes use of the non-linear inverting negative current mirror of Figure 8.3a. Its operation can conveniently be explained using the MESFET drain current expression of Equation 8.1. The input voltage in Figure 8.12 defines the Vgs of M1. Since M2 and M4 have the same drain current as M1, it follows that Vo- is equal to -Vin and that the Vgs of M2 is equal to Vin. The Vgs of M2 is inverted by the current mirror when applied to M3 and since M5 shares the same drain current as M3, it follows that Vo+ is equal to Vin. This circuit achieves a linear voltage-to-voltage transfer characteristic, inspite of the non-linear voltage-to-current transfer chracteristic of the MESFET and the non-linear current-to-current transfer characteristic of the current mirror used.

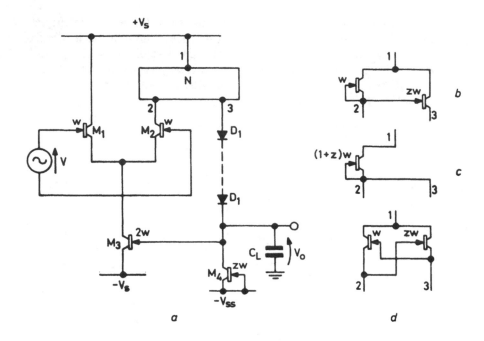

Figure 8.13   Differential to single-ended converter
a Basic architecture b Subcircuit using source follower
c Subcircuit using current source d Subcircuit using current mirror

8.5   Differential to Single-ended Converter

A general architecture for a differential to single-ended converter circuit is shown in Figure 8.13a [16,17]. The network 'N' is designed to ensure that the drain current of MESFET M2 is constant. Thus the small-signal currents in M1 and M3 are the same, leading to an overall voltage gain for the converter of the order of 0.5. This gain is acceptable for many applications, but a modification to allow higher gains has been presented [17]. For the network 'N' in Figure 8.13a, three configurations have been investugated and these are shown in Figures 8.13b, c and d.   Figure 8.13b, used in converter configuration 1, is basically a current source connected MESFET and a source follower. Figure 8.13c, used in configuration 2, is simply a larger current source. Configuration 3 makes use of the basic voltage following current mirror of Figure 8.3a.

The simulated gain responses of the three configurations are shown in Figure 8.14 together with MESFET sizing and circuit conditions. It can be seen that configuration 3, using the voltage following current mirror, avoids the undesirable tendancy towards instability and peaking of configuration 1, while providing a higher gain than configuration 2. The greater stability factor is attributed to the 'single node', or impedance transforming,

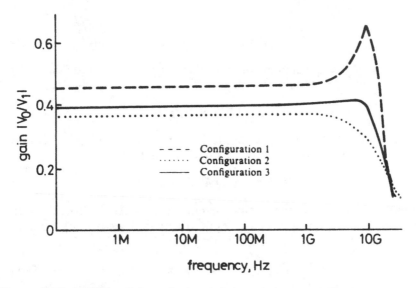

Figure 8.14 Differential-to-single ended converter amplitude responses
Gate width W=50 μm; Z=2; CL=39 fF (=typical Cgs);Vs=5V,Vss=7.5V

property of the current mirror. This property gives the added benefit that the gain is less dependent on the geometries of the devices, giving considerably greater freedom of design. Although the current mirror used has a non-linear current transfer characteristic, its non-linearity does not significantly affect the converter linearity because the mirror is within an internal feedback loop.

The differential to single ended converter using the current mirror can be used in conjunction with single ended input amplifiers, of the type to be described, for the realisation of high performance composite amplifiers with differential input capability [13,17].

8.6 Linear Transconductors

8.6.1 Linearisation Principles

We see in Chapters 4, 9, 10 and 17 that the linear transconductor, or voltage controlled current source, is a very powerful circuit building block for high frequency applications, including filters, amplifiers and components for optical communication systems. We now present a general approach for the design of linear transconductors using GaAs MESFET's, which is a generalisation of the approach in [18] developed for MOS technology.

We begin by assuming that the GaAs MESFET drain current in the saturation region can be described by a general expression of the form of Equation 8.1. Consider now the current function denoted Io which is defined as

$$Io = W F(Vgs) - W F(-Vgs) \qquad (8.14)$$

We now express F as the sum of components which are even and odd in Vgs, which we denote as Fe and Fo. We thus obtain

$$Io = 2 W Fo(Vgs) \qquad (8.15)$$

For the system to be linear, Io should be proportional to Vgs. Denoting the constant of proportionality Gm, this requires that

$$Fo(Vgs) = (Gm/2W) Vgs \qquad (8.16)$$

The simplest model for the drain current of a GaAs MESFET in the saturation region ignoring channel length modulation is given by Equation 8.7, which may be written in even and odd parts as

$$Id = (KW/2) [Vgs^2 + Vt^2] - K W Vt Vgs \qquad (8.17)$$

The odd part of this expression satisfies the requirement of Equation 8.16 and gives a transconductance of

$$Gm = 2 K W (-Vt) \qquad (8.18)$$

where -Vt is positive for depletion mode devices. We now consider implementation of GaAs transconductors based on this linearisation principle.

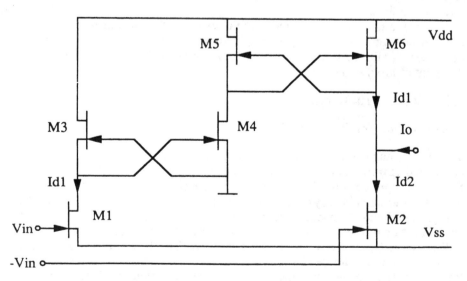

Figure 8.15 GaAs transconductors using non-inverting current mirror

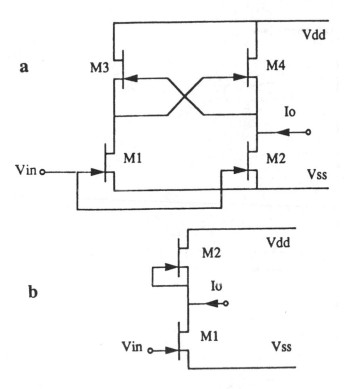

Figure 8.16 Alternative GaAs transconductors
a Using inverting current mirror b Classical inverter-type circuit

8.6.2 *Linear Transconductor Implementation*

8.6.2.1 *Circuit Using Non-inverting Current Mirror*

Consider the circuit in Figure 8.15. M1 and M2 are driver MESFET's and M3 to M6 comprise the linear non-inverting current mirror of Figure 8.9. This arrangement implements the required relationship in Equation 8.14 because the output current is the difference Id1 - Id2 and M1 and M2 are driven by complementary Vgs inputs. This arrangement has the disadvantage that linearity depends on the input signals being truly complementary, which may be difficult to achieve in practice. The circuit in Figure 8.12 could in principle be used to generate these signals but, in fact, a much more attractive transconductor circuit is possible.

8.6.2.2 *Transconductor Using Inverting Current Mirror*

We now consider the transconductor architecture in Figure 8.16a. The input voltage is applied directly to MESFET's M1 and M2. MESFET M3 shares the same drain current as M1 and, provided its gate width equals that of M1

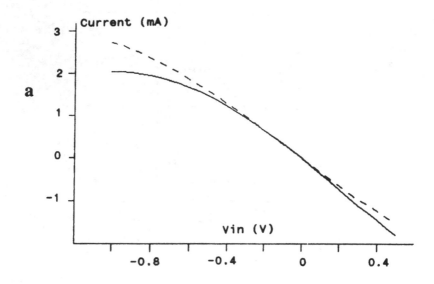

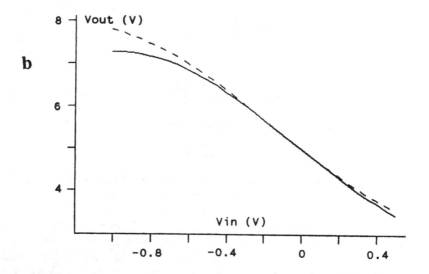

**Figure 8.17   Simulated transconductor DC transfer characteristics
a Short-circuit load   b RL = 1144 Ohm**

(W3 = W1), its Vgs equals Vin. MESFET M4 is cross-coupled to M3, and so
the Vgs of M4 is equal to -Vin. Assuming that the gate widths for M2 and
M4 are equal (W4 = W2), the output current Io is the difference Id2 - Id4,
which satisfies Equation 8.14, the fundamental requirement for linearity,
with W equal to W2.

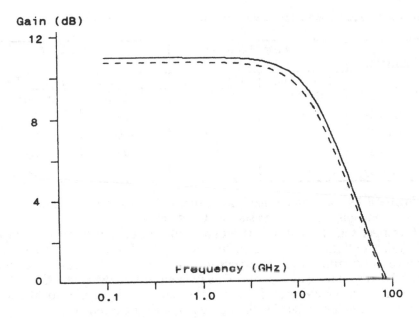

Figure 8.18   Transconductor frequency responses
_____ **Figure 8.16a   - - - - Figure 8.16b**

An alternative, but less elegant way of viewing this circuit, is that it permits the non-linear current transfer characteristic of the current mirror (Equation 8.10) to compensate for the non-linearity of the MESFET voltage-to- current transfer characteristic (Equation 8.7).

Figure 8.16b shows a conventional inverter circuit comprising a MESFET M1 fed from a current source connected MESFET M2. This circuit will be used as a benchmark for evaluation of the transconductor in Figure 8.16a. For comparison, the gate widths are the minimum 16 microns for the circuit in Figure 8.16a and 32 microns for the inverter in Figure 8.16b, required in order to achieve the same transconductance because the inverter does not realise a push-pull output.

Figure 8.17a shows the simulated DC current transfer characteristics of the trancoductors in Figure 8.16. HSPICE was used with the level 2 MESFET parameters of Table 8.2. The outputs are short circuit and the input voltage ranges from cut-off (Vgs = Vt = -1V) to Schottky breakdown point (Vgs = 0.5V).

Figure 8.17b shows a comparison of the simulated voltage-to-voltage transfer characteristics of the same two transconductors for a load resistance of 1144 Ohm, giving a gain of approximately 11 dB. The results in Figure 8.17 show that even for a fairly sophisticated MESFET model, the new transconductor achieves greater linearity. The small signal frequency responses of the two transconductors in 11 dB gain configuration are shown

Table 8.3   Transconductor distortion figures (Vin = 0.1 Vpeak)

Harmonic	Fo = 100MHz		Fo = 1 GHz	
	Fig. 8.16b	Fig. 8.16a	Fig. 8.16b	Fig. 8.16a
Fo (Vpeak)	0.3294	0.3511	0.3290	0.3503
2Fo (dB)	-37	-53	-37	-47
3Fo (dB)	-60	-62	-60	-61
4Fo (dB)	-99	-93	-98	-90
THD (%)	1.3	0.2	1.3	0.4

in Figure 8.18, confirming that both transconductors have a similar small signal bandwidth for a load resistance of 1144 Ohm.

A comparison of simulated distortion results for the transconductors in 11 dB gain configuration is shown in Table 8.3. The new transconductor achieves a significant improvement in total harmonic distortion (THD) at 100 MHz by a factor of about 6. This is reduced somewhat at 1 GHz. The reason for the greater variation of distortion with frequency for the new circuit is being investigated with improved modelling techniques.

The results in Figures 8.17 and 8.18 indicate that the transconductor in Figure 8.16a achieves improved linearity and lower distortion compared with existing circuits without any discernable penalty in terms of high frequency performance.

8.7   High Gain Operational Transconductance Amplifiers

8.7.1   General

The operational amplifier is a centrally key component for the realisation of many analogue circuits. For high frequency sampled data applications,

including GaAs switched capacitor filters and analogue-to-digital and digital-to-analogue converters [14], the requirements differ in key aspects from those for general purpose amplifiers [8,19-21]. For example, a high output impedance is preferred (current output instead of voltage output) in view of the capacitive loads. Settling time is a very important parameter since it

determines the maximum speed of circuit operation. DC gain must be above specified limits which are typically 60 dB, or 40 dB is permissible if finite gain insensitive switched capacitor circuits are adopted [12].

8.7.2   Circuits Using Diode Level-shifting

A basic structure for a moderate gain inverting amplifier is shown in Figure 8.19a; M1 is the driver transistor and M2 the cascode transistor. Assuming

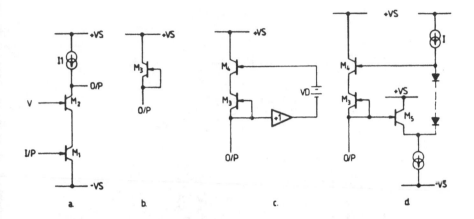

Figure 8.19 Basic amplifier techniques
a Basic inverter architecture b Single MESFET current source
c Cascode current source architecture d Cascode current source

an ideal current source, the voltage gain is $gm_1gm_2/(go_1go_2)$ which would have a typical value of 400. For the realisation of the current source, a single MESFET connected as in Figure 8.19b may in principle be used. However, this would reduce amplifier voltage gain to a low value of approximately gm_1/go_3; the typical low value of gm/go of about 20 makes it necessary to develop a high impedance current source using cascode techniques. This is shown in Figures 8.19c and d. In Figure 8.19c, the drain-source voltage of M3 is kept relatively constant (independent of Vo) by means of the buffer amplifier, DC voltage source (VD) and source follower M4. A realisation of the buffer and DC source is as in Figure 8.19d, using the source follower M5, diode chain and associated current sources.

The realisation of a complete amplifier is shown in Figure 8.20. The devices M1 to M5 are as identified in Figure 8.19d; M6 realises the upper current source in Figure 8.19d; the lower current source in Figure 8.19d is realised in two parts, namely M7 and M8; finally, M9, M10 and M11 are cascode transistors which improve source follower performance. DC gain for this amplifier is typically $(gm/go)^2/4$ (about 40dB). For the bias MESFET's M5 - M11 in Figure 8.20, the gate widths (indicated as factors multiplying the minimum gate width) may be equal to the minimum, ie unity. The gate width in the input/output chain comprising M1 - M4 determines the amplifier transconductance and slew rate and the gate widths of these devices are defined as an amplifier scaling factor k multiplied by the minimum gate width.

SPICE simulation of a unit scaled amplifier using the typical MESFET model parameters of Table 8.1, indicates a DC gain of 41 dB, a gain-bandwidth product of 3.6 GHz and a phase margin of 70 degrees, when used

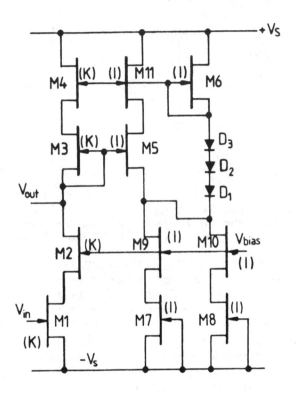

Figure 8.20   40 dB op-amp using diodes; Vs=5V, Vbias=-3V

with an effective capacitative load of 80 fF. The simulated performance figures are given in Table 8.4.

8.7.3   Circuits Using Device Width Ratioing

We illustrate biassing techniques based on device width ratioing by the design of a pushpull amplifier. A basic architecture for a pushpull amplifier is in fact realised by the transconductor architecture already shown in Figure 8.16a. The advantage of this push-pull arrangement is that it is possible to reduce the widths of M1 and M3 below those of M2 and M4 without affecting gain or transconductance, yielding a given transconductance and maximum output current for about half the effective area and power consumption compared with a simple (non push-pull) inverter (Figure 8.16b). However, the gain of the amplifier in Figure 8.16a is only of the order of gm/(2go), which is the same as for the inverter of Figure 8.16b, and it is thus necessary

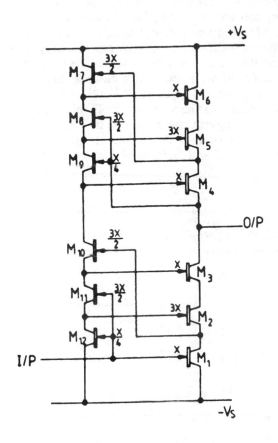

Figure 8.21 High gain double cascode amplifier; Vs=5V

Table 8.4 Operational amplifier performance parameters

Amplifier circuit	Minimum settling time (ps)	Load capacitance (pF)	DC gain (dB)	GB product (GHz)	Phase margin (deg)
Fig. 8.20	200	0.08	41	3.6	70
Fig. 8.21	457	0.8	63	3.1	64
Fig. 8.22	230	0.8	61	3.2	71
Fig. 8.24	215	0.8	65	3.1	70

to introduce double cascode techniques for each MESFET. The circuit in Figure 8.16a is now   considered to consist of two parts; a current mirror consisting of M3 and M4, and a dual driver circuit consisting of M1 and M2. A double cascode version of the current mirror has already been shown in Figure 8.7b. This circuit is also used with modified connections to the gates and sources of M1 and M2 to form the double cascode dual driver required, and the final amplifier is shown in Figure 8.21 [2,13].

The simulated performance of the amplifier in Figure 8.21, with a 0.8 pF load capacitance, using the device parameters as in Table 8.1, gives the key performance parameters tabulated in Table 8.4. It can be seen that the low frequency voltage gain is 63 dB and that a minimum settling time of 457 ps is achieved.

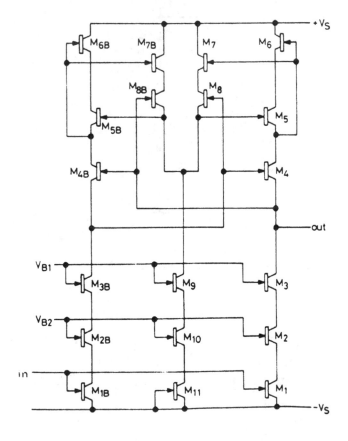

Figure 8.22   Amplifier using current mirror with composite MESFET's (VS=5V;VB1=-2.2V;VB2=-3.6V;all devices 100 μm except M7, M8 and M9 200 μm   and M5, M6, M10, M5B, M6B, M10B 500 μm)

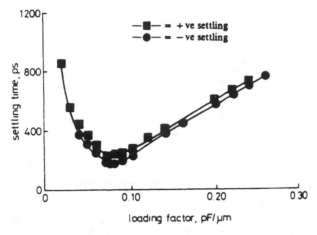

Figure 8.23   Settling time characterisation curve for amplifier in Figure 8.22

8.7.4   *Fast Settling Architectures*

The minimum settling time of the amplifier in Figure 8.21 permits its use in sampled data sytems with maximum switching rates up to 400 MHz [22]. However, a second generation of amplifier architectures with even lower settling times aimed at higher frequency systems is possible, based on both non-pushpull and pushpull architectures [13,15]. We now concentrate on fast pushpull architectures which make use of the previously described current mirrors.

The current mirror using composite MESFET's of Figure 8.8 can be used to realise a fast settling push-pull amplifier and the circuit is shown in Figure 8.22. The current mirror comprises M4 - M8 and M4B - M8B, and also M9 - M11 which implement the required current source. M1 - M3 and M1B - M3B implement the double cascode dual input driver also required. For evaluation of the speed capability of operational amplifiers, a charcterisation technique has been developed in which settling time is simulated as a function of load capacitance. For the amplifier in Figure 8.22, such settling time characterisation curves, for both positive and negative output voltage steps, are shown in Figure 8.23. In order to facilitate use of these curves for high speed circuit design [14,23], the load capacitance on the horizontal axis is divided by the amplifier's driver device gate width and referred to as loading factor. From Figure 8.23, the minimum settling time reaches 230 ps. It can be seen that the circuit features excellent symmetry of positive and negative settling. The simulated performance parameters for this amplifier are summarised in Table 8.4.

The double cascode current mirror using seperate bias chain of Figure 8.6 can be used to realise a high gain, fast settling amplifier as shown in Figure 8.24 with drive circuitry similar to that already utilised in Figure 8.22. The

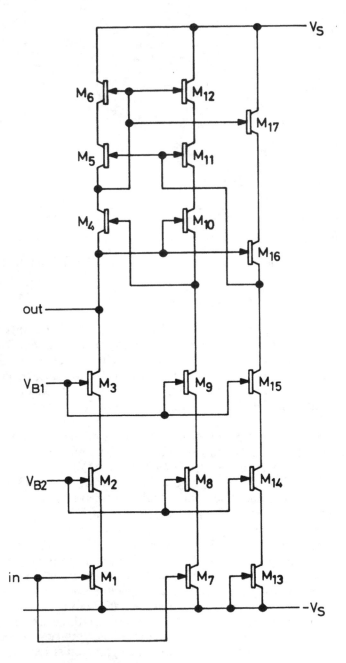

Figure 8.24 Amplifier using current mirror with seperate bias chain (Vs=5V;VB1=-2.2V;VB2=-3.6V all devices 100 μm except M5,M6,M11,M16,M17 500μm)

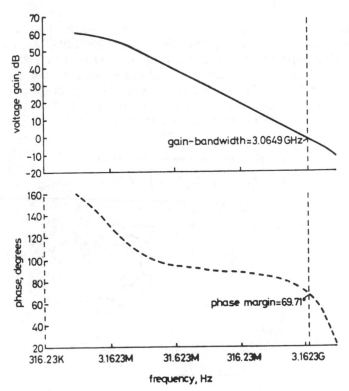

Figure 8.25   Gain and phase versus frequency responses for amplifier in Figure 8.24

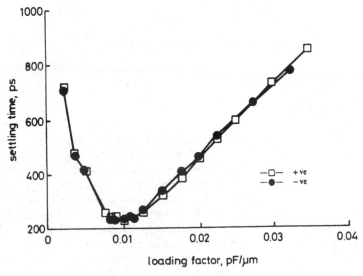

Figure 8.26   Settling time characterisation for amplifier in Figure 8.24

simulated gain and phase versus frequency responses for this amplifier are shown in Figure 8.25. The settling time characterisation curve is shown in Figure 8.26 (Key performance parameters are summarised in Table 8.4). The minimum settling time for positive and negative steps is the lowest achieved figure of 215 ps. The achievement of such fast settling amplifiers is an encouraging necessary step towards the development of switched capacitor filters with switching frequencies approaching 1 GHz.

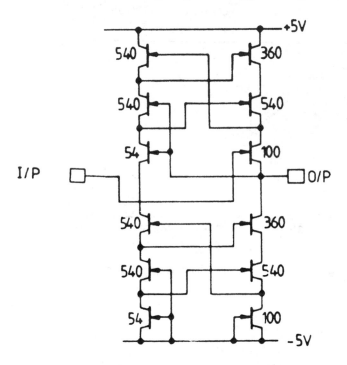

Figure 8.27   High quality unity gain buffer amplifier

8.8   Buffer Amplifiers

In integrated circuit design, on-chip impedance levels are usually much higher than off-chip impedance levels. It is thus necessary to introduce, into the chip output ports, buffer amplifiers which can tolerate off-chip loading impedances. Such buffers usually require a gain of unity, high input resistance, low input capacitance and ability to drive relatively high load capacitances. A suitable circuit for implementation in GaAs MESFET technology is shown in Figure 8.27. It is derived from the high gain double cascode amplifier of Figure 8.21 by a number of operations. We connect the original input terminal to the negative supply and create a new input terminal from the disconnected gate of MESFET M4. This results in a buffer amplifier with a relative gain error from unity of the order of 1/A, where A is the gain of the original amplifier, giving in this case a gain error of 0.1 %. The circuit in Figure 8.27 has a bandwidth of at least 100 MHz driving a signal of 1 Vrms into a 2 pF load.

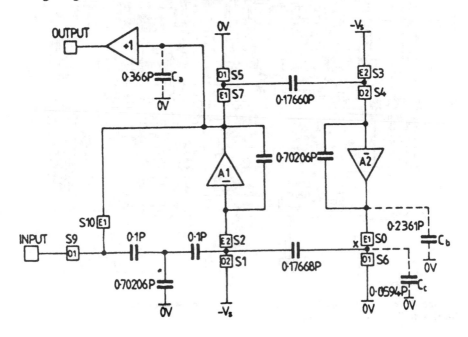

Figure 8.28 2nd order SC filter for implementation

8.9 Applications

8.9.1 Switched Capacitor Filter Example

A 2nd order switched capacitor bandpass filter designed to realise a Q-factor of 16, midband gain of unity and midband frequency of 1/25th of the switching frequency is shown in Figure 8.28 [11,14]. The design maximum switching frequency is 250 MHz using the single-stage, double-cascode, double-level- shifting operational amplifier design of Figure 8.21. When switched capacitor circuits are implemented in GaAs technology, switch control circuits are required to prevent forward biassing of the switch device gate-channel Schottky diodes [14]. In this filter example, the switch control circuit used is that of [24], which features low power consumption and low signal dependence of clock feedthrough. The switched capacitor circuit is designed for high frequency operation using an optimisation technique [11,14] based upon the amplifier settling time characterisation curves. In this approach, amplifier load capacitances are introduced to maintain acceptable settling behaviour and stability in all switching states.

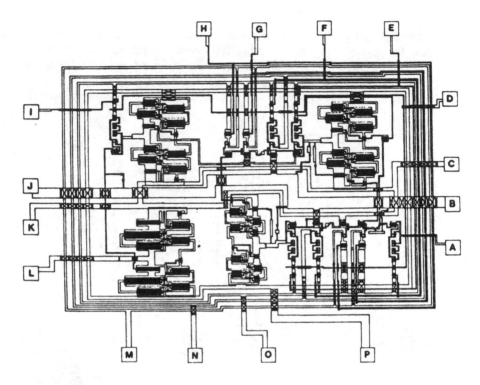

Figure 8.29 Layout plot for GaAs switched capacitor filter; total size 3.3 mm x 2.6 mm

The capacitors Ca, Cb and Cc in Figure 8.28 have been introduced as a result of the optimisation procedure.

The circuit was layed out using MAGIC with a specially written technology file for the Anadigics 0.5 microns GaAs process and a layout plot is shown in Figure 8. 29. The size of the chip, which was fabricated by Anadigics, is 3.3 mm x 2.6 mm and the power consumption is 440 mW.

The filter was found to operate very accurately up to beyond its design maximum switching frequency of 250 MHz. The amplitude/frequency response was measured for switching frequencies of 300 MHz, 350 MHz and 400 MHz. The responses for 300 MHz and 400 MHz are shown in Figures 8.30a and b, respectively. Expanded passband plots for the above switching frequencies are shown in Figures 8.31a and b. Peak gains, -3 dB frequencies and calculated midband frequencies and Q-values are tabulated in Table 8.5. Midband frequency accuracies range from less than 1 % to about 2 % and represent a considerable improvement over the previous results presented in [25]. The Q- factor accuracy is reasonable for a 300 MHz switching frequency but the Q- factor reduces for higher switching frequencies. The peak noise at the output of the filter measured for a measuring bandwidth of 10 kHz is at -75 dB relative to the maximum signal

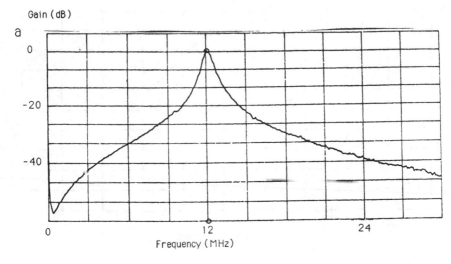

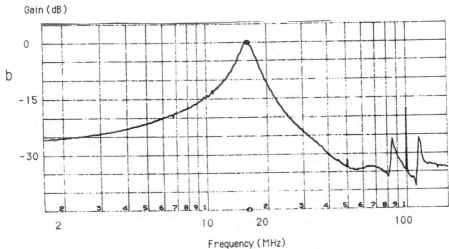

**Figure 8.30   Measured amplitude/frequency responses of filter
a 300 MHz switching frequency   b 400 MHz switching frequency**

Table 8.5   Measured Filter performance data

| Switching Frequency (MHz) | Peak Gain (dB) | -3dB Frequencies | | Fo $=(F1F2)^{1/2}$ (MHz) | Q-factor $=Fo/(F_2-F_1)$ |
		F1 (MHz)	F2 (MHz)		
300	-0.15	11.7	12.5	12.1	15
350	-0.20	13.5	14.7	14.1	12
400	-0.15	14.6	16.9	15.7	7

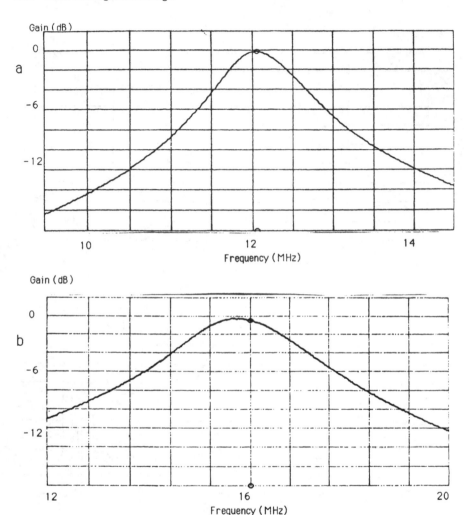

**Figure 8.31   Measured expanded passband responses
a 300 MHz switching frequency   b 400 MHz switching frequency**

level. Observation of the filter output signal at the resonant frequency
indicates a DC offset voltage of less than 50 mV and reasonably low levels of
clock feedthrough.

Inspite of the results achieved, the operational amplifier used in the design
is relatively slow compared with more recent designs, such as those of
Figures 8.22 and 8.24, and advanced designs implemented using state-of-
the-art technology [26]. Thus future opportunities are opened up for SC
systems operating with GigaHertz clock rates.

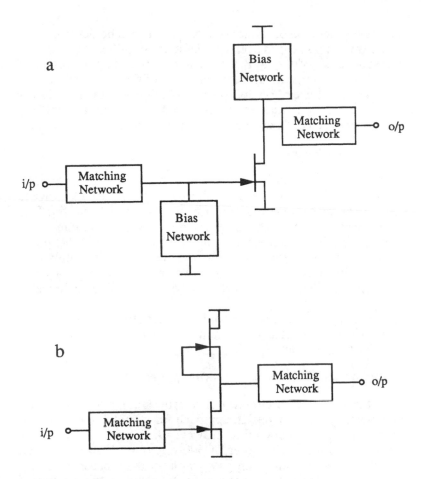

**Figure 8.32   Microwave amplifier architectures
a Single MESFET   b Inverter structure**

8.9.2   *Microwave Applications*

8.9.2.1 *Microwave Amplifiers*

Traditional designs for microwave amplifier stages are based on either the single GaAs MESFET of Figure 8.32a or the classical inverter of Figure 8.32b [27,28].   In either case, the non-linear nature of the Id vs Vgs characteristic of the GaAs MESFET (eg Equation 8.7) introduces a significant   source of non-linearity.   In many microwave systems, high linearity is essential and this means that conventional circuits often have to be operated in a 'backed-off' condition, that is with AC signals which are only a small fraction of the DC quantities [29].

However, a new generation of microwave amplifiers can be developed by replacing the active components in such conventional amplifier stages by the linearised transconductor of Figure 8.16a. The results in Figures 8.17 and 8.18 and also Table 8.3 indicate that this basic amplifier architecture will allow more linear circuit operation and permit AC signals which are larger proportions of the DC quantities resulting in improved signal handling capability, lower distortion and reduced chip area and power consumption.

8.9.2.2 Microwave Combiners

Apart from microwave amplifiers, a number of other microwave applications for advanced GaAs circuit designs are envisaged. Microwave combiners are important components in many microwave systems [30]. The successful achievement of a linearised transconductor as in Figure 8.16a could allow the realisation of combiners in monolithic microwave integrated circuit form by directly interconnecting the outputs of a number of transconductors with seperate input signals. The high input impedance of the transconductors facilitates easy interfacing with preceeding analogue or digital circuitry.

The achievement of low intermodulation distortion in this application may require the introduction of cascode techniques, as described in Section 8.3.3.2, within the transconductors [13].

8.9.2.3 New Approach to Integrated Microwave System Design

The adoption of more complex circuits for microwave applications could bring about a radically different design approach for monolithic microwave integrated systems. In conventional microwave IC design, typical GaAs MESFET gate widths are of the order of 300 microns. However, much smaller devices close to the minimum gate width for the process, typically 16 microns, could have decisive advantages. Such a twentyfold reduction of the scale of the devices and their interconnections in comparison with the signal wavelength would significantly increase the frequency range for which low frequency circuit techniques are useable. Bulky transmission lines for interconnections between components could in many cases be avoided and circuits can be envisaged which consist of 'islands' of high density circuitry, with interconnections between islands being conventional transmission lines.

One interesting possibility for high circuit density microwave systems is to exploit the wide bandwidth capability of current amplification [31], as described in Chapter 4. The linear current mirror in Figure 8.9 could be used as an amplifying component to achieve a good compromise between high gain and high bandwidth beyond that achievable with conventional architectures.

8.9.2.4 Microwave Buffers

Such a novel approach to monolithic microve integrated system design introduces a requirement to interface the high impedance on-chip microwave circuitry with low impedance off-chip microwave systems, thus leading to the concept of a microwave buffer. It is proposed that such a buffer could be realised using the transconductor of Figure 8.16a operating into a 50 Ohm load resistance. By making the output devices M2 and M4 about 100 microns wide, a transconductance of 20 milli-Siemens is achieved giving a unity voltage gain with reasonable linearity and efficiency.

8.9.3   Optical Applications

8.9.3.1   Driver Circuits for Wideband Optical Systems

In optical communication systems, information is contained in the optical signal in the form of frequency modulation or intensity modulation [32]. Both optical frequency and intensity are proportional to the current in a laser diode, which is non-linearly related to laser diode voltage. The requirement for linearity, which applies to digital as well as to analogue systems, makes it desirable that the driver circuit should have an output current which is linear with modulating signal. In present optical communication systems, with typical bandwidths of 0.5 GHz, the driver and detector circuits are realised using bipolar transistors.

The growth of communications is creating a demand for wideband optical systems operating up to around 15 GHz, necessitating the use of III-V semiconductor materials, and in particular Gallium Arsenide [14]. Thus there is a requirement for a range of GaAs building block circuits, not only for driver and detector circuits but also for non-regenerative optical-electronic-optical repeater circuits required while all-optical repeaters [32] are being developed.

Two possible applications of the GaAs circuits developed in this Chapter within the context of optical systems are shown in Figure 8.33. Figure 8.33a shows a proposal for a driver circuit for a laser diode. The current source Ib defines the quiescant diode current. The transconductor is the linearised transconductor developed in Section 8.6 and shown in Figure 8.16a. The high impedance voltage input provides convenient interfacing with both analogue and digital preceeding circuits. The transconductor can be operated with larger signal swings, mininising power consumption and chip area and optimising system performance.

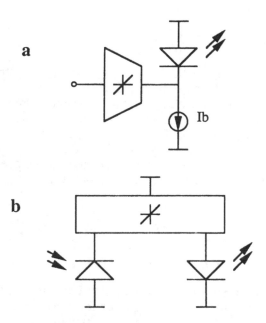

Figure 8.33   Circuits for optical applicatiions
a Linear driver circuit   b Non-regenerative repeater

8.9.3.2   Optical Repeater Circuits

The circuit in Figure 8.33b is a proposed non-regenerative linear repeater. The current mirror block is the linear negative current mirror developed in Section 8.3.4 (see Figure 8.9). The circuit achieves a linear current gain given by Equation 8.13. By making the gate width of M4 ten times that of the other MESFET's (which may be minimum size), a current gain of 20 dB can be achieved which is acceptable in some applications. The quiescant current of the laser diode is dependant on the gate width of M4 and an additional current source may be added. This circuit could be used as a photodetector circuit with the laser diode replaced by a resistor and voltage buffer.

8.9.4   Continuous-time Filtering

Filters play a major role in communication systems. GaAs technology has been used to realise integrated switched capacitor filters at frequencies up to 20 MHz, as described in Section 8.9.1, and it is anticipated that frequencies up to 100 MHz should be possible with this technique. Continuous-time filters realised using transconductors and capacitors implemented in GaAs technology could provide a means of realising integrated precision filters at higher frequencies.

The linearised transconductor developed in Section 8.6. is a candidate component for high frequency filter realisation. However, for filter realisation, a number of transconductor features are necessary which this transconductor does not yet have. These include the possibility of realising a negative transconductance and the electronic control of the transconductance value. The successful achievement of these requirements could lead to the realisation of filters which are viable for frequencies beyond 1 GHz.

8.10 Conclusions

Gallium Arsenide technology is maturing rapidly with the achievement of higher yields, lower costs and faster turnarounds. Current developments in modern wideband communication systems are creating a firm need for circuit and system components realised in this technology. In this Chapter, we have presented some design techniques specifically developed to yield optimum high frequency performance, while overcoming the material technological difficulties of Gallium Arsenide. It is hoped that the building blocks presented in this Chapter, ranging from low level subcircuits such as current mirrors, to experimentally confirmed system components such as high precision filters, will help to bring about the realisation of the possibilities that undoubtedly lie ahead. As in other technologies, the use of current is an essential aspect of Gallium Arsenide circuit techniques if the fastest circuit performance is to be achieved.

8.11 Acknowledgements

The authors gratefully acknowledge the invaluable help of numerous colleagues. In particular, Kevin Steptoe, Heloisa Facanha, Paul Radmore, Ahmad Khanifar and Marco Federighi of University College London (Kevin Steptoe is now with Cadence Ltd), Jack Sewell and Brian Wishart of Glasgow University, and Owen Richards and Graham Johnson of Imperial College. Also, Liza Wong and Steve Newett of the SERC Rutherford Appleton Laboratory, Shiva Carver, David Goldsmith and Steven Hart of Interactive Solutions Ltd, Charlie Suckling and Jim Arnold of Plessey, and Robert Bayruns of Anadigics Inc gave valuable support. The financial support of the SERC is gratefully acknowledged.

8.12 References

[1]   P E Allen and C M Breevoort, An analogue circuit perspective of GaAs technology, Procs 1987 IEEE ISCAS (Philadelphia), May 1987

[2]   C Toumazou and D G Haigh, Design of a high-gain, single-stage operational amplifier for GaAs switched-capacitor filters, Electronics Letters, vol 23, no 14, 2nd July 1987, pp 752-754.

[3]   Larson L E, K W Martin and G C Temes, Comparison of amplifier gain enhancement techniques for GaAs MESFET analogue integrated circuits,Electron Lett, 1986, 22, pp 1138-1139

[4] C Camocho-Penalosa and C S Aitchison, Modelling frequency dependence of output impedance of a microwave MESFET at low frequencies, Electronics Letters, 1985, vol 21, pp 528-529

[5] B L Hart and W J Barker, Modified current mirror with a voltage following capability, ibid, vol 18 1982, pp 970-972

[6] B Gilbert, Wideband negative current mirror, Electronics Letters, vol 11, no 6, March 1975, pp 126-127

[7] W J Barker and B L Hart, Negative current mirror using NPN transistors, ibid, vol 13, 1977, pp 311-312

[8] N Scheinberg, Design of high speed operational amplifiers with GaAs MES FET's, Procs 1987 IEEE ISCAS (Philadelphia), May 1987, pp 193-198

[9] C Toumazou and D G Haigh, Design and application of GaAs current mirror circuits, to be published IEE Procs Part G (CDS), vol 137, no 2, April 1990

[10] E L Larson, K W Martin and G C Temes, GaAs switched capacitor circuits for high speed signal processing, IEEE J Solid State Circs, SC-22,pp 971- 981,Dec 1987

[11] D G Haigh, C Toumazou, S J Harrold, J I Sewell and K Steptoe, Design and optimisation of a GaAs switched capacitor filter, IEEE 1989 International Symposium on Circuits and Systems, Portland, Oregon, USA, June 1989

[12] D G Haigh, A K Betts, K Steptoe and J T Taylor, The design of switched capacitor filter circuits for GaAs MSI technology, Procs 1989 European Conference on Circuit Theory and Design, Brighton (UK), 5th - 8th September 1989

[13] C Toumazou and D G Haigh, Design of GaAs operational amplifiers for analogue sampled data applications, to be published IEEE Trans Circs and Systems, vol 37, no 7 or 8, July or August 1990

[14] D G Haigh, C Toumazou and A K Betts, Chapter 10 in 'GaAs technology and its impact on circuits and systems' (Edited by D G Haigh and J K A Everard, Published by Peter Peregrinus, London 1989

[15] D G Haigh and C Toumazou, Fast settling high gain GaAs operational amplifiers for switched capacitor applications, Electronic Letters, vol 25, no 1, 25th May 1989, pp 734 - 736

[16] C Toumazou and D G Haigh, Level-shifting differential to single-ended convertor circuits for GaAs MESFET implementation, Electronics Letters, vol 23, no 20, 24th September 1987, pp 1053-1055.

[17] C Toumazou and D G Haigh, Analogue design techniques for high speed GaAs operational amplifiers, Procs 1988 IEEE ISCAS (Helsinki), June 1988

[18] Z Wang, Novel linearisation technique for implementing large-signal MOS tunable transconductor, Electronics Letters, vol 26, no 2, 18th January 1990, pp138-139

[19] L E Larson et al, A 10 GHz operational amplifier in GaAs MESFET technology, Digest 1989 IEEE Int Solid State Circs Conf, pp 72-73, February 1989

[20]   A A Abidi, Gain bandwidth enhancement in GaAs MESFET wide band amplifiers, Procs 1988 IEEE ISCAS (Helsinki), pp 1465-1468, June 1988

[21]   S I Katsu, M Kazumura and G Kano, Design and fabrication of a GaAs monolithic opertaional amplifier, IEEE Trans on Electron Devs, vol 35, no 7, pp 831-838 July 1988

[22]   C Toumazou, D G Haigh et al, Design and testing of a GaAs switched capacitor filter, Procs IEEE 1990 Int Symp Circs & Systs (New Orleans), 1st - 3rd May 1990

[23]   C Toumazou and D G Haigh, Some designs and a characterisation method for GaAs operational amplifiers for switched capacitor applications, Electronics Letters, vol 24, no 18, 1st September 1988, pp 1170-1172 .pa

[24]   S J Harrold, A switch-driver circuit suitable for high order switched capacitor filters implemented in GaAs, Electronics Letters, vol 24, no 15, pp 982-984, July 1988

[25]   S J Harrold, I A W Vance and D G Haigh, Second-order switched-capacitor bandpass filter implemented in GaAs, Electronics Letters, Vol 21, no 11, pp 494-496, 23rd May 1985.

[26]   L E Larson, C S Chou and M J Delaney, An ultra high speed GaAs MESFET operational amplifier, IEEE Jnl Slid State Circs, vol 24, no 6, December 1989, pp 1523 -1528

[27]   J A Archer et al, A GaAs monolithic low-noise broadband amplifier, IEEE J Solid State Circs, SC-16, no 6, 1981, pp 648-652

[28]   Gallium Arsenide IC technology, N Sclater, TAB Books, 1988

[29]   Applications of GaAs MESFET's, edited by R Soares J Graffeuil J Obregon, Artech House, 1983

[30]   R Olshansky, Microwave subcarrier multiplexing: New approach to wideband lightwave systems, IEEE Circuits and Devices Magazine, vol 4, no 6, Nov 1988, pp 8-14

[31]   C Toumazou, F J Lidgey and C A Makris, Extending voltage mode op-amps to current mode performance, IEE Proc G (CDS), vol 137, no 2, April 1990

[32]   M J O'Mahoney, Semiconductor laser optical amplifiers for use in future fibre systems, IEEE Jnl Lightwave Technology, vol 6, no 4, April 1988, pp 531- 544

Chapter 9

Continuous-Time Filters

Rolf Schaumann and Mehmet Ali Tan

9.1 Introduction

A filter is a twoport that shapes the spectrum of the input signal in order to obtain an output signal with the desired frequency content. Thus, a filter has passbands where the frequency components are transmitted to the output and stopbands where they are rejected. Traditionally, such circuits working in the continuous-time domain have been designed as resistively terminated lossless LC filters[1, 2, 3] where resonance could be employed to achieve complex poles and the desired steep transition regions between passbands and stopbands. With the growing pressure towards microminiaturisation, inductors were found to be too bulky so that designers started to replace passive RLC filters by active RC circuits where gain, obtained from operational amplifiers (op amps), together with resistors and capacitors in feedback networks, was used to achieve complex poles. Active filters based on the ubiquitous op amp have found wide acceptance throughout the industry over the last two or three decades and their mature development and technology guarantee their continued popularity in many applications.

One disadvantage of op amp-based active RC filters is the limited frequency range over which these circuits can be used: the finite bandwidth of op amps usually constrains the applications to be below 100 kHz, with performance deviations becoming increasingly worrisome and difficult to control as operating frequencies increase[3, 4, 5].

Analog filters, even on-chip in fully integrated form, for operation in the megahertz range are beginning to appear for use in video signal processing applications and computer disk drives. The main design challenges are (i) reliable high-frequency performance, which makes op amp-RC filters less and less desirable, and (ii) automatic on-chip tuning against fabrication tolerances and changing operating conditions or even to achieve some measure of adaptive behaviour. These tolerance problems of integrated continuous-time filters are addressed by on-chip automatic control circuitry to be discussed briefly below.

Thus, as applications in communication circuits and systems tend to call for more high-frequency and even fully integrated designs, engineers began to search for a more suitable active element to provide the unavoidable gain without imposing severe frequency limitations. At the same time, this element should have simple circuitry, be realisable in any IC technology and lead to easy and methodical synthesis procedures for active filters. The solution was found in voltage-to-current converters where the output current rather than the output voltage is proportional to an input voltage:

$$I_{out} = g_m V_{in} \qquad (9.1)$$

g_m is the transconductance parameter provided by the active devices. The design of such circuits is the topic of Chapters 3 through 5 and 8 in this book and need not be discussed in any detail here. Suffice it to say that for application in continuous-time filters, transconductances should satisfy the following main properties:

(i)   the circuits must be simple, linear and have a wide frequency response;

(ii) they must have large input and output impedances to prevent undesirable interactions and simplify circuit design;

(iii) they should preferably work with low-voltage power supplies to conserve power and to be compatible with the prevalent digital technologies on the same IC chip; and

(iv) their transconductance parameter must depend on some dc bias voltage or current to facilitate electronic tuning against environmental or processing variations.

The literature contains many useful designs in different technologies, see e.g. references 3, 6-13. Depending on the technology chosen, the frequency range of transconductance circuits extends to > 50 MHz (CMOS), > 500 MHz (bipolar) or even to >1 GHz (GaAs) so that the design of high-frequency continuous-time telecommunication circuits becomes feasible (see chapter 8). Further, because transconductances and capacitors are the only components required for realising a filter (a resistor can be simulated if necessary - see Figure.9.1 and (9.3)), transconductance-C filters can readily be implemented in fully integrated form, compatible with the remaining, often digital, system in any desired technology. In many useful active simulations of filters, we may even insist that all transconductors are identical and all capacitors grounded for especially simple IC layout and processing.[3, 17]

We shall in this chapter present a brief overview of a few important synthesis and implementation methods for current-based continuous-time filters. The designs do not stay exclusively in the current domain; instead, because transconductances are voltage driven with a current output, the circuits are constructed such that each transconductance "sees" a (generally frequency-dependent) load impedance that converts the output current into a voltage which in turn becomes the input for the next transconductance stage. Thus, the signals alternate quite naturally between voltages and currents.

A simple example of this approach is indicated in Figure.9.1a which shows the usual symbol for a voltage-to-current converter with differential-input, the so-called operational transconductance amplifier, or OTA, realizing

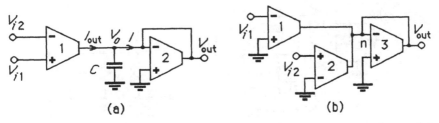

(a)   (b)

Figure 9.1   Fundamental transconductance building blocks to perform (a) **Integrating,** (b) **Voltage summing**

$$I_{out} = g_{m1} (V_{i1} - V_{i2}) \qquad (9.2)$$

I_{out} drives the impedance $1/(sC)$ to produce the voltage $V_o = I_{out}/(sC)$ which forms the input of OTA 2 with transconductance g_{m2}. Its output in turn is connected back to its inverting input terminal so that, as the reader may show, OTA 2 realizes a grounded resistor of value

$$R = V_o/I = 1/g_{m2} \qquad (9.3)$$

Consequently, the circuit in Figure.9.1a realizes a differential-input lossy integrator (lossless if OTA 2 is absent),

$$V_{out} = V_o = \frac{g_{m1}}{sC + g_{m2}} (V_{i1} - V_{i2}) \qquad (9.4)$$

which constitutes an important fundamental building block for active filters as we shall find shortly. Evidently the integrator can be inverting or non-inverting depending on which input is used. Note that, for $C = 0$, the circuit is an amplifier with positive or negative gain g_{m1}/g_{m2}:

$$V_{out} = \frac{g_{m1}}{g_{m2}} (V_{i1} - V_{i2}) \qquad (9.5)$$

In an analogous way, the operation of scaled voltage summing can be implemented by the circuit in Figure 9.1b which realises

$$Vout = - \frac{gm1}{gm3} Vi1 + \frac{gm2}{gm3} Vi2 \qquad (9.6)$$

Observe especially that current summing is "free" with this technique because it is given by Kirchhoff's current law at any node (e.g., node n in Figure.9.1b where $g_{m2}V_{i2}$ is added to $-g_{m1}V_{i1}$). This free operation of summing often simplifies current-based circuit design significantly when compared to the better known approaches relying on op amps.

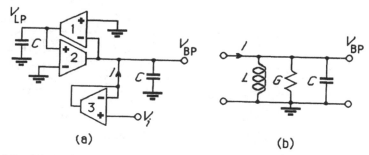

(a)   (b)

Figure 9.2 (a) Active OTA-C simulation of the passive RLC resonance circuit in (b)

As a further simple example, the reader may wish to verify that the circuit in Figure.9.2a realises the bandpass and lowpass functions

$$\frac{V_{BP}}{V_i} = \frac{s g_{m3} C}{s^2 C^2 + s g_{m3} C + g_{m1} g_{m2}} \tag{9.7a}$$

and

$$\frac{V_{LP}}{V_i} = -\frac{g_{m1} g_{m3}}{s^2 C^2 + s g_{m3} C + g_{m1} g_{m2}} \tag{9.7b}$$

respectively.   Note that, formally, (9.7a) is identical to

$$\frac{V_B}{I} = \frac{1}{sC + 1/sL + G} = \frac{sL}{s^2 LC + sLG + 1}$$

the transfer function realized by the passive RLC circuit in Figure 9.2b. Thus, the g_m-C filter in Figure.9.2a is simply an active simulation of the passive prototype: it implements the conductance G via $G = g_{m3}$ and the inductor via the gyrator[3] formed from the combination of OTA 1 and OTA 2 as $L = C/(g_{m1} g_{m2})$. The input current I is obtained from $I = g_{m3} V_i$.

The two examples should have given the reader an indication of the simplicity with which continuous-time transconductance-C filters can be realized. In the next section we shall establish a more formal base for some of those design procedures which have been shown to be practically useful.

9.2 Design of Continuous-Time Transconductance-C Filters

The reader familiar with the active filter literature[1, 2, 3] will recognise that fundamentally almost all active filter topologies consist of suitable combinations of integrators and/or gyrators and summers. Because both of these building blocks can be implemented in the current-domain with OTAs and capacitors, see Figures.9.1 and 9.2, we can expect that most popular and traditional design procedures, with appropriate modifications, still apply.

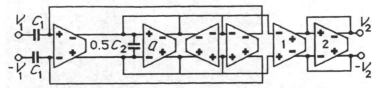

Figure 9.3 Fully differential transconductance-C bandpass section

Indeed, we can design cascade filters provided that cascadable biquadratic sections (e.g. Figure. 9.2a) can be built. If lowest passband sensitivities are of importance[1, 2, 3], we can attempt to simulate LC ladder filters: we can select either the element substitution method provided that active inductors, i.e. gyrators, can be obtained with OTAs and capacitors (see Figure. 9.2a) or, we can implement an operational simulation of an LC ladder if integrators and summers are available (see Figure. 9.1). Details of these methods and some examples will be discussed in the following sections.

9.2.1 Cascade Design

The oldest and probably best known method of active filter design starts by factoring a prescribed high-order transfer function H(s) into a product,

$$H(s) = \prod H_i(s) \qquad (9.8)$$

where (usually) the factors $H_i(s)$ are second-order sections or biquads. These sections are then realised by any of the numerous available methods based on op amps[1, 2, 3] and are connected in cascade to obtain the indicated product. The assumption is, of course, that the sections are cascadable, i.e., that their output impedances are very low (ideally zero) so that loading effects can be neglected.

We discussed earlier that these biquads can also be implemented with transconductances (OTAs) as is shown, for example in Figure. 9.2. If better power supply rejection, immunity from noise and from parasitic signals is important [3, 11, 14], designers find it often preferable to build the circuits in fully differential, balanced form. Conversion procedures from single-ended circuits to fully differential form are relatively straightforward [3] and need not be discussed in this chapter. An example of such a balanced circuit is shown in Figure. 9.3: assuming that the three unlabeled OTAs are identical with value g_m and that the remaining ones have the values g_{mQ}, g_{m1} and g_{m2}, the reader should verify that the filter realises the bandpass function

$$H_{BP}(s) = \frac{g_{m1}}{g_{m2}} \frac{s g_m / C}{s^2 + \dfrac{g_{mQ} - g_m}{C} s + \left(\dfrac{g_m}{C}\right)^2} = H_B \frac{s \omega_o}{s^2 + \dfrac{s \omega_o}{Q} + \omega_o^2} \qquad (9.9)$$

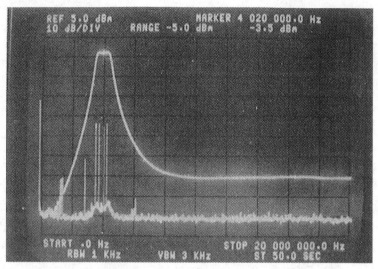

Figure 9.4   Experimental performance of a fully integrated CMOS eighth-order cascade transconductance-C bandpass filter

Note that the two transconductances g_{m1} and g_{m2} are needed only to be able to set the gain constant H_B. As soon as suitable biquad structures and their transfer functions are identified, the designer may use any convenient means to determine the component values (g_{mi} and C_i) such that the prescribed filter coefficients, such as Q, ω_o and H_B in (9.9), are realized. Paying attention to sensitivity problems and dynamic range scaling[1, 3], the biquadratic sections are then connected in cascade in order to obtain the required high-order function. Note, that some care must be exercised when cascading OTA-C sections because their output impedances are normally not small and loading may cause problems. However, g_m-C or OTA-C biquads can always be constructed such that their input impedances are very large (see Figure.9.2a) so that interactions between the circuit blocks are minimised.

A published example of such a high-frequency transconductance-C cascade design is available in reference 15. Using four biquads similar to those in Figure.9.3, but in single-ended operation, an eighth-order bandpass filter was designed to have a Chebyshev transfer characteristic with bandcenter at 4 MHz and a 0.5 dB equiripple bandwidth of 800 kHz. The filter was realized in fully integrated form in a 3 μm CMOS technology; its experimental performance is shown in Figure. 9.4. Numerous design details, including the all-important automatic tuning scheme to eliminate the effects of fabrication tolerances and temperature drifts, can be found in reference 15. The circuit with on-chip automatic tuning was found to perform as required over the temperature range $0° \leq T \leq 60°$.

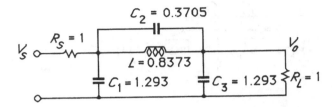

Figure 9.5 Normalised third-order elliptic LC lowpass ladder filter

9.2.2 Element Substitution Method

Sensitivity of circuit performance to element tolerances is of great practical concern in the design of active filters. Although cascade filters behave relatively well, their passband sensitivities tend to become too large for many applications, especially as the filter order increases. Since it is known, on the other hand, that appropriately designed LC ladders have very low sensitivities to component tolerances, it is natural that designers have looked for various methods to simulate passive LC ladders in active form such that their low sensitivity behaviour is maintained. One such method simply attempts to replace the offending component, the bulky inductor, by an integrable active circuit. We have encountered one such technique in Figure. 9.2 where an inductor is implemented by a capacitively loaded gyrator. Naturally, this procedure can be extended to filters of higher order, as we shall illustrate next.

A third-order elliptic LC lowpass ladder filter with transfer function

$$H(s) = \frac{0.28163\,(s^2 + 3.2236)}{(s + 0.7732)(s^2 + 0.4916s + 1.1742)} \qquad (9.10)$$

is shown in Figure.9.5. This filter is difficult to miniaturise in the given form because of the floating inductor. However, using two gyrators of the form given in Figure.9.2 in a back-to-back connection permits us to realise a floating inductor[1, 3]. The resulting circuit, for illustration purposes in single-ended form, is shown in Figure.9.6a. The inverter symbol depicts a single-input - single-output transconductance element, as is indicated in the inset. Figure 9.6b shows the fully differential form of the same circuit. Note especially, that all transconductances are identical, a convenience in practice.

To complete the design, let us assume that the lowpass filter must have a bandwidth of 4.5 MHz; we then denormalize the frequency in (9.10) by $\omega_0 = 2\pi \cdot 4.5 \cdot 10^6$ rad/s. Further, let us choose the transconductance value as $g_m = 200\ \mu$S. With these values we obtain[8, 3]

$$C_1 = C_3 = 9.146\ \text{pF} \qquad C_2 = 2.621\ \text{pF} \qquad C_L = 9.923\ \text{pF}$$

where we used that $C_L = Lg_m^2$.

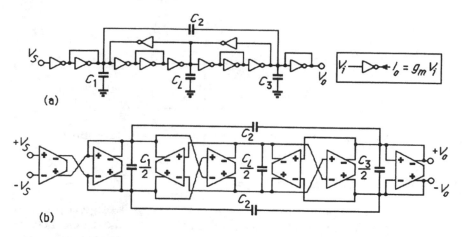

(a)

(b)

Figure 9.6   (a) Active simulation of the passive LC filter in Figure 9.5 (Inset: Definition of the transconductance symbol) (b) fully differential form of the circuit in (a)

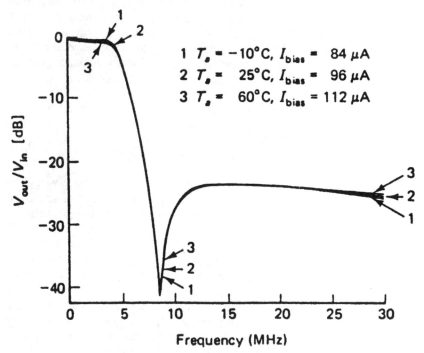

1 $T_a = -10°C$, $I_{bias} = 84 \mu A$
2 $T_a = 25°C$, $I_{bias} = 96 \mu A$
3 $T_a = 60°C$, $I_{bias} = 112 \mu A$

Figure 9.7   Experimental performance of the integrated filter of Figure 9.6b[8]. (© IEEE, June 1988)

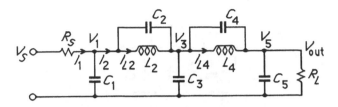

Figure 9.8   Fifth-order elliptic LC lowpass filter

The design of this filter in CMOS technology was discussed in reference 8 with the experimental performance depicted in Figure 9.7. Again, in order to maintain filter performance in the face of fabrication tolerances and temperature changes, the integrated circuit chip contains an automatic control or tuning circuit whose design details are discussed in the original paper.

9.2.3 The Operational Simulation Method

A very successful and popular way of implementing an LC filter with active components uses an operational simulation of the equations describing the circuit. It is based on the powerful signal-flow graph (SFG) method[1, 3, 16] which, after impedance scaling, treats all voltages and currents in the passive circuit as voltage signals [1] and realises the effect of both inductors and capacitors via integrators. For example, the voltage V_1 in Figure 9.8 is obtained by integrating the difference between I_1 and I_2:

$$V_1 = \frac{1}{sC_1} (I_1 - I_2)$$

We also note as a further advantage of the SFG method the possibility of scaling the element values such that the circuits have the maximum possible dynamic range[1, 3]. This scaling is not normally available in the element substitution procedure (see Chapter 10).

Because integrators can be realised with transconductances and grounded capacitors (see Figure. 9.1a) it stands to reason that g_m-C design methods can be adapted to the SFG procedure. Specifically, it has been shown [17, 3] that ladder filters with quite arbitrary branches can be implemented as g_m-C circuits where furthermore all transconductors, with the exception of possibly one, are identical[2] and all capacitors are grounded. As was

---

[1] After impedance normalisation with an arbitrary resistor R, the currents become R I with the unit [volt].

[2] For LC ladders with unequal source and load resistors one transconductance value is different.

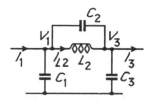

Figure 9.9   Typical ladder section with a floating capacitor

mentioned before, this approach is very desirable because it results in simple processing technologies and easy and methodical IC layout and design. The insistence on only grounded capacitors, however, entails a penalty: the realisation requires many active devices (transconductors) because floating capacitors, such as C_2 in Figure.9.8, result in equations with terms proportional to s,

$$I_2 = (sC_2 + \frac{1}{sL_2})(V_1 - V_3)$$

which with integrators must be realised in the form 1/(1/s) [3, 17, 18]. A very efficient realisation with savings of a large number of transconductors can be achieved, however, if floating capacitors are acceptable[18]. Consider the typical ladder section in Figure.9.9. The relevant equations describing this circuit can be written as

$$V_1 = \frac{1}{sC_1}[(I_1 - I_{L2}) - sC_2(V_1 - V_3)] \qquad I_{L2} = \frac{1}{sL_2}(V_1 - V_3) \qquad (9.11a)$$

To convert the currents into voltages for active SFG simulation, we multiply both equations in (9.11a) by a normalising resistor R to obtain the expressions

$$V_1 = \frac{1}{sC_1R}[(RI_1 - RI_{L2}) - sC_2R(V_1 - V_3)] \qquad RI_{L2} = \frac{R}{sL_2}(V_1 - V_3) \qquad (9.11b)$$

which are realised by the circuits in Figure.9.10 as the reader may verify. In the Figure, we have labelled: $V_{I1} = R\,I_1$ and $V_{I2} = R\,I_{L2}$. For the remaining equations describing Figure.9.8 we use the analogous procedures.

As an example, let us realise a lowpass filter with the following specifications: Passband attenuation ≤ 0.0005 dB in $f \leq 3.4$ MHz; stopband attenuation ≥ 26.5 dB in $f \geq 4.6$ MHz; equal source and load resistors of value 2 kΩ. From filter tables[19] we find that a fifth-order elliptic filter, Figure. 9.8, with

$$R_S = R_L = 2\ k\Omega, \quad L_2 = 85.91\ \mu H, L_4 = 53.74 \mu H, \text{ and}$$

$$C_1 = 8.962\ pF, C_2 = 3.476\ pF, C_3 = 24.81\ pF, C_4 = 12.98\ pF, C_5 = 3.246\ pF$$

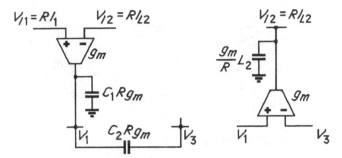

Figure 9.10   Realisations of Equ. (9.11b)

satisfies the specifications. The prescribed load resistors specify the value of the transconductance: $g_m = (2 \text{ k}\Omega)^{-1} = 500 \text{ μS}$; let us now (arbitrarily [3]) set R = $1/g_m$. From Figure.9.10, the capacitors representing the inductors are then obtained by $C_L = g_m^2 L$ to give

$$C_{L2} = 21.48 \text{ pF} \quad C_{L4} = 13.44 \text{ pF}$$

Because of the choice R = $1/g_m$, the remaining capacitor values do not change. Thus, combining the appropriate blocks identified in Figure.9.10 in the way specified via Figure.9.8 gives the g_m-C simulation of the fifth-order elliptic lowpass ladder in Figure.9.11. For ease of reference, we have in Figure.9.11 labelled those signals that represent the currents and voltages in the passive prototype Figure.9.8. Observe that the realisation is very efficient, it uses only seven capacitors and seven OTAs for the seven reactances in the original ladder, including the source and load resistors that were simulated via R = $1/g_m$.

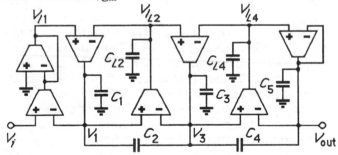

Figure 9.11   Transconductance-C SFG simulation of the circuit in Figure 9.8

---

[3] Note that the choice of R permits scaling the element values

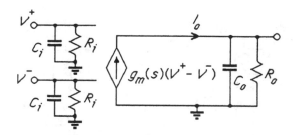

Figure 9.12   Practical OTA circuit model

9.2.4 Effects of OTA Nonidealities

We have assumed until now the transconductances to be ideal voltage-controlled current sources with their operation described by (9.2). In practice, there are, of course, a number of nonidealities to be considered which may adversely affect the performance of a g_m-C circuit. Thus, as for all active devices, OTAs have finite input and output impedances and a frequency-dependent gain parameter, leading to the more realistic circuit model given in Figure.9.12. The capacitors C_i and C_o may be assumed to model not only the parasitic device input and output capacitances but also those contributed by wiring (layout and routing). For well-designed OTAs, the input resistor R_i is so large (>10 MΩ upto >100 MΩ in CMOS circuits) that its effect can be neglected. The output resistor, on the other hand, is usually so small (of the order of 100 kΩ or less) that its effect on circuit performance must be included. Finally, the frequency dependence of g_m can be modelled via a dominant pole or excess phase shift

$$g_m(j\omega) \approx \frac{g_{m0}}{1+j\omega/a} \approx g_{m0}\, e^{-j\phi(\omega)} \,, \quad \phi(\omega) \approx \tan^{-1}\left(\frac{\omega}{a}\right) \qquad (9.12)$$

similar to that of the gain of an op amp, where $\phi(\omega)$ is found to cause Q-enhancement [1, 3] in filter circuits. In g_m-C filters, however, this effect is noticeable only at fairly high operating frequencies because the dominant pole is located at much higher frequencies: transconductances can be designed readily to have constant gain g_{m0} until 100 MHz or higher. An additional advantage of transconductance-C filters is that their topologies often consist of mostly "integrators driven by integrators", i.e., most OTAs have circuit capacitors connected to both their inputs and outputs (see e.g. Figures.9.2a, 9.6 and 9.11). Consequently, the parasitics C_i and C_o can be absorbed by predistortion in the circuit capacitors and the main effect is to cause capacitor-losses whose influence on the performance of LC ladder filters is well known.[20, 1, 2, 3] Notice also, that the effects of lossy capacitors and those of excess phase oppose each other and partially cancel. We may conclude, therefore, that g_m-C filters are generally affected to a far smaller degree by device nonidealities and circuit parasitics than their op amp-based counter parts.

Phase detector

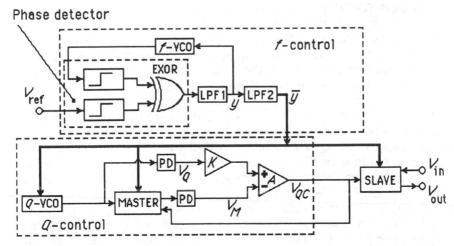

Figure 9.13 General MASTER-SLAVE tuning scheme for integrated filters

9.3 Tuning Procedures

It is known that analog filters generally require post-fabrication tuning or adjustments in order to set the critical frequencies exactly. This is even more true in integrated filters because the g_m/C ratios that set the frequency parameters are likely to be obtained very inaccurately in IC fabrication due to the high tolerances (> ±25%) of absolute component values. Since tuning in the usual sense is clearly impossible or impractical in a filter on an IC chip, designers have generally adopted an automatic on-chip tuning or control scheme that maintains the accuracy of the design values of important filter parameters by locking them to a suitable stable reference signal.[3, 8, 14, 15, 21] Therefore, before attempting to realise an IC continuous-time filter, the designer is well advised to make certain that the filter is amenable to some automatic tuning scheme as is depicted in the block diagram in Figure.9.13. Clearly, this requires that filter components can be changed electronically: in our case the transconductances are functions of bias currents or voltages.

The operation of an actual tuning scheme depends on the details of a filter circuit and can, therefore, be discussed only in connection with specific cases. The references contain many examples[3, 8, 14, 15, 21]; for this presentation, let it be sufficient to illustrate the principle of automatic tuning via the "master-slave" approach with the aid of the block diagram in Figure.9.13.

All analog circuits need to have their frequency parameters stabilised because of the large absolute value tolerances of IC processes. This is accomplished usually with a frequency control ("f-control" in Figure.9.13) circuit based on a digital or analog phased-locked loop (PLL). f-VCO is an on-chip oscillator whose output frequency is compared to a stable reference, V_{ref}, such as a system clock. f-VCO, the "master", is designed to model well the relevant behaviour of the main filter, the "slave", including its deviations arising from whichever source. This means that the oscillation frequency of the "master" f-VCO should depend on the same type of components and parasitics as the pole frequencies of the "slave". Before reaching the EXOR gate, the signals are sent through hard limiters to remove any amplitude dependence of the PLL. The dc output of the "f-control" circuit, obtained after lowpass filtering (LPF2) is then applied to the control terminals of the appropriate transconductances to change g_m(bias)/C in a way that minimises the frequency errors.

Assuming that this scheme corrects all frequency errors on the chip, the only defects that can alter the desired frequency response are errors in pole quality factors, Q_i. Q, a dimensionless quantity, is set by ratios of like components, a forte of IC processing. However, at least at high frequencies and medium-to-high values of Q, tuning is still necessary because of its well-known high sensitivity to excess phase shifts in the active devices.[3] Thus, in these cases a "Q-control" scheme is advisable whose function is to "tune the transfer function shape" of the filter. As indicated in Figure.9.13, Q-control works by comparing the response of the "master", which again is designed to model well the relevant behaviour of the main filter, the "slave", with the Q-VCO output itself and thereby creates an error voltage V_{QC}. PD are two identical peak detectors and K is an amplifier of dc gain K. The voltage V_{QC} is applied to the "master" in a closed loop which lets us recognise the Q-control circuitry as a magnitude locking scheme [4]. V_{QC} is then also applied to the "slave" and thereby corrects any remaining transfer function shape problems.[3]

The examples discussed earlier as well as all other cases that appeared in the literature use tuning schemes along the lines of Figure.9.13. Their performance proved quite satisfactory; the remaining work that needs to be done on the development of tuning methods consists of

(i) design of "masters" which are better models of their "slaves"
(ii) simplification of the tuning circuitry which now consumes a significant portion of the chip area and the total chip power.

9.4 Discussion and Summary

With the growing demand for placing as many components of electronic systems as possible on an IC chip, analog continuous-time circuits need to be combined with digital circuitry for lower cost and improved reliability. At the same time, many applications call for operations at increasingly higher frequencies so that simple and fast devices are needed that permit a methodical design of integrated signal processing systems. Current-based,

---

[4] This simple method works because, as one can show, with the previously adjusted and now assumed correct frequency parameters, Q errors manifest themselves in magnitude errors[3].

transconductance-C designs satisfy almost all the requirements and begin to emerge as the approach of choice.

We have in this Chapter discussed the principles of g_m-C continuous-time filters and presented the most successful design procedures. Their main advantages are extremely simple and transparent design methodologies, IC layout and processing possibilities. With designs based on identical transconductances and grounded capacitors, implementation of integrated analog filters based on analog gate arrays appears a distinct possibility. In addition, the wider useful bandwidth of transconductances coupled with the reduced effects of circuit and device parasitics on filter performance result in far higher operating frequencies at which the circuits can function. A problem to be paid attention to, which appears to plague all continuous-time analog circuitry, is that of reliability and stability of filter parameters in the face of fabrication tolerances and changing environmental conditions. On-chip automatic tuning and control schemes appear to yield practical solutions to these difficulties.

9.5 References

[1]   Sedra, A. S. and P. Brackett, Filter Theory and Design: Active and Passive, Matrix Publishers, 1978.

[2]   Temes, G. C. and G. LaPatra, Introduction to Circuit Synthesis and Design, McGraw-Hill 1977.

[3]   R. Schaumann, M. S. Ghausi and K. R. Laker, Design of Analog Filters: Passive, Active RC and Switched Capacitor, Prentice Hall, 1990.

[4]   Brackett, P. O. and A. S. Sedra, "Active Compensation for High-Frequency Effects in Op-Amp Circuits with Applications to Active RC Filters," IEEE Trans. Circuits Syst., Vol. CAS-23, pp 68-72, 1976.

[5]   Bruton, L. T. and A. I. A. Salama, "Frequency Limitations of Coupled Biquadratic Active Ladder Structures," IEEE J. Solid-State Circuits, Vol. SC-9, pp 70-72, 1974.

[6]   Nedungadi, A. and T. R. Viswanathan, "Design of Linear CMOS Transconductance Elements," IEEE Trans. Circuits Syst., Vol. CAS-31, pp 891-894, 1984.

[7]   Park, C. S. and R. Schaumann, " A High-Frequency CMOS Linear Transconductance Element," IEEE Trans. Circuits Syst., Vol. CAS-33, pp 1132-1138, 1986.

[8]   Krummenacher, F. and N. Joel, "A 4 MHz CMOS Continuous-Time Filter with On-Chip Automatic Tuning," IEEE J. Solid-State Circuits, Vol. SC-23, pp 750-758, 1988.

[9]   Gray, P. R. and R. G. Meyer, Analysis and Design of Analog Integrated Circuits, Wiley, 1984.

[10] Toumazou, C. and D. G. Haigh, "Analogue Design Techniques for High-Speed GaAs Operational Amplifiers," Proc. IEEE Int. Symp. Circuits Syst., pp 1453-1456, 1988.

[11] Gregorian, R. and G. C. Temes, Analog MOS Integrated Circuits for Signal Processing, Wiley-Interscience, 1986.

[12] Wu, P. and R. Schaumann, "A High-Frequency GaAs Transconductance Circuit and Its Applications,"Proc. IEEE Int. Symp. Circuits Syst., 1990.

[13] Wu, P., R. Schaumann and S Szczepanski, "A CMOS OTA with Improved Linearity Based on Current Addition," Proc. IEEE Int. Symp. Circuits Syst., 1990.

[14] Banu, M. and Y. Tsividis, "An Elliptic Continuous-Time CMOS Filter with On-Chip Automatic Tuning," IEEE J. Solid-State Circuits, Vol. SC-20, pp 1114-1121, 1985., 1988.

[15] Park, C. S. and R. Schaumann, "Design of a 4 MHz Analog Integrated CMOS Transconductance-C Bandpass Filter," IEEE J. Solid-State Circuits, Vol. SC-23, pp 987-996, 1988.

[16] Martin, K. and A. S. Sedra, "Design of Signal-Flow Graph (SFG) Active Filters, IEEE Trans. Circuits Syst., Vol. CAS-25, pp 185-195, 1978.

[17] M. A. Tan and R. Schaumann, "Simulating General-Parameter LC-Ladder Filters for Monolithic Realizations with Only Transconductance Elements and Grounded Capacitors, "IEEE Trans. Circuits Syst., Vol. CAS-36, pp 299-307, 1989.

[18] Tan, M. A. and R. Schaumann, "A Reduction in the Number of Active Components Used in Transconductance Grounded Capacitor Filters, "Proc. IEEE Int. Symp. Circuits Syst., 1990.

[19] Zverev, A. I., Handbook of Filter Synthesis, Wiley, 1967.

[20] Temes, G. C., "Fist-Order Estimation and Precorrection of Parasitic Loss Effects in Ladder Filters," IEEE Trans. Circuit Theory, Vol. CT-9, pp 385-400, 1967.

[21] Schaumann, R. and M. A. Tan, "The Problem of On-Chip Automatic Tuning in Continuous-Time Integrated Filters," Proc. IEEE Int. Symp. Circuits Syst., pp. 106-109, 1989.

Chapter 10

Continuous-time and Switched Capacitor Monolithic Filters Based on LCR Filter Simulation using Current and Charge Variables

David Haigh

10.1 Introduction

An important area of application of analogue circuits is for the realisation of high precision integrated circuit filters. In Chapter 9, design techniques have been presented for high frequency state-of-the-art continuous-time filters implemented in CMOS technology. This present Chapter attempts to generalise on the realisation of monolithic filters, using a common framework based on the transconductor element, which encompasses many switched capacitor and integrated continuous-time filter realisation architectures. We focus particularly on the distinction of whether voltages or currents are used as the simulating variables, or charges in the switched capacitor case. Although the adoption of current processing will almost certainly eventually lead to filters with very high operating frequencies [1,2], and we are beginning to see this process starting already, in this Chapter we confine our scope to examining the benefits of current processing from the points of view of minimising sensitivity of response to component parameter variations and also maximising signal handling capability.

Current conveyors have already been used to realise biquadratic active RC filter sections [3]. However, for high order monolithic filters, with which we are concerned in this Chapter, the method of cascading biquadratic sections leads to unacceptably high sensitivity of the response to component parameter variations. Consequently, we use the general method of simulating low sensitivity LCR filters. The simulating filters derived will be described and analysed using two alternative approaches which have specific features.

In one approach, we derive a signal flow graph representation of the LCR filter, and alternative grouping of the signal flow graph branches yields alternative realisations including voltage-based and current-based structures [4,5].

In the other approach, we derive a state-matrix type of representation for the prototype LCR filter, and view the various realisations as alternative ways of realising required matrix elements [6,7]. This latter approach constitutes a general formulation which allows comparison of the properties

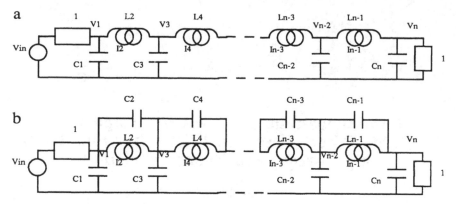

Figure 10.1 LCR lowpass filter networks
a Polynomial b Elliptic

of the various structures, including sensitivity and signal handling capability, within a common framework.

10.2 LCR Prototype Filters and State Matrix Formulations

In this paper we shall be concerned primarily with lowpass filters. A typical section of an equally-terminated LCR network for polynomial lowpass filters, which are those not possessing transmission zeros at finite frequencies, is shown in Figure 10.1a. Network analysis using the inductor currents and capacitor voltages as variables yields a set of equations which may conveniently be expressed in state matrix form, as follows:

$$
\begin{vmatrix}
sC_1 & 0 & & & & & & & & \\
0 & sL_2 & 0 & & & & & & & \\
 & 0 & sC_3 & 0 & & & & & & \\
 & & 0 & sL_4 & 0 & . & & & & \\
 & & & 0 & . & & & & & \\
 & . & & & & & 0 & & & \\
 & & & & & 0 & sL_{n-3} & 0 & & \\
 & & & & & & 0 & sC_{n-2} & 0 & \\
 & & & & & & & 0 & sL_{n-1} & 0 \\
 & & & & & & & & 0 & sC_n
\end{vmatrix}
\begin{Vmatrix}
V_1 \\
I_2 \\
V_3 \\
I_4 \\
. \\
. \\
I_{n-3} \\
V_{n-2} \\
I_{n-1} \\
V_n
\end{Vmatrix}
=
$$

$$
\begin{vmatrix}
-1 & 1 & 0 & . & & & & & & \\
1 & 0 & -1 & 0 & . & & & & & \\
0 & 1 & 0 & -1 & 0 & . & & & & \\
. & 0 & 1 & 0 & -1 & . & . & & & \\
 & . & 0 & 1 & . & . & . & . & & \\
 & & . & . & . & . & 0 & -1 & 0 & \\
 & & . & . & . & 1 & 0 & -1 & 0 & \\
 & & . & & & 0 & 1 & 0 & -1 & 0 \\
 & & . & & & & 0 & 1 & 0 & -1 \\
 & & . & & & & & 0 & 1 & -1
\end{vmatrix}
\begin{Vmatrix}
V_1 \\
I_2 \\
V_3 \\
I_4 \\
. \\
 \\
I_{n-3} \\
V_{n-2} \\
I_{n-1} \\
V_n
\end{Vmatrix}
+
\begin{vmatrix}
1 \\
0 \\
0 \\
0 \\
0 \\
0 \\
0 \\
0 \\
0 \\
0
\end{vmatrix}
V_{in}
\qquad (10.1)
$$

A general section of an odd order equally-terminated elliptic lowpass LCR ladder filter is shown in Figure 10.1b, where the transmission zeros have been implemented by capacitors connected in parallel with the inductors of the corresponding polynomial filter. Analysis of the prototype filter in Figure 10.1b yields a set of equations which may be conveniently expressed in state space-type form as in follows:

$$
\begin{vmatrix}
sC_1' & 0 & k_{13} & 0 \\
0 & sL_2 & 0 & \\
k_{31} & 0 & sC_3' & 0 & k_{35} & 0 \\
& & 0 & sL_4 & 0 \\
& & & \cdot & & \\
& & & & \cdot & \\
& & & & 0 & sL_{n-3} & 0 \\
& & & k_{n-2,n-4} & 0 & sC_{n-2}' & 0 & k_{n-2,n} \\
& & & & & 0 & sL_{n-1} & 0 \\
& & & & & k_{n,n-2} & 0 & sC_n'
\end{vmatrix}
\begin{vmatrix}
V_1 \\
I_2 \\
V_3 \\
I_4 \\
\cdot \\
\cdot \\
I_{n-3} \\
V_{n-2} \\
I_{n-1} \\
V_n
\end{vmatrix} =
$$

$$
\begin{vmatrix}
-1 & 1 & 0 & \cdot \\
1 & 0 & -1 & 0 & \cdot \\
0 & 1 & 0 & -1 & 0 & \cdot \\
\cdot & 0 & 1 & 0 & -1 & \cdot & \cdot \\
& \cdot & 0 & 1 & \cdot & \cdot & \cdot & \cdot \\
& & \cdot & \cdot & \cdot & \cdot & 0 & -1 & 0 \\
& & & \cdot & \cdot & \cdot & 1 & 0 & -1 & 0 \\
& & & & & 0 & 1 & 0 & -1 & 0 \\
& & & & \cdot & & 0 & 1 & 0 & -1 \\
& & & & & & & 0 & 1 & -1
\end{vmatrix}
\begin{vmatrix}
V_1 \\
I_2 \\
V_3 \\
I_4 \\
\cdot \\
\\
I_{n-3} \\
V_{n-2} \\
I_{n-1} \\
V_n
\end{vmatrix}
+
\begin{vmatrix}
1 \\
0 \\
0 \\
0 \\
0 \\
0 \\
0 \\
0 \\
0 \\
0
\end{vmatrix} V_{in}
$$

$$(10.2)$$

where

$$C_1' = C_1 + C_2 \quad C_i' = C_{i-1} + C_i + C_{i+1} \quad C_n' = C_{n-1} + C_n$$

and

$$k_{i,i+2} = C_{i+1}/C_i, \quad k_{i,i-2} = C_{i-1}/C_i.$$

The techniques described in this Chapter for odd order filters can be similarly developed for the even order cases. The prototype filter matrix descriptions of Equations 10.1 and 10.2 can be conveniently translated into corresponding active realisations using signal flow graphs.

10.3 Signal Flow Graph Representations and Realisations

The relationships between the state variables for a polynomial lowpass filter contained in Equation 10.1 can be represented in the form of a signal flow

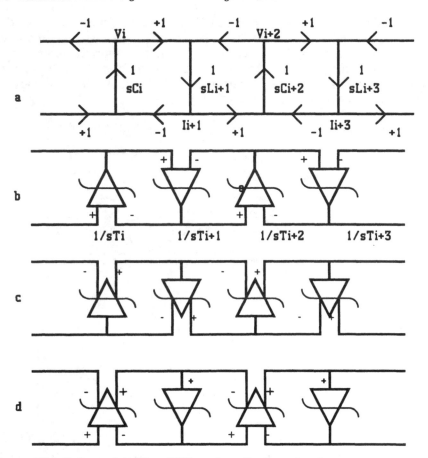

Figure 10.2 Polynomial filter SFG and realisation structures
a Signal flow graph   b Realisation using voltage integrators
c Realisation using current integrators   d Realisation using mixed variables

graph (SFG) [8] as shown in Figure 10.2a. At this stage, we ignore the source and load terminations, showing that part of the signal flow graph corresponding to the reactive elements in the prototype. We note that the branch weights in the signal flow graph are either +1 or -1 unity branches or are of the integrator form $1/sT_i$, where T_i is the value of an inductor or capacitor in the LCR network and s is the complex frequency variable. We now combine the +1 and -1 branches, which occur in pairs, at the inputs or outputs of the $1/s$ type integrator branches.

By combining the unity branch pairs at the inputs of the integrator branches, we have an implentation in terms of differential input integrators, as shown in Figure 10.2b. When they are combined at the integrator outputs we have a realisation in terms of differential output integrators as shown in Figure 10.2c. A third realisation is shown in Figure 10.2d, where we treat alternate integrators differently; for one set, pairs of unity branches are

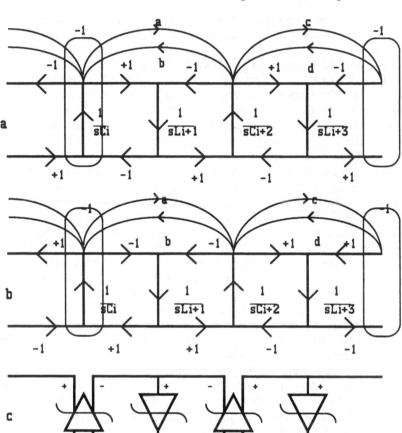

Figure 10.3   Elliptic filter SFG and realisation structure
a Signal flow graph   b Scaled signal flow graph
c Realisation structure (zero-producing branches not shown)

associated with both the input and output of the integrator branches to form a differential input/differential output integrator; the other alternate set of integrators have no unity branches associated with them and are single input/single output.

The implementations in Figure 10.2b, c and d imply certain assumptions about the type of simulating variable used. In the structure in Figure 10.2b the output of each integrator drives a pair of integrator inputs. In order that loading does not alter the integrator transfer functions, the integrators could have a very low output impedance and/or a very high input impedance. Thus for the structure in Figure 10.2b it is most convenient if the signals are voltages. Similarly, the structure in Figure 10.2c relies on direct connection of pairs of integrator outputs together to achieve addition at each integrator input, which requires that the signals are currents. By similar arguments, the structure in Figure 10.2d is most conveniently realised if a combination of currents are used for the upper variables and voltages for the lower variables.

For the elliptic lowpass LCR network of Figure 10.1b, the state matrix description of Equation 10.2 may be realised in signal flow graph form as in Figure 10.3a. The transmission zeros are realised by cross coupled branches (shown dashed) with branch weights which are positive constants [9]. Such constant branches can be realised simply [9], by means of capacitors, in the active realisation, provided the branch weights are negative. The signs of these branches can be changed using contour scaling [10], in which the weight of each incoming branch crossing a contour is multiplied, and that of each outgoing branch divided, by an arbitrary scaling factor. Using the contours shown in Figure 10.3a with a scaling factor of -1 leads to the signal flow graph in Figure 10.3b. This operation satisfies the requirement that the transmission zero producing branches are negative but also changes the signs of some of the unity branches. In fact, if we look at the branches derived from the inductors of the LCR prototype filter, we see that now both unity branches at the output have the same sign and both input branches have the same sign. Therefore, these integrators can not be realised in canonic differential input or differential output form. We observe, however, that the constraints on the signs of the branches of the signal flowgraph can in fact be met by adopting the mixed-variable architecture of Figure 10.2d, as shown for the elliptic case in Figure 10.3c [11].

Having derived signal flow graphs for polynomial and elliptic lowpass LCR filters and identified some general types of realisation using block diagram integrator components with voltage variables, current variables and mixed voltage and current variables, we now consider the realisation of the required integrator blocks.

10.4   Integrator Subcomponents

In this Chapter, we assume that integrators are realised from two subcomponents. The first consists of an operational amplifier with a feedback capacitor as shown in Figure 10.4a. Its input-output behaviour is defined by

$$V_0 = - (1/sC) \, I_{in} \qquad (10.3)$$

and it may be described as an integrating current-to-voltage converter. It is defined by the single parameter C. Note that although the output variable is a voltage, the input variable is a current.

The second subcomponent is shown in general form in Figure 10.4b and has the input-output relation

$$I_{o1} = -I_{o2} = G_m \, (V_{in1} - V_{in2}) \qquad (10.4)$$

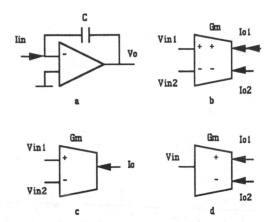

Figure 10.4   Sub-components for integrator realisation
a integrating voltage-to-current converter   b transconductor
c differential input transconductor   d differential output transconductor

It may be described as a differential input/differential output voltage-to-current converter and is defined by the single parameter G_m, its transconductance. It is often referred to as a transconductor. Note that although the input variables are voltages, the output variables are currents. Two special cases of this general transconductor are of interest. In the first, one of the output terminals is grounded or not present so that there is only one output as shown in Figure 10.4c; this will be referred to as a differential input transconductor. In the second version (Figure 10.4d), there is only one input terminal and this will be referred to as a differential output transconductor. We will now examine the use of such components to realise various types of integrator.

10.5   Integrator Types

10.5.1   Voltage Integrators

Figure 10.5a shows a differential input voltage integrator formed from a cascade of the differential input transconductor of Figure 10.4c and the integrating current to voltage converter of Figure 10.4a. It is defined by

$$V_o = (G_m/sC) (V_{in1} - V_{in2}) \tag{10.5}$$

It may be used directly in the polynomial filter structure of Figure 10.2b. Note that there is a one-to-one correspondence between each integrator time constant (C/G_m) and the value of an inductor or capacitor in the LCR filter.

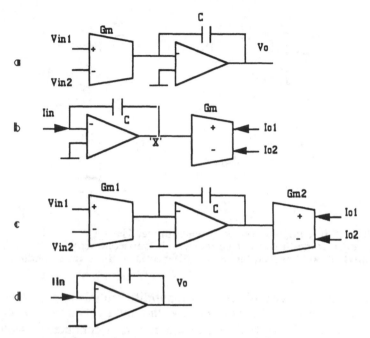

Figure 10.5   Realisation   of alternative integrators
a voltage integrator   b current integrator
c voltage/current integrator   d current/voltage integrator

10.5.2   Current Integrators

A differential output current integrator may be realised as shown in Figure 10.5b by cascading an integrating current to voltage converter with a differential output transconductor. The transfer function is

$$I_{o1} = - I_{o2} = (G_m/sC) \, I_{in} \tag{10.6}$$

This integrator may be used in the polynomial filter structure of Figure 10.2c. The integrator has a time constant C/G_m. The state variables in the SFG are now being represented in the circuit by currents.

10.5.3   Mixed Variable Integrators

By forming a cascade of a differential input transconductor, an integrating current-to-voltage converter and a differential output transconductor, as shown in Figure 10.5c, we obtain a differential input/differential output transconductor, as required for the polynomial and elliptic filter realisations in Figures 10.2d and 10.3c. For this integrator, the input variables are voltages and the output variables are currents. The input output relationship is

$$I_{o1} = -I_{o2} = (G_{m1}G_{m2}/sC)(V_{in1} - V_{in2}) \qquad\qquad 10.7)$$

yielding a time constant $C/(G_{m1}G_{m2})$.

The filter realisations in Figures 10.2d and 10.3c also require a single input/single output integrator, and this may be simply realised by the integrating current to voltage converter, shown again in Figure 10.5d.

10.5.4   Integrator Canonicity

In order that monolithic filters have reasonably good response accuracies, it is important that the sensitivity of response to component parameter variations is minimised. An LCR filter designed for maximum power transfer at the frequencies of minimum loss in the passband has zero sensitivity of the response to changes in inductor and capacitor values at these frequencies and the sensitivity is low throughout the passband. For the three structures in Figures 10.2b, c and d, there is a one-to-one correspondence between the time constant of each integrator and the value of a reactive element in the LCR network. Furthermore, we have now established a direct link between the integrator time constant and the values of circuit capacitances and transconductances. Hence, the three structures in Figure 10.2 do preserve the low sensitivity property. Integrators such as those in Figure 10.5, which are defined by a single time constant, are referred to as canonic.

10.6   Signal Flow Graph Scaling

Modern CMOS integrated circuit processes which are used for the implementation of monolithic analogue filters tend to have low power supply voltages. In view of the large peaks in the values of the voltages and currents in an LCR filter at frequencies in or around the passband edges, it is important to be able to apply scaling to the voltages within the simulating filter at the design stage in order that signal levels within such filters are not limited to unaccepatably low levels. Let us assume that the variables in the SFG of Figure 10.2a are to be simulated by voltages and carry out scaling on the nodal variables. The SFG is shown again in Figure 10.6a with a contour around each node with which a desired scaling factor is associated. The result of applying contour scaling [10] is shown in Figure 10.6b. In general, each node will require scaling by a different factor and this leads as shown in Figure 10.6b to a situation where the original +1 and -1 unity branch pairs at the integrator inputs are no longer equal. Hence, realisation in terms of canonic differential input voltage integrators as in Figure 10.2b becomes impossible. Thus, we see that using voltage integrators, the requirements for minimum sensitivity (use of canonic integrators) and good signal handling capability (ability to apply nodal voltage scaling) are incompatible.

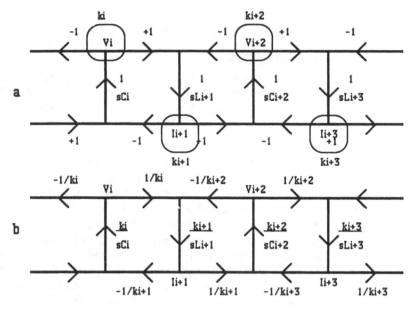

Figure 10.6   Application of voltage scaling to SFG
a SFG showing scaling contours   b SFG after scaling transformation

10.7   Generalised Filter Realisation Architecture

10.7.1   The Need for a Generalisation

We have seen that current is a useful concept in the realisation of low sensitivity filters with good signal handling capability. The basic building blocks which we have used for filter realisation, namely the integrators of Figure 10.5, were derived by combining an integrating charge to voltage converter (Figure 10.4a) with special cases of the general differential input/differential output transconductor of Figure 10.4b having an input or output terminal grounded. This suggests that filter structures realised using the integrators in Figure 10.5 may be regarded in a more general way as special cases of a general realisation architecture based simply on the general transconductor of Figure 10.4b in conjunction with the integrating current-to-voltage converter of Figure 10.4a.

Although a classification of filter structures as based on use of the different types of integrators considered (voltage, current and mixed variable) has allowed us to make some fairly general statements about sensitivity and signal handling capability, this classification is not entirely satisfactory. Consider the realisation in Figure 10.7 of the polynomial lowpass filter signal flow graph in Figure 10.2a. This circuit may be interpretted in several ways. We may consider the transconductors to be associated in pairs at the inputs of the amplifiers in which case we see the circuit as realised using differential input voltage integrators. We may also associate the transconductors in pairs at the

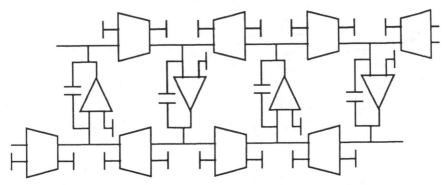

Figure 10.7   Filter realisation using non-canonic transconductors

outputs of the amplifiers, forming differential output current integrators. The circuit may also be interpretted as a mixed variable structure of the kind represented in Figure 10.2d. These ambiguously interpretted 'integrator entities' are non-canonic and therefore do not have the minimum sensitivity property. Nevertheless, the structure in Figure 10.7 is important, and a popular approach for realising elliptic switched capacitor filters [12] is based on it.

Apart from the difficulty in classifying structures like that in Figure 10.7, there is another problem. This is that a mathematical description of structures using integrators with current variables may not explicitly contain important information about the voltages in the circuit. Voltages are still important in such structures because, generally, the filter input and output variables have to be voltages and also because the nodal voltage peaks determine filter signal handling capability and have to be scaled. Thus a more general classification of filter structures which took account of the fact that for all structures both voltages and currents are relevant at different points in the circuit could provide a non-ambiguous means of classifying circuits [13] and, at the same time, provide a basis for a uniform mathematical treatment of sensitivity and nodal voltage scaling [14]. Such an approach could also, in principle, lead to new realisation structures not previously envisaged.

10.7.2   Generalised Architecture

The basis of the generalised architecture to be presented is a set of integrating current-to-voltage converters, as shown in Figure 10.8a. The n inputs of the integrating current-to-voltage converters are fed from the n outputs of a generalised transconductor network NG, as shown in Figure 10.8b. The n + 1 inputs of NG are the n outputs of the current to voltage converters in Figure 10.8a and also the filter input voltage V_i. This system can be described by,

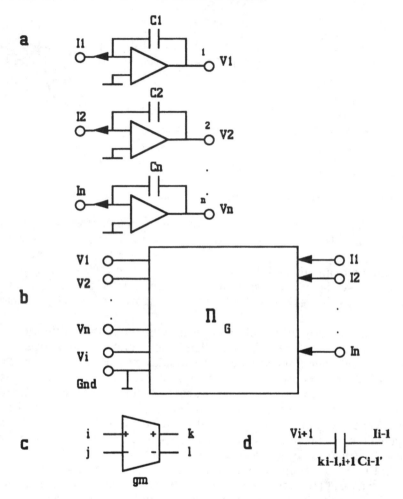

Figure 10.8   Generalised filter realisation
a integrating current-to-voltage converter network   b transconductor
network   c transconductor   d capacitor arrangement

$$
s \begin{vmatrix} C_1 & 0 & & 0 \\ 0 & C_2 & & . \\ . & . & & \\ . & & 0 & \\ 0 & . & 0 & C_n \end{vmatrix} \begin{vmatrix} V_1 \\ V_2 \\ . \\ . \\ V_n \end{vmatrix} = G \begin{vmatrix} V_1 \\ V_2 \\ . \\ . \\ V_n \end{vmatrix} + \begin{vmatrix} g_{01} \\ g_{02} \\ .. \\ .. \\ g_{0n} \end{vmatrix} V_{in}
$$

$$(10.8)$$

where **G** is an n x n transconductance matrix. The transconductor network
NG is assumed to consist of an interconnection of a number of general
transconductors with differential input and differential output as shown in
Figure 10.8c. The inputs to each transconductor (i and j) may be connected

to any of the $n + 1$ input terminals of NG or ground and the transconductor output terminals (k and l) to any of the n output terminals of NG or ground. Comparing Equation 10.8 with Equations 1 and 2, we see that $g_{02}, g_{03} \cdots g_{0n}$ must be zero and that the conductance matrix G is required to have a bi-diagonal form. The $+1$ and -1 entries required will be realised as conductance terms and the general form for G is

$$
G = \begin{vmatrix}
-g_{11} & -g_{12} & 0 & & & & & & \\
g_{21} & 0 & -g_{23} & 0 & & & & & \\
0 & g_{32} & 0 & -g_{34} & 0 & & & & \\
& 0 & g_{43} & 0 & \cdots & & & & \\
& & 0 & \cdot & \cdot & \cdot & \cdot & \cdot & \\
& & & \cdot & \cdot & 0 & -g_{n-3,n-2} & 0 & \\
& & & 0 & g_{n-2,n-3} & 0 & -g_{n-2,n-1} & 0 & \\
& & & & 0 & g_{n-1,n-2} & 0 & -g_{n-1,n} & \\
& & & & & 0 & g_{n,n-1} & -g_{nn} &
\end{vmatrix}
$$

$$(10.9)$$

Comparing Equation 10.8 with Equation 10.2 for the elliptic case, we see that we need to realise additional constant terms of the form $k_{i-1,i+1}$ in the left hand side matrix. Such terms may be simply realised by a capacitor (Figure 10.8d) connected between the input of amplifier i-1 in Figure 10.10a and the output of amplifier 1+1.

The way in which the transconductance terms in Equation 10.9 are built-up as the transconductances of transconductor elements defines various types of realisation which will be described in Sections 10.10 and 10.11. However, we first consider the general properties of such realisations.

10.8   Sensitivity

An LCR filter designed for maximum power transfer at the frequencies of minimum loss in the passband has zero sensitivity of the response to changes in inductor and capacitor values at these frequencies and the sensitivity is low throughout the passband. It can be seen from Equation 10.1, that the component parameters in a unity resistance terminated odd order polynomial lowpass filter occur on the leading diagonal of the left hand side matrix in the state equation. Comparison of Equation 10.1 with the matrix describing the general realisation architecture in Equation 10.8 indicates that the integrating capacitors in the realisation are in corresponding positions in the matrix equation and therefore they must have the same low sensitivities. Looking now at the right hand sides of Equations 1 and 8, we see an important difference. In Equation 10.1, for the LCR prototype network, the state matrix consists of $+1$ and -1 entries. In Equation 10.8, on the other hand, the state matrix, G, which has the general form of Equation 10.9, contains the transconductances of the transconductors in the realisation. Additional sensitivities appear as a finite sensitivity of zero frequency response to

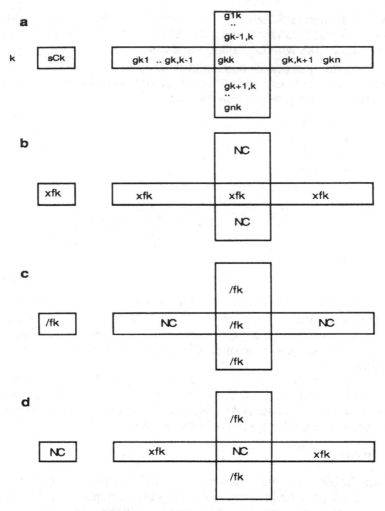

Figure 10.9   Matrix scaling operations
a elements prior to scaling   b elements after row scaling
c elements after variable scaling   d alternative variable scaling scheme

transconductance values. Setting $s = 0$ in Equation 10.8, with G given by
Equation 10.9, and solving for the transfer function, yields

$$\frac{X_n}{X_{in}} = g_{01} \left| \begin{array}{cc} g_{11} g_{23} g_{45} .. g_{n-1,n} & + & g_{12} g_{34} .. g_{n-2,n-1} g_{nn} \\ g_{21} g_{43} .. g_{n-1,n-2} & & g_{32} g_{54} .. g_{n,n-1} \end{array} \right|^{-1}$$

(10.10)

Since for $s = 0$ the description for the elliptic case reduces to the same form,
this equation can be used for elliptic filters also. It will be used to compare
the sensitivity of alternative realisations in section 10.11.

10.9   General Scaling Operations

The tendency of modern CMOS integrated circuit processes to have low power supply voltages makes it desirable to scale the voltages in the simulating filter in order to maximise the signal handling capability. In Section 10.5, we scaled the voltages within the simulating filter by applying contour scaling to the signal flow graph. We now apply general scaling operations to the state matrix equation.

Since an equation is unaffected if both sides are multiplied by the same factor, we can multiply any row in the state variable description of Equation 10. 8 by a factor without altering the transfer function or any variables. The effect of such row scaling on the terms in the affected coefficients in Figure 10.9a is shown in Figure 10.9b. This operation can be interpretted as scaling of the integrating capacitor value and the transconductance terms in the corresponding row by the same factor.

Now let a voltage variable, say V_k, in the state equation of Equation 10.8 be multiplied by a factor f_k. The overall transfer function V_n/V_{in} is left unaltered by this scaling if the coefficients of V_k in Equation 10.8 are divided by f_k as shown in Figure 10.9c. The k'th integrating capacitor and the transconductance terms in the k'th column of the transconductance matrix are both divided by f_k.

An alternative method of variable scaling is derived by applying row scaling by a factor f_k to the k'th row in Figure 10.9c to obtain the result in Figure 10.9d. Here, the only changes are to respectively multiply and divide the corresponding row and column of the transconductance matrix by the factor f_k.

10.10   Transconductor Configurations

A general transconductor, as shown in Figure 10.8c, in general contributes four elements to the transconductance matrix $\mathbf{G}$:

$$
\mathbf{G} = \begin{array}{c}
\\ k \\ \\ \\ l \\ \\
\end{array}
\left|
\begin{array}{ccccccc}
& i & & & j & & \\
& 0 & & & 0 & & \\
.. \; 0 & g_m & 0 & .. \; 0 & -g_m & 0 & .. \\
& 0 & & & 0 & & \\
& . & & & . & & \\
& 0 & & & 0 & & \\
.. \; 0 & -g_m & 0 & .. \; 0 & g_m & 0 & .. \\
& 0 & & & 0 & &
\end{array}
\right|
$$

$$(10.11)$$

If one of the output terminals, say l, is grounded, we have only two terms which occur on a single row,

$$G = k \begin{array}{c} \\ \end{array} \begin{array}{ccccccccc} & i & & & j & \\ | & 0 & & & 0 & & | \\ | \,.. & 0 & g_m & 0 \,.. & 0 & -g_m & 0 \,.. & | \\ | & 0 & & & 0 & & | \end{array} \tag{10.12}$$

The corresponding relationship between the state variables is

$$V_k = (g_m/sC_k)\,(V_i - V_j) \tag{10.13}$$

Thus this transconductor configuration corresponds to the realisation of the differential input voltage integrator of Figure 10.5a.

If one of the input terminals of the general transconductor, say j, is grounded, we also have two terms in **G** but now in a single column,

$$G = \begin{array}{c} \\ k \\ \\ \\ l \\ \\ \end{array} \begin{array}{ccccc} & & i & & \\ | & & 0 & & | \\ | \,.. & 0 & g_m & 0 \,.. & | \\ | & & 0 & & | \\ | & . & & & | \\ | & & 0 & & | \\ | \,.. & 0 & -g_m & 0 \,.. & | \\ | & & 0 & & | \end{array} \tag{10.14}$$

The corresponding state variable relationship is

$$V_k C_k = -Vl_1 C_l = (gm/s)\,V_i \tag{10.15}$$

This transconductor configuration corresponds to the realisation of the differential output current integrator of Figure 10.5b.

If one input terminal and one output terminal of the general transconductor are grounded (say terminals j and l), then each transconductor contributes only one element to the transconductance matrix,

$$G = k \begin{array}{ccccc} & & i & & \\ | & & 0 & & | \\ | \,.. & 0 & g_m & 0 \,.. & | \\ | & & 0 & & | \end{array} \tag{10.16}$$

The corresponding state variable relationship is

$$V_k = (g_m/sC_k)\,V_i \tag{10.17}$$

This transconductor configuration corresponds to the integrators used in the non canonic filter realisation in Figure 10.7.

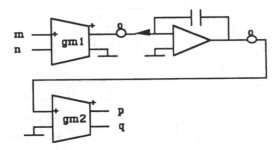

Figure 10.10   Transconductor arrangement

Finally, consider two transconductors, one differential input and one differential output, connected as in Figure 10.10. In this case, the transconductance matrix has four entries arranged as follows,

$$
\begin{array}{c}
\\
\end{array}
$$

$$(10.18)$$

The four entries lie on a vertical and a horizontal line whose intersection corresponds to the output node of the integrating voltage to current converter. The state variable relationships are

$$V_o = (g_{m1}/sC_k)\,(V_m - V_n) \qquad (10.19)$$

and

$$V_p\,C_p = -V_q\,C_q = (g_{m2}/s)\,V_o \qquad (10.20)$$

corresponding to a differential input /differential output integrator using voltage and current variables as in Figure 10.5c.

10.11   Classification of Filter Realisations and Properties

10.11.1   General

In this section we use the general realisation architecture approach to define and classify various realisation structures. We also use our general analysis of sensitivity and nodal voltage scaling to predict their properties.

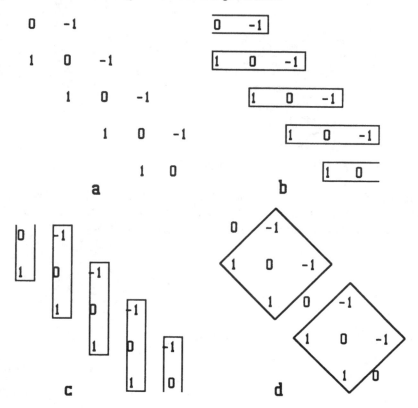

Figure 10.11 **Transconductance matrix realisation** **a required matrix form** **b row-type realisation** **c column-type realisatio** **d cross-type realisation**

10.11.2 Row-type realisation

The required bidiagonal terms in the transconductance matrix of Figure 10.11a may be realised row-by-row as shown in Figure 10.11b. From Equation 10.12, the terms in the transconductance matrix are related according to $g_{k,k-1} = -g_{k,k+1}$. In the filter gain expression of Equation 10.10, such pairs of transconductance terms will precisely cancel. Therefore the sensitivity of gain to transconductance values is zero. This type of realisation corresponds to the structure using differential input voltage integrators as in Figure 10.2b.

Nodal voltage scaling has the effect of scaling terms in the same column, as shown in Figure 10.9c. This will make the transconductance pairs in each row unequal and therfore not realisable as a row-type realisation.

10.11.3 Column-type realisation

The required terms in the **G** matrix can alternatively be realised column-by-column as shown in Figure 10.11c. Equation 10.14 shows   that the transconductance terms are now related by $g_{k-1,k} = -g_{k+1,k}$. In Equation 10.10, such transconductance pair terms will also precisely cancel and therefore the sensitivity of gain to transconductance value is zero. This realisation corresponds to the structure using differential output current integrators as in Fig 10.2c.

Since nodal voltage scaling affects terms in columns, this will leave the transconductance term pairs in each column equal and therefore still realisable in column form. Thus, nodal voltage scaling is compatible with column-type simulation.

10.11.4   Non-canonic realisation

In this case, each term in the G matrix is realised by a seperate transconductor, which has one input terminal and one output terminal grounded (see equation 10.16). It follows that transconductance terms in Equation 10.10 cannot cancel and therefore each transconductance will affect the gain. Therefore sensitivity will be relatively high. This type of realisation has already been illustrated in Figure 10.7.

Nodal voltage scaling can, however, be carried out freely as there are no constraints on the equality of the transconductance matrix terms.

10.11.5   Cross-type simulation

In this case, the required terms in Figure 10.11a are realised as in Figure 10.11d. Each set of four entries, in the form of a cross, are related according to Equation 10.18 by $g_{po} = -g_{po}$, $g_{om} = -g_{on}$. These terms will cancel in the gain expression of Equation 10.10 and therefore the sensitivity of gain to transconductance values is zero. This realisation corresponds to the structure using mixed variables as in Figure 10.2d.

Nodal voltage scaling can be applied to the nodes corresponding to the centres of the 'crosses' in Figure 10.11d because this leaves all pairs of equal terms still equal. It cannot be applied to nodes corresponding to the diagonal (zero) elements between the crosses because this will make the row pairs in adjacent 'crosses' unequal. Thus, cross-type simulation allows nodal voltage scaling to be carried out only at certain nodes.

10.11.6   'Z'-type simulation

Figure 10.12a shows that the G matrix may be realised as a combination of 'Z'-shaped groups. Figure 10.12b shows the realisation of each group as a rectangular block of four terms and an additional pair of terms in a row. The block of four terms corresponds to a transconductor with no input or output terminals grounded, and the pair of terms correspond to a differential input transconductor. Since cancelled terms are in a position in the conductance matrix of Equation 10.9 where no terms exist, the gain expression of Equation 10.10 cannot be used to evaluate sensitivity to these terms

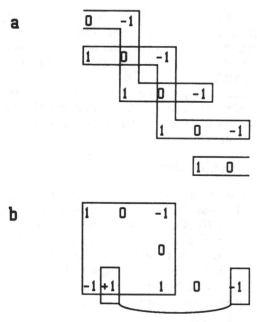

**Figure 10.12 Alternative transconductance matrix realisation
a matrix decomposition b realisation of matrix element group**

However, analysis indicates that filter gain is independent of the transconductance of the transconductors involved.

The presence in Figure 10.12b of quads of equal terms and pairs of row terms, means that it is impossible to carry out nodal voltage scaling, which affects values in the corresponding column, while still maintain the equality of terms necessary for realisation. Thus, we see that 'Z'-type simulation does not permit nodal voltage scaling to be carried out.

10.11.7 Elliptic Filters

Elliptic filter realisation is based on the architecture of Figure 10.3c. The constraints on the signs of the transconductance entries are best satisfied using the 'cross-type' of realisation. Thus the comments in Section 10.11.5 on sensitivity and nodal voltage scaling apply also to elliptic filters.

10.12 Examples

10.12.1 Polynomial Filters

The general differential input/ differential output transconductor of Figure 10.8c may be realised in continuous time form using a number of circuits [15] and Chapter 5 contains an excellent review of available approaches and techniques. In this paper, we shall concentrate on switched capacitor implementations. In this case, the transconductor of Figure 10.8c can be

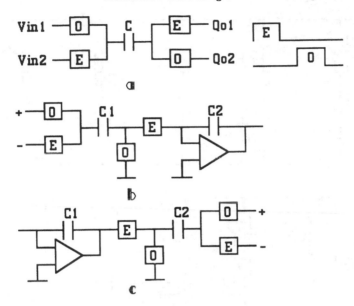

Figure 10.13 Switched capacitor polynomial filter realisation
a switched capacitor 'transconductor'
b differential input SC voltage integrator
c differential output SC charge integrator

realised very simply as shown in Figure 10.13a by means of a single capacitor and four switches operated from a 2-phase non-overlapping clock as shown (E - even, O - odd). Note that for switched capacitor realisation, the output variable is charge instead of current.

From the basic switched capacitor transconductor building block, we may derive canonic voltage and canonic charge integrators as in Figures 10.5a and b and these are shown in Figures 10.13b and c.

In this section, we consider realisation of a 5th order polynomial lowpass filter, for which case Equation 10.9 becomes

$$
G = \begin{vmatrix}
-g_{11} & -g_{12} & 0 & 0 & 0 \\
g_{21} & 0 & -g_{23} & 0 & 0 \\
0 & g_{32} & 0 & -g_{34} & 0 \\
0 & 0 & g_{43} & 0 & -g_{45} \\
0 & 0 & 0 & g_{54} & -g_{55}
\end{vmatrix}
$$

$$(10.21)$$

The gain expression, Equation 10.10, becomes

$$
\frac{X5}{X_{in}} = g_{01} \left| \frac{g_{11} g_{23} g_{45}}{g_{21} g_{43}} + \frac{g_{12} g_{34} g_{55}}{g_{32} g_{54}} \right|^{-1}
$$

$$(10.22)$$

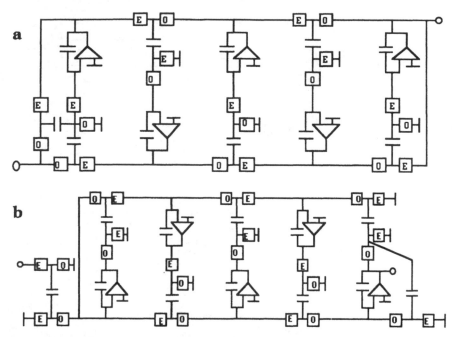

Figure 10.14   SC polynomial 5th order lowpass filter realisations
a row-type realisation   b column-type realisation

For simplicity, we shall ignore switching delays in the switched capacitor systems and make a continuous-time approximation of transconductor operation with transconductance assumed equal to the capacitor value C; all the circuits generated are amenable to the standard exact design procedures for discrete-time systems.

For a row type realisation, the terms in Equation 10.21 are realised as follows:

$$g_{11} = g_1, g_{12} = g_2, g_{21} = g_{23} = g_3,$$
$$g_{32} = g_{34} = g_4, g_{43} = g_{45} = g_5, g_{54} = g_{55} = g_6$$

where gi is the transconductance of the i'th transconductor. The gain, from Equation 10.22, is $X_5/X_{in} = g_{01}/(g_1 + g_2)$. We can choose $g_{01} = g_1 + g_2$, as suggested in [7], to obtain $X_5 = X_{in}$ independent of all transconductances and hence independent of all switched capacitor values. The row-type realisation structure is shown in Figure 10.14a. As mentioned above, nodal voltage scaling cannot be carried out in this realisation.

For a column type realisation, the terms in Equation 10.21 are realised as:

$$g_{11} = g_{21} = g_1, g_{12} = g_{32} = g_2, g_{23} = g_{42} = g_3,$$
$$g_{34} = g_{54} = g_4, g_{45} = g_5, g_{55} = g_6$$

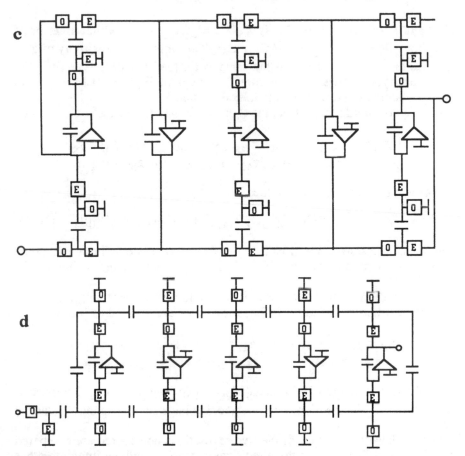

Figure 10.14 (Continued)
c cross-type realisation d non-canonic realisation

The gain, from Equation 10.22, is $X_5/X_{in} = g_{01}/(g_5 + g_6)$. In terms of capacitor values, we have $X_5/X_{in} = C_0/(C_5 + C_6)$. The worst case sensitivity of gain to capacitor value is 2. This was previously attributed to a voltage-to-charge conversion at the input (via C_0) and charge-to-voltage conversion at the output. The column-type realisation structure is shown in Figure 10.14b. Nodal voltage scaling can be carried out for all nodes by scaling capacitor values.

For a 'cross'-type realisation, the terms in Equation 10.21 are realised as:

$$g_{11} = g_{01} = g_1, \ g_{12} = g_{32} = g_2, \ g_{21} = g_{23} = g_3,$$
$$g_{34} = g_{54} = g_4, \ g_{43} = g_{45} = g_5, \ g_{55} = g_6$$

The gain, from Equation 10.22, is $X_5/X_{in} = g_1/(g_1 + g_6)$, which in terms of capacitor values is $X_5/X_{in} = C_1/(C_1 + C_6)$. The worst case sensitivity of gain to capacitor value is 1, which is the same as for the LCR prototype filter [8]. The cross-type realisation structure is shown in Figure 10.14c. Nodal voltage scaling can be applied only to nodes 2 and 4.

For a non-canonic realisation, the terms in Equation 10.21 are realised as:

$$g_{01} = g_1, g_{11} = g_2, g_{12} = g_3, g_{21} = g_4, g_{23} = g_5, g_{32} = g_6,$$
$$g_{34} = g_7, g_{43} = g_8, g_{45} = g_9, g_{54} = g_{10}, g_{55} = g_{11}$$

The gain in terms of capacitor values is

$$X_5/X_{in} = C_1/[C_2C_5C_9/(C_4C_8) + C_3C_7C_{11}/(C_6C_{10})] \qquad (10.23)$$

The worst case sensitivity of gain to capacitor value is 6. The non-canonic realisation structure is shown in Figure 10.14d. Nodal voltage scaling can be carried out.

For realisation of the 5th order polynomial lowpass filter using the Z-type realisation structure, the terms in Equation 10.21 are realised as follows:

$$g_{01} = g_{11} = g_1, g_{12} = g_2, g_{21} = g_{43} = g_{23} = g_3, g_{41} = g_3 - g_4$$
$$g_{45} = g_4, g_{32} = g_{34} = g_{54} = g_5, g_{25} = g_5 - g_6, g_{55} = g_6$$

The gain, from Equation 10.22, is $X_5/X_{in} = g_1/(g_1 + g_2)$ yielding a worst case sensitivity of 1. As mentioned above, nodal voltage scaling cannot be carried out.

Some relevant properties of the five realisations considered are compared in Table 10.1. We see that the circuit using canonic voltage integrators has minimum sensitivity, but nodal voltage scaling cannot be applied. For the circuit using charge integrators, nodal voltage scaling can be applied but there is a worst case sensitivity of 2 due to the voltage/charge conversions at the filter input and output. Nodal voltage scaling is also possible in the realisation using non-canonic voltage integrators (circuit 3) but there is a considerable increase in sensitivity. Cross-type simulation allows a very close approach to minimum sensitivity but nodal voltage scaling can only be carried-out partially. Z-type simulation provides moderate sensitivity and only a partial capability to carry out nodal voltage scaling. Apart from the non-canonic realisatiom, the other realisations all require the same components, namely five integrating charge to voltage converters and six transconductors, and Table 10.1 illustrates a wide variety of circuit properties for the different transconductor configurations. This example shows that the adoption of charge integrators seems to represent an optimum solution as it provides a close approach to minimum sensitivity while fully permitting nodal voltage scaling.

Table 10.1 Comparison of realisation properties

Realisation	Worstcase DC sensitivity	Nodal voltage scaling
Row-type	0	No
Cross-type	1	Partial
Column-type	2	Yes
Non-canonic	6	Yes
Z-type	1	No

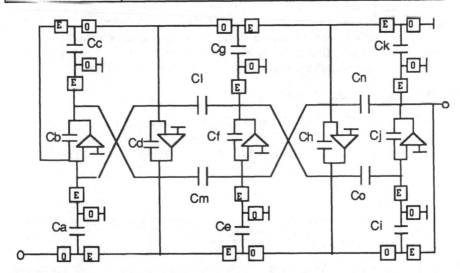

Figure 10.15 5th order elliptic switched capacitor filter example

10.12.2 Elliptic Filters

We consider the realisation of a 5th order elliptic lowpass filter using the basic architecture of Figure 10.3c with cross-type integrator realisation and the resulting circuit is shown in Figure 10.15. The source and load termination loops have been realised in two different ways, leading to a DC gain for the complete filter of $1/[1 + C_k/C_a]$ (DC gain is independent of all other capacitor values). For the typical case of $C_k = C_a$ the DC sensitivities of C_k and C_a are $-1/2$ and $+1/2$, respectively, and the summed DC sensitivity is 1, as it is for the LCR prototype filter. For a state space realisation, the summed DC sensitivity would be zero, but the capacitor spread would be prohibitively large[11]. Thus the proposed approach based on integrators using both voltage and charge variables makes a contribution to reducing sensitivity under the practical constraint of avoiding excessive capacitor spread.

10.13   Conclusions

We have attempted to make a generalised study of the use of voltage and current (or charge) variables in the realisation of continuous-time and switched capacitor monolithic filters. For polynomial lowpass filters, the use of current (or charge) integrators allows us to carry out nodal voltage scaling while at the same time obtaining low sensitivity, which is not possible using voltage integrators alone. A formulation in terms of generalised transconductor element configurations provides a framework within which to classify the various realisations possible and allows a systematic mathematical treatment of sensitivity and nodal voltage scaling, confirming the expected properties of the structures. For elliptic filters, use of a judicious combination of voltages and currents (or charges) provides the lowest sensitivity of any approach with acceptable component parameter spread. We conclude that in order to obtain optimum implementations, we cannot afford to ignore the use of current (or charge) processing.

10.14   Acknowledgements

The author would like to thank Bhajan Singh of Plessey Research, and John Taylor and Paul Radmore, both of University College London for collaboration and for important contributions to the work described in this Chapter.

10.15   References

[1] C.Toumazou F J Lidgey and PK Cheung, Current actuated analogue signal processing circuits: Review and recent developments. Procs 1988 IEEE Int Symp Circs and Systems (Portland, USA), pp 1572-1575, June 1989

[2] C Toumazou and F J Lidgey,and C.A Makris, Extending voltage-mode opamps to current-mode performance, IEE Proceedings Part G (ECS), vol 137, no 2, pp 116-130, April 1990

[3] C Toumazou and F J Lidgey, Universal active filters using current conveyors, Electronics Letters, vol 22, pp 662-664, 1986

[4] D G Haigh, B Singh and J T Taylor, Low Sensitivity Switched Capacitor Filters with Maximum Signal Handling Capability, Procs 1987 IEEE Int Symp on Circs and Systems (Philadelphia), May 1987, pp 1110-1113.

[5] D G Haigh, J T Taylor and B Singh, Low sensitivity switched capacitor simulation of elliptic lowpass LCR ladder filters, Electronics Letters, vol 24, no 1, 7th January 1988, pp 52 - 54

[6] Roberts G W, Snelgrove W M and Sedra A S, Switched capacitor state-space filters using intermediate function synthesis, Procs 1986 IEEE Int Symp Circs and Systems, San Jose, May 1986, pp 614-617

[7] Kunieda H and Ohshimo A, State space approach to the design for low passband sensitivity switched capacitor filters, Procs 1984 IEEE Int Symp Cics and Systems, Montreal, May 1984, pp 292-295

[8] D G Haigh and J T Taylor, Continuous time and switched capacitor monolithic filters based on current and charge simulation, Procs 1989 IEEE Int Symp on Circs and Systems (Portland, Oregon), May 1989, pp 1580 - 1583

[9] Jacobs G M, Allstot D J, Brodersen R W and Gray P R: Design techniques for MOS switched capacitor ladder filters, IEEE Trans Circuits Syst, vol CAS-25, no 12, Dec 1978, pp 1014-1021

[10] Perry D, Scaling transformation of multiple feedback filters, IEE Proceedings Part G, vol 128, no 4, pp 176-179, August 1981

[11] Haigh D G and Taylor J T, State space description and realisation of elliptic lowpass LCR ladder filters, Electronics Letters, 1987, 23, pp 1032-1034

[12] K Martin, Improved circuits for the realisation of switched capacitor filters, IEEE Trans Circuits Syst, vol CAS-28, pp 85-92, February 1981

[13]   D G Haigh, J T Taylor and B Singh, Continuous-time and switched capacitor monolithic filters based on current and charge simulation, IEE Procs Part G (CDS), vol 137, no 2, pp 147-155 , April 1990

[14]   D G Haigh and P Radmore, Some properties of continuous-time and switched capacitor filters based on current charge and voltage simulation, Procs 1990 IEEE Int Symp on Circs and Systems, New Orleans, 1st -3rd May 1990

[15] P J Ryan and D G Haigh, Novel fully differential MOS transconductor for integrated continuous-time filters, Electronics Letters, Vol 23, no 14, 2nd July 1987, pp 742-743..

Chapter 11

Switched-Current Filters

John.B.Hughes

11.1 Introduction

Analogue sampled-data signal processing has been dominated for the past decade by the switched-capacitor technique. It has proved to be a reliable workhorse in a wide range of signal processing applications, both linear and non-linear, over a frequency range spanning audio and video. So why should a new sampled-data technique such as switched-currents be needed?

Following its inception in 1972,[1] switched capacitors gained favour as a technique for implementing active filters capable of greater precision and compactness than earlier active-RC filters, especially in low frequency applications (e.g. telephony). Its repertoire has grown to include more general signal processing[2] and at the height of its popularity it was used to implement complete sampled-data subsystems[3]. But by the mid-1980's, systems which had previously been made using switched-capacitors were starting to find solutions using digital signal processing. Although often more dissipative and less silicon-area-efficient, the digital approach offered easier computer-aided design with shorter time-to-market. Test methods were simpler and as VLSI feature sizes diminished, complexity rose (and continues to rise) so that complete systems can now be integrated on a single silicon chip. This has changed the role of switched-capacitors. Single chip digital systems require analogue circuits at their interface with the outside world and switched-capacitors are now commonly found performing A-D and D-A conversion , sample and holding, filtering etc. at these interfaces.

Switched-capacitors have never fitted standard VLSI processing. Their need for linear floating capacitors has led to special process options, e.g. double polysilicon, being added to the digital VLSI process. In their hey-day, when they occupied complete chips, this extra expense was not serious. However in their present role, where they may occupy no more than 10-20 of the total chip area, the extra expense has become more difficult to justify.

In the future, as feature sizes shrink further into the sub-micron region, digital systems with higher circuit density will be possible but this wiil lead to higher power dissipation. Further, the smaller dimensions will give higher device electric fields resulting in poorer MOSFET performance. To contain this situation, the industry has proposed lowering the standard supply from 5V to 3.3V [4]. Inevitably, the VLSI process will be optimised for digital rather than analogue performance and, while threshold voltages will be lowered, logic gate leakage will prevent them being reduced far enough for optimum switched-capacitor performance. Analogue performance will suffer doubly: reduced voltage swings will have a direct impact on dynamic range, and sub-optimal threshold voltages will worsen the performance of MOSFET switches. Of course, extra threshold options, beyond those needed for the digital circuits, would help but would make processing still more expensive. So, switched-capacitor performance is set to degrade[5] and the expense of its extra processing option(s) will make it less attractive.

It was against this background that a new technique, called switched-currents[6], was proposed. By overcoming the aforementioned difficulties, switched-currents has aspirations to supersede switched-capacitors. In this Chapter, the technique is fashioned to the needs of filtering although, like switched-capacitors, it may be applied to many other signal-processing functions (see Chapters 7, 13 and 14). The chapter starts with a brief review of the switched-capacitor technique and then introduces the concepts of sampled-current signal-processing and current memory. Circuit modules are developed which are suitable for implementing filters and this is illustrated by the design of a 6th order low-pass filter. The causes of analogue error are outlined and circuit enhancements are given for improving performance and for rendering the modules amenable to design-automation techniques.

11.2   Switched-capacitor Background

A good starting point for introducing a new technique such as switched-currents is to review the state-of-the-art alternatives. In the case of a mature technique such as switched-capacitors a complete review is not needed as many books and review papers are available[7,8,9]. The intention here is merely to highlight the major features of the technique so that switched-currents may be viewed from the right perspective.

Most switched-capacitor filter structures have resulted from the substitution of an active-RC filter's continuous-time integrators by switched-capacitor counterparts. This approach has been applied to state-variable filters[10,11] and to filters which simulate the nodal voltages of lossless ladder prototypes[12]. Three of the principal switched-capacitor integrator building blocks are shown in Figure 11.1. The non-inverting integrator (Figure 11.1(a)) operates as follows. On phase ϕ_2 of the non-overlapping clock period (n-1), the charge on capacitor C holds the output voltage at $v_0(n-1)$ while capacitor $\alpha_1 C$ is charged to $v_1 (n-1)$. The next clock phase is ϕ_1 of period (n), and capacitor $\alpha_1 C$ is discharged into capacitor C causing the output voltage to charge to $v_0(n)$. It is easily shown that

$$V_o(n) = V_o(n-1) + \alpha_1 V_1(n-1)$$

(11.1)

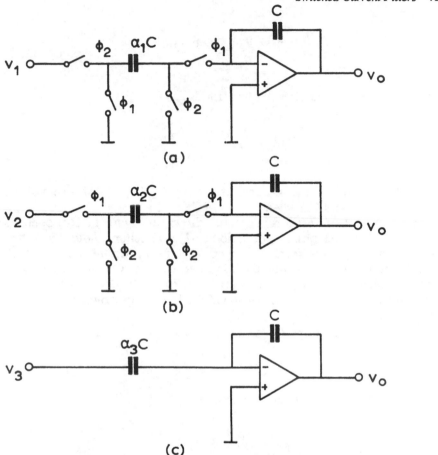

Figure 11.1 **Switched-capacitor circuits (a) non-inverting lossless integrator (b) inverting lossless integrator (c) inverting feed-forward integrator**

which gives the z-domain transfer function

$$H_1(Z) = \frac{V_o(Z)}{V_1(Z)} = \frac{\alpha_1 Z^{-1}}{1 - Z^{-1}}$$

(11.2)

This is the Forward Euler z-transform ($s \rightarrow (1 - z^{-1})/Tz^{-1}$) of a non-inverting integrator ($H(s) = 1/sRC$) where $\alpha_1 = T/RC$

The inverting integrator shown in Figure 11.1(b) operates as follows. On clock phase ϕ_2 of period (n-1), capacitor C holds the output voltage at $v_0(n-1)$ while capacitor $\alpha_2 C$ is discharged. On the next phase ϕ_1 of period (n), capacitor $\alpha_2 C$ is charged to $v_2(n)$ and capacitor C charges to $v_0(n)$. It is easily shown that

$$V_o(n) = V_o(n - 1) - \alpha_2 V_2(n)$$

(11.3)

which gives the z-domain transfer function

$$H_2(Z) = \frac{V_o(Z)}{V_2(Z)} = -\frac{\alpha_2}{1 - z^{-1}}$$

(11.4)

This is the backward Euler z-transform ($s \rightarrow (1 - z^{-1})/T$) of an inverting integrator ($H(s) = -1/sRC$) where $\alpha_2 = T/RC$.

The inverting feed-forward integrator shown in Figure 11.1(c) operates as follows. On clock phase ϕ_2 of period (n-1), capacitor C holds the output at $v_0(n-1)$ and capacitor $\alpha_3 C$ is charged to $v_3(n-1)$. On the next ϕ_1 clock phase of period (n), capacitor $\alpha_3 C$ is charged to $v_3(n)$ and in so doing transfers a charge of magnitude $(v_3(n) - v_3(n-1))\alpha_3 C$ onto capacitor C causing the output voltage to change to $v_0(n)$. It is easily shown that

$$V_o(n) = V_o(n - 1) - \alpha_3(V_3(n) - V_3(n - 1))$$

(11.5)

which gives the z-domain transfer function

$$H_3(Z) = \frac{V_o(Z)}{V_3(Z)} = -\alpha_3$$

(11.6)

This feed-forward has resulted from applying the derivative of the input signal to the integrator loop formed by capacitor C and the op.amp. with the result that the input is fed-forward with gain and inversion. It should be noted that all the circuits shown in Figure 11.1 are d.c. unstable, as indeed is the continuous-time integrator, and would never be used in isolation.

The virtual earth of the op.amp. is a current summing node and this enables any number of the input branches of the circuits of Figure 11.1 to be combined. This is shown in the generalised integrator of Figure 11.2(a). In this arrangement, the three input branches previously described are combined with a fourth branch ($\alpha_4 C$) which is driven from the output, v_0, having the same switch phasing as the second branch ($\alpha_2 C$). By superposition,

$$V_o(n) = V_o(n - 1) + \alpha_1 V_1(n - 1) - \alpha_2 V_2(n) - \alpha_3(V_3(n) - V_3(n - 1)) - \alpha_4 V_o(n)$$

(11.7)

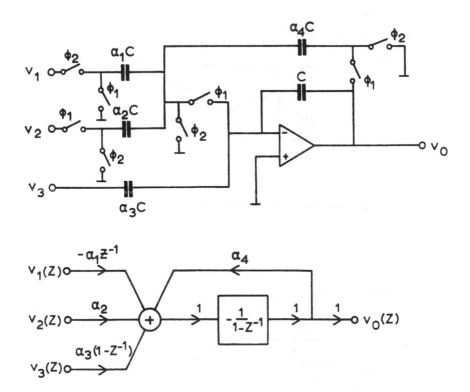

Figure 11.2   Generalised switched-capacitor integrator (a) circuit topology (b) z-domain block diagram

which, after some manipulation and z-transformation, gives

$$V_o(Z) = A_1 \frac{Z^{-1}}{1 - BZ^{-1}} V_1(Z) - A_2 \frac{1}{1 - BZ^{-1}} V_2(Z) - A_3 \frac{1 - Z^{-1}}{1 - BZ^{-1}} V_3(Z)$$

(11.8)

where

$$A_1 = \frac{\alpha_1}{1 + \alpha_4} \ , \ \ A_2 = \frac{\alpha_2}{1 + \alpha_4} \ , \ \ A_3 = \frac{\alpha_3}{1 + \alpha_4} \ , \ \ B = \frac{1}{1 + \alpha_4}$$

(11.9)

The z-domain block diagram is shown in Figure 11.2(b). Clearly, the extra feedback branch (α_4 C) has introduced damping into the integrator. The generalised integrator is a frequently used building block for even-order state-variable filters. Each biquadratic section comprises a combination of lossless or damped integrators in a feedback loop, and can be made to execute

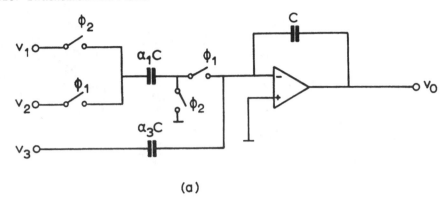

(a)

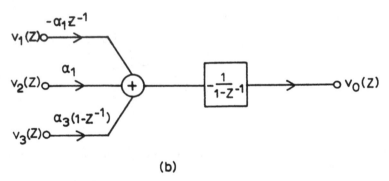

(b)

Figure 11.3   Switched-capacitor subtracting integrator   (a) circuit topology (b) z-domain block diagram

the bilinear z-transform (s --> $2(1 - z^{-1})/(1 + z^{-1})T$) which enables 'exact' filter design.

A second useful building block is the subtracting integrator shown in Figure 11.3(a).This operates as follows.On clock phase ϕ_2 of period (n-1), capacitor αC is charged to $v_1(n-1)$ while capacitor C holds the output voltage at $v_0(n-1)$. On phase ϕ_1 of the next clock period (n), capacitor $\alpha_1 C$ is charged to $v_2(n)$ and in so doing transfers a charge with magnitude $\alpha_1 C$ $(v1(n-1) - v2 (n))$ to capacitor C. The branch $\alpha_3 C$ behaves as described earlier. The resulting output voltage is given by

$$V_o(n) = V_o(n - 1) + \alpha_1(V_1(n - 1) - V_2(n)) - \alpha_3(V_3(n) - V_3(n - 1))$$

(11.10)

which, after z-transformation, gives

$$V_o(Z) = \frac{\alpha_1}{1 - Z^{-1}} (V_1(Z)Z^{-1} - V_2(Z)) - \alpha_3 V_3(Z)$$

(11.11)

This is expressed in the z-domain block diagram shown in Figure 11.3(b). The circuit may be regarded as a special case of the generalised integrator of Figure 11.2(a) with $\alpha_4 = 0$ and $\alpha_2 = 1$ but avoiding mismatch errors resulting from separate branches (α_1 C and α_2 C) . It finds particular use in filters designed by simulation of lossless ladder prototypes.

In summary, the switched-capacitor technique may be characterised by the following features :

o   Switched-capacitor systems are usually made by substitution of switched-capacitor integrators in well-established active-RC circuit structures. This enables them to retain their modularity and low sensitivity to component spreads.

o   The integrators execute algorithms, defined by their difference equations, involving the manipulation of past and present voltage samples. Low supply voltage operation necessarily implies degraded performance.

o   Manipulation of voltage samples is accomplished by transferring charge between floating capacitors. For linear operation these capacitors must also be linear and so switched-capacitor circuits are not truly VLSI compatible.

o   Building blocks have been defined which are versatile (generalised integrator), self-contained (only capacitors within the integrator require critical matching) and which have low design interaction (loading of a module by other modules does not significantly change the module's performance). These are necessary attributes for enabling hierarchical design leading to design-automation.

11.3   Switched-current Systems

A switched-current system may be defined as a system using analogue sampled-data circuits in which signals are represented by current samples. This is in contrast with switched-capacitor circuits which use voltage samples. The applications for switched-current systems will be much the same as for switched-capacitors viz. filters, A-D and D-A converters, general signal processing etc. but it will be shown that unlike switched-capacitors, switched-current circuits can be implemented using a standard VLSI CMOS process. Linear floating capacitors are not needed in switched-current circuits and, in principle, voltage swings need not be large as signals are represented by currents, giving a potential for low supply voltage operation.

To be clear on this last point, any circuit which handles signal currents must develop internal voltage swings.   However, the voltages in switched-

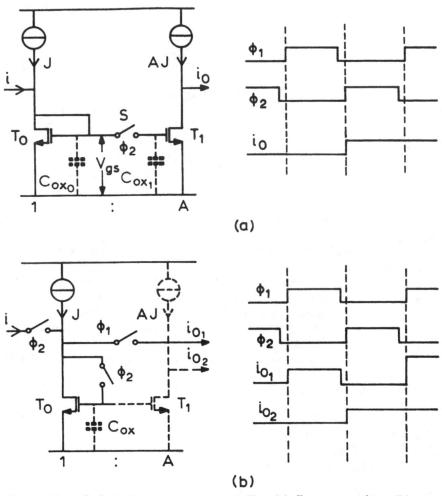

Figure 11.4 Switched-current memory cells   (a) first-generation   (b) second-generation

current circuits are incidental to the signal-processing and are not used to represent signals. By the same token, switched-capacitor circuits pass currents internally in the act of charging their capacitors but these currents are also incidental, not representing signals either.

A switched-current counterpart of the generalised switched-capacitor integrator shown in Figure 11.2(a) would have the following difference equation

$$i_o(n) = \frac{i_o(n-1)}{1+\alpha_4} + \frac{\alpha_1 i_1(n-1)}{1+\alpha_4} - \frac{\alpha_2 i_2(n)}{1+\alpha_4} - \frac{\alpha_3(i_3(n) - i_3(n-1))}{1+\alpha_4}$$

$$(11.12)$$

Taking z-transforms gives

$$i_o(Z) = A_1 \frac{Z^{-1}}{1 - BZ^{-1}} i_1(Z) - A_2 \frac{1}{1 - BZ^{-1}} i_2(Z) - A_3 \frac{1 - Z^{-1}}{1 - BZ^{-1}} i_3(Z)$$

$$(11.13)$$

where A_1, A_2, A_3 and B are as defined in equation (11.9). A switched-current generalised integrator described by the algorithm above could be used directly to make switched-current systems employing structures which are duals of those already used in active-RC and switched-capacitor systems. Such a switched-current system should inherit the desirable attributes of switched-capacitor systems (modularity, component insensitivity), just as switched-capacitors did from earlier active-RC circuits.

11.4   Switched-current Integrators

11.4.1   Current Memories

Current memory can be achieved in two ways as shown in Figure 11.4. The memory circuit shown in Figure 11.4(a) is simply a current mirror with a switch S separating its input and output transistors. Clock phases ϕ_1 and ϕ_2 are defined by non-overlapping voltages and it is assumed that the switches turn on when their control voltage is high. On phase ϕ_2 switch S is closed and both oxide capacitances, (C_{oxo} and C_{ox1}), are charged to V_{gs} where

$$V_g \cong V_T + \sqrt{\frac{J + i}{\frac{\mu C_{o0}}{2} \frac{W}{L}}}$$

$$(11.14)$$

By normal current mirror action, $i_0 = -Ai$ and i_0 is available simultaneously with the input sample. On phase ϕ_1, switch S opens and isolates the input from the output. A voltage close to V_{gs} is held on C_{ox1} and sustains a current close to $-A.i$ at the output.

The arrangement in Figure 11.4(b) can achieve memory within a single transistor, T_0. On phase ϕ_2 T_0 is diode connected and conducts current $J + i$ and as before, V_{gs} is stored on the oxide capacitance. On phase ϕ_1, T_0 maintains its current $J + i$, and so $i_{01} = -i$. To achieve scaling by a factor A, an extra output stage (T1) can be used to give $i_{02} = -A i$. As T_0 is used alternately as an input diode and an output transistor, i_{01} is available only during phase ϕ_1. Output current i_{02} is available for the whole period as with the other memory cell.

Both arrangements are displaying a current memory property. Of course, in reality this results from the voltage V_{gs} being stored on C_{ox1} and this highlights a crucial distinction from switched-capacitors. Switched-capacitors require linear floating capacitors to perform linear transfer of voltages. The switched-current memory cell requires only a grounded capacitor, which need not be linear, to hold V_{gs} at the value imposed by the current J + i. With ideal MOSFET's, the 'stored' current is transferred linearly from input to output without the need for special linear capacitors.

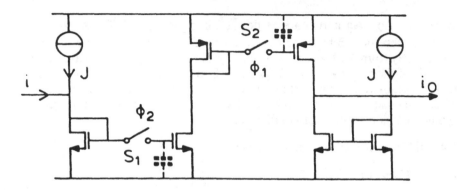

Figure 11.5 **First-generation switched-current delay cell**

11.4.2 First-generation Switched-current Integrators

The first attempt to make a switched-current integrator[6], made use of the current memory cell of Figure 11.4(a). Figure 11.5 shows a simple delay circuit formed by cascading n-channel and p-channel memory cells of this type. The n-channel memory is clocked on phase ϕ_2 and the p-channel memory on phase ϕ_1. During phase ϕ_2 (the end of period (n-1)) the input current i(n-1) and the bias current J enter the input diode of the n-channel memory cell. During phase ϕ_1 (the start of period (n)) this current is stored in the n-channel memory cell and fed to the p-channel memory and, since switch S2 is closed, a current i_0 (n) equal to i(n-1) flows in the output. The input current, i(n), does not propagate to the p-channel memory on this phase as switch S1 is open. On the next phase, ϕ_2, i_0(n) is sustained at the value i(n-1), so the output is clearly the input signal delayed by one clock period.

A switched-current integrator is shown in Figure 11.6(a). It is formed from a delay circuit with two output stages weighted A and B, the output from the B weighted transistors being fed back to the input summing node. The signal, i_0, from the A weighted transistors is the integrator output signal. The circuit operates as follows. The output signal is established as soon as switch S2 closes at the beginning of phase ϕ_1 and is held until the end of phase ϕ_2. During phase ϕ_2 of the (n - 1)[th] clock period the output signal is i_0 (n-1) and the value of i_f is B i_0 (n - 1)/A . The input current is i_1(n - 1) and the total current in T_1 is I_1 where

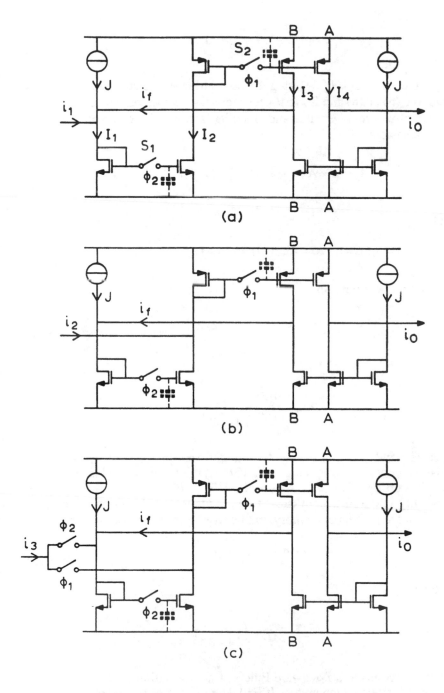

Figure 11.6 First-generation switched-current integrators (a) non-inverting (b) inverting (c) feed-forward

$$I_1 = J + i_1(n-1) + \frac{B}{A} i_o(n-1)$$

(11.15)

S_1 is closed so $I_2 = I_1$. During the next phase ϕ_1 the currents i_1, i_0 and i_f change to $i_1(n)$, $i_0(n)$ and $B\ i_0(n)/A$ respectively. However, since S_1 is open, I_2 is held at its previous value and since S_2 is closed I_2 is mirrored into the output transistors. So,

$$I_4 = AI_1 = A(J + i_1(n-1) + \frac{B}{A} i_o(n-1))$$

(11.16)

and

$$i_o(n) = I_4 - AJ = Ai_1(n-1) + Bi_o(n-1)$$

(11.17)

Expressed in the z-domain,

$$i_o(Z) = Ai_1(Z)\ Z^{-1} + Bi_o(Z)\ Z^{-1}$$

(11.18)

Therefore

$$H_1(Z) = \frac{i_o(Z)}{i_1(Z)} = \frac{AZ^{-1}}{1 - BZ^{-1}}$$

(11.19)

This corresponds to a Forward Euler z-transformation (s -->(1 - z $^{-1}$) / Tz^{-1}) of a damped non-inverting integrator. If B=1, the integrator is lossless.

The integrator shown in Figure 11.6(b) differs from that of Figure 11.6(a) only in that the input current, i_2 , is applied to the summing node at the input of the p-channel memory. In this case, it is readily shown that

$$i_o(n) = Bi_o(n-1) - Ai_2(n)$$

(11.20)

This gives

$$H_2(Z) = \frac{i_o(Z)}{i_2(Z)} = -\frac{A}{1 - BZ^{-1}}$$

(11.21)

which corresponds to a Backward Euler z-transformation (s--> (1 - z $^{-1}$) /T) of a damped inverting integrator. If B=1, the integrator is lossless.

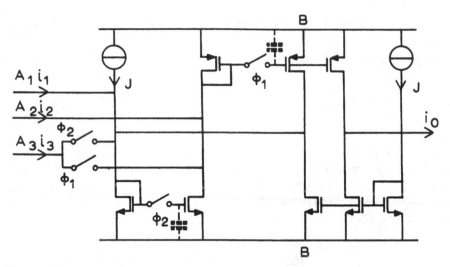

Figure 11.7   **Generalised first-generation switched-current integrator**

The feed-forward integrator of Figure 11.6(c) has its input current, i_3, connected to the summing node of the n-channel memory on phase ϕ_2 and to the summing node of the p-channel memory on phase ϕ_1 . It is readily shown that

$$i_o(n) = Bi_o(n - 1) - A(i_3(n) - i_3(n - 1))$$

(11.22)

Expressed in the z-domain

$$H_3(Z) = \frac{i_o(Z)}{i_3(Z)} = - A \frac{1 - Z^{-1}}{1 - BZ^{-1}}$$

(11.23)

This corresponds to the damped integration of the derivative of the input signal, $i_3(n) - i_3(n-1)$. With B=1, $H_3(z) = -A$ and the input signal is merely scaled and inverted.

If the three integrators of Figure 11.6 are combined as shown in Figure 11.7 then a generalised switched-current integrator results. The gain factor A, defined by the weight (W/L ratio) of the output transistor in the previous integrators, has been replaced by individual gains, A_1 , A_2 ,and A_3 , applied to the input currents. In practice, this is accomplished by scaling the weights of the output stages supplying the input currents but it is included here to maintain equivalence with the generalised switched-capacitor integrator. By superposition,

$$i_o(n) = Bi_o(n - 1) + A_1 i_1(n - 1) - A_2 i_2(n) - A_3(i_3(n) - i_3(n - 1))$$

(11.24)

or, expressed in the z-domain

$$i_o(Z) = A_1 \frac{Z^{-1}}{1 - BZ^{-1}} i_1(Z) - A_2 \frac{1}{1 - BZ^{-1}} i_2(Z) - A_3 \frac{1 - Z^{-1}}{1 - BZ^{-1}} i_3(Z)$$

(11.25)

Comparison with equation (11.8) shows that the generalised switched-current integrator of Figure 11.7 is the dual of the generalised switched-capacitor integrator of Figure 11.2.

It can be seen from equation (11.25) that if $i_2 = - i_1 = i$, $i_3 = 0$ and $A_1 = A_2 = A$, then

$$H(Z) = \frac{i_o(Z)}{i(Z)} = A \frac{1 + Z^{-1}}{1 - BZ^{-1}}$$

(11.26)

This corresponds to the bilinear z-transformation $(s \rightarrow 2(1 - z^{-1})/(1 + z^{-1})T)$ of a damped or lossless (B=1) integrator.

The generalised switched-current and switched-capacitor integrators have been shown to perform identical algorithms. The only distinction is that the switched-current integrator employs sampled currents and its coefficients are determined by current mirror ratios while the switched-capacitor integrator uses sampled voltages and its coefficients are determined by capacitor ratios. In practice, these ratios are both subject to errors which cause distortion of the ideal transfer function. While the magnitudes of these two types of error may be comparable in practice[13], their effect on the transfer function is quite different, as will now be demonstrated.

If a continuous-time damped integrator transfer function, $H(s)=a_0/(1+(s/\omega_0))$ is transformed using the Forward Euler s-z transformation to produce a non-inverting z-domain transfer function then it will be of the form given in equation (11.19) where the low frequency gain, a_0 , and the cut-off frequency, ω_0 , are defined by

$$a_o = \frac{A}{1 - B}$$

(11.27)

and

$$\omega_0 = \frac{2}{T} \left(\frac{1 - B}{1 + B} \right)$$

(11.28)

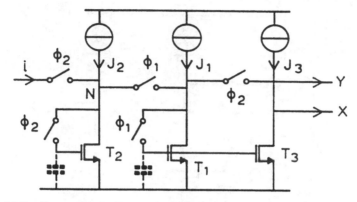

Figure 11.8   Second-generation switched-current delay cell

The sensitivities of a_0 and ω_0 to variations in the coefficient B are given by

$$S_B^{a_0} = \frac{B}{1 - B}$$

$$\tag{11.29}$$

and

$$S_B^{\omega_0} = -\frac{2B}{1 - B^2}$$

$$\tag{11.30}$$

Clearly the low frequency gain and cut-off frequency of integrators with very low damping (B --> 1) are very sensitive to variations in B. In practice, this makes this circuit structure ill-suited to filter designs having a clock frequency which is very much higher than the filter cut-off frequency, or having a high Q-factor.

11.4.3   *Second-generation Switched-current Integrators*

A switched-current integrator, which overcomes the aforementioned shortcoming of the earlier integrator, is based on the alternative current memory shown in Figure 11.4(b). First, consider the delay cell shown in Figure 11.8 which comprises two of these current memories connected in cascade. As before, the switches are operated by a two-phase non-overlapping clock, ø1 / ø2 , where ø1 precedes ø2 . On phase ø2 of sampling period (n-1) the input signal current, i(n-1), and the bias current, J_2 , sum at node N and a current, $J_2 + i(n-1)$, flows into T_2 which is diode-connected. On the next phase ø1 (of sampling period (n)), the current stored in T_2 and the bias currents, J_1 and J_2 , flow into T_1 which is diode-connected and so the current in T_1 is given by $I_1 = J_1 + J_2 - (J_2 + i(n-1)) = J_1 - i(n-1)$. The current at output X is $J_3 - I_3 = J_3 - I_1 = J_3 - (J_1 - i(n-1))$. For unity gain, $J_1 = J_2 = J_3$

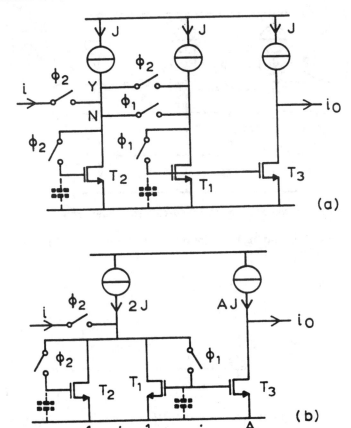

Figure 11.9   Second-generation switched-current integrator
(a) delay cell with feedback   (b) simplified integrator structure

and so the output current at X is $i_0(n) = i(n-1)$, the input current from the previous sampling period.   On the next phase ϕ_2 , a similar current flows at output Y.

In Figure 11.9(a), a non-inverting integrator is formed by connecting the delay cell's output Y to the input summing node, N.   By inspection, the parallel combination of the ϕ_1 and ϕ_2 switches may be replaced by a short circuit to yield the structure shown in Figure 11.9(b).   The transfer function may be found as follows.   On phase ϕ_2 of clock period $(n-1)$, transistor T_2 is diode-connected and passes current I_2 where

$$I_2 = 2J + i(n-1) - I_1 = J + i(N-1) + \frac{i_0(n-1)}{A}$$

(11.31)

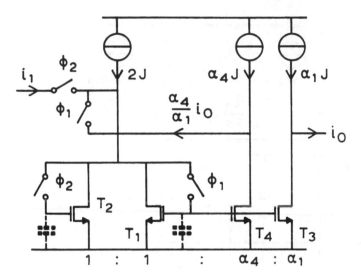

Figure 11.10 Second-generation non-inverting damped integrator

On the next phase ϕ_1 of clock period (n), transistor T1 is diode-connected and passes current I_1 where

$$I_1 = 2J - I_2 = J - i(n-1) - \frac{i_o(n-1)}{A}$$

(11.32)

Therefore

$$i_o(n) = A(J - I_1) = i_o(n-1) + Ai(n-1)$$

(11.33)

Taking z-transforms gives

$$H(Z) = \frac{AZ^{-1}}{1 - Z^{-1}}$$

(11.34)

which corresponds to a Forward Euler z-transformation of a lossless integrator.

This integrator may be damped as shown in Figure 11.10. It contains an extra feedback stage (T4 and current source) which is weighted α_4, and the output stage is weighted α_1. On phase ϕ_1, T1 and T4 are connected in parallel and the current they receive is shared between them. On phase ϕ_2, these currents are stored but only that current stored in T1 is fed-back to the summing node. In this way, the loop-gain is made equal to $1 / (1 + \alpha_4)$ and the z-domain transfer response becomes

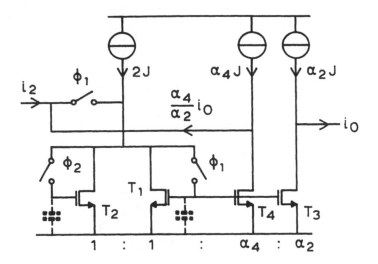

Figure 11.11   Second-generation inverting damped integrator

$$H_1(Z) = \frac{i_o(Z)}{i_1(Z)} = \frac{A_1 Z^{-1}}{1 - BZ^{-1}}$$

$$(11.35)$$

where $A_1 = \alpha_1 / (1 + \alpha_4)$ and $B = 1 / (1 + \alpha_4)$. Clearly, this is the same as that given in equation (11.19), i.e. it is a Forward Euler damped integrator.

Figure 11.11 shows an inverting integrator which differs from that shown in Figure 11.10 only in that the input current is sampled on phase ϕ_1 instead of phase ϕ_2 . On phase ϕ_1 , the input current enters T_1 (being diode-connected) and the output current, i_0 , is mirrored immediately. Analysis of this circuit gives the following z-domain transfer characteristic

$$H_2(Z) = \frac{i_o(Z)}{i_2(Z)} = - \frac{A_2}{1 - BZ^{-1}}$$

$$(11.36)$$

where $A_2 = \alpha_2 / (1 + \alpha_4)$, which corresponds to the Backward Euler z-transform of a damped integrator.

Figure 11.12 shows a feed-forward integrator in which the input current, i_3 , is fed directly to the summing node. On phase ϕ_2 , the input current has value $i_3(n-1)$ and flows in T_2 , while on phase ϕ_1 it has value $i_3(n)$ and flows in T_1 . Of course, on phase ϕ_1 , the stored current from T_2 also flows into T_1 giving an effective input signal in T_1 of $i_3 (n) - i_3 (n-1)$ and so the damped

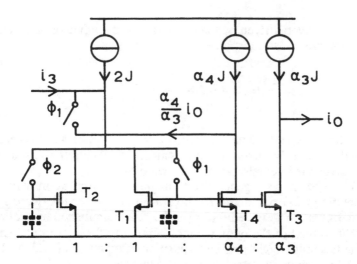

Figure 11.12 Second-generation feed-forward damped integrator

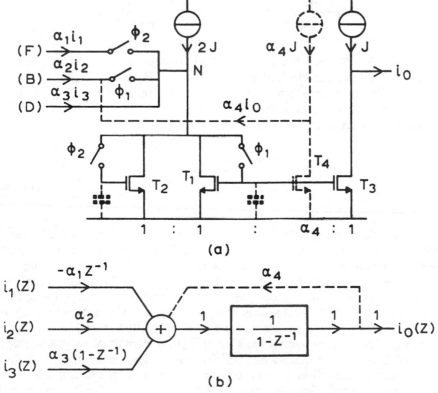

Figure 11.13 Second-generation generalised integrator (a) circuit topology (b) z-domain block diagram

integrator is effectively driven by the derivative of the input signal. Analysis gives the z-domain transfer response

$$H_3(Z) = \frac{i_o(Z)}{i_3(Z)} = -A_3 \frac{(1 - Z^{-1})}{1 - BZ^{-1}}$$

(11.37)

where $A_3 = \alpha_3 / (1 + \alpha_4)$. Note that when $B = 1 / (1 + \alpha_4) = 1$ (lossless), the input signal is merely inverted and scaled. Clearly, the direct input connection gives the integrator its feed-forward capability.

Figure 11.13(a) shows a generalised integrator configuration made from the superposition of Forward and Backward Euler and feed-forward inputs. The input currents are weighted α_1, α_2 and α_3, accomplished by scaling the weights (W/L ratios) of the output stages supplying these currents, and the output stage has unit weight. By superposition of equations (11.35), (11.36) and (11.37), the z-domain output current is given by

$$i_o(Z) = \frac{AZ^{-1}}{1 - BZ^{-1}} i_1(Z) - \frac{A_2}{1 - BZ^{-1}} i_2(Z) - \frac{A_3(1 - Z^{-1})}{1 - BZ^{-1}} i_3(Z)$$

(11.38)

where A_1, A_2, A_3 and B are as defined in equations (11.35), (11.36) and (11.37). The z-domain block diagram is shown in Figure 11.13(b).

A bilinear z-transform integrator may be formed by setting $i_1(z) = -i_2(z) = i(z)$, $i_3(z) = 0$ and $A_1 = A_2 = A$. The z-domain transfer response then becomes,

$$H_4(Z) = \frac{i_o(Z)}{i(Z)} = \frac{A(1 + Z^{-1})}{1 - BZ^{-1}}$$

(11.39)

The second-generation switched-current integrator comprises a lossless integrator feedback loop (transistors T_1 and T_2, Figure 11.13) with extra feedback (transistor T_4) to produce damping. This is in contrast with the first-generation switched-current integrator (Figure 11.7) which combined the feedback paths. It is more akin to the switched-capacitor integrator (Figure 11.2) which has a lossless integrator (op. amp. and capacitor C) to which is added extra feedback (switched-capacitor, $\alpha_4 C$) to produce damping. The current memories, T_1 and T_2 of the lossless integrator loop, both sink and source current with single transistors and so random errors due to transistor mismatch in current mirrors do not arise; they influence only the accuracy of coefficients α_1, α_2, α_3 and α_4.

If a continuous-time damped integrator transfer function, $H(s) = a_0/(1 + s/\omega_0)$ is transformed using the Forward Euler s-z transformation to produce a non-inverting z-domain transfer function, as performed earlier for the first-generation integrator (equations (11.27)-(11.30)), then the low frequency gain, a_0, and the cut-off frequency, ω_0, are defined by

$$a_o = \frac{\alpha_1}{\alpha_4}$$

(11.40)

and

$$\omega_0 = \frac{2}{T} \left(\frac{\alpha_4}{2 + \alpha_4} \right)$$

(11.41)

The sensitivities of a_0 and ω_0 to variations in the coefficient α_4 are given by

$$S_{\alpha_4}^{a_o} = -1$$

(11.42)

and

$$S_{\alpha_4}^{\omega_0} = \frac{2}{2 + \alpha_4}$$

(11.43)

Clearly the low frequency gain and cut-off frequency of integrators with very low damping ($\alpha_4 \to 0$) are no longer highly sensitive to variations in α_4. This makes the second-generation integrator circuit structure well-suited to filter designs having a clock frequency which is very much higher than the filter cut-off frequency, or having a high Q-factor.

11.5 Switched-current Differentiators

A novel approach to high-pass and band-pass ladder filters was reported recently[14] using switched-capacitor differentiators. These circuits were free from d.c. stability problems and offered improved component sensitivities. Equivalent differentiator building blocks can be implemented with switched-current techniques.

Figure 11.14 shows a Backward Euler differentiator module. The operation is as follows. For $i_2 = i_3 = 0$, on phase ϕ_2 of period (n-1), transistor T_2 is diode-connected and its current is $J + \alpha_1 i_1 (n - 1)$. On the next phase ϕ_1 of period (n), transistor T_1 is diode-connected and it receives current from both $\alpha_1 i_1$ and from the current stored in T_2. The resulting

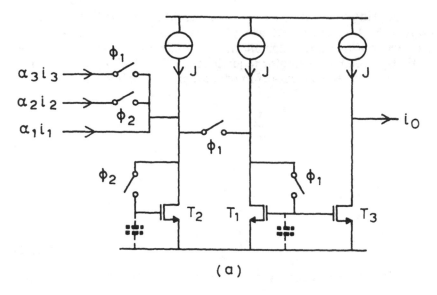

(a)

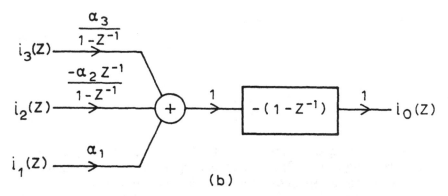

(b)

Figure 11.14 Second-generation inverting lossless differentiator (a) circuit topology (b) z-domain block diagram

current in transistors T_1 is $J + \alpha_1 (i_1 (n) - i_1 (n-1))$ and this is mirrored into T_3 giving an output current $i_0 (n) = - \alpha_1 (i_1 (n) - i_1 (n-1)$. Therefore, the output current is the inverted derivative of the input current. With $i_1 = i_3 = 0$, on phase ϕ_2 of period $(n-1)$, transistor T_2 is diode-connected and conducts current $J + \alpha_2 i_2 (n-1)$. On the next phase ϕ_1 of period (n), transistor T_1 is diode-connected and conducts current $J - \alpha_2 i_2(n1)$ and the output current is $\alpha_2 i_2 (n-1)$. The output current is merely the input current delayed by one clock period. With $i_1 = i_2 = 0$, on phase ϕ_1 of period (n), transistor T_1 is diode-connected and conducts current $J + \alpha_3 i_3(n-1)$. This is mirrored into T_3 to give an output current of $- \alpha_3 i_3(n)$. By superposition, the z-domain output current for the complete circuit is given by

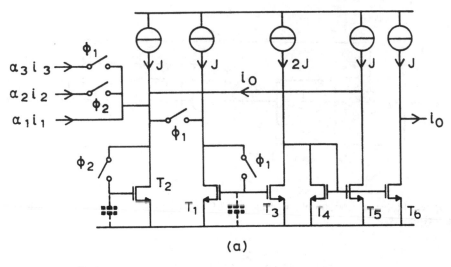

(a)

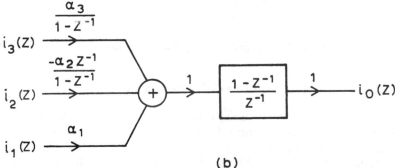

(b)

Figure 11.15 **Second-generation non-inverting lossless differentiator** (a) **circuit topology** (b) **z-domain block diagram**

$$i_o(Z) = -\alpha_1 i_1(Z)(1 - Z^{-1}) + \alpha_2 i_2(Z)Z^{-1} - \alpha_3 i_3(Z)$$

$$(11.44)$$

The first term corresponds to the Backward Euler mapping of a lossless differentiator, the second term to a delay and the third term to inversion and feed-forward. These are represented by the z-domain block diagram of Figure 11.14(b).

Figure 11.15(a) shows a Forward Euler differentiator module. It comprises the Backward Euler differentiator (transistors T_1, T_2 and T_3) the output of which is inverted to produce two output currents each of value i_0. One output is fed back to the summing node, producing a positive feedback loop with unity gain. In isolation, the module is unstable, as is its switched-capacitor counterpart, and can only be used in circuits which render it stable by applying negative feedback. It can be shown that the output current is given by

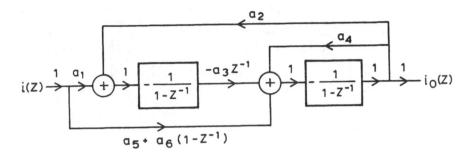

Figure 11.16   z-domain block diagram of an integrator-based biquadratic section

$$i_o(Z) = \alpha_{11} i_i(Z) \frac{1 - Z^{-1}}{Z^{-1}} - \alpha_{22} i_i(Z) + \frac{\alpha_{33} i_i(Z)}{Z^{-1}}$$

(11.45)

This is represented by the z-domain block diagram shown in Figure 11.15(b).

11.6   Switched-current State-variable Filter Synthesis

Having established second-generation circuit topologies for both integrators and differentiators which perform either Forward Euler (non-inverting) or Backward Euler (inverting) s-z transforms, it is now possible to define suitable biquadratic sections. As the integrators and differentiators have direct counterparts in the switched-capacitor technique it follows that switched-current biquadratic sections can be derived from well known switched-capacitor sections either by direct module substitution[15] or by using its z-domain block diagram. Of course, there are many known biquadratic sections and any of these may be used to derive a corresponding switched-current topology. Here, to illustrate the approach, just one integrator-based biquad and one differentiator-based biquad are described.

11.6.1   Integrator-based Biquadratic Sections

Figure 11.16 shows the z-domain block diagram of a well known biquadratic section[8]. The transfer function is given by

$$H(Z) = \frac{i_o(Z)}{i(Z)} = -\frac{(a_5 + a_6)Z^2 + (a_1 a_3 - a_5 - 2a_6)Z + a_6}{(1 + a_4)Z^2 + (a_2 a_3 - a_4 - 2)Z + 1}$$

(11.46)

The s-domain biquadratic transfer function is given by

Table 1. Coefficients for use with integrator-based biquad section. T is the clock period.

Coefficient	Value
$a_1 a_3$	$4k_0 T^2/D$
$a_2 a_3$	$4\omega^2_0 T^2/D$
a_4	$4\omega_0 T/QD$
a_5	$4k_1 T/D$
a_6	$(4k_2 - 2k_1 T + k_0 T^2)/D$
D	$\omega^2_0 T^2 - 2(\omega_0/Q)T + 4$

$$H(s) = -\frac{k_2 s^2 + k_1 s + k_0}{s^2 + \left(\dfrac{\omega_0}{Q}\right)s + \omega_0^2}$$

(11.47)

Applying the bilinear z-transform the z-domain transfer function becomes

$$H(Z) = \frac{\left(\dfrac{4k_2 + 2k_1 T + k_0 T^2}{D}\right)Z^2 + \left(\dfrac{2k_0 T^2 - 8k_2}{D}\right)Z + \left(\dfrac{4k_2 - 2k_1 T + k_0 T^2}{D}\right)}{\left(\dfrac{\omega_0 T^2 + \dfrac{2\omega_0 T}{Q} + 4}{D}\right)Z^2 + \left(\dfrac{2\omega_0^2 T^2 - 8}{D}\right)Z + 1}$$

(11.48)

where $D = \omega_0^2 T^2 - (2\omega_0 T)/Q + 4$. Comparing coefficients between equations (11.46) and (11.48) gives the weights which must be used to produce a biquadratic section with an 'exact' response, i.e. one which performs with a bilinear z-transform. These coefficients are given in Table 11.1 and correspond to capacitor ratios in a switched-capacitor implementation or to current mirror ratios in a switched-current implementation.With the integrator blocks in Figure 11.16 replaced by corresponding switched-current integrator modules (Figure 11.7), the biquadratic section is as shown in Figure 11.17.

11.6.2 Filter Design Example

Using the integrator-based biquadratic section described in 11.6.1 (or any other biquad that may be derived from a known integrator-based switched-

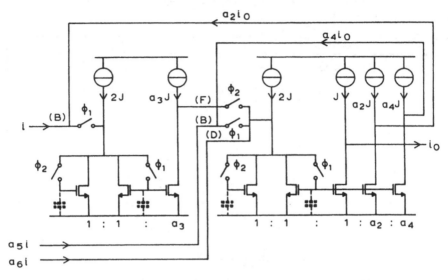

Figure 11.17   Biquadratic section using second-generation switched current integrators

Table 2.   MOSFET aspect ratios (W/L) for a 6th order Chebyshev low-pass filter (0.5dB ripple), f_{co} = 5 MHz, f_{cK} = 20 MHz

W/L	1st Biquad	2nd Biquad	3rd Biquad
a_1	1.984504	1.913393	0.984467
a_2	2.102105	1.913393	0.984467
a_3	1	1	1
a_4	0.165887	0.716175	1.894220
a_5	0	0	0
a_6	0.496126	0.478348	0.246117

capacitor biquad) it is possible to design complete filters which execute the bilinear z-transform. To demonstrate this a complex filter is designed and then tested with a switched-capacitor CAD package. The chosen specification is that of a 6th order low-pass filter with a Chebyshev response (0.5dB equiripple), cut-off frequency of 5MHz and a clock frequency of 20MHz. After pre-warping the specification and applying the relationships of Table 11.1, the coefficients for the three section design are as given in Table 11.2.

This filter's transfer characteristic was simulated with a switched-capacitor analysis program using the current memory model shown in

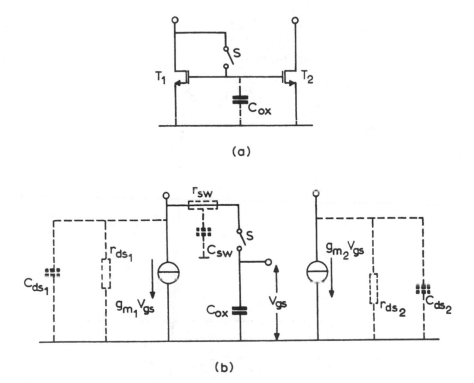

Figure 11.18 Modelling of switched-current cells (a) simple current memory (b) equivalent circuit

Figure 11.18 (the parasitics are shown dashed), with ideal switches and no parasitics. The simulated characteristic of the filter (the sinx/x effect was removed by the simulator) is shown in Figure 11.19 (solid line). The response is exact (apart from a small computational error) indicating that the circuits are executing their algorithms exactly as intended. Choosing suitable bias currents and device dimensions, typical parasitics for transistors and switches were computed and then included in the filter model. The dashed line shows the resulting simulated transfer characteristic. The distortion of the response was less than 0.2dB, indicating that the switched-current technique has much potential for high frequency filters.

11.6.3 Differentiator-based Biquadratic Section

Figure 11.20 shows the z-domain block diagram of a differentiator-based biquadratic section[16] for a known switched-capacitor implementation. The z-domain transfer function is given by

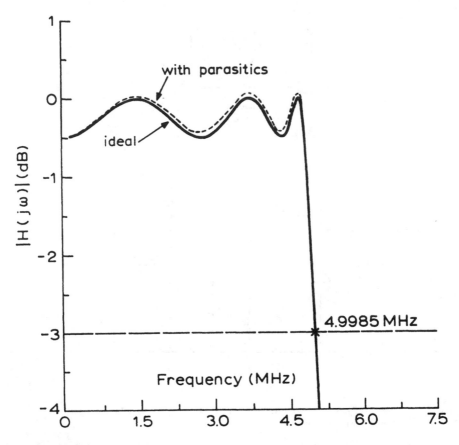

Figure 11.19   Simulated transfer characteristic of a 6th. order low-pass filter (Chebyshev   response, 0.5dB ripple, 5MHz cut-off frequency, 20MHz clock frequency)

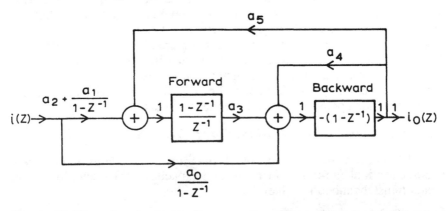

Figure 11.20   z-domain block diagram of a differentiator-based biquadratic section

**Table 3.   Coefficients for use with differentiator-based biquad section.
T is the clock period.**

Coefficient	Value
a_0	k_0/ω^2_0
a_1a_3	$k_1/T\omega^2_0$
a_2a_3	$(k_0 - 2k_1/T + 4k_2/T^2)/4\omega^2_0$
a_4	$1/QT\omega_0$
a_5a_3	$(4/T^2 + 2\omega_0/QT + \omega^2_0)/4\omega^2_0$

$$H(Z) = -\frac{(a_2a_3 + a_1a_3)Z^2 + (a_0 - a_1a_3 - 2a_2a_3)Z + a_2a_3}{a_3a_5Z^2 + (1 + a_4 - 2a_3a_5)Z + (a_3a_5 - a_4)}$$

$$(11.49)$$

Despite using a non-causal Forward Euler differentiator, the biquad is stable (as evidenced by the relative magnitudes of the coefficients of z^2 and z^0 in the denominator of H(z)) because of the negative feedback applied by the Backward Euler differentiator. Comparing coefficients between equations (11.48) and (11.49) gives the weights which must be used to produce a biquadratic section with an 'exact' response, i.e. one which performs with a bilinear z-transform. These coefficients are given in Table 11.3. The z-domain block diagram of Figure 11.20 may be used to derive a switched-current circuit configuration by just the same means that were used to derive the integrator-based topology.

11.7   Analogue Errors

All sampled-data circuits suffer degraded performance (accuracy, dynamic range, linearity, offsets, etc) through analogue errors resulting from device imperfections. Switched-capacitor circuits suffer from capacitor-ratio mismatch errors, from op. amp. finite gain-bandwidth effects, from op. amp. and switch noise sources and from switch charge-injection effects. Second-generation switched-current circuits are no exception to this rule; in practice the integrator and differentiator topologies presented earlier do not perform accurately. They were only intended as primitive cells giving easy explanation of circuit operation. Device imperfections produce errors in the following ways.

1. Variations   in   transistors'   threshold   voltages   (V_T)   and   gains ($\beta=\mu C_{OX}W/L$) produced by processing spreads give errors in current

mirror ratios and hence filter coefficient values. Systematic errors can be substantially eliminated by employing arrays of unit transistors in like-manner to the practice in switched-capacitors, and by ensuring that drain voltages are equalised. Random errors in V_T and β can be reduced by increasing the area of the transistors, and their influence on current ratio-matching can be reduced by choosing a large saturation voltage[17] A unity gain current mirror made in a VLSI CMOS process (1.6 μm feature size) with transistors having aspect ratios of 40 μm/ 40 μm and a saturation voltage of 1V has a current-matching standard deviation of about 0.118 which is comparable with switched-capacitor ratio matching for capacitors of similar area.

2.   Channel length modulation through drain voltage variation produces errors in both the direct and mirrored outputs of the current memory (i.e. i_{01} and i_{02} of Figure 11.4). This can be greatly improved with extra circuits to maintain constant drain voltage e.g. with cascode transistors (see Section 11.8) or servo-amplifiers (see Chapter 13).

3.   Charge injected from the memory switch into the oxide capacitance of the memory transistor produces an error in the gate-source voltage which gives a corresponding error in the drain current. The error voltage may be reduced by increasing the oxide capacitance (by increasing the area of the memory transistor) and decreasing the injected charge (by reducing switch dimensions and using balanced switch arrangements [19]). The resulting memory current error may be reduced by choosing a memory transistor with a large saturation voltage. In the second-generation switched-current integrators, charge injection errors are partially cancelled by the combined operation of the two memory cells forming the lossless integrator loop [20]

4.   Incomplete charging of the memory transistor's oxide capacitance when it is diode-connected will lead to memory current errors. However, so long as the memory transistor's output conductance is sufficiently low, switch resistance has little influence and the charging time-constant is simply C_{ox}/g_m and this should be made sufficiently low compared with the clock period. In switched-capacitor integrators, charging of its sampling capacitors is more demanding because they are completely discharged every clock period, and the settling accuracy is influenced by both the op. amp. gain-bandwidth product and the sampling switches' channel resistance. This gives second-generation switched-current circuits an advantage when the clock frequency is high.

5.   The memory transistor generates both thermal and 1/f noise and these cause memory current errors. In second generation switched-current circuits the thermal noise is under-sampled and high frequency noise components are aliased into the passband. However, 1/f noise is   over-

sampled and an inherent correlated double sampling (CDS) mechanism in the current memory shifts this noise out of the passband. Switch noise is not significant so long as the memory transistor's output conductance is sufficiently low [21]. Switched-capacitor integrators require special chopper amplifier arrangements [22] to achieve CDS, and thermal noise generated in the integrator's sampling switches (KT/C noise) can be very significant. These two factors give second-generation switched-current integrators an unexpected advantage over their switched-capacitor counterparts.

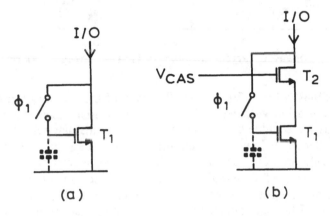

Figure 11.21 **Current memory cells (a) simple memory cell (b) cascoded memory cell**

11.8 CURRENT MEMORY CIRCUIT ENHANCEMENTS

To make switched-current circuits competitive with switched-capacitor circuits they must have both comparable accuracy and be equally amenable to design-automation techniques [23]. The simple current memory cell, comprising only a transistor and a switch (Figure 11.21(a)), produces unacceptably large analogue errors and gives a ratio between output and input resistance which is not large enough to enable carefree interconnection of modules. The input resistance of the diode-connected transistor ($1/g_m$) and its output resistance (r_{ds}) produce a ratio ($g_m r_{ds}$) of only about 100. Cascoding, as shown in Figure 11.21(b), increases the output resistance by approximately $g_m r_{ds}$ giving a ratio of output to input resistance ($g_m^2 r_{ds}^2$) of about 10,000 which is high enough. However, the need to connect cells in cascade can cause non-saturated operation in the cascode transistors with consequent loss of output resistance.

Alternative memory cells which avoid this problem are shown in Figure 11.22 (a) and (b). The conveyor memory cell, so-called because of its similarity to the current conveyor (see Chapter 3), operates as follows. On phase ϕ_1, the current conveyor loop is formed, comprising transistors T_1 - T_5, and the input node is forced to the conveyor reference voltage, V_{CON}.

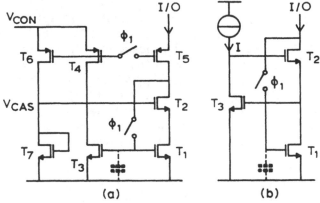

Figure 11.22   Enhanced current memory cells   (a) conveyor (b) regulated cascode

The cascode bias voltage (V_{CAS}) sets the drain of the memory transistor T_1 near to its saturation voltage.   Ideally, the input resistance is zero but in practice it has a low value determined by transistor mismatches. On phase ϕ_2 , the current conveyor loop is broken and the current stored in T_1 is fed via T_2 and T_5 to the output.   The cascode bias V_{CAS} is sustained and so the voltage on the drain of T_1 remains unchanged, a necessary requirement for accurate operation. Cascading cells presents no problem because the voltage presented to the I/O node is close to V_{CON} on both phases of the clock.   The cell is particularly useful when many signals are summed at the cell input node.

The regulated cascode [24] , so-called because of its similarity with the regulated cascode current source [25] , operates as follows. The drain voltage of T_1 is forced to a fixed value, V_{gs3} , by the action of the negative feedback loop comprising transistors T_3 and T_2 . On phase ϕ_1 , T_1 is diode-connected and the cell's input resistance is $1/g_m$. On phase ϕ_2 , the output resistance is approximately $g_m r_{ds}$ times larger than that of the cascode cell (i.e.$g_m{}^2 r_{ds}{}^3$) due to the extra voltage gain of transistor T_3 .   With the output forced to a voltage low enough to operate T_2 in non-saturation, the output resistance is still as high as that of the cascode cell.   This makes cascade connection to other cells straightforward.   Alternatively, T_1 may be operated in non-saturation which lowers output resistance to a value similar to that of the cascode cell but gives the opportunity to improve switch charge-injection errors..

11.9   Interface Circuits

Sampled-data circuits require continuous-time pre- and post-filters. If the system's signals are represented by voltages, the switched-current circuit will need voltage-to-current and current-to-voltage converters. These interface functions must be integrated with the switched-current circuit using

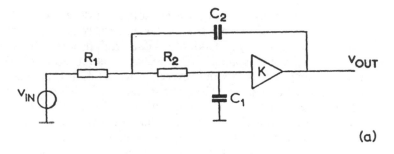

(a)

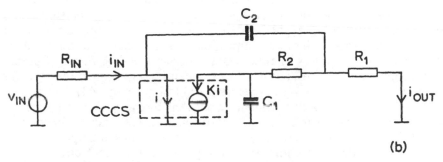

(b)

Figure 11.23 Input interface circuit (a) Sallen-Key circuit (b) Corresponding CCCS-based adjoint circuit including voltage-to-current converter

low voltage techniques requiring no special processing options and in a manner that incurs minimal circuit overhead.

A good approach is to first convert the signals into currents and then perform continuous-time filtering in the current domain. Well known voltage amplifier circuits can be readily transformed into the current domain using the interreciprocal property of linear networks (see Chapter 3). The resulting current-domain filter uses a current controlled current source (e.g. current conveyor) and this allows the input voltage-to-current conversion to be simply implemented with a resistor. Figure 11.23 shows such an input interface circuit derived from a Sallen-Key filter configuration; a similar approach may be adopted for the output interface. The filters' passive components may be formed from the VLSI process's conducting layers (e.g. polysilicon or aluminium) and dielectrics. Capacitors so-formed can be of good quality but have large bottom plate parasitic capacitance. This presents no problem for filter configurations of the type shown when the capacitors are connected with their bottom plates on the low impedance nodes.

11.10 Conclusions

Switched-Currents is a new sampled-data technique with aspirations to supersede switched-capacitors. By operating in the sampled- current domain it offers several advantages over switched-capacitors: it does not need linear floating capacitors and will operate from reduced supply voltage, and so it can be implemented using only a standard digital VLSI process. It has wide ranging applications but the emphasis in this chapter is on filtering. To this end a generalised integrator configuration was introduced. It is an exact dual of a well known switched-capacitor integrator, and is capable of performing integration with Backward Euler (inverting), Forward Euler (non-inverting), or bilinear z-transformation, and has a feed-forward capability. The second-generation integrator topology was shown to be simpler, less sensitive to coefficient errors and to have improved analogue performance. To demonstrate correct algorithmic operation of the integrators, a 6th order 5MHz low-pass filter was designed and simulated with a switched-capacitor CAD package. Inclusion of parasitics distorted the ideal response by less than 0.2dB, indicating a good video frequency capability. Generalised differentiator topologies were introduced and these can be used as alternatives to integrators in either state-variable or ladder filters. Sources of analogue error were outlined and second-generation circuits offer prospect of higher frequency of operation and lower 1/f and KT/C noise than switched-capacitor circuits. Enhanced current memory circuit structures were given offering better analogue performance. Generalised integrators or differentiators using these enhanced current memories are highly suitable for automated design techniques, making switched-current circuits truly competitive with switched-capacitors.

11.11 ACKNOWLEDGEMENTS

The author wishes to acknowledge the invaluable contributions to this work of N.C.Bird, I.C.Macbeth and D.M.Pattullo.

11.12 REFERENCES

1.   D.L. Fried, "Analog Sample-data Filters," IEEE Journal. Solid-State Circuits,Vol.SC-7, Aug. 1972, pp. 302-304.

2.   R. Gregorian, K.W. Martin, G.C. Temes, "Switched-Capacitor Circuit Design", Proceedings of the IEEE 1984, Vol. 71, No. 88, pp. 671-678.

3.   J.B. Hughes, N.C. Bird, R.S. Soin, "A Receiver IC for a 1+1 Digital Subscriber Loop", IEEE Journal of Solid State Circuits, June 1985,   Vol. SC-20, No.3, pp. 671-678.

4.   Solid State Products Engineering Council, " JEDEC Standard No.8, Standard for reduced operating voltages and interface levels for   integrated circuits", Electronic Industries Association, Washington   D.C., Dec. 1984.

5.   P.R.Grey, R.Castello, "Performance Limitations in Switched-Capacitor Filters", Design of MOS VLSI Circuits for Telecommunications,   Prentice-Hall, 1985, pp. 314-332.

6. J.B.Hughes, N.C.Bird, I.C.Macbeth, "Switched Currents - A New Technique for Analogue Sampled-Data Signal Processing", IEEE   International Symposium on Circuits and Systems, 1989,   pp. 1584-1587.

7. G.C. Temes, ed.,"Special Section on Switched-Capacitor Circuits", Proc. IEEE, Vol.71, pp. 926-1005, Aug. 1983.

8. R.Gregorian, G.Temes, "Analog MOS Integrated Circuits for Signal Processing", J.Wiley ,pp.280-284

9. A.S.Sedra, "Switched-capacitor filter synthesis", from "Design of MOS VLSI circuits for Telecommunications", Chapter 9, Prentice Hall, 1985.

10. K.Martin, A.S.Sedra, "Exact Design of Switched-Capacitor Bandpass Filters using Coupled-Biquad Structures", IEEE Trans. Circuits and Systems, Vol. CAS-27, No. 6, June 1980, pp.469-475.

11. J.C.M.Bermudez, B.B.Bhattacharyya, "A Systematic Procedure for the Generation and Design of Parasitic Insensitive SC Biquads", IEEE Transactions on Circuits and Systems, Vol. CAS-32, No. 8, August, 1985, pp. 767-783.

12. G.M.Jacobs, D.J.Allstot, R.W.Broderson, P.R.Gray, "Design Techniques for MOS Switched-Capacitor Ladder Filters", IEEE Trans. Circuits and Systems, Vol. CAS-25, No. 12, Dec. 1978, pp. 1014-1021.

13. J-B Shyu, G.C.Temes, F.Krummenacher, "Random Errors Effects in   Matching MOS Capacitors and Current Sources", IEEE Jnl. Solid-State Circuits, Vol. SC-19, No. 6, Dec. 1984, pp. 984-955.

14. C-Y Wu, T.C. Yu, "The Design of High-Pass and Band-Pass Ladder Filters Using Novel SC Differentiators", ISCAS89 Proceedings, pp. 1463-66.

15. J.B.Hughes, I.C.Macbeth, D.M.Pattullo, "Switched-Current Filters ", IEE Proceedings Part G, No. 2, April, 1990.

16. C-Y Wu, Private communication.

17. K.R.Lakshmikumar, R.A.Hadaway, M.A.Copeland, "Characterisation and Modeling of Mismatch in MOS Transistors for Precision Analog Design", IEEE Jnl. of Solid-State Circuits, Vol.SC-21, No.6, Dec. 1986, pp.1057-1066.

18. M.J.M.Pelgrom, A.C.J.Duinmaijer, A.P.G.Welbers, "Matching Properties of MOS Transistors", IEEE Jnl. of Solid-State Circuits, Vol.24, No.5, Oct. 1989, pp.1433-1439.

19. G.Wegman, E.A.Vittoz, "Basic Principles of Accurate Dynamic Current Mirrors", IEE Proceedings Part G, No.2, April, 1990.

20. J.B.Hughes, I.C.Macbeth, D.M.Pattullo, "Second-Generation Switched-Current Signal Processing", IEEE International Symposium on Circuits andSystems,1990.

21. S.J.Daudert, D.Vallencourt, "Operation and Analysis of Current Copier Cells", IEE Proceedings PartG, No.2, April, 1990.

22. K.C.Hsieh, P.R.Grey, D.Senderowicz,D.G.Messerschmitt, "A Low-Noise Chopper-Stabilized Differential Switched-Capacitor Filtering Technique", IEEE Jnl. Solid-State Circuits, Vol. SC-16, No.6, Dec.   1981, pp.708-715.

23.   J.B.Hughes, I.C.Macbeth, D.M.Pattullo, "Switched-Current System   Cells", IEEE International Symposium on Circuits and Systems, 1990.

24.   C.Toumazou, J.B.Hughes, D.M.Pattullo, "A Regulated Cascode   Switched-Current Memory Cell", Electronics Letters, No.5, 1st March,   1990.

25.   E.Sackinger, W.Guggenbuhl, "A Versatile Building Block, The CMOS Differential Difference Amplifier", IEEE Journal of Solid-State   Circuits, Vol. SC-42, No. 2, April 1987, pp.287-294.

Chapter 12

Analog Interface Circuits For VLSI

Evert Seevinck

12.1 Introduction

In present-day information systems, signal processing is increasingly being carried out by digital VLSI integrated circuits. Broad fields of importance are ASICs and RAMs. Interface functions are required between the "real world" and the silicon system. The primary information acquired from the real world is usually in the form of time-continuous analog signals and must be interfaced to digital circuitry. The result of the digital processing must likewise be converted to back to analog form. In addition, the interfacing will frequently convert from the voltage domain outside the chip to current domain inside the chip. This is favourable for dynamic range in low supply voltage VLSI circuits and for high speed in the capacitance dominated chip interior.

The performance of an information system as a whole is determined to a large degree by the quality of the analog peripheral circuits. This is because digital processing can be performed to as high a resolution as is desired by simply increasing the word length. In addition, the speed of digital processing is steadily increasing by the evolution of submicron VLSI technology. The analog parts generally form the weak link in the system thus defining performance limits. The required interface functions are increasingly being realized on the same chip as the digital processing, i.e. in VLSI technology. The latter is determined by the digital demands. Submicron CMOS and, to a lesser degree, BiCMOS, will be technologies of choice. Pure bipolar will remain for niche applications.

A typical system interface is illustrated in Figure 12.1. Amplification, filtering and signal conditioning are required before converting to the digital format. After output signal D-A conversion, again filtering is needed. Finally, the smoothed output signal must be amplified to the appropriate power level to achieve the desired effect. The limit of monolithic integration is steadily progressing outward from the digital processing and memory core, towards both the input and output interface functions.

This chapter will address the design of interface circuits compatible with VLSI technologies: CMOS, bipolar and BICMOS. The discussion will concern the following time continuous analog functions: input interfacing, filters, signal conditioning, internal interfacing, output interfacing and,

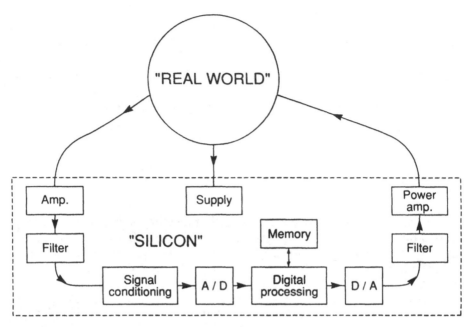

Figure 12.1 Typical information system

finally, power supply interfacing. Circuit examples selected from recent work, illustrating the techniques, will be presented. It is not intended to strive for completeness; rather, a personal viewpoint will be pursued. A broader overview can be found in the literature [1, 2]. The important topics of analog digital conversion and switched capacitor techniques are not addressed here (Chapters 10, 11, 13 and 14).

Since VLSI CMOS technology developed for digital applications will be the dominant medium for realization, it is important to take note of its strengths and weaknesses.

On the negative side, short channel MOS transistors suffer from hot-carrier degradation. To avoid this, the supply voltage must be restricted. At present the supply voltage is 5V; it will possibly have to be decreased to between 3V and 4V in future. In addition, only enhancement transistors are usually available with Vth = 0.8 V. This means that stacking of more than two transistor gate source voltages can hardly be applied within the available supply voltage since a considerable margin is needed for sufficient gate source drive. Additional weaknesses of short-channel transistors are mobility reduction, channel length-modulation and pronounced body-effect. This results in low voltage gain and low current drive capability when compared with bipolar circuits. In addition, the noise properties are inferior. Of course, using BiCMOS allows exploiting the strengths of both bipolar and MOS technologies, but at a considerable price.

Strong points of modern short channel CMOS are the available speed, the high packing density and the low power dissipation of standard logic circuitry. For example, the f_T of a MOS transistor can be roughly expressed as

$$f_T \cong \frac{\mu}{2\pi L^2} (V_{GS} - V_{th})$$

(12.1)

with μ the mobility, L the channel length and V_{th} the threshold voltage. For an n channel device having $L = 0.7$ μm, this predicts f_T of 16 GHz per volt of gate source drive potential.

Finally, there is the important issue of transistor matching which has always been much worse for MOS compared to bipolar. However, the present technological drive to submicron VLSI circuits is expected to have a spin off of improved large geometry MOS device matching.

Standard CMOS processes intended for digital applications, only provide n-and p-channel enhancement transistors. All the other required components have to be improvised from transistors. Resistors can only be made from poly interconnect, n- or p- wells, or source/drain implants. Capacitors of limited values can be realized using the gate oxide or as a sandwich consisting of metal, field-oxide and poly. Since these improvised components have various weaknesses such as non-linearity, parasitics and large spread, they can best be avoided entirely.

12.2 Input interface circuits

The input signal available to the monolithic system is usually low-level, often contaminated by noise and distortion, and containing redundant information in an unnecessarily large dynamic range, and bandwidth. The signal must first be amplified linearly to avoid further degradation and intermodulation. A good approach is to first convert to the current domain using a linear transconductor. Subsequent linear current amplification, filtering and signal conditioning can then be implemented with simple bipolar or CMOS circuits. Also, instrumentation amplifier functions are easy to implement with V-I converter building blocks or transconductors. Various V-I convertor techniques will be presented. A CMOS ECL-compatible input buffer circuit will also be introduced.

12.2.1 Linear CMOS voltage-current converters

Linear V-I converter circuits can be designed based on the first-order square-law characteristic of MOS transistors biased in saturation. This topic is also addressed in Chapter 5 where circuits based on the triode-region of

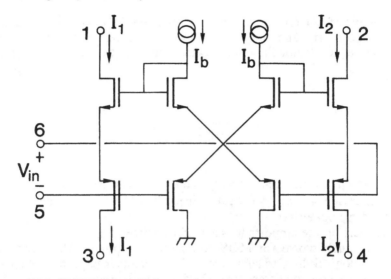

Figure 12.2 CMOS linear V-I converter circuit or transconductor

the MOS-characteristic are discussed as well. An example of the circuits to be discussed here is shown in Figure 12.2.

This circuit linearly converts a differential input voltage V_{in} into a differential output current $I_{out} = I_1 - I_2$. The current subtraction can be performed by a current-mirror (not shown here). The circuit provides two separate outputs: at nodes 1 and 2, and at nodes 3 and 4. Its operation is based on the following principle [3]:

Consider two matched MOS transistors operating in saturation and characterized to first order by

$$I_{DI} = \frac{\beta}{2} (V_{GS1} - V_{tr})^2$$

(12.2a)

$$I_{D2} = \frac{\beta}{2} (V_{GS2} - V_{tr})^2$$

(12.2b)

We find for the differential output current, $I_{out} = I_{D1} - I_{D2}$ and so

$$I_{out} = \beta/2 \, (V_{GS1} - V_{GS2}) \, (V_{GS1} + V_{GS2} - 2 \, V_{th})$$

(12.3)

If the first term is made equal to the applied differential input voltage V_{in} and the last term is kept constant, say $2V_b$, we will have a linear voltage-to-current conversion

$$I_{ar} = \beta \, V_b V_{in}$$

(12.4)

Analysis of the circuit shown in Figure 12.2, along the lines sketched above, reveals class AB operation, yet furnishing a linear transconductance given by [4].

$$I_{out} = 2\sqrt{2\beta_{eq} I_b} \cdot V_{in}$$

(12.5)

$$\text{with } \beta_{eq} = \frac{\beta_n \beta_p}{(\sqrt{\beta_n} + \sqrt{\beta_p})^2}$$

(12.6)

The measured large-signal distortion is 0.2% and the DC to -3 dB bandwidth is 20 MHz for $I_b = 100 \, \mu A$, using a 5 μm CMOS process [4].The distortion is due mainly to mobility reduction. Larger bandwidthis possible by not connecting the n-wells to the sources as was done for this realization.

Note that the transconductance can be varied via I_b. In addition to the linear function, this versatile circuit also provides a square-law function. Summing I_1 and I_2 by interconnecting nodes 1 and 2, or 3 and 4, gives

$$I_1 + I_2 = 2 I_b + \beta_{eq} V_{in}^2$$

(12.7)

The linear, variable transconductance (5) enables application in the fields of amplifiers, gain control, filters and analog multipliers [4]. Figure 12.3 shows an instrumentation amplifier constructed with two V-I converters of Figure 12.2. The voltage gain is given by $\sqrt{I_{b1}/i_{b2}}$ and can therefore be easily varied. The unused outputs can be conveniently used for other purposes, for example signal squaring or adaptive biasing, using (12.7).

A drawback of the circuit of Figure 12.2 is the stacking of two gate-source voltages, thus restricting the minimum supply voltage. This problem is solved by the circuit of Figure 12.4. The circuit operates according to the same principle as discussed before, performing a linear, variable V-I conversion [5]. In addition, it has the interesting property of constant

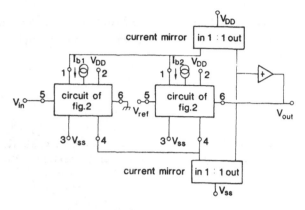

Figure 12.3 Variable-gain instrumentation amplifier application

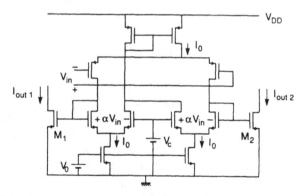

Figure 12.4 Linear V-I converter with variable transconductance and constant bandwidth

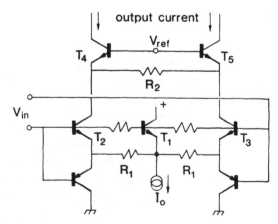

Figure 12.5 Large-signal bipolar V-I converter

bandwidth (a feature of current-mode design we have seen in other Chapters) and phase-shift when the transconductance is varied.

In Figure 12.4 all transistors operate in saturation. M_1 and M_2 are the square-law transistors that perform the linear transconductor function. Two proportional copies of V_{in} are formed across the lower differential pairs. These voltages are added to and subtracted from transconductance control voltage V_c.

Thus it follows that $V_{GS1} - V_{GS2} = 2\,\alpha\,V_{in}$, with $\alpha = \sqrt{\beta_p/\beta_n}$ and $V_{GS1} + V_{GS2} = 2\,V_c$. Therefore, from (12.3)

$$I_{\alpha t1} - I_{\alpha t2} = 2\,\beta_1 \sqrt{\beta_p/\beta_n} \cdot (V_c - V_{t}) \, V_{in}$$

$$(12.8)$$

Summing the output currents provides a square-law function, similar to (12.7):

$$I_{\alpha t1} + I_{\alpha t2} = \beta_1 (V_c - V_{t})^2 + (\beta_1 \beta_p/\beta_n) \, V_{in}^2$$

$$(12.9)$$

12.2.2 *Large-signal bipolar V-I converter*

Input stages designed to accommodate signals of large dynamic range often produce excessive offset and noise and this is the result of inevitable component mismatches and large bias currents. Adaptive biasing or class AB operation can minimize these problems. The bipolar V-I converter of Figure 12.5 combines a very large input signal handling capability (limited only by the supply voltages) with low offset voltage and noise [13].

The circuit features adaptive biasing provided by transistor T_1 progressively cutting off for increasing input voltage, thus steering more of the bias current I_0 toward T2 or T3. This class AB operation is further enhanced by the pnp transistors becoming active for large V_{in} and thus supplementing the bias current. In addition the (lateral) pnp transistors protect T2 and T3 against base-emitter breakdown. A measure of linearity-connection is provided by T4, T5 and R2.

The circuit was integrated as the input stage for an offset cancelling circuit, using an 18 V bipolar process [13]. The maximum symmetrical input voltage amplitude is $36V_{pp}$, and the offset voltage is approx. 0.5 mV. The transconductance is constant within $\pm 8\%$.

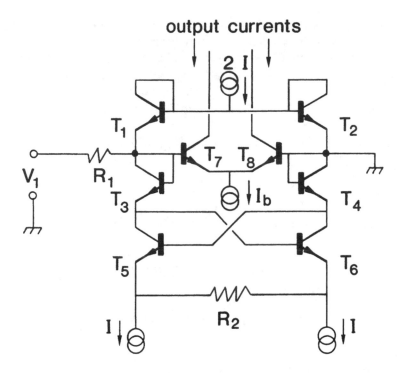

Figure 12.6 Linear large bandwidth gain-controlled bipolar V-I converter

12.2.3   *High-frequency bipolar V-I converter*

A wide-band V-I converter circuit with gain-control capability and excellent linearity is shown in Figure 12.6.

The single-ended input voltage V_1 is converted to a signal current by resistor R_1 which is connected to a low-impedance node (emitter of T_1). The resulting signal current $V_1/R1$ modulates the bias currents of diode-connected transistors T1 and T2. The translinear gain-cell T1, T2, T7,T8 transfers the input current to the output nodes and permits gain-control via I_b. The basic circuit is simple and provides a transparent high-frequency signal path. All circuit nodes are low-impedance with small voltage swings, thus conserving bandwidth.

The impedance at the emitter of T1, although low, is non-linear and therefore will cause linearity errors. The network T3-T6 and R2 (R2 = R1) constitutes a linear negative resistance precisely equal to - R1. This negative resistance exactly compensates the nonlinearity caused by T1 and T2 without affecting the favourable high-frequency properties of the V-I converter.

The circuit was applied as input stage for a monolithic high-frequency RMS-DC converter, using a standard 300 MHz bipolar process [6]. The measured performance for R1 = R2 = 1k Ω, I = 150 μA and I_b = 300 μA is as follows. Low-frequency nonlinearity error is less than 0.5% up to 90% FSR. This figure becomes + 7% when the error correction resistor R2 is removed, thus confirming the effectiveness of the nonlinearity compensation technique. The small-signal DC to -3 dB bandwidth is 85 MHz and the -1% bandwidth is 25MHz.

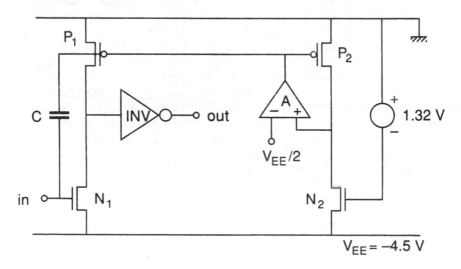

Figure 12.7 ECL-compatible CMOS input buffer principle

12.2.4 CMOS subnanosecond ECL-compatible input buffer

It is well known that fast digital systems are often implemented in emitter-coupled logic (ECL) technology. It is therefore desirable for gate-arrays, logic chips and memory chips intended for fast systems to have ECL-compatible interface circuits.

ECL is based on bipolar technology which has the disadvantage of high power dissipation, making it unsuitable for VLSI circuits. CMOS technologies on the other hand, offer high density and low power dissipation. Consequently, a combination of CMOS VLSI circuits with ECL-compatible input and output buffers would allow fast, high-density systems, with relatively low power dissipation. One possible approach is the use of BiCMOS technology, which is however more complex and costly than CMOS ECL compatible interface circuits implemented in CMOS-only circuits would have obvious advantages. ECL-compatible input and output buffers suitable for submicron CMOS have recently been developed. The principle of the input buffer is shown in Figure 12.7. The design of the ECL output buffer is described in section 12.6.3.

The circuit has to convert an ECL input signal to a CMOS signal with as short a delay as possible. Because the typical high and low ECL 100k voltage levels are at -0.95 V and -1.7V, the switch-level of the input buffer should be at -1.32 V. The first stage, formed by N1 andP1, is a capacitively coupled CMOS inverter. The second stage is a standard CMOS inverter. A replica-bias feedback circuit, consisting of differential amplifier A, P2 and N2, adjusts the gate voltage of P1 such that the switch-level of the input stage is at -1.32 V. The operation is based on the fact that the replica-bias circuit P2, N2 differs from the first stage only by a scale factor. Differential amplifier A drives the gates of P2 and P1 such that the inputs of A are kept about equal. It follows that when the input voltage is at the same level as the on-chip reference voltage (-1.32 V), the output of the first stage will be at about VEE/2 which is approximately the switch-level of the inverter forming the second stage. ECL-levels at the input will therefore result in CMOS-levels at the output. The accuracy of the switch level mainly depends on the accuracy of the reference voltage (-1.32 V) and on the matching of the input stage and the replica stage. The capacitive input stage coupling is the key to obtaining a very high speed. The sum of the gate-source bias voltages of N1 and P1 is more than 6V, considerably higher than the supply voltage of 4.5V.

The significance of this is as follows: The small-signal transconductance of a CMOS inverter with both transistors biased in saturation, is given by

$$g_m = \beta (V_{GSn} + V_{GSp} - V_{tn} - V_{tp})$$

$$(12.10)$$

This can be easily calculated, using the first order transistor model (12.2), and assuming ßn = ßp = ß. It follows from (12.10) that the transconductance and therefore the speed, will be optimized by maximizing the gate-source voltage sum, other things being equal.

Figure 12.8 shows the complete circuit implementation of the input buffer. The differential amplifier of Figure 12.7 is implemented here by differential pair NM6/NM7 and current mirror PM4/PM5. The combination PM6/NM8 can be considered as an inverter with output connected to input. It thus establishes a voltage at the inverting input of the differential amplifier equal to the switch level of the second stage.

As noted before, capacitor C forms a short-circuit for fast input signals, almost doubling the dynamic gain and speed of the first stage. In addition, it constitutes the frequency compensation (together with "resistor" NM9/PM7) for the bias feedback circuit.

The delay for an on-chip load capacitance of 50 fF is about 0.6 ns when implemented in an 0.7 µm CMOS process.

12.3   Monolithic filters

The amplified input signal usually must be filtered to remove redundant information and noise, to restrict the bandwidth (anti-aliasing) and to shape the signal spectrum [7]. With current-domain signals, filtering is relatively

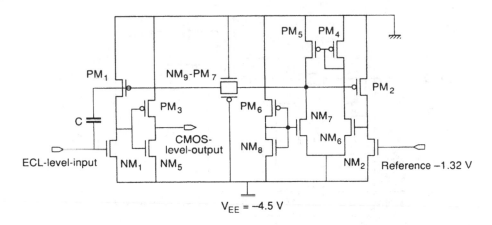

Figure 12.8 ECL 100k-compatible CMOS input buffer circuit

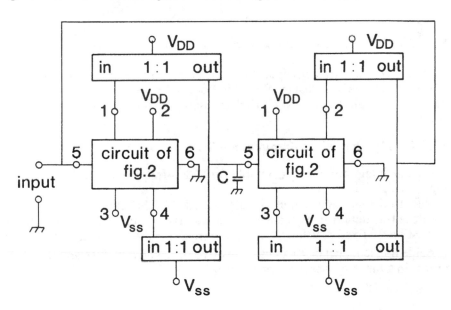

Figure 12.9 Tunable gyrator application of CMOS transconductor

easy since a capacitor driven by a current signal provides an integrating function. More generally, arbitrary transfer functions can be realized with transconductor-capacitor integrators. For example, Figure12.9 shows a gyrator constructed with two CMOS transconductor circuits of Figure 12.2. The circuit synthesizes a tunable inductor seen at the input port. For I_b = 100 μA and C = 100 nF, an inductance of 1.6 H was measured at 500 Hz, with a Q-factor of 900.

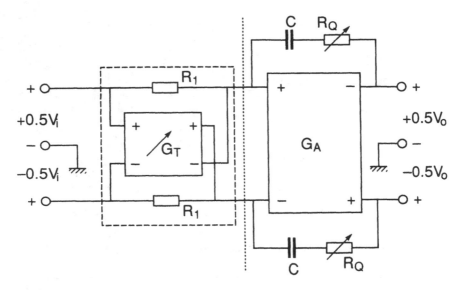

Figure 12.10 Active-RC integrator concept

High-frequency transconductance-C filter techniques are being developed at present [7]. Also, Chapter 9 of this book reviews the development of a number of high frequency CMOS transconductance-C continuous time filters.

12.3.1   Bipolar video-frequency filters

A bipolar integrator concept suitable for video frequencies is illustrated in Figure 12.10 [8].

Balanced input and output signals are used. The transconductor block (enclosed in dotted lines) consists of a passive nominal transconductance consisting of fixed resistors R1, and a variable transconductance G_T. The variable transconductance can have positive or negative values and therefore results in a tunable transconductor of value

$$G = \frac{1}{R1} + G_T$$

(12.11)

The nominal value of G_T is zero; it is always considerably smaller than the passive part since it only needs to correct fabrication spreads of the unity gain frequency $f_0 = 1/(2\pi R_1 C)$. This has the advantage of reducing the adverse effects of noise, nonlinearity and excess phase shift added by the active part G_T.

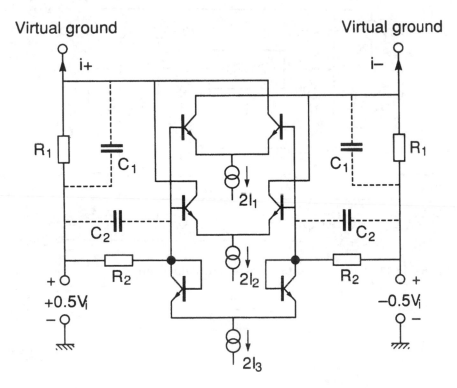

Figure 12.11 Circuit implementation of tunable transconductor

A transconductance amplifier (OTA) having transconductance G_A is used as gain block. The unity-gain frequency of the complete integrator is given by approximately

$$f_o = \frac{1 + G_T R_1}{2\pi R_1 C}$$

. (12.12)

and is tunable by G_T

The resistors R_Q control a high-frequency zero caused by the finite G_A. This zero is at frequency

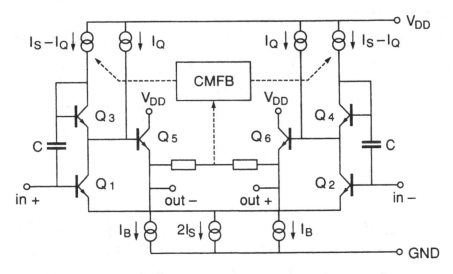

Figure 12.12 OTA with feedback path and output buffers

$$f_z = \frac{1}{2\pi C \left(\dfrac{1}{G_A} - R_Q\right)}$$

$$(12.13)$$

For $R_Q = 0$, the zero is in the right half plane. If R_Q is made equal to $1/G_A$, the zero is placed at infinity.

When R_Q is made greater than $1/G$ the zero is moved into the left half plane where it can compensate for the excess phase caused by parasitic poles and zeros of the circuit. Automatic Q-tuning can be applied by making R_Q variable.

The schematic of the tunable transconductor is shown in Figure 12.11 The variable part consists of a double translinear current amplifier. The overall transconductance is given by

$$G = \frac{1}{R_1} + \frac{I_1 - I_2}{R_2 I_3}$$

$$(12.14)$$

and can be tuned around $1/R_1$ by the difference of control currents $(I_1 - I_2)$. Capacitors C_1 and C_2 compensate for the phase shift of the resistors.

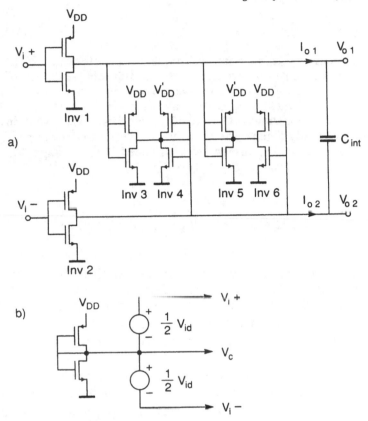

Figure 12.13 (a) VHF transconductance - C integrator (b) Definition of input signals

The OTA circuit is shown in Figure 12.12. It consists of a simple differential pair Q_1, Q_2 buffered at the output by emitter followers Q_5, Q_6. The diode connected transistors, Q_3, Q_4 form the variable resistors R_Q, controlled by currents I_Q. For $I_Q = 0$, R_Q will be very close to $1/G_A$, thus placing the zero near infinity. For $I_Q > 0$, R_Q becomes greater than $1/G_A$ and the zero is moved into the left-half plane. This enables Q-tuning. It is interesting to note that the diodes provide large-signal compensation of the signal-dependent base-emitter voltages V_{BE1}, V_{BE2} by V_{BE3}, V_{BE4}. This reduces distortion.

The tunable transconductor has good high frequency properties up to 70 MHz and less than 1% linearity error for input signals up to $2V_{pp}$. Simulations of a fifth order elliptic lowpass filter at 5 MHz indicate video-frequency capability. Techniques for on-chip tuning of cut-off frequency and quality factor are being investigated [9].

12.3.2   CMOS VHF filters

A transconductance - C integrator technique compatible with CMOS VLSI technology and suitable for frequencies beyond 100 MHz is shown in Figure 12.13 [10]. These frequencies become possible by using a linear transconductance element without internal nodes [11].

The transconductor shown in Figure 12.13 consists of six CMOS inverters which are essentially equal. The basic V-I conversion is performed by two inverters Inv_1, and Inv_2 driven by a balanced input voltage V_{id}. The differential output current is given by

$$I_{od} = I_{o2} - I_{o1} = V_{id}(V_{DD} - V_{tn} - V_{tp}) \sqrt{\beta_n \beta_p}$$

(12.15)

This result is based on the first-order saturated MOS model (12.2). Note that a linear transconductance is obtained, even with nonlinear individual inverters, i.e. if $\beta_n = \beta_p$. This is an interesting and attractive property. However, matched β_n and β_p is advantageous since this reduces the common-mode output currents. The transconductance can be tuned by means of V_{DD}.

The common-mode level of the output voltages V_{01} and V_{02} is controlled by the four inverters Inv_3 - Inv_6. Inv_4 and Inv_5 are shunted as resistors 1/gm4 and 1/gm5. For common signals the V_{01}-node is loaded by a resistance 1/(gm5+gm6) and the V_{02} - node by a resistance 1/(gm3 +gm4). For differential signals, however, these nodes are loaded by resistances 1/(gm5 - gm6) and1/(gm4-gm3) respectively. Thus the network Inv_3 - Inv_6 forms a low-ohmic load for common signals and a high-ohmic load for differential signals, resulting in a controlled common-mode voltage level of the outputs.

The DC-gain of short-channel CMOS inverters is low. This is improved by loading the differential inverters by a negative resistance for differential signals. By choosing gm3 > gm4, gm5 = gm4 and gm6 = gm3 this negative resistance, 1/Δgm = 1/(gm4-gm3) = 1/(gm5 - gm6), is simply implemented, without adding nodes to the circuit. In operation, the DC-gain of the integrators is fine-tuned with a separate supply voltage V'_{DD} for Inv_4 and Inv_5 to implement automatic on-chip Q-tuning. Thanks to the feedback loops in the filter the integrators will remain stable, even if the net output resistance of the transconductor becomes negative.

A third order elliptic low-pass filter was realized in a 3 μm CMOS process [10]. For a supply voltage of 5V the filter cut-off frequency was measured as 71 MHz and the low-frequency THD was less than 1% for input signal of 1

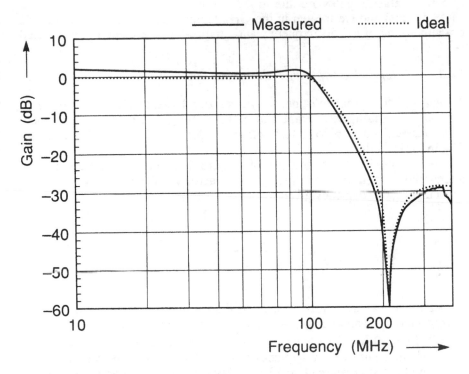

Figure 12.14 Measured and ideal response of CMOS 3rd order elliptic filter (V_{DD} = 10 V)

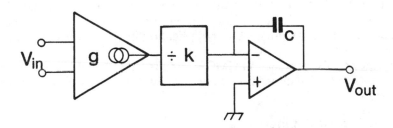

Figure 12. 15 Realization of large on-chip time-constant

V_{pp}. For a supply voltage of 10V the cut-off frequency was 110 MHz and low- frequency THD was 0.5% for input signal of 2 V_{pp}. In Figure 12.4 the measured filter response is compared with the ideal response.

This filter technique is felt to offer much promise for high-frequency applications. It owes its large bandwidth capability to the absence of internal nodes, utilizing the high-speed potential of VLSI to best advantage. The effective parasitic poles are due to the finite transit-time of carriers in the MOS channel and are located in the GHz region. Techniques for automatic on-chip tuning of cut-off frequency and quality-factor are being investigated [12].

12.3.3 *Integration of large time-constants*

A long- standing problem is that of synthesizing large time-constants on-chip without also multiplying noise and offset voltage. A recently developed solution, realized in bipolar technology, is depicted in simplified form in Figure 12.15 [13].

The class AB transconductor of Figure 12.5 is used, together with attenuating current mirrors to realize the factor k. The resulting integrator has a unity-gain frequency given by

$$f_{ug} = \frac{g}{2\pi kC}$$

(12.16)

Offset and noise are hardly multiplied thanks to the class AB input stage having a large dynamic range. A monolithic implementation with $R_1 = R_2 = 50$ kΩ, $I_o = 30$ μA for Figure 12.5 and $k = 6000$, $C = 50$ pF for Figure 12.15, yielded a unity gain frequency below 5 Hz and input offset voltage of 2mV for chip temperature up to 120 °C. By comparison, a passive R-C low-pass section with $C = 50$ pF and cut-off frequency of 5 Hz would require a resistance of 700 MΩ. Inconventional IC-technology this would need a diffused resistor of width 5 μm and length 20 meters!

The circuit has been used as an offset cancelling circuit for monolithic audio amplifiers [13]. Its use allows the elimination of up to two large external electrolytic capacitors. Wider application in low-frequency control loops is envisaged.

Implementing this technique in CMOS circuits should enable even larger time-constants to be synthesized owing to the low leakage currents and zero gate current.

12.4 Signal conditioning

The band-limited input signal frequently requires linearity correction and preliminary analog signal processing to remove redundant information and to improve the efficiency and speed of the subsequent A-D conversion and digital processing. Examples are, signal normalization [14], folding and interpolation [15], gain-control, and amplitude-compression. Examples of gain-control circuits have already been presented in section 12.2. For bipolar (and BICMOS) circuits, the translinear principle offers process- and temperature-insensitive algebraic signal conditioning [16]. This circuit

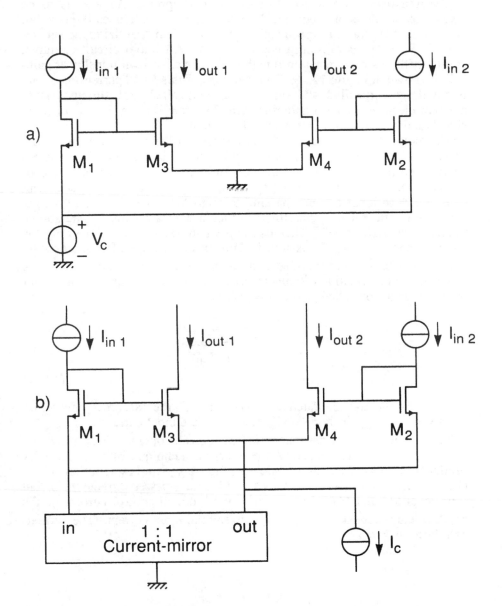

Figure 12.16 (a) Voltage-controlled current gain cell (b) Current-controlled gain cell

principle is based on exploiting the strictly exponential current-voltage characteristic of bipolar transistors. Its application requires current-signals; these are provided by V-I converters.

The translinear principle and bipolar circuit applications are discussed elsewhere in this book (chapter 2) , as well as in references [6,16,17,18]. Extension of the translinear principle to CMOS circuits, utilizing the square-law MOS model, is presently being studied.   An initial circuit example, similar both in structure and in function to the well-known translinear gain-cell, is shown in Figure 12.16. Two versions of this MOS current-gain cell, one voltage-controlled and one current-controlled, are shown.   Both perform electronically variable differential current scaling, linear for large signals and having a gain-independent bandwidth.

The gain-cell circuits require input currents of the form provided by the linear CMOS voltage-current converters discussed in section 12.2.1. It can be recalled that the linear V-I conversion was based on the principle of deriving currents in such a way that the sum of gate-source voltages is kept constant. The input currents passing through M1 and M2 of the voltage-controlled gain-cell will therefore produce a constant sum of gate-source voltages for M1 and M2. This also applies to M3 and M4 since only the common control voltage V_c is added. This means that the differential output current will be a linearly scaled copy of the differential input current. A more detailed calculation [19] reveals that the differential current-gain factor for the voltage-controlled gain-cell is given by

$$A_i = 1 + \frac{V_c}{\sqrt{2I_{in}/\beta}}$$

(12.17)

where ß is the transconductance parameter of the (assumed identical) transistors, and I_{ino} is the zero-signal value of the input currents.

The current-controlled version (Figure 12.16b) operates according to a related principle. The linear V-I conversion technique of section 12.2.1 produces currents, the sum of which contains a square-law function (12.7), (12.9). It has been shown [19] that when this sum is passed through the output transistors M3 and M4, together with an additional control current I_c, the result is again a linear scaling of the differential input current. The current-gain factor is given by

$$A_i = \sqrt{1 + \frac{I_c}{2I_{in}}}$$

(12.18)

where I_{ino} is again the zero-signal value of the input currents.

These MOS gain-cell circuits, unlike their bipolar counterparts, have a constant (gain-independent) bandwidth. This favourable property is owing

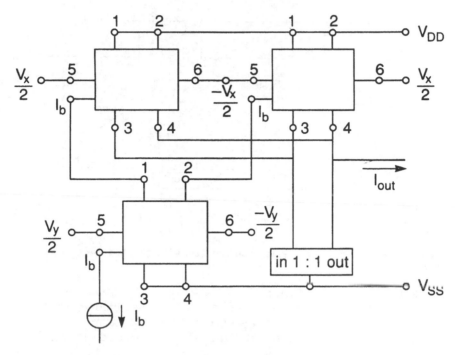

Figure 12.17 A CMOS four-quadrant multiplier. The blocks are the V-I converters of Figure 12.2

to the fact that the transconductance of a MOS transistor can be varied without affecting its gate-source capacitance, in contrast to the bipolar case. Expressed in another way, these circuits have gain-bandwidth products a factor A_i larger than for a simple current mirror.

Measured performance of a breadboarded voltage-controlled gain-cell using CA 3600 E transistor arrays produced current-gain of 8 for $V_c = 5$ Volts and THD < 0.5% for signal-amplitude of 95% of its maximum value. For 50% signal-amplitude the THD dropped to about 0.1%.

Nonlinear MOS circuit techniques for signal conditioning and function synthesis have been proposed [20,21]. In addition, the versatile CMOS V-I converter circuits described in section 12.2.1 can be configured to implement useful functions. Figure 12.17 depicts a four-quadrant analog multiplier circuit based on the V-I converter circuit of Figure 12.2.

Using two V-I converter blocks and modulating their bias currents by a third block yields a multiplier with transfer function given by

$$I_{at} = 4\,\beta_{eq}\,V_x V_y$$

(12.19)

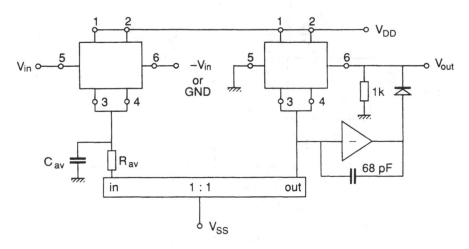

Figure 12.18 RMS-DC converter constructed with the V-I converters of Figure 12.2

Using a square-law function such as (12.7) or (12.9), a simple RMS-DC converter circuit can be realized. In Figure 12.18, each block represents the circuit of Figure 12.2 and computes the square of its respective input voltage. The filter $R_{av}C_{av}$ removes the AC-part. The transfer function is therefore

$$V_{out} = \sqrt{av(V^2in)}$$

(12.20)

The diode and resistor at the output eliminate a latch-up condition and the 68 pF capacitor ensure high-frequency stability. The measured linearity error at 10 kHz was within 0.5% of full-scale value for square-wave, sine-wave, and triangle-wave signals. The -1% bandwidth was 1 MHz and the-3dB bandwidth was 6 MHz.

Omitting the low-pass filter $R_{av}C_{av}$ from the circuit of Figure 12.18 yields a low-frequency precision rectifier or absolute-value circuit. This application is based only on circuit symmetry and matching, and therefore does not require accuracy of the square-law function. The maximum signal amplitude is restricted only by the supply voltage.

12.5 Internal interfacing

Frequently, circuits located far apart on a VLSI chip have to communicate at high speed or in a multiplexed manner. An example of the former situation is found in static RAMs where the data in the fragile memory cells have to be

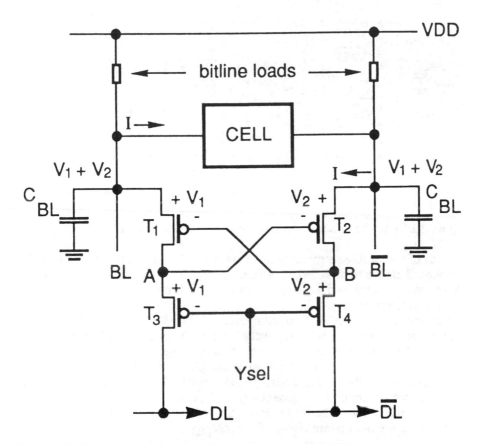

Figure 12.19 Current sense-amplifier circuit for SRAMs

accessed and transported to the output circuits via long lines and yet at high speed. Multiplexed information transfer, on the other hand, can be applied in other situations where not high speed, but rather economy of interconnecting lines is of paramount importance.

12.5.1 Current sensing in SRAMs

Designing fast, high-gain and robust sense-amplifier circuits is a challenge, particularly for large SRAMs realized in submicron CMOS technology. This is because firstly, large memories use long bitlines which present a large capacitance load for the memory cells, thus causing extra signal delay. Secondly, the small-signal voltage gain provided by submicron transistors decreases as the channel length becomes smaller, due to channel length modulation. The latter point presents a problem for conventional voltage sense-amps.

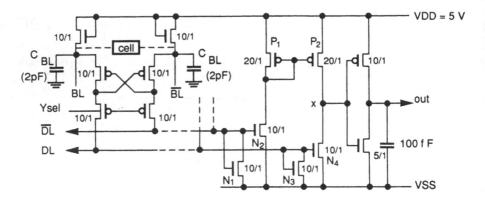

Figure 12.20 Current sense-amp followed by output section

Both of the problems mentioned above are relieved when current signals are sensed rather than voltage signals. Figure 12.19 shows a new type of sense-amp circuit which directly measures the cell read-current and transfers it to the output circuits [22].

The circuit consists of four equal sized PMOS transistors in a cross-coupled configuration. Since it is very simple, the circuit can in many cases fit in the column pitch, thus again reducing propagation delay. The sense-amp is selected by grounding the Y_{sel} node. Currents will then flow through the transistors via the bitline loads. The drains of T3 andT4 are connected to data-lines which are close to ground-level. This means that these transistors operate in saturation. The bitline loads are low-ohmic to ensure that during read-access, the bitlines are always close toV_{DD}

The circuit operates as follows. Suppose the cell is accessed and draws current I as shown in Figure 12.19. The gate source voltage of T1 will be equal to that of T3 since their currents are equal and both are in saturation. This voltage is represented by V1. The same applies to T2 and T4. Their gate-source voltages are represented by V2. It follows that since Y_{sel} is grounded, the left bitline will have voltage V1+V2, and the right bitline will also have voltage V1+V2. Therefore the potentials of the two bitlines will be equal independent of the current distribution. This means that there exists a virtual short-circuit across the bitlines. Since the bitline voltages are equal, the bitline-load currents will also be equal. As the cell draws current I, it follows that the right-hand leg of the sense-amp must pass more current than the left leg. In fact the difference between these currents is I, the cell current. The drain currents of T3 and T4 are passed to current-transporting datalines.The differential data-line current is therefore equal to the cell current. Thus we obtain current sensing. The cross-coupled structure is actually a flip-flop configuration, but sufficient margin from unwanted latching behaviour is provided by the body-effect and the low output resistance of short-channel transistors.

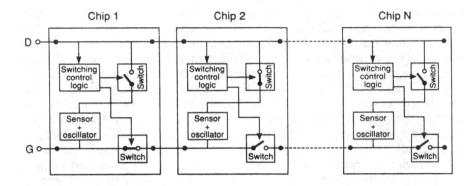

Figure 12.21 Schematic diagram of a two-lead multiplex array

Since the bitline voltages are kept equal, the sense-amp possesses intrinsic equalizing action. This removes the need for bitline equalization. The sensing delay is insensitive to bitline capacitance since no capacitor discharging is required to sense the cell data. A second speed enhancing feature is provided by the common-mode discharge current pulses from the bitline capacitors effectively precharging the sense-amp (in particular precharging nodes A and B, and the data lines) as soon as Y_{sel} is grounded. This is a kind of dynamic biasing which is very favourable for speed and which does not increase the current consumption. The output of this current sense-amp circuit can take the form of a current as in Figure 12.19 or, alternatively, the differential voltage signal at nodes A and B can be taken as output, to be further amplified by a second stage, as shown in Figure 12.20.

Simulations for a submicron CMOS process and supply voltage of 5V predict a sensing delay of 2 ns for a typical bitline capacitance of 2 pF. This is three times faster than a conventional voltage sensing sense-amp which requires more than three times the current. In order to verify the effectiveness of the low-impedance current sensing, the bitline capacitance was changed over a wide range. The sensing delay remains 2ns and only starts increasing for C_{BL} larger than 15 pF.

12.5.2 Two-lead multiplexing system

A multiplexing information transfer principle needing only two leads for both supplying power to and extracting information from any number of identical signal sources, is illustrated in Figure 12.21.

The array depicted in the figure consists of N identical smart sensor chips connected in parallel by two leads, the D (data) line an the G (ground) line. This particular system constitutes a temperature sensor array which was developed for the measurement of tissue temperature profiles during hyperthermia treatment of cancer patients [23]. This multiplexing principle was originally devised for the detonation of explosions by sequential control in gold mines [24]. Although the array shown here as an example consists of separate chips, the multiplexing principle can also be applied within a VLSI circuit where the identical signal sources can consist of separate circuit blocks. Each chip shown in Figure 12.21 contains a multiplexer (switching control logic) that controls which sensor has to be read out. The supply voltage is also derived from the D line. A switching pulse on the D line causes the sensor which is connected to the D line to be disconnected by opening the appropriate switch and simultaneously causes the sensor on the next chip to be connected to the D line by closing the ground-line switch. In this way all the sensors of the array are connected to the D line sequentially, after which the array is reset and the sensor on chip 1 is connected to the D line again. In the example, the sensor of chip 2 is being read out. Since the sensor array is controlled sequentially by means of switching pulses, by counting these pulses it is always possible to know which of the identical sensors is being read out. The D line will also carry the sensor signal, in this case as a frequency-modulated current whose frequency is proportional to the chip temperature.

Thus the D lead transports supply power and control pulses to the chips and modulated sensor information from the connected sensor. The switching control logic is simple, consisting mainly of some switches and a set-reset flip-flop.

12.6 Output interface circuits

The output interfacing function invariably involves power-handling since the low-impedance "real world" has to be manipulated. Two examples will be presented, the first a bipolar circuit technique for analog applications and the second a CMOS technique for digital applications. In addition, the design of bipolar power transistors for extended safe operating area will be briefly discussed.

12.6.1 Bipolar power amplifier

The design of monolithic power amplifiers is no simple matter. Major problems are the linearity and stability of the class AB output stage. Figure 12.22 shows the circuit principle of a recently developed bipolar power amplifier suitable for high-quality audio applications for up to 40 Watt output power [25]. The output transistors are driven in push-pull by the differential stage T_3, T_4. This eliminates the customary quasi-PNP compound transistor, thus improving stability. The output transistor bias currents are feedback-controlled by the translinear bias control circuit at the top of Figure 12.22. Note that at the collectors of T_3 and T_4 we have common-mode drive for the class AB biasing and differential-mode drive for the signal.

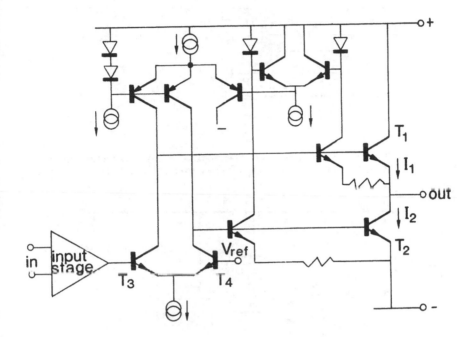

Figure 12.22 Circuit principle of high-quality power amplifier

The bias control loop forces the output transistor currents to conform to a harmonic-mean control law

$$\frac{I_1 I_2}{I_1 + I_2} = \text{constant}$$

(12.21)

rather than the geometric-mean law common for conventional class AB output stages. The significance of (12.21) is seen when an output current is delivered. I_1 and I_2 will then become very different. It follows from (12.21) that the nondriven output transistor can never cut off in contrast to the conventional case. In fact, the residual output transistor current stabilizes at one-half the zero-signal quiescent value, when the other output transistor is driven very hard. This is a nearly ideal class AB behaviour. The measured THD was around -80 dB at 20 kHz, for a 7 Ω load. Possible application of these, or similar, biasing techniques to the design of CMOS power amplifiers is presently being studied.

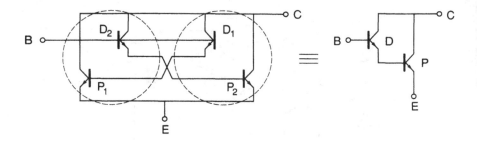

Figure 12.23 Thermally cross-coupled Darlington power transistor to extend safe operating area. Driver D_1 (D_2) is thermally coupled to power transistor P_2 (P_1).

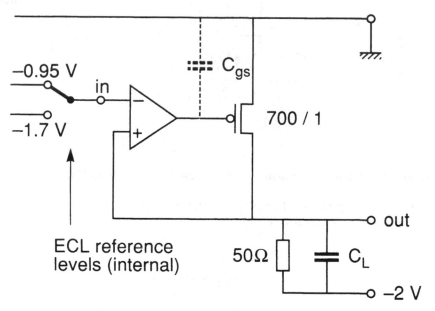

Figure 12.24   Feedback-controlled ECL output buffer

12.6.2 Bipolar power transistor design for extended safe operating area

It is well known that bipolar transistors suffer from secondary breakdown. This phenomenon is the result of a thermal runaway mechanism which causes the current being passed by a multi-emitter power transistor to be crowded into a small section of the transistor. This concentrates the power dissipation in one small part, thus leading to destructive overheating. Emitter ballast resistors are often applied, but this increases the saturation voltage and power dissipation. A better solution is to partition the

Darlington transistor into thermally cross-coupled sections as shown in Figure 12.23.

This technique of using the driver of one power transistor section to sense the temperature of the other transistor section is a simple, yet effective method to equalize the current distribution, thereby extending the safe operating area (SOAR) [26].

The operating mechanism is based on creating an electrothermal negative feedback loop within the power transistor structure. This electrothermal feedback controls the thermal runaway characteristic inherent to bipolar transistors, resulting in thermal stability up to significantly larger temperature gradients.

12.6.3 CMOS subnanosecond true-ECL output buffer

In section 12.2.4 an ECL-compatible CMOS input buffer circuit was described. The complement of this function, the ECL output buffer, is much more difficult to implement in CMOS only technology due to the combination of required precision, high speed, stability and pulse-response for variable load capacitance. In fact, this circuit constitutes the key to realizing ECL-compatible VLSI circuits in CMOS-technology.

The technique to be described below results in the first CMOS only output buffer circuit compatible with standard ECL 100k systems and not needing external components or additional supply voltages. High speed (0.9 ns delay), sufficient precision and good pulse-response are achieved through use of a new circuit principle [27]. Simply replacing the emitter follower transistor of bipolar ECL by an n- or p- MOS transistor is not viable due to the output level accuracy required.

Figure 12.24 illustrates the basic approach used. A feedback amplifier driving a large PMOS output transistor is connected as a unity-gain buffer. The ECL levels can be switched to the buffer input by CMOS logic. Since the ECL reference sources are not loaded they can be generated on-chip by a suitable voltage reference (e.g. bandgap). The feedback circuit has two major time-constants, one due to load capacitance and one due to gate-source capacitance of the large output transistor. The amplifier contributes additional delay, thus resulting in total phase shift at high frequency of possibly more than 180°. This means that the negative feedback could change to positive, leading to an unstable loop which will oscillate. This is a classic problem and is usually solved by phase compensation which, however, sacrifices speed.

The new technique combines stability and precision without sacrificing speed, Figure 12.25. We start with a high-speed, low-gain differential amplifier circuit, comprising M_1-M_5. Since almost no overshoot of the pulse-response is allowed, a phase-margin of more than 90° is needed. Therefore we decrease the gain further by adding resistor R. Because of the low gain, the precision will now be insufficient. This is corrected by a

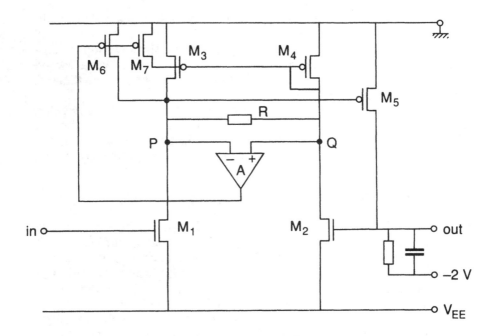

Figure 12.25 Principle of high-speed output buffer with error-correcting feedback loop

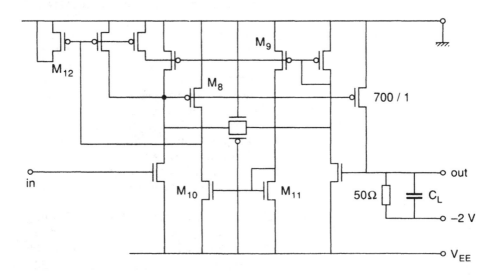

Figure 12.26 Complete ECL-compatible output buffer circuit

slower error-correction circuit comprising differential amplifier A, M_6 and M_7.

The speed of the basic amplifier is not affected. This feedback loop corrects the input-output error caused by unbalanced drain voltages of M1 and M2. Amplifier A detects a voltage difference between nodes P and Q and relatively slowly drives M6 and M7 to change the common-mode bias currents of current-mirror M3-M4. As a result, the potential of node Q is driven towards that of node P. This new differential-mode to common-mode bias feedback (DCMFB) principle thus forces almost equal drain voltages for M1 and M2. Since M1 and M2 pass equal currents this results in a small input-output error.

The complete buffer circuit is shown in Figure 12.26. The error-correcting amplifier A consists of M8-M11 and is loaded by capacitor M12. Prototypes have been fabricated in an 0.7 µm CMOS process. Acceptable precision, unconditional stability and good pulse-response for variable load capacitance were confirmed. The measured delay is 0.9ns for a standard ECL-load of approximately 3 pF, increasing to 2.5ns for 50 pF and 3.6ns for 100 pF.

12.7 Power supply interfacing

As was mentioned in section 12.1, submicron CMOS circuits need a reduced supply voltage to protect the transistors against hot carrier degradation. The standard system supply voltage is 5V, while the internal circuits require a supply of about 3.5 V. At first sight designing a suitable on-chip supply voltage converter would appear to be an easy, if not trivial, task. However, a closer look reveals that this presents a surprisingly challenging problem for SRAMs, for example.

A major section of the SRAM-market requires a very low standby current specification, in the order of 1 µA. The leakage current of a large SRAM makes up a fair part of this current value. A supply voltage converter circuit consisting of a voltage reference and a voltage regulator therefore is required which consumes less than 1 µA, yet is able to deliver the maximum SRAM supply current of tens of mA within an insignificant delay. In addition, stability for highly capacitive loads must be guaranteed (the internal chip represents a capacitance of some nF).

A topic which turns out to be closely related to supply voltage down-conversion is that of on-chip supply current testing. This so-called I_{DD}-testing relies on the principle that in the quiescent state a full-CMOS device should draw no supply current (zero I_{DD}) apart from the very small leakage currents. By measuring the I_{DD} just prior to activating or clocking a device it is possible to detect processing defects such as shorts and opens that cause a higher than normal leakage current. The advantage of this method with respect to conventional testing is the possibility to detect early failure states that will cause fatal errors after a longer period of usage.

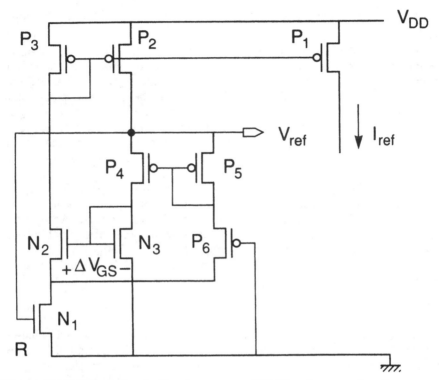

Figure 12.27 Principle of ultra-low power CMOS reference circuit

12.7.1 "Nanopower" CMOS voltage and current reference

The reference voltage generator needed for the supply voltage down-convertor function discussed above is only allowed to consume a very small current. The reference circuit principle shown in Figure 12.27 was devised for a submicron process and requires only about 100nA. The circuit generates a reference voltage as well as a reference current, both of which are temperature-insensitive. The approach is similar to the well-known bipolar bandgap reference technique but operation is based only on well-characterized CMOSproperties.

The feedback-loop consisting of "resistor" R (implemented by transistor N1), N2, P3, P2, P4 and N3 forms a current-stabilizer modelled on the bipolar PTAT) generator. For the moment P5 and P6 are ignored. The transistors are dimensioned such that for small currents the loop-gain is greater than unity. The currents will then increase until the voltage drop across R (ΔV_{GS}) decreases the loop-gain to unity. R is made independent of V_{DD} by connecting N1's gate to the stabilized reference voltage V_{ref}. N2 and N3 are designed such that they operate in weak inversion (sub-threshold) [28]. Then ΔV_{GS} will be proportional to absolute temperature (PTAT):

$$\Delta V_{GS} \; \alpha \; \frac{n.kT}{q}$$

(12.22)

with kT/q the thermal voltage ($\cong$ 26 mV at room temperature).

The stabilized loop current is then given by $\Delta V_{GS}/R$, with R expressed from the triode-region MOS-model as

$$R \cong \frac{1}{\beta \, (V_{ref} - V_{th})}$$

(12.23)

The temperature dependence of ß is determined by the mobility $\mu \; \alpha \; T^{-3/2}$. If V_{ref} is constant, $V_{ref} - V_{th}$ will increase slightly with temperature. It follows from (12.23) that R will vary slightly more than PTAT. Recalling (12.22), we finally conclude that I_{ref} will remain almost constant with temperature. For example, simulations predict I_{ref} to decrease from 45nA to 44nA for 0-70°C temperature increase. However, I_{ref} is process-dependent, mainly through ß in (12.23).

In order to realize these very small currents, R must be very large. The drain connection of P6 reduces the required value to some MΩ in this case. This resistance can be obtained by choosing the W/L of N1 about 1/400.

The reference voltage V_{ref} is produced by the sum of the gate-source voltages of P5 and P6. These are determined by the stabilized current flowing through them and by their threshold voltages. The latter are made free of body-effect by connecting the n-wells to the sources of P5 and P6 respectively. P5 and P6 are dimensioned such that they operate in strong inversion (long channels). It follows that V_{ref} is given by

$$V_{ref} = 2 \, V_{thp} + 2 \sqrt{ 2I_p/\beta_p }$$

(12.24)

with I_p the part of the stabilized current $\Delta V_{GS}/R$ flowing through P_5 and P_6. Using (12.22) and (12.23), and solving for V_{ref}, results in

$$V_{ref} = 2V_{th} + k\Delta V_{GS}$$

(12.25)

where the constant k is mainly determined by β_n/β_p and the current-division ratio of P4/P5. Since V_{th} and ΔV_{GS} have opposite temperature coefficients, proper choice of k will result in a temperature-stable reference voltage, similar to the bandgap-reference technique.

The circuit was designed for $V_{ref} = 3.6$ V. Simulations predict a variation of less than ± 50 mV between 0 °C and 70 °C. However, V_{ref} is process-dependent, mainly through V_{th} in (12.25). This level of performance, although probably adequate for SRAMs, is not nearly as good as that obtainable from bipolar bandgap-references. The latter can be realized in CMOS technology, but require process compatible bipolar transistors with guaranteed DC performance [29,30]. An important feature of the new CMOS reference circuit presented here is its ultra-low power dissipation, comparable with the normal chip leakage. This factor may prove to be more important than the inferior precision when compared to the bipolar bandgap technique which requires a much larger supply current.

Finally, in common with many reference circuits, provision for proper starting is needed. A simple solution is to connect a long-channel diode-connected MOS transistor across P2.

12.7.2 Regulator circuit with adaptive biasing having low standby current

The principle of a regulator circuit with standby current below 1 µA is shown in Figure 12.28. The load is represented by the chip capacitance C_c (some nF) and a current sink. Rather than relying on an activation signal or clock to switch the circuit from a standby state to a condition for delivering a large load current, this circuit automatically adapts its bias currents to changes in the required load current. This makes the circuit more generally applicable.

The regulator consists of two parts: a reference buffer (P1-P4) which consists of a modified Wilson current mirror, and a two-stage current amplifier (N1-N4, P5-P7). The reference buffer provides a low-impedance node A at a constant voltage of $V_{ref} + V_{GS}(P2)$, independent of varying current drawn by P5. This circuit is based on a universal adaptive biasing technique [31]. P2 passes the constant small current I_{ref} (approx. 100nA) and P4 provides the current required by the current amplifier part. The latter contains two ratioed current mirrors N1-N4 and P6-P7 and has a total current gain of about 50.

The capacitor C_k (approx. 2 pF) has a double function. First, it acts as a "kick-starter" to obtain sufficient response speed. The standby bias currents

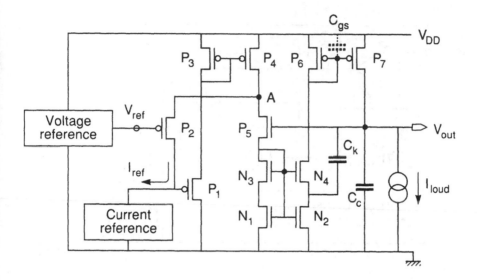

Figure 12.28 Principle of regulator circuit with low standby current

are far too small to charge the gate-source capacitance C_{gs} of P7 (size 800/1) at the required rate. If the required chip current increases suddenly from leakage levels to tens of mA, the output voltage V_{out} will dip down sharply. C_k then produces a current pulse which passes N4 and quickly charges C_{gs}. The second function of C_k is to stabilize the feedback loop around the current amplifier by providing a zero in the forward signal path. A second "kick-start" capacitor (not shown) is required to enhance the response speed of the reference buffer part.

Simulations predict a total standby current (including the reference circuit) of 600 nA, a maximum output current of 60 mA (10^5 times the standby current), and output voltage regulation of 0.25 V.

12.7.3 On-chip I_{DD} monitor circuit

A supply current monitor circuit for I_{DD}-testing is required to measure the supply current while keeping the internal chip voltage constant. The circuit principle is depicted in Figure 12.29.

The core of the monitor circuit is formed by current mirror transistors T1 and T2. Transistor T1 provides the current which is drawn by the digital circuit being tested. In order to have a fixed internal supply voltage V'_{DD}, close to V_{DD}, a differential amplifier A1 drives the gate of T1. The output voltage V'_{DD} is fed back to A1 and is thus forced to be equal to V_{ref}. It can be seen that the combination of A1 and T1 constitutes a voltage regulator. Therefore, a supply voltage down-converter can conveniently be used to

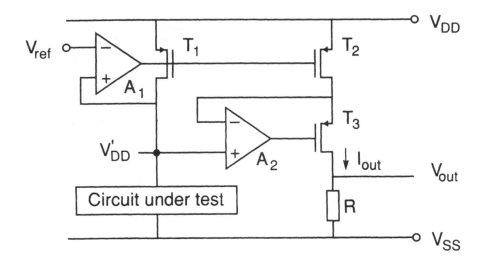

Figure 12.29 Principle of regulator circuit with low standby current

provide an additional function, that of I_{DD}-testing. If A1 is allowed to consume a reasonable standby current, many circuit implementations are possible which allow V_{ref} to be very close to V_{DD} [32].

Amplifier A2, T2 and T3 from a scaled copy of the supply current being measured. The feedback arrangement equates the drain voltages of T1 and T2 and thereby ensures accuracy even when T1 is nonsaturated. The output current can be further processed on-chip by converting it to a voltage, which can be measured or compared with a threshold value. The monitor circuit was realized as part of an SRAM in a standard 1.6 μm CMOS process [32]. Supply currents down to 2 μA can be measured at a test-rate of 1 MHz.

12.8   Conclusions

This chapter has addressed the design of time-continuous analog interface circuits required between the "real world" and digital VLSI silicon systems. Circuit techniques compatible with the dominant monolithic technologies CMOS, bipolar and BICMOS were discussed. The interface functions must prepare the real-world analog signals for digital processing and vice versa. In addition, the interfacing will frequently convert from the voltage-domain outside the chip to current-domain inside the chip. This is favourable for dynamic range in low-supply voltage VLSI circuits and for high speed in the capacitance-dominated chip interior. The important topic of analog-digital interference on a VLSI chip has not been addressed here. Some information is provided in [2].

The following interface functions were considered: input interfacing, filters, signal conditioning, internal interfacing, output interfacing and power supply interfacing. Some topics which were discussed were: linear voltage-to-current conversion (transconductors), video and VHF filter techniques, nonlinear function synthesis, SRAM sense-amplifiers, two-lead multiplexing, CMOS subnanosecond ECL-compatible input and output buffers, power amplifiers, bipolar power transistor design for extended safe-operating-area, ultra low-power CMOS voltage and current references, and on-chip I_{DD}-testing.

The circuits which were presented make use of electronic principles, old and new, such as negative feedback, error compensation, class AB operation, translinear principle, and differential-mode to common-mode feedback. In many cases exploiting the intrinsic monolithic transistor properties results in simple, high-speed or low-power circuits. It is expected that the development of new and improved circuit structures will continue, notwithstanding the tapering-off of process-technology innovation

12.9 Acknowledgments

Much of the work described in this chapter was carried out at the University of Twente, Enschede and Philips Research Laboratories, Eindhoven. I should like to thank my former and present co-workers, students and colleagues for their collaboration and their contributions.

12.10 References

[1] E. Habekottaa, B. Hoefflinger, H.W. Klein and M.A. Beunder, "State of the Art in the Analog CMOS Circuit Design", Proc. IEEE, Vol. 75, pp.816-828, June 1987.

[2] E.A. Vittoz, "The Design of High)Performance Analog Circuits onDigital CMOS Chips", IEEE J. Solid)State Circuits, Vol. SC)20, pp.657-665, June 1985.

[3] A. Nedungadi and T.R. Viswanathan, "Design of linear CMOStransconductance elements", IEEE Trans. Circuits Syst., Vol. CAS-31,pp. 891-894, October 1984.

[4] E. Seevinck and R.F. Wassenaar "A Versatile CMOS LinearTransconductor/Square)Law Function Circuit", IEEE J. Solid)StateCircuits, Vol. SC)22, pp. 366)377, June 1987.

[5] E. Klumperink, E. v.d. Zwan and E. Seevinck, "CMOS variable transconductance circuit with constant bandwidth", Electron. Lett.,Vol. 25, pp. 675)676, 11 May 1989.

[6] R.F. Wassenaar, E. Seevinck, M.G. van Leeuwen, C.J. Speelman and E.Holle, "New Techniques for High-Frequency RMS-to-DC Conversion Based on a Multifunctional V-to-I Converter", IEEE J. Solid-State Circuits,Vol. SC)23, pp. 802)814, June 1988.

[7] J.O. Voorman, "Analog integrated filters or continuous-time filters for LSI and VLSI", Revue de Physique Appliquaae, Vol. 22, pp. 3-14,January 1987.

[8]   W.J.A. de Heij, E. Seevinck and K. Hoen, "Transcon-ductor andintegrator circuits for integrated bipolar videofrequency filters",Internat. Symp. on Circuits and Systems Proceedings (ISCAS 1989), pp.114-117, May 1989.

[9]   W.J.A. de Heij, K. Hoen and E. Seevinck, "Accurateautomatic tuningcircuit for bipolar integrated filters", Internat. Symp. on Circuits and Systems Proceedings (ISCAS 1990), May 1990.

[10] B. Nauta and E. Seevinck, "A 110 MHz CMOS transcon- ductance)C low*pass filter", Dig. Tech. Papers, European Solid)State Circ. Conf.(ESSCIRC 1989), pp. 141-144, September 1989.

[11] B. Nauta and E. Seevinck, "Linear CMOS transconductance element forVHF filters", Electron. Lett., Vol. 25, pp. 448-450, 30 March 1989.

[12] B. Nauta and E. Seevinck, "Automatic tuning of quality factors for VHF CMOS filters", Internat. Symp. on Circuits on Systems Proceedings(ISCAS '90), May 1990.

[13] R.J. Wiegerink, E. Seevinck and W. de Jager, "Offset CancellingCircuit", IEEE J. Solid-State Circuits, Vol. SC-24, pp. 651-658, June1989.

[14] B. Gilbert, "A monolithic 16channel analog array normalizer", IEEEJ. Solid-State Circuits, Vol. SC 19, pp. 956-963, December 1984.

[15] R.J. van de Plassche and P. Baltus, "An 8-bit 100 MHz Full-Nyquist Analog-to-Digital Converter,", IEEE J. Solid)State Circuits, Vol. SC-23, pp. 1334-1344, Dec. 1988.

[16] B. Gilbert, "Translinear circuits: a proposed classification",Electron. Lett., Vol. 11, pp. 14-16, Jan. 1975, and "Errata", ibid.,p. 136.

[17] E. Seevinck, "Synthesis of nonlinear circuits based on thetranslinear principle", IEEE Internat. Symp. on Cir- cuits and SystemsProceedings, pp. 370-373, May 1983.

[18] E. Seevinck, Analysis and Synthesis of Translinear IntegratedCircuits, Elsevier Science Publishers, Amsterdam 1988.

[19] E.A.M. Klumperink and E. Seevinck "MOS Current Gain Cells withElectronically Variable Gain and Constant Bandwidth", IEEE J. Solid-State Circuits, Vol. SC)24, pp. 1465-1467, Oct. 1989. 7

[20] J.W. Fattaruso and R.G. Meijer, "MOS Analog Function Synthesis", IEEEJ. Solid)State Circuits, Vol. SC-22, pp. 1056-1063, Dec. 1987.

[21] K. Bult and H. Wallinga, "A class of analog CMOS cir- cuits based onthe square)law characteristic of an MOS transistor in saturation",IEEE J. Solid-State Circuits. Vol. SC)22, pp. 357-365, June 1987.

[22] E. Seevinck, "A current sense amplifier for fast CMOS SRAMs", Dig. Tech. Papers, Symp. VLSI Circuits (Hawaii), June 1990.

[23] A. Kaalling, S. Koomen, P. Bergveld, and E. Seevinck, "Two lead multiplex system for sensor array applications", Sensors andActuators, Vol. 17, pp. 623)628, 1989.

[24] T.C. Verster, E. Seevinck and R.F. Greijvenstein, "Detonation of explosions by electronic sequential control", South African patent application, 1978.

[25] E. Seevinck, W. de Jager and P. Buitendijk, "A Low)Distortion OutputStage with Improved Stability for Monolithic Power Amplifiers", IEEEJ. Solid-State Circuits, Vol. SC)23, pp. 794-801, June 1988.

[26] W. Smulders, U.S. patent 3 952 258, 1974.

[27] H. J. Schumacher, J. Dikken and E. Seevinck, "CMOS SubnanosecondTrue)ECL Output Buffer", IEEE J. Solid-State Circuits, Vol. SC-25,Feb. 1990.

[28] E. Vittoz and J. Fellrath, "CMOS analog integrated circuits based onweak inversion operation", IEEE J. Solid)State Circuits, Vol. SC)12,pp. 224-231, June 1977.

[29] M.G.R. Degrauwe, O.N. Leuthold, E.A. Vittoz, H.J. Oguey and A.Descombes, "CMOS Voltage Reference using Lateral BipolarTransistors", IEEE J. Solid)State Circ., Vol. SC)20, pp. 1151-1157,Dec. 1985.

[30] E. Holle, "A CMOS bandgap reference with reduced offset sensitivity",Dig. Tech. Papers, European Solid)State Circ. Conf. (ESSCIRC '88), 7p. 207-210, September 1988.

[31] E. Seevinck, R.F. Wassenaar and W. de Jager, "Univer- sal adaptivebiasing principle for micropower amplifiers", Dig. Tech. Papers,European Solid-State Circ. Conf. (ESSCIRC '84), pp. 59-62, September1984.

[32] A. Welbers, B. Verhelst, E. Seevinck and K. Baker, "A built)in CMOSI=D=D quiescent monitor circuit", submitted to European Solid-State Circ. Conf. (ESSCIRC '90), Sept. 1990.

Chapter 13

Current Mode A/D and D/A Converters

C. Andre T. Salama, David G. Nairn and Henry W. Singor

13.1 Introduction

The continued proliferation of mixed analog/digital VLSI systems has and will ensure that the need for small size, high speed analog-to-digital and digital to analog converters (ADCs and DACs) fabricated using commonly available digital processes will continue to grow. Initially, the widespread use of MOS technology, with its unique ability to accurately store and transfer voltages or charge packets [1], led to the development of analog integrated circuit techniques in which voltage was used as the signal. Although these techniques are quite successful in many applications, reductions in the available supply voltage and the deterioration in the performance of the analog components caused by the move to ever smaller geometries, is likely to limit their performance [2]. To address this issue, converters that do not require high quality analog components and still offer good performance with a minimum of chip area must be developed.

Current mode techniques (in which the signal is essentially processed in the current domain) offer a number of advantages. Generally current mode circuits do not require amplifiers with high voltage gains thereby reducing the need for high performance amplifiers [3]. At the same time current mode circuits generally do not require either high precision resistors or capacitors and when capacitors are used to store the signal, the capacitors need not display either good ratio matching or good linearity [2, 4, 5]. Consequently, current mode circuits can be designed almost exclusively with transistors making them fully compatible with most digital processes.

This chapter discusses the application of current mode techniques to the design of a specific set of analog to digital and digital to analog converters.

13.2 Analog to digital converters

Although current mode techniques may reduce the need for high performance analog components, the choice of architecture for the ADC has also a significant impact on the converter's performance. Presently, for voltage mode ADCs, the algorithmic or cyclicconversion technique [6] generally results in the smallest circuit size for a given resolution [7, 8, 9].

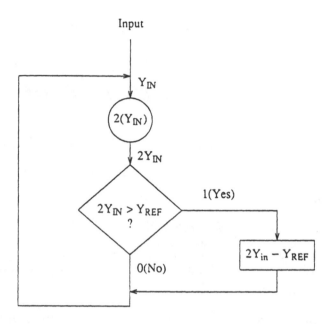

Figure 13.1 Flow chart for an algorithmic analog-to-digital converter.

Consequently current mode design techniques using an algorithmic architecture should offer the designer a small size ADC that is fully compatible with most digital VLSI processes. The focus here will be on current mode algorithmic ADC's. The basic design considerations and limitations of the techniques are discussed.

13.2.1   Algorithmic analog-digital conversion

The algorithmic analog-to-digital conversion technique [6] has been known for many years as a conversion method that can take advantage of relatively simple hardware to produce ADCs. An algorithmic conversion is perfomed as shown in Figure 13.1. The input signal, Y_{IN} which can take on any value between zero and the reference, Y_{ref}, is first doubled to create $2Y_{IN}$. The new signal, $2Y_{IN}$ is then compared with the reference. If $2Y_{IN}$ is less than the reference, the digital output is set to zero and $2Y_{IN}$ becomes the new Y_{IN}. If $2Y_{IN}$ exceeds the reference, the digital output is set to one. For this case, the reference is then subtracted from $2Y_{IN}$ to create a new Y_{IN} The new Y_{IN} can either be fed back to the input as shown in Figure 13.1 or on to a following identical cell which will perform exactly the same function and generate another bit of resolution. This process is repeated as many times as necessary to obtain the desired resolution.

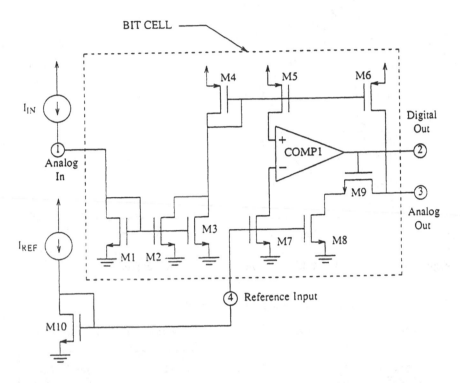

Figure 13.2 A bit cell that implements a one-bit algorithmic conversion.

13.2.2 *Basic mirror analog-to-digital converters*

To implement an algorithmic ADC using current mode techniques and basic current mirrors, the circuit shown in Figure 13.2 can be used [10, 11]. The circuit performs a one bit algorithmic analog to digital conversion in the following manner. The input current, I_{IN}, is first multiplied by two using the current mirror composed of M1, M2, and M3. Following the multiplication, the signal, $2I_{IN}$, is mirrored from M4 through M5 to the comparator, COMP1, and through M6 to the output. COMP1 is used to compare $2I_{IN}$ (from M5) with I_{REF}, the reference current (from M7). If $2I_{IN}$ is less then I_{REF} the digital output goes low and M9 remains off, resulting in an output current of $2I_{IN}$ (from M6). On the other hand, if $2I_{IN}$ exceeds I_{REF}, the digital output will be high causing M9 to be on. With M9 on, I_{REF} (from M8) will be subtracted from $2I_{IN}$(from M6) resulting in an output current of $2I_{IN}$-I_{REF}. This completes the one bit algorithmic conversion.

To produce an N-bit converter, N bit cells are cascaded with the analog output of one cell connected to the analog input of the following cell as

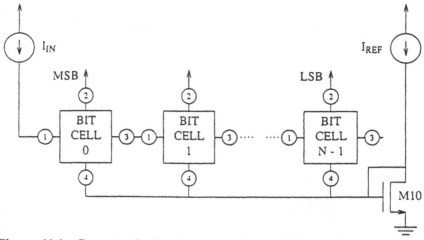

Figure 13.3   Cascade of bit cells for an N-bit converter.

Table 13.1   Current Mode ADC Characteristics

Parameter	Basic Mirror	Active Mirror	Current Matching
Technology Circuit Area	3μm CMOS 0.45 mm^2	3μm CMOS 0.74 mm^2	3μm CMOS 0.32 mm^2
Power Supply Reference Current Power Dissipation	+5V 25μA 4.65 mW	+5V 100 μA 63 mW	+5V 100μA 3.5 mW
Resolution Conversion Time	6 bits 5.0 μsec	8 bits 2.0 μsec	10 bits 40 μsec
Gain Error @ full scale	1.11 LSB	-0.06 LSB	-2.29 LSB
Offset Error	0.33 LSB	0.565 LSB	0.03 LSB
Differential Nonlinearity maximum DNL minimum DNL missing codes	0.35 LSB -0.79 LSB 0	0.46 LSB -0.45 LSB 0	0.10 LSB -0.87 LSB 0
Integral Nonlinearity maximum INL minimum INL	0.16 LSB -0.62 LSB	0.52 LSB -0.47 LSB	0.92 LSB -0.92 LSB

illustrated in Figure 13.3. Transistor M10 is shared by all the bit cells. The resulting linear sequence of bit cells does not require control signals. Therefore, this configuration will result in a very compact circuit that can be easily modified for different resolutions.

To achieve good resolution, the current mirrors used in the algorithmic current mode ADC must display excellent current matching. Therefore devices that exhibit a high output resistance and display good matching are essential. To improve the device's output resistance, long channel length devices must be used. To improve the device matching, both the channel length and the channel width must be significantly larger than the minimum feature size permitted by the technology [12, 13]. A current mode 6-bit ADC based on this principle was fabricated using a 3µm CMOS process. The resulting circuit had a total area of 0.45mm^2 [10]. The circuit itself displayed full 6-bit resolution with a maximum sampling rate of 200 kHz and a power consumption of less than 5 mW as shown in Table 13.1. The converter's speed is primarily limited by the settling time of the current mirrors and can be improved by using shorter channel length devices or by using bipolar devices. The circuit's resolution is limited by the finite output resistance of the current mirrors.

Due to the mirror's finite output resistance, errors are generated by the subtraction operation at the bit cell's output. The most significant error for the converter occurs at the 1000 to 1001 transition which corresponds to an input slightly larger than half the full scale input. In this case, the output of the first section will be very small (ie. $2I_{IN}\text{-}I_{REF}$), causing the voltage at the output to be pulled down to the threshold voltage of the n-channel device present at the input stage of the following bit cell. The low output voltage causes the reference mirroring device M8, in Figure 13.2 to come out of saturation resulting in a significant current error. This error is then amplified in successive stages of the converter leading to conversion errors and possibly missed codes.

To solve the subtraction problem, the voltage at the input to each bit cell must be kept as high as possible even for extremely low currents. Both objectives can be met by using cascoded current mirrors, as shown in Figure 13.4. Such a solution would not only increase the minimum input voltage from V_T to $2V_T$, where V_T is the device threshold voltage, but would also increase the output resistance of the current mirrors. Unfortunately, such a solution will also reduce the maximum gate voltage for the mirroring devices, thereby increasing the current mirror's errors due to V_T mismatches [12].

To achieve greater resolutions, a current mirror that does not suffer from either the V_T mismatches or the device's finite output resistance must be used. The effects of the V_T mismatches can be reduced by operating the mirroring transistors with the highest possible gate voltage. Simultaneously, the drain to source voltage of the mirroring devices must be buffered from differences between the mirror's input and output voltage thereby ensuring that the devices' finite output resistance does not degrade the signal.

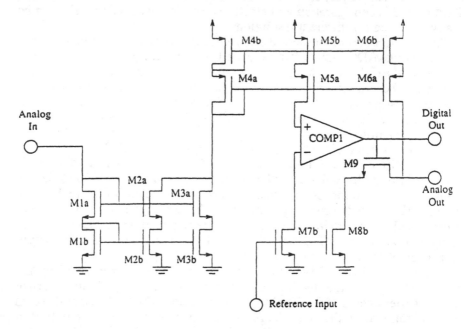

Figure 13.4 Schematic of the bit cell using cascoded mirrors.

13.2.3 Active mirror analog-to-digital converters

To implement a current mode ADC in which the converters resolution is not degraded by either a reduction in the device's gate voltage or the finite output resistance of the mirroring devices, the basic current mirror shown in Figure 13.5a can be replaced with the active current mirror shown in Figure 13.5c [14]. The active current mirror which is a simplified current conveyor [15] has the primary advantage of allowing its input voltage to be fixed independent of the mirroring device's gate voltage. If the active current mirror is operated such that the output terminal is at the same potential as the input terminal, the detrimental effects of the device's finite output resistance will be eliminated.

To illustrate the operation of the active current mirror, consider the basic current mirror shown in Figure 13.5a and its small signal model shown in Figure 13.5b. The output current source of transistor M1 drives the parallel combination of r_o, the device's output resistance, and the mirror's effective input resistance, $1/gm$, causing current division to occur. Consequently, the current mirror's input current, i_{in_2}, can be expressed as

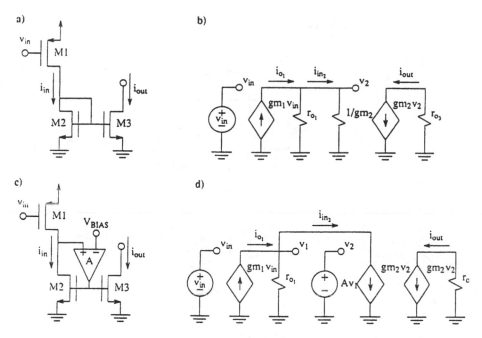

Figure 13.5 Current mirror structures and their small signal models: a) the basic current mirror, b) the small signal model of the basic current mirror, c) the active current mirror, d) the small signal model fo the active current mirror.

$$\frac{i_{in_2}}{i_{o_1}} = \frac{r_o}{r_o + 1/gm} \tag{13.1}$$

where i_{o_1} is the output current of the input current source M1. Therefore, to obtain a current transfer ratio closer to unity, either r_o must be increased, or $1/gm$ must be decreased. When the current mirror in Figure 13.5a is replaced with an active current mirror, as illustrated in Figure 13.5c, the small signal model of Figure 13.5b is modified as shown in Figure 13.5d resulting in the following current transfer function

$$\frac{i_{in_2}}{i_{o_1}} = \frac{r_o}{r_o + 1/Agm} \tag{13.2}$$

where A is the amplifier's gain. From equation (13.2), it is evident that the amplifier reduces the mirror's effective input resistance by a factor equal to the amplifier's gain leading to a current transfer ratio closer to unity. Therefore the active current mirror reduces current mismatches by reducing the current mirror's input resistance.

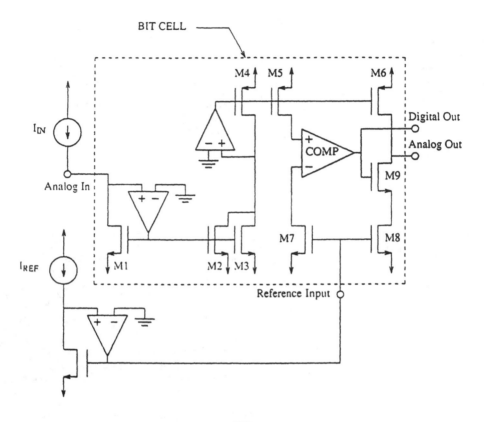

Figure 13.6 Bit cell for a current mode ADC using active current mirrors.

To determine the required gain of the active current mirror's amplifier, the desired reduction in the current mismatch should be considered. This mismatch can be expressed as

$$\frac{i_{o_1} - i_{in_2}}{i_{o_1}} = \frac{1/Agm}{r_o + 1/Agm} \tag{13.3}$$

and can be simplified to

$$\frac{i_{o_1} - i_{in_2}}{i_{o_1}} = \frac{1}{Agmr_o} \tag{13.4}$$

provided $r_o \gg 1/Agm$. Consequently an amplifier consisting of a simple differential pair with a gain of 100V/V will improve the current matching by two orders of magnitude, thereby virtually eliminating the detrimental effects of the current mirror's finite output resistance.

A bit cell designed using the active current mirror is illustrated in Figure 13.6. The operation of this bit cell is identical to the one presented in Figure 13.2. To maximize the circuit's performance, both active mirrors must be biased at the same potential and this potential must be equal to the threshold voltage of the comparator. Such a biasing arrangement ensures that the input and output devices for each current mirror will have the same drain to source potential. Therefore, the effect of the devices' finite r_o is minimized both within the bit cell and when the bit cell is connected to other bit cells.

A complete N-bit converter is obtained by cascading N bit cells as was done for the previous circuit. Using the active current mirrors, an 8-bit ADC was implemented in a $3\mu m$ CMOS technology [14]. The circuit occupies $0.74mm^2$ of active area and exhibits a maximum sampling rate of 500 kHz with a power dissipation of 63 mW. The circuit's performance is summarized in Table 1. The circuit's sampling rate was limited by the amplifier's settling time. In view of the relatively low gain required by the amplifier, it is expected that significant improvements in the sampling rate can be achieved by optimizing the amplifier's performance. The converter's resolution was limited by the device mismatches and hence can only be improved by reducing the device mismatches.

13.2.4 Current matching analog-to-digital converters

To implement high resolution ADCs in a cost effective manner, the reliance on the close matching of device parameters must be avoided. In most current mode circuits though, good device matching is relied upon to generate closely matched currents. To avoid this device parameter dependence, an arrangement that uses the actual input current itself to set the device's current level is more appropriate. Once the device's current level has been set, the resulting gate voltage can then be stored and recalled whenever the current is required at a future time.

An algorithmic ADC can be implemented using this current matching concept with the circuit shown in Figure 13.7 [16, 17]. For this circuit, only one bit cell and some control circuitry as shown in Figure 13.8 are required for a complete N-bit converter instead of the usual N bit cells.

To perform an algorithmic conversion, the converter's switches are controlled by the clock signals shown in Figure 13.9. The clock sequence is essentially the same for each bit conversion except during the acquisition of the input signal for the conversion of the most significant bit. The conversion of the most significant bit is initiated by closing switches S1, S2, and S3, causing the current in N1 to be set to I_{IN}. Once the gate voltage of N1 has settled, S2 and S3 are opened while S4 and S5 are closed to set the current in N2 to I_{IN}. Then by opening S1 and S5 and by closing S2, S4, S6, and S7, the current stored in N1 and N2 are summed, generating a current equal to $2I_{IN}$ without the need for well matched devices. The resulting current is then loaded into P1. Once P1 is set, S2, S4, and S7 are opened while S6 and S8 are

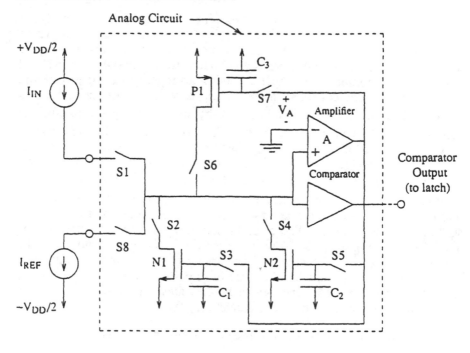

Figure   13.7   Algorithmic ADC using current matching.

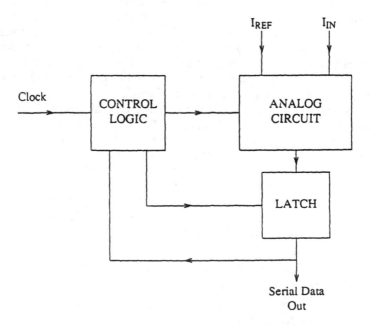

Figure   13.8   Block diagram of the complete ADC.

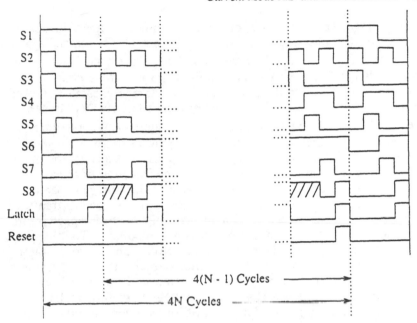

4(N - 1) Cycles

4N Cycles

Figure 13.9 Clock signals for the ADC shown in Figure 13.7.

closed, thus allowing the comparator to sense the current imbalance and hence determine if the signal, $2I_{IN}$ is greater than the reference, I_{REF}. If the signal exceeds the reference, the output for the MSB will be a "1" otherwise it will be a "0". This completes the conversion of the MSB.

The remaining (N-1) bits are then converted in the following manner. The signal, which is now stored in P1, is loaded into N1 by closing S6, S2, and S3. If the output of the preceding bit was a "1", S8 is also closed to subtract the reference from the signal. If the output was a "0", S8 is opened so that the signal remains unchanged. Once N1 is set, N2 is then set by closing S6, S4, and S5 (and S8 if the previous bit was a "1"). The signal is then doubled and stored on the gate of P1 by closing S2, S4, S6, and S7. Finally, the signal is compared with the reference by closing S6 and S8. This sequence is repeated until the desired resolution has been achieved.

The total conversion time depends on the clock rate and the converter's resolution. As can be seen in Figure 13.9, each bit conversion requires four clock cycles hence, an N-bit conversion requires 4N clock cycles and the sampling rate is equal to the clock rate divided by 4N. Alternatively, due to the sampled nature of the circuit, a pipelined architecture could be used to achieve a sampling rate equal to one quarter of the clock rate, independent of the converter's resolution.

When implementing the current matching ADC shown in Figure 13.7, careful layout and device sizing are important to minimize the detrimental effects of noise on the circuit's performance. To maximize the circuit's dynamic range, the gain constant of device P1 should be approximately twice the gain constant of devices N1 and N2. The size of the storage capacitors, C_1, C_2 and C_3, are determined by the maximum acceptable level of switch induced charge injection and kT/C noise. To minimize coupling of the power supply noise into the signal path, the storage capacitors should be physically and electrically close to their associated transistors. In particular, note that C_3 is connected to the positive power supply and not to the negative supply or ground, thereby ensuring that the gate-to-source voltage of P1 is independent of fluctuations in the power supplies. It is also important to use a lot of substrate connections to minimize the injection of noise from the substrate into the circuit. Although the size of the storage capacitors will affect the amount of noise in the circuit, they are not used to transfer and store precise amounts of charge as in the case of switched-capacitor circuits and hence their charge-voltage linearity is not important. To minimize the effects of offset voltages between the amplifier and the comparator, the amplifier can be used as both the amplifier and the comparator thus completely eliminating the detrimental effects of the amplifier's offset voltage on the converter's performance. Following these guidelines, allows one to achieve the high resolution capabilities inherent with the current matching technique.

Although the current matching ADC requires control logic and uses storage capacitors, the resulting circuit offers a significant reduction in the total circuit area over that of the previous designs due to a better utilization of the analog components. For example, a prototype ADC based on this technique, using a $3\mu m$ CMOS process, achieved 10-bit resolution at a sampling rate of 25 kHz with a power consumption of only 3.5 mW [16]. The analog portion of the circuit occupies a chip area of $0.18mm^2$ while the control logic requires a further $0.14mm^2$. The performance of this circuit is summarized in Table 13.1. In this circuit, switch induced charge injection was found to limit the resolution while the amplifier's settling time determined the maximum clock rate and hence the maximum conversion rate.

13.2.5   *Summary*

As seen in Table 1, the current mode algorithmic ADCs have a number of features that make them attractive for VLSI systems. Their extremely favourable speed/area trade-off makes them well suited for use in large signal processing systems where it is important that the ADC not consume a large portion of the chip area. The basic current mirror ADC is well suited for built in analog self testing purposes due to its small size, low power consumption and lack of control circuitry. The basic mirror and the current matching based converters are expected to be useful for battery powered circuits in which their low power requirements are a distinct advantage. At the same time these circuits can be implemented without the need for high gain amplifiers and if capacitors are required, their matching and linearity are not significant factors in the circuit's performance.

While the converters discussed above are suitable for many applications, further circuit refinements should permit the designer to achieve both better speed and better resolution. In the case of the active mirror and the current matching based converter, it is expected that the speed can be improved by at least a factor of two by improving the settling time of the amplifiers. In the case of the current matching ADC, the use of a pipelined architecture will allow one to further increase the sampling rate by an order of magnitude. To achieve higher resolutions, higher output resistance current mirrors (Chapter 6) will be required for the basic mirror ADC while better device matching will improve the resolution of the active current mirror based ADC. In the case of the current matching ADC, which does not require matched components, an increase in the resolution to at least the 13-bit range can be expected through the use of a better charge injection cancellation scheme.

The use of new and different technologies is also expected to enhance the performance of current mode ADCs. In particular, the use of a bipolar process may significantly increase the speed of the ADCs due to the significant reduction in voltage swing required for a given change in current level. In a similar manner, the use of a BiCMOS process should lead to a significantly improved active current mirror with ideally zero input current and a very fast response time.

13.3 Digital to Analog Converters

Current-scaling digital-to-analog converters have the advantage of offering both high linearity and high switching speed. However, current scaling circuits implemented in a CMOS process often require additional processing steps for thin-film resistor fabrication and laser trimming. An alternate technique, capable of high resolution without additional processing, is based on the multiple current source approach [18]. With this approach, weighted current sources are created using multiple unit current sources. If mismatches between unit current sources are uncorrelated, then the relative accuracy of a weighted current source improves with the square root of the number of unit current sources used to create it [14]. Thus, highly accurate weighted current sources can be created using a large number of relatively inaccurate unit current sources. However, when a large number of unit current sources are used, the area occupied by each unit current source becomes very significant. As a result high resolution converters require unit current sources with both good matching and small physical dimensions.

Multiple current source type D/A converters based on NMOS transistor unit current sources have been previously reported [19, 20]. However, with MOS transistors there is a tradeoff between the area required to implement a current source and the matching that can be obtained with it [21]. Unfortunately this tradeoff is also a function of the current level in the device and is particularly unacceptable at low current levels.

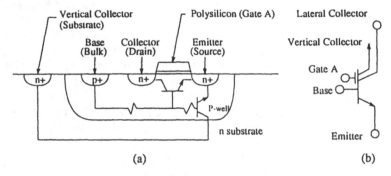

Figure 13.10 Npn lateral bipolar transistor: (a) device cross section, (b) device symbol.

The converter discussed here uses CMOS-compatible split-collector lateral bipolar transistors as current sources. Unlike MOS devices, the mismatch between bipolar devices does not increase at low current levels. Thus, at low current levels, bipolar transistors can achieve better matching than MOS devices while occupying a smaller area. Furthermore, by using multiple collector lateral bipolar transistors, even smaller well matched unit current sources can be obtained. In addition to highly linear current sources, the D/A converter presented here uses improved current switches which exhibit high speed and low switching noise at low current levels.

13.3.1   Split collector lateral bipolar transistors in CMOS

Lateral bipolar npn transistors are available in a standard P-well CMOS process [22]. Single-collector versions of these transistors have been previously used for low-noise amplifier inputs [22, 23, 24] voltage references [22, 25] variable gain amplifiers [26] and digital-to-analog converters [27]. A simple cross section of a single-collector device and a commonly used circuit symbol are shown in Figs. 13.10(a) and 13.10(b) respectively. The lateral bipolar transistor has a structure identical to that of a normal NMOS transistor. To operate the structure in the lateral bipolar mode requires forward biasing the source-bulk (emitter-base) junction and reverse biasing the drain-bulk (collector-base) junction. In addition the voltage at the gate must be biased slightly negative with respect to the source to eliminate subthreshold conduction in the inherent MOS transistor. Activating the lateral npn transistor however, also activates a parasitic vertical npn transistor. This transistor has its collector tied to the substrate and depending on the process used can carry a significant portion of the total emitter current. For this reason the symbol, shown in Figure 13.10(b), includes both vertical and lateral collectors. Minimizing the vertical current requires using short channel length devices with small geometry emitters and lateral collectors completely surrounding the emitters. The layout of a single collector device is shown in Figure 13.11(a). Here the collector and emitter are created from a single n^+ implant by using a polysilicon layer to divide the implant into two separate regions. In a similar manner, polysilicon may be used to divide the collector into two separate regions as shown in Figure 13.11(b). Here a polysilicon tab extended past the field region divides the collector into two separate parts. Because of the symmetric nature of the two collector transistor, good matching between collectors can be expected.

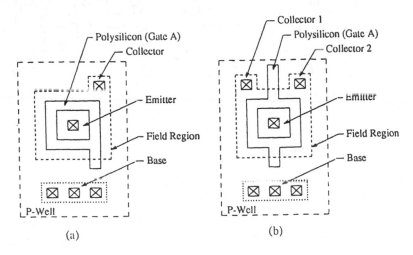

Figure 13.11 Lateral bipolar transistor topologies: (a) single collector device, (b) two collector device.

13.3.2 DAC Configuration

A simplified schematic of the 10-bit converter circuit is shown in Figure 13.12. This converter may be divided into three separate parts: the weighted current sources, the differential current switches, and the control circuit which biases the weighted current sources. The weighted current sources Q_9 to Q_2 generate the first eight bit currents of the converter. These bit currents are switched to either I_{out} or $\overline{I_{out}}$ using the NMOS differential current switches M_{9-1}, M_{9-2} to M_{2-1}, M_{2-2}. The current switches are driven such that one device operates in the saturation region carrying the entire bit current while the other device operates in the cutoff region carrying none of the bit current. Operating the switches in the saturation region increases the converters output impedance and also isolates the switches highly capacitive common-source node from the outputs.

The final two bit currents are generated by dividing the collector current of transistor Q_{10} into three part with the ratio 2:1:1 between parts. The first two of these parts are used as the final two bit currents while the last part,

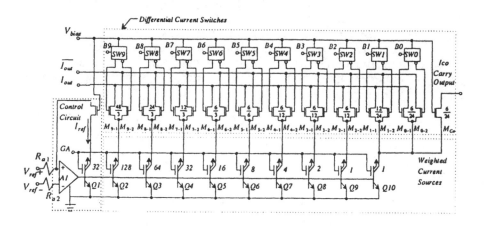

Figure   13.12   Simplified converter schematic.

called the carry output (I_{co}), is discarded. Current division is accomplished directly within the differential current switches. Here, the collector current of Q_{10} is split between three transistors consisting of M_{1-1} or M_{1-2} in parallel with M_{0-1} or M_{0-2} and M_{c_0}. Since each transistor operates in the saturation region with the same gate-source voltage, the ratio of their drain currents is equal to the ratio of their widths. Current division within the switches has several advantages when compared with using a secondary current divider. First, the absence of an explicit secondary current divider allows both transistors Q_9 to Q_{10} to operate with the same collector voltage. This eliminates nonlinearities due to the output resistance of Q9 and Q10. Second, transients at the common source node of the least significant bit switch are recharged not only by the transistor carrying the LSB current, but also by the transistor carrying the 2nd LSB current and the transistor M_{c_0}.

The control circuit for the converter of Figure13.12 consists of the control amplifier A1 and the weighted current source Q_1. Together Q_1 and A_1 form a feedback arrangement which generates a reference current

$$I_{ref} = \frac{V_{ref+} - V_{ref-}}{R_{a1}}$$

(13.5)

in the collector of transistor Q_1. This reference current is mirrored by transistors Q_2 to Q_{10} to provide the converter with a full scale output $I_{FS} = 8I_{ref}$.

13.3.3 Weighted Current Source Implementation

The weighted current sources , in the converter of Figure13.12, are created by interconnecting 288 unit current sources in the ratios shown. The unit current sources are created from an array of 144 two-collector transistors by using each collector as a separate unit current source. The unit current source array is divided into six smaller sub-arrays to minimize voltage drop along the emitter metal lines. Each sub-array consists of two rows of devices with 16 devices per row. Of the 16 devices in each row, only 12 are used in creating weighted current sources. The four remaining devices (two at each end of the row) operate as dummy transistors which carry the same current but whose collectors are connected to the positivesupply rail instead.

Weighted current sources are created by interconnecting unit current sources into weighted groups. Furthermore, a symmetric interconnection scheme is employed whereby a weighted current source is created using an equal number of unit current sources from each portion of the array. Although a symmetric interconnection scheme requires more area to implement, it substantially reduces mismatches between weighted current sources that are associated withprocessing gradients across the wafer surface.

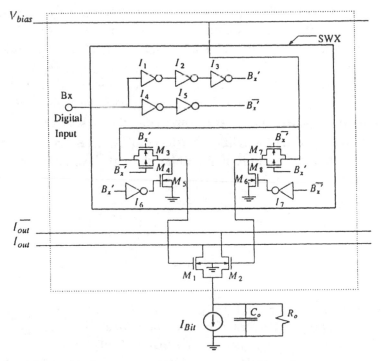

Figure 13.13 Fully differential current switch.

13.3.4 High-Speed Current Switch Implementation

High-speed fully-differential current switches are used to obtain a fast settling time in the converter of Figure 13.12. Fully-differential switches achieve high switching speeds by eliminating steady state voltage changes at the weighted current source outputs[28-30]. Although steady-state voltage changes are eliminated, switching transients can still occur. Previously, these transients were forced to settle quickly by operating the switches at high current levels[28].

In the present work, improved fully-differential current switches are developed to achieve high switching speeds at low current levels. By reducing the switching transients at the weighted current source outputs, smaller switching transients and faster settling times are achieved at the output of the converter.

A circuit diagram for the differential current switches is shown in Figure 13.13. Here the weighted current source I_{Bit} is shown along with its finite output resistance r_o and parasitic output capacitance C_o . The bit current I_{Bit} is switched to either I_{out} or $\overline{I_{out}}$ using an NMOS differential pair consisting of transistors M_1 and M_2. To switch the current to I_{out} for instance requires driving the gate of transistor M_1 to a positive potential V_{bias} while pulling the gate of transistor M_2 to ground. Driving the gate potential of M_1 to V_{bias} is done with the transmission gate composed of transistors M_3 and M_4 while the gate of M_2 is pulled to ground using a single N-channel device M_6 .

The transmission gate is driven by a pair of complementary signals B_x' and $\overline{B_x'}$. These signals are derived from the digital input B_x by using two inverter strings I_1, I_2, I_3 and I_4, I_5. By equalizing the delay through each inverter string, the rising and falling transitions of B_x' and $\overline{B_x'}$ can be made to overlap perfectly. Furthermore if the delay through inverter i equals T_i, then choosing

$$T_2 = T_5 \tag{13.6}$$

and

$$T_1 + T_3 = T_4 \tag{13.7}$$

ensures that the rising and falling transitions of B_x' and $\overline{B_x'}$ occur simultaneously, even in the face of process variations[31]. To satisfy (13.6) and (13.7), device sizes are chosen such that each of the inverters I_1 to I_5 drive the same load capacitance. The inverters I_1, I_2, I_3 and I_5 are chosen to be the same size with Wp/Lp = 36/3 and Wn/Ln = 15/3, while the inverter I_4 is chosen to have twice the channel length Wp/Lp = 36/6, Wn/Ln = 15/6. The pull-down transient at the gate of M_2 is controlled by the device M_6 driven by inverter I_7. The addition of inverter I_7 delays the pulldown transient at the gate of M_2 until the voltage at the gate of M_1 has risen sufficiently to turn M_1 on. Both inverters I_6 and I_7 have Wp/Lp = 6/3, Wn/Ln = 15/3. The resulting low logic threshold voltage means that I_6 will turn off M_5 quickly, allowing the gate voltage of M_1 to charge up. The low logic threshold voltage also means that I_7 will further delay the turn on of M_6.

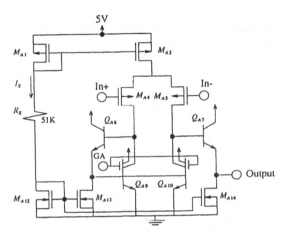

Figure 13.14 Control amplifier A1.

13.3.5 Control Amplifier Implementation

The reference circuit of the converter in Figure 13.12 consists of the current source Q_1 along with the control amplifier A_1 of Figure 13.14. The control amplifier consists of a single differential gain stage with a buffered output. The differential gain stage consists of the PMOS differential pair MA4 and MA5 which drive the active loads QA9 and QA10. Lateral bipolar transistors are used as the active load because their good matching at low current levels results in a low input offset voltage. The active load is buffered by the vertical bipolar transistor Q_{A6}. Q_{A6} is biased with M_{A13} to allow it to operate with a higher collector current and hence a higher f_T. The output of the differential stage is buffered using the vertical bipolar transistor Q_{A7} which is biased by M_{A14}.

The bias current generator consists of the diode connected transistors MA1 and MA2 and the external resistor R_x. When Rx equals 51 KΩ, a bias current Ix of 50 μA is generated. This current is mirrored by M_{A2} to bias the input stage at $50\mu A$. Similarly the bias current flowing in M_{A12} is mirrored by M_{A13} and M_{A14} to bias each of the buffer transistors QA6 and QA7 at a current of $50\mu A$.

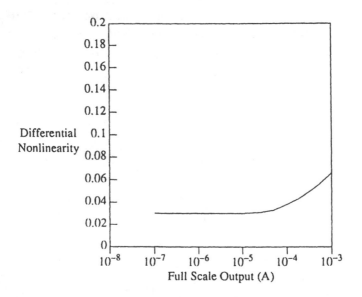

Figure   13.15   Differential nonlinearity vs. full scale output.

13.3.6   *Experimental Results*

The 10-bit converter of Figure 13.12 was implemented in a standard $3\mu m$ P-well CMOS process. For the two collector transistors, current matching between the collectors in a single device was typically ±0.7%. Similar mismatches were observed between the collectors of separate transistors.

From a group of six converters tested, the worst case differential nonlinearity was, ±0.05%FS or ±1/2LSB. The worst case differential nonlinearity for a typical converter is shown as a function of full scale current in Figure 13.15. This plot shows that the converter achieves low differential nonlinearity over a wide range of operating currents.

The integral nonlinearity for the converters tested was better than ±0.04%FS while the average gain error was ±0.04%FS. Power dissipation was only 5 mW when operated at a full scale current of $100\mu A$.

The 10 bit converter displays a high output impedance and a high voltage compliance. The output impedance, measured at a full scale current of $100\mu A$, was in excess of $15M\Omega$, while the voltage compliance was 1 to 5V. Thus, 4V signal swings could be supported at the converter's output. Although a signal swing at the converter's output will introduce an integral nonlinearity into the converters response[28], even with a signal swing of 4V, this integral nonlinearity is less than ±0.035%FS. The small size of this nonlinearity eliminates the need for cascoded current sources to increase the converter's output impedance. The absence of the cascode connection has the advantage of increasing the converter's voltage compliance.

The converter achieves very high switching speed even at low current levels. The full scale response of the converter, operating at $100\mu A$ *FS* and driving a 100Ω load exhibits a setting time of 30 ns Test results are given in Table 13.2.

Table 13.2 Current Mode DAC Characteristics

Parameter		10 Bit Converter
Resolution		10 Bits
Acrive Area		13 mm2
Nonlinearity	Differential	0.05% FS
	Integral	0.04% FS
	Gain Error	0.04% FS
Power Supply		5 V
Power Dissipation*		5 mW
Setting Time*		30 ns
Output Impedance*		> 15 MΩ
Output Voltage Compliance		1V-5V

* with full scale output = 100 μA

13.3.7 Summary

A 10-bit current-scaling D/A converter has been implemented in a standard $3\mu m$ P-well CMOS process. This converter uses an array of two-collector lateral bipolar transistors to create its weighted current sources. By using lateral bipolar transistors rather than MOS devices, the converter achieves much better linearity at low full scale current levels. The converter is also capable of operating over a wider range of full-scale currents than either MOS transistor based converters or R/2R ladder based converters. In addition, by using split-collector lateral bipolar transistors, the converter achieves high resolution and high linearity in a very small active area.

By comparison, the 10-bit converter presented here occupies less than one third the area of a typical 10-bit R/2R ladder based DAC [1], and substantially less area than previously reported 8-bit MOS unit current source based DACs[28, 32].

The switching speed, which is considerably faster than other general purpose converters, is comparable to that of special purpose video speed

---

[1]Comparison based on the AD-7520 10-bit CMOS DAC offered by Analog Devices.

DACs [32, 33]. In addition, the 10-bit converter developed here features higher resolution and much lower power dissipation than these converters.

13.4 Acknowledgements

This work was supported by the Natural Sciences and Engineering Research Council, Northern Telecom Electronics and the Information Technology Research Centre of Ontario. The authors wish to thank the Canadian Microelectronics Corporation for fabrication of the integrated circuits.

13.5 References

[1] J. L. McCreary and P. R. Gray,: "All MOS Charge Redistribution Analog-to-Digital Conversion Techniques -PART I", IEEE Journal of Solid-State Circuits, vol. SC-10, pp. 371-379, 1975.

[2] J. B. Hughes, N. C. Bird, and I. C. Macbeth, "Switched Currents - A New Technique For Analog Sampled-Data Signal Processing", 1989 International Sympo-sium on Circuits and Systems", Proceedings, pp. 1584-1587, Portland, 1989.

[3] B. Gilbert, "A New Wide-Band Amplifier Technique", IEEE Journal of Solid-State Circuits, Vvl. SC-3, pp. 353-365, 1968.

[4] W. Groeneveld, H. Schouwenaars, and H. Termeer, "A Self Calibration Technique for Monolithic High-Resolution D/A Converters", ISSCC Digest of Technical Papers, vol. 32, pp. 22-23, New York, 1989.

[5] D. Vallancourt, and Y. P. Tsividis, "Sampled-Current Circuits", Proceedings of the 1989 International Symposium on Circuits and Systems", pp. 1592-1595, Portland, 1989.

[6] B. D. Smith, "An Unusual Electronic Analog-to-Digital Conversion Method", IRE Transactions on Instrumentation, vol. PGI-5, pp. 155-160, 1956.

[7] H. Onodera, T. Tateishi, and K. Tamarur, "A Cyclic A/D Converter That Does Not Require Ratio-Matched Components", IEEE Journal of Solid-State Circuits, Vol. SC-23, pp. 152-158, 1988.

[8] C. C. Shih and P. R. Gray, "Reference Refreshing Cyclic Analog-to-Digital and Digital-to-Analog Converters", IEEE Journal of Solid-State circuits, vol. SC-21, pp. 544-554, 1986.

[9] P. W. Li, M. J. Chin, P. R. Gray and R. Castello, "A Ratio-Independent Algorithmic Analog-to-Digital Conversion Technique", IEEE Journal of Solid-State Circuits, Vol. SC-19, pp. 828-836, 1984.

[10] G. Nairn and C. A. T. Salama, "An Algorithmic Analog-to-Digital Converter Based on Current Mirrors", Electronics Letters, vol. 24, pp. 471-472, 1988.

[11] D. G. Nairn and C. A. T. Salama, "A Current Mode Algorithmic Analog-to-Digital Converter", International Symposium on Circuits and Systems, Proceeding, pp. 2573-2576, Helsinki, 1988.

[12] K. R. Lakshmikumar, R. A. Hadaway, and M. A. Copeland: "Characterization and Modeling of Mismatchin MOS Transistors For Precision Analog Design", IEEE Journal of Solid-State Circuits, vol. SC-21, pp. 1057-1066, 1986.

[13] J. B. Shyu, G. C. Temes, and F. Krummenacher,: "Random Error Effects in Matched MOS Capacitors and Current Sources, IEEE Journal of Solid-State Circuits, vol. SC-19, pp. 948-955, 1984.

[14] D. G. Nairn, and C. A. T. Salama, "High-Resolution Current Mode A/D Converters Using Active Current Mirrors", Electronics Letters, vol. 24, pp. 1331-1332, 1988.

[15] A. S. Sedra and K. C. Smith, "A Second Generation Current Conveyor and Its Applications", IEEE Transaction on Circuit Theory, vol. CT-17, pp. 132-134, 1970.

[16] D. G. Nairn and C. A. T. Salama, "Ratio-Independent Curren Mode Algorithmic Analog-to-Digital Converters", International Symposium on Circuits and Systems, Proceedings, pp. 250-253, Portland, 1989.

[17] D. G. Nairn and C. A. T. Salama, "Current Mode Analog-to-Digital Converters", International Symposium on Circuits and Systems Proceedings, pp. 1588-1591, Portland, 1989.

[18] P.H. Saul and A.J. Jenkins, "A 10-bit monolithic tracking A/D converter", ISSCC Digest of Technical Papers, pp. 138-139, 1978.

[19] H. Post and K. Waldschmidt, "A High-Speed NMOS A/D Converter with a Current Source Array", IEEE J. Solid State Circuits, vol. SC-15, pp. 295-300, 1980.

[20] H.J. Schouwenaars, W. Groeneveld, and H. Termeer, "A Stereo 16b CMOS D/A Converter for Digital Audio", ISSCC Digest of Technical Papers, pp. 200-201, 1988.

[21] K.R. Lakshmikumar, R.A. Hadaway, and M.A. Copeland , "Characterization and Modeling of Mismatch in MOS Transistors for Precision Analog Design", IEEE J. Solid State Circuits, vol. SC-21, pp. 1057-1066, 1986.

[22] E.A. Vittoz, "MOS Transistors Operated in the Lateral Bipolar Mode and Their Application in CMOS Technology", IEEE J. Solid State Circuits, vol. SC-18, pp. 273-279, 1983.

[23] S. Gustafsson, R. Sundblad, and C. Svensson, "Low-Noise Operational Amplifiers Using Bipolar Input Transistors in a Standard Metal Gate CMOS Process", Electronics Letters, vol. 20, pp. 563-564, 1984.

[24] C.A. Laber, C.F. Rahim, S.F. Dreyer, G.T. Uehara, P.T. Kwok, and P.R. Gray, "Design Considerations for a High-Performance 3-um CMOS Analog Standard-Cell Library", IEEE J. Solid State Circuits, vol. SC-22, pp. 181-189, 1987.

[25] E.A. Vittoz, "CMOS Voltage References Using Lateral Bipolar Transistors", IEEE J. Solid State Circuits, vol. SC-20, pp. 1151-1157, 1985.

[26] T. Pan and A.A. Abidi, "A 50-dB Variable Gain Amplifier Using Parasitic Bipolar Transistors in CMOS", "IEEE J. Solid State Circuits, vol. SC-24, pp. 951-961, 1989.

[27] H.W. Singor and C.A.T. Salama, "A High Performance CMOS Compatible 8-bit Current Scaling D/A Converter", accepted for publication in Proceedings of the IEE, Part G, 1989.

[28] T. Miki, Y. Nakamura, M. Nakaya, S. Asai, Y. Akasaka, and Y. Horiba, "An 80-MHz 8-bit CMOS D/A Converter", IEEE J. Solid State Circuits, vol. SC-21, pp. 983-988, 1986.

[29] R.B. Craven, "An Integrated Circuit 12-Bit D/A Converter", ISSCC Digest of Technical Papers, pp. 40-41, 1975.

[30] J.A. Schoeff, "An Inherently Monotonic 12b DAC", IEEE J. Solid State Circuits, vol. SC-14, pp. 904-911, 1979.

[31] M. Shoji, "Elimination of Process Dependant Clock Skew in CMOS VLSI", IEEE J. Solid State Circuits, vol. SC-21, pp. 875-880, 1986.

[32] K.R. Lakshmikumar, R.A. Hadaway, M.A. Copeland, and M.I.H. King , "A High Speed 8-Bit Current Steering CMOS DAC", Custom Integrated Circuits Conference, Proceedings, pp. 156-159, 1985.

[33] P.H. Saul, D.W. Howard, and C.J. Greenwood, "An 8b CMOS Video DAC", ISSCC Digest of Technical Papers, pp. 32-33, 1985

Chapter 14

Applications of current-copier circuits

David Vallancourt and Steven J. Daubert

14.1 Introduction

In Chapter 7, a technique for sampling, holding, and making practically identical copies of a current was described. A number of research groups [1-4] developed the idea independently, giving the class of circuits names that often reflect the application that the inventors envisioned. The term used by Eric Vittoz in Chapter 7, "dynamic current mirror" is suggestive of the historical continuity of the principle within his body of research in dynamic techniques; thus, the emphasis here is on the general utility of the circuit in any design situation to which improvement of performance through dynamic techniques may apply (some examples of these are given in Chapter 7). The term "current copier" was first used [5] by one of us, for whom analog signal processing was the main focus. In this case, the name arose from the system-level requirement for a circuit capable of distributing multiple copies of an input, output, or state variable in current form within a current-mode discrete-time analog filter. Groeneveld et al. use the term "calibration circuit", indicative of the function performed by the cell in their monolithic current-mode D/A converter. The algorthmic A/D converter reported by Nairn and Salama employs a "current matching" circuit as its essential feature, ideally eliminating dependence on the matching or ratioing of any circuit components.

These multiple geneses suggest that the time has come for this circuit, for reasons of both technological feasibility and usefulness of function. Further, the variety of names is a clear indication of the variety of systems to which a cell capable of copying a current can be applied. In this chapter, we review applications of "current copier" circuits, and suggest new ones. Deferring a discussion of scientific nomenclature to one more capable [6] we will use the name "current copiers" in this chapter because (a) it is the name used in the first publication on this subject [1] and (b) we feel it is the most comprehensively descriptive of the circuit operation and use.

14.2 Current Copier Concept Review

A simple current copier cell is shown in Figure 14.1. The transistor need not have well-controlled geometric or electrical characteristics, and the capacitor value is similarly non-critical (the capacitor may even be

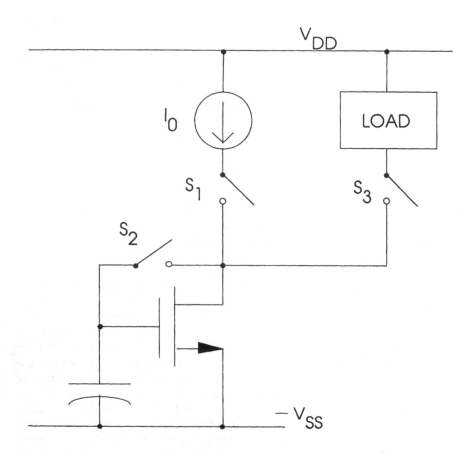

Figure 14.1 Basic current copier

nonlinear). To copy the current I_0 into the cell, switches S_1 and S_2 are closed; the capacitor will then charge to the gate voltage needed by the transistor in order to achieve a drain current equal to I_0 . The switches are then opened, disconnecting the transistor and capacitor from the original current source. Thereafter, the transistor and capacitor combination is capable of sinking a current I_0 when connected to another circuit via S_3. Note that under appropriate conditions, an explicit capacitor may not be necessary; the gate capacitance C_{gs} of the transistor itself may be sufficient.

Clock feedthrough, leakage, and channel length modulation are among the problems associated with the simple copier. Analyses of these effects [1,4,7-9 , Chapter 7,] show that some form of corrective circuitry is usually required. Common circuit techniques such as cascoding, placing the copier in an opamp feedback loop, using dummy switches to cancel feedthrough, and so forth, have been applied successfully in these cases. The difference between I_0 and its copy for one such modified copier has been reported to be in the range of 100ppm [4]. Further, in applications such as D/A converters, the difference between two copies derived from the same reference I_0 may be more important than the absolute copy fidelity; this difference can be reduced to the range of tens of ppm [1]. High-performance current copiers are presented in detail in Chapter 7.

Assuming that appropriate circuit techniques can be applied to reduce the deterministic error sources mentioned above, the remaining noise will set a lower limit on usable current. An assessment of the noise current magnitude is critically dependent on the specific copier circuit used, and on the application. As an example [8] analysis of the simple copier of Figure 14.1 used as an impulse sample/zero-order hold cell predicts about 7 nA rms noise from 1 to 25 MHz, assuming a 10 µA input current and reasonable device sizes and process parameters. The reader is again referred to Chapter 7 for a detailed treatment of this topic.

14.3 Applications of Current Copiers

The following two sections constitute a survey of published applications of current copiers; some have been fabricated and tested, others are as yet proposals. Included also are some previously unpublished systems. Unless noted otherwise, a reference given in the title of a section implies that all significant designs, analyses, and experimental results included in the section are from that reference.

14.4 Examples of Function Blocks

14.4.1 Ratio-Independent Integer Current Multipliers

One obvious application of current copiers is the integer current multiplier, an example of which is shown in Figure 14.2. If the input current I_0 is constant,then an output current equal to an integer multiple of I_0 can be obtained as follows. The switches S_{1i}, S_{2i} are closed in pairs, so only one copier cell at a time is connected in the opamp feedback loop. In the figure, the second copier is being "loaded", that is, the current I_0 is being copied into it. The copying process is repeated sequentially until all cells are loaded with I_0. Any or all of the cell outputs can then be summed; the output of the ith cell will be included in the sum if switch S_{3i} is closed. Ideally, the sum of the outputs of the cells is therefore an integer multiple of I_0 achieved without relying on element matching. The system described above makes use of class-A bidirectional copiers in anticipation of a time-varying, bidirectional input current I_0, as would be encountered in a general-purpose integer analog multiplier. In applications with known, fixed I_0 such as D/A converters, simpler cells would suffice.

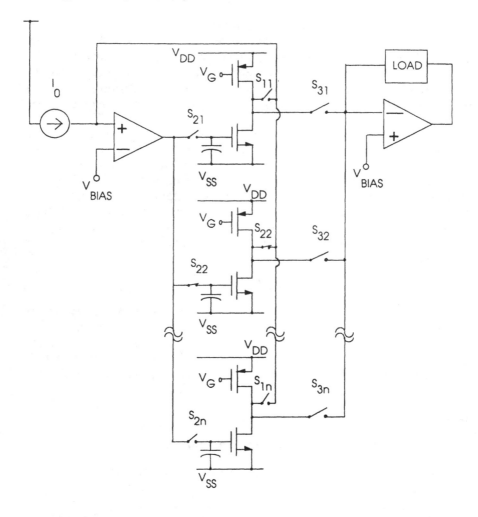

Figure 14.2 Integer multiplier constructed using a   parallel current sample/hold/summer

14.4.2 A Ratio-Independent Voltage Multiplier/Divider [10]

Integer current multipliers such as the one described above might be used as the basis for implementation of a ratio-independent, non-integer coefficient voltage multiplier, as shown in Figure 14.3. In the figure, the two current sources, nI_0 and mI_0 , are schematic representations of the outputs of two integer current multipliers; thus, m and n, are integers. The proposed circuit operates as follows:

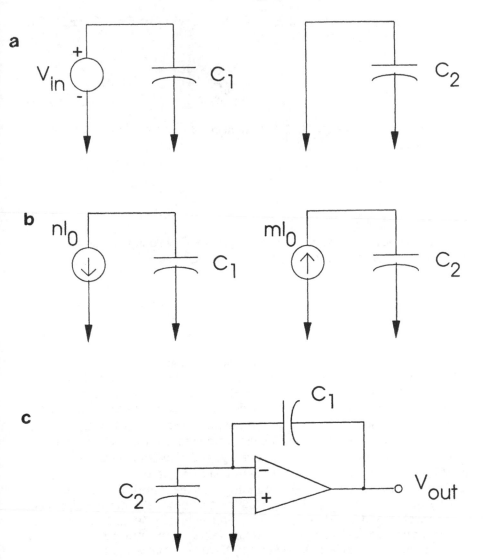

Figure 14.3   Ratio-independent voltage multiplier/divider
(a) Step #1   (b) Step #2   (c) Step #3

Step #1 (Figure.14.3a): The voltage signal sample, V_{in} is stored on a capacitor of arbitrary value, C_1. Another capacitor of arbitrary value C_2 is initially discharged.

Step #2 (Figure 14.3b): The current nI_0 is used to ramp the voltage on C_1 down to zero; while this is happening, the current mI_0 is used to charge C_2.

Once the voltage on C_1 reaches zero, the charging of C_2 is halted. The amount of time τ taken to discharge C_1 is

$$\tau = \frac{V_{in} C_1}{n I_o}$$

(14.1)

therefore, the total charge Q_{C2} accumulated on the top plate of C_2 during this operation is

$$Q_{C2} = \tau m I_o = \frac{m}{n} V_{in} C_1$$

(14.2)

Step #3 (Figure14.3c): All of the charge Q_{C2} on C_2 is dumped onto C1. The magnitude of the voltage V_{C1} appearing across C_1 is then

$$|V_{C1}| = |V_{out}| = \frac{Q_{C2}}{C_1} = \frac{m}{n} V_{in}$$

(14.3)

The multiplication factor realized, m/n, is a rational number which can be selected digitally by appropriate programming of the switches in the current multiplier discussed previously. The algebraic sign of the factor can be changed by either changing the direction of the current source $m I_0$ or by swapping the plates of C_2 in the charge dumping operation (Step #3).

Since the charging currents $n I_0$ and $m I_0$ are obtained using current copiers which do not require component matching, and since the values of I_0, C_1, and C_2 are irrelevant (although capacitor linearity is essential), the circuit achieves digitally programmable analog voltage multiplication without relying on component matching. Note that some of the non-ideal effects which must be addressed in a practical implementation (such as accuracy of the circuit used to detect zero-crossings in Step #2) are similar to those found in integrating A/D converters, where they have been successfully treated.

14.4.3 A Ratio-Independent Current Divider [11]

A proposed algorithmic method for dividing a given current I_{REF} by an integer factor without depending on device matching or even linearity is depicted in Figure 14.4, for the specific case of division by a factor of 2. This divider was discussed in Chapter 7; a somewhat different explanation of its operation will be given in this section. Four current copiers ("CC1" - "CC4") are required. In the figure, a copier engaged in sampling a current is

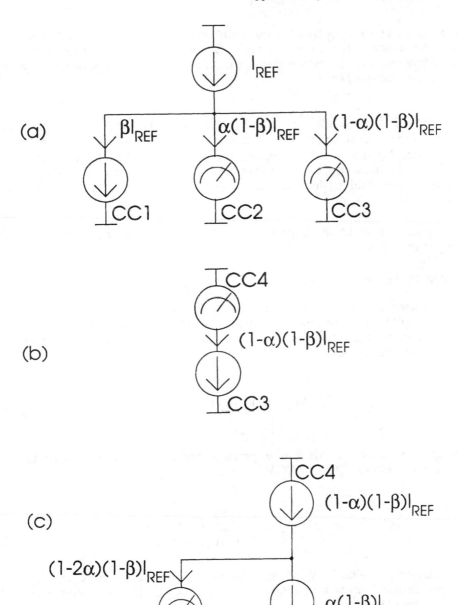

Figure 14.4 Ratio-independent current divider
(a) Step #1   (b) Step #2   (c) Step #3

indicated schematically by a "meter" symbol, and a copier configured to act as a current source is indicated as such. Several iterations of the following procedure will force the steady-state value of the current in CC3, denoted by I_{CC3}, to equal $I_{REF}/2$:

Step #1 (Figure 14.4a): $I_{REF} - I_{CC1}$ is copied into the parallel combination of CC2 and CC3. Since the latter do not match, the split will not be even; in the figure, this is shown by assigning a fraction $0< \alpha <1$ of the current to CC2, and the remaining $(1-a)$ of the current to CC3. The current in CC1 is assumed to be equal to βI_{REF} initially, where β is a function of the mismatch parameter α.

Step #2 (Figure 14.4b): I_{CC3} is copied into CC4 (in order to change the current direction).

Step #3 (Figure.14.4c): $I_{CC4} - I_{CC2}$ is copied into CC1. Return to step #1.

If we assume that some steady state has been reached, then the current values in each copier will be the same after step #3 as they were in step #1. We may thus choose to equate the current in CC1 from step #1 (Figure 14.4a) and step #3 (Figure. 14.4c):

$$\beta I_{REF} = (1 - 2\alpha)(1 - \beta I_{REF}$$

$$(14.4)$$

This may be solved to yield an expression for β in terms of the current mismatch α between CC2 and CC3:

$$\beta = \frac{1}{2}\frac{1 - 2\alpha}{1 - \alpha}$$

$$(14.5)$$

Substituting this value for β into the expression for the current in CC3 in step #1 (which must be its asymptotic value if steady state has indeed been reached), we obtain:

$$I_{CC3} = (1 - \alpha)(1 - \beta) = (1 - \alpha)(1 - \frac{1}{2}\frac{1 - 2\alpha}{1 - \alpha}) = \frac{1}{2}I_{REF}$$

$$14.6)$$

Similarly, substitution of (14.5) into the expressions for the currents in CC1 and CC2 given in Fig. 14.4a shows that the sum of these currents is also equal to I $_{REF}$/2, independent of the mismatch parameter α

For α in the range 0 <α< 0 .5, it canbe shown [11 Chapter 7] that the system response is overdamped, and all copier currents are unidirectional. Thus, the simple copier of Figure 14.1 may be used to implement all cells (with the p-channel version implementing CC4) as long as the transistors in CC2 and CC3 are properly sized. Such a circuit is shown in Figure 7.16 of Chapter 7. If α is instead in the range0.5<α<1, the system will become underdamped, and some currents may change sign while the system is converging. In this case, the roles of CC2 and CC3 could be swapped if switch programming allows, or bidirectional copiers could be used. The latter may also be used if I$_{REF}$ is allowed to be bidirectional, as in a general purpose analog multiplier.

As described in [11], the system above may be modified to implement ratio-independent division by integer factors greater than two, at the expense of additional hardware. It is also possible to provide a continuous output current, alternating between CC3 and the sum of CC1 and CC2. This property is used to advantage in the pipelined A/D converter described in section 14.5.3.

The accuracy of the divider is limited by the same non-ideal effects as those that limit the performance of the basic current copier. However, if dividers are used in a D/A converter (where their function would be to provide the continuous fixed currents that are summed to produce the analog output), then switch feedthrough could be reduced by increasing the hold-capacitor size. No speed penalty would be incurred, since the constant copier currents need only be incrementally refreshed once they have been established, in order to compensate for leakage. Further, the area penalty may even be avoided if stacked capacitors are available, since the usual non-linearity associated with these devices does not influence the operation of the copiers.

14.5 Examples of Systems

14.5.1 A Self-Calibrating D/A Converter [2,9]

Modified current copiers are used to provide the 6 MSB's of a 16-bit self-calibrating digital audio DAC, shown in block diagram form in Figure 14.5. Each copier contributes a nominal 10 μA current to the output sum or dumps it's current to ground, depending on the word to be converted. The output of one copier is divided by a 10-bit resolution network to provide the LSB's. To compensate for leakage, the 64 copiers are periodically refreshed; a 65th copier is therefore added which replaces the copier being refreshed at any given time, allowing continuous operation of the converter.

In this segmented design, it is apparant that the copier output currents must match to within 10 bits in order to achieve overall 16 bit performance.

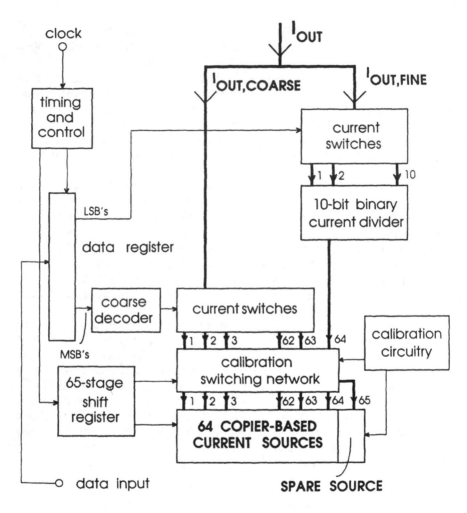

Figure 14.5 16-bit self-calibrating D/A converter

Analysis of the basic copier of Figure 14.1 showed that it could just marginally meet this specification, and then only when an unreasonably fast refresh rate was assumed. The accuracy requirement was met using the copier shown in Figure 14.6, designed to reduce errors caused by switch leakage, clock feedthrough, and channel length modulation. The complete circuit will be referred to below as a current "cell"

In the cell, a standard current mirror provides about 90 percent of the reference current. The actual "copier" portion, M1, handles the remaining 10 percent, in effect "calibrating" the standard mirror. Since the current handled by M1 is only about 0.1 as much as would have been the case without augmentation by the mirror, the transconductance of M1 is decreased by a

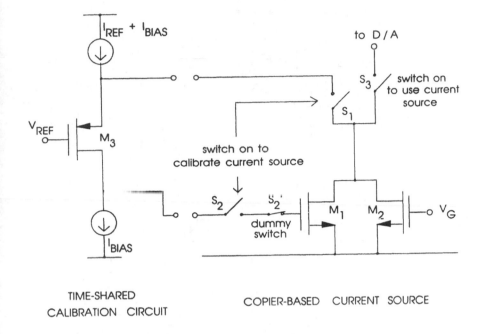

$I_{REF} + I_{BIAS}$

to D/A

S_3 / switch on to use current source

S_1

V_{REF} M_3

switch on to calibrate current source

S_2 / S_2'

dummy switch

M_1 M_2 V_G

I_{BIAS}

TIME-SHARED
CALIBRATION CIRCUIT

COPIER-BASED CURRENT SOURCE

Figure 14.6 One of sixty-five current cells in the D/A converter of Fig. 14.5

factor of $(10)^{1/2}$. The reduction in required current handling also allows optimization of M1's aspect ratio W/L, resulting in an overall decrease in transconductance by a factor of 8. Thus, errors resulting from leakage and feedthrough, which contaminate the stored gate voltage of M1, are reduced by the same factor. Channel length modulation effects were suppressed by using a non-critically biased level-shifting transistor M3 during the refresh operation; this device helps maintain the drain voltage of M1 and M2 at the same value during refresh as that which will be experienced when the cell is in use by the converter.

The addition of a fixed main current source M2 derived from mirrors is advantageous with regard to lowering the transonductance of M1, but the current variation from mirror output to mirror output, which may be as much as 30 percent,causes a larger mismatch in transconductance of the copiers from cell to cell. A portion of the beneficial effect is therefore lost. This was recovered by reducing the absolute amount of feedthrough at the outset. Dummy switch S 2' performs this task in the usual manner.

The converter was fabricated in 1.6 μm CMOS, occupies 2.8 mm2 (active area), and draws 20 mW from a single 5V supply. The measured integral nonlinearity is +/-1 LSB, appearing almost exclusively at the range extremes; at intermediate codes, the linearity is considerably better. A 94 dB dynamic range and 92 dB S/(N + THD) were achieved.

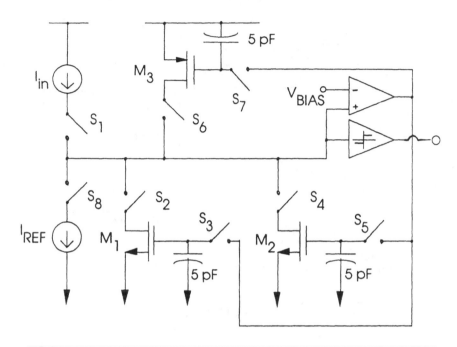

Figure 14.7 Ratio-independent algorithmic A/D converter

14.5.2 A Ratio-Independent Algorithmic A/D Converter [3]

An integer current multiplier is the essential element in the ratio-independent algorithmic A/D converter shown in Figure 14.7. The algorithm used is well-established, and was recently used in a switched-capacitor implementation [12], but the signals in the new converter are currents rather than voltages. The conversion proceeds from MSB to LSB, with the output appearing in serial form, according to the procedure below.

Step #1: An input sample is taken directly from the signal to be converted.

Step #2: The sample is multiplied by 2, and compared to the reference. If the reference is smaller, the present output bit is set to "1", and the reference is subtracted from the doubled sample. The result is stored, and will become the sample for the next bit determination. If instead the reference is larger, the present output bit is set to "0 ",and the doubled sample itself is stored; it will become the sample for the next bit determination.

Step #3: Go back to step (2).The process continues until the desired number of bits of resolution have been obtained.

In Figure 14.7, M1 and M2 in conjunction with the opamp perform the doubling function, as discussed in section 14.4.1. Transistor M3 is used to store the doubled sample minus the reference (if the present out-put bit is "1") or just the doubled sample (if the present output bit is "0 "). Generation of each bit requires four clock phases ϕ_1 - ϕ_4; in the figure, the switches that are closed during each phase are indicated accordingly. The circuit shown is the entire processing unit; it is used recursively m times for m bits of resolution.

The circuit was integrated in 3 μm CMOS technology, occupying only 0 .2 mm^2. Using a single 5 volt supply and reference current of 100μA, 10 bit resolutionwas achieved in 40 μS total conversion time. Worst case differential nonlinearity was measured at 0.87 LSB, and worst case integral nonlinearity was 0.92 LSB. Switch feedthrough limited the accuracy; other than the use of 5 pF hold capacitors, no means to reduce feedthrough were implemented.

The speed of the algorithmic converter could be increased at the expense of additional hardware by the well-known pipelining technique, in which a cascade of processing units similar to the one above are used. One such system has been proposed [13], in which the basic processing unit can accomodate bidirectional input currents and the algorithm is optimized for the pipelined architecture. Theoretical analysis in [13] taking into account switch feedthrough and noise in a chosen 2 μm process yields a predicted accuracy of 12 bits at a throughput rate of 6 μS. The estimated die area is between 1.5 and 2mm^2 not including any circuits that may be required to convert the input signal or the reference into current form.

14.5.3 A Pipelined A/D Converter Using Current Division [15]

One of the drawbacks of the conversion algorithm given in section 14.5.2 involves the signal doubling required in step #2. As the conversion progresses, errors are amplified by this factor of two for each bit computed, eventually limiting the converter accuracy.

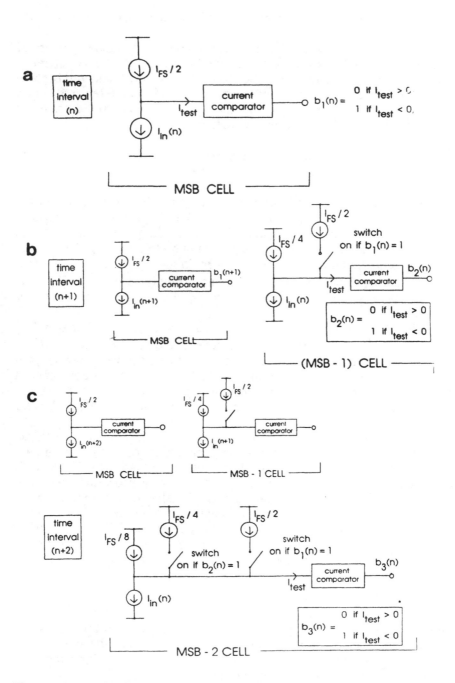

Figure 14.8 Pipelined successive-approximation   A/D   conversion
(a) sample period (n)   (b) sample period (n+1)   (c) sample period (n+2)

An alternative procedure [14] avoids this problem by pipelining the standard successive approximation approach. Here, the input signal sample is not processed through a cascade of doubling cells; instead, it is compared in its original form to progressively finer reference values. Figures 14.8a-14.8c show the time development of the conversion process for an input current sample $I_{in}(n)$ taken in time interval n. Only the first three MSB'sof the conversion are depicted. The circuit realization ofthe current sources shown will be discussed shortly. In the figure, I_{FS} represents the full-scale current; b_1 is the MSB, b_2 is the next most significant bit,and so forth; and all quantities X associated with an input sample taken in time interval n are indicated as X(n).

Figure 14.8a: The signal sample $I_{in}(n)$ is compared with the half- scale current. If it is larger, the MSB $b_2(n)$ is set to "1", otherwise, it is set to "0" Digital logic, not shown, stores the MSB.

Figure 14.8b: The signal sample $I_{in}(n)$ is now observed by the next cell, which determines the next most significantbit $b_2(n)$. The previously computed MSB is used to control a switch in such a way that if $I_{in}(n)$ was found to be greater than $I_{FS}/2$ by the previous stage(that is, $b_1(n) = 1$), then a comparison to $I_{FS}/2 + I_{FS}/4$ will now be made. If on the other hand $b_1(n) = 0$, then a comparison to $I_{FS}/4$ will be made. The result of this comparison determines $b_2(n)$. While the second stage is determining $b_2(n)$,the first (MSB) stage is at work processing the next input signal sample, $I_{in}(n+1)$.

Figure 14.8c: The third stage now compares $I_{in}(n)$ to

$b_1(n)I_{FS} 2^{-1}+b_1(n)I_{FS} 2^{-2}+I_{FS} 2^{-2}$ The result determines $b_3(n)$ the process continues through succeeding stages (not shown) until the desired resolution is achieved.

An implementation based on the use of current copiers and the ratio-independent current divider presented in sec-tion 14.4.3 has been proposed [15]. In this system, the bit currents $2^{-m} I_{FS}$ are generated from one reference current I_{FS} using a cascade of divide-by-2 current dividers, and multiple copies (beyond those available from the current dividers themselves) are provided by additional current copiers.

As described previously, the current divider is com-posed of four current copiers CC1-CC4, and requires several iterations of a three step procedure in order to achieve a steady state condition in which the desired currents are available. It is important to note that not all copiers CC1-CC4 are in use during each step in the procedure;further, the divided output is available at various times from CC3, CC4, and the parallel combination of CC1 and CC2.

The proposed A/D converter begins operation once the divider currents have settled. The algorithm that generates the divider currents is merged with the conversion algorithm itself, so that no new phases are needed. Briefly, this is accomplished by using for other purposes the outputs of any of CC1-CC4 not actively involved in current division during a given step. For example, since CC4 is not used during step #1 of the division procedure, the copy of its stage's bit current which CC4 contains is available to feed the next divider in the cascade. A theoretical analysis in [15] predicts 12 bit performance with a 10 0µA reference current in a chosen 3µm CMOS process at a rate of 2µS per conversion (not including latency), limited mainly by the comparison speed.

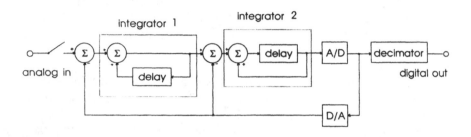

Figure 14.9 Second-order sigma-delta modulator

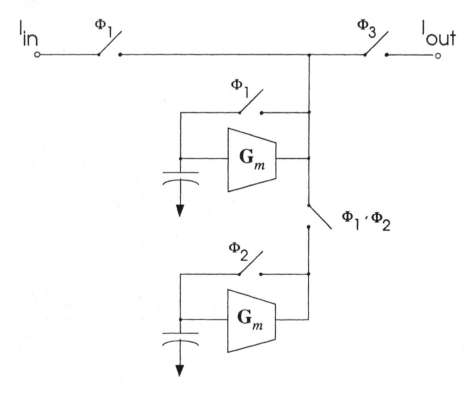

Figure 14.10 Current-copier based integrator

14.5.4 A Sigma Delta Front End

Another promising application of current copiers is in the implementation of sigma delta A/D converters. The block diagram of one such converter [16] is shown in Figure 14.9. At present, switched-capacitor circuits are usually used to implement the integrators, making linear capacitors necessary. While these may be obtained by adding a second level of poly to a digital CMOS process, the use of current copiers could eliminate this requirement.

A copier-based implementation of one of the integrators(#2) may appear as in Figure 14.10. As compared to a switched-capacitor realization, the copier-based approach requires one additional clock phase, used to facilitate the delay function; with appropriate timing, the complete second-order loop may be executed using a total of four phases. Given the high oversampling rate necessary for sigma-delta conversion, it is undesirable to allow any further expansion in the number of clock phases. Thus, if any noise-shaping is to be included (not shown in Fig.14.9), the non-unity branch gains required could not be implemented using copier-based current multipliers or dividers. The divider used in the D/A converter presented in section 14.5.1 [9,17] represents one possible solult ion to this problem, as it is compatible with digital processes (whereas a simple passive resistor divider isnot). Depending on the achievable linearity, the weighted current mirror approach [18, Chapter 11] might also be chosen.

14.5.5 Digitally Programmable Analog Filters [5,10]

Returning to the circuit shown in Figure 14.2, if I_0 is considered to be a time-varying input, an analog transversal filter could be obtained. In the proposed filter, the parallel current S/H stage shown in the figureis used as a cyclic delay line. Ratio-independent integer tap weights are implemented using the current multiplying approach in section 14.4.1; non-integer ratio-independent weights might be obtained using timing control. Briefly, to implement a fractional weight γ for the current sample in cell i, switch S_{3i} is closed periodically, with duty cycle γ; this chopped signal is then smoothed by a low pass filter or periodically reset integrator, producing the desired weighted output. Details of this process are given in [5]. Since the duty cycles of the switches can be controlled digitally, the resulting transversal filter has a digitally programmable transfer function. Simple modifications [5,10] allow voltage-in, voltage-out operation without re-introducing dependence on element ratios.

A recursive filter can be implemented with the same hardware if some of the copiers in the parallel current S/H are used to store samples of the output. Since the weights of the feedback paths in a recursive filter cannot be scaled arbitrarily, the timing-control method mentioned above must be applied with care. A proposed ratio-independent solution is discussed in [5].

Timing-controlled digitally-programmable analog filters using standard voltage sample/hold stages have been success-fully fabricated [5,19];

however, the process of generating coefficients using duty cycles is a relativelyslow one for reasonable resolutions. Nevertheless, the simplicity of the system, compact circuit realization, and powerful programming capabilities make this option attractive in low-bandwidth, low-cost applications. If higher operating speeds are necessary, then faster dividers based on standard resistors or linearized MOSFETs [20] may be considered, at the expense of ratio-independence.

14.6 References

[1] S.J. Daubert, D. Vallancourt, and Y.P. Tsividis, "Current Copier Cells", Electronics Letters, vol. 24, no. 25, pp. 1560-1562, 8 Dec. 1988.

[2] W. Groeneveld, H. Schouwenaars, and H.Termeer, "A Self- Calibration Technique for Monolithic High-Resolution D/A Converters", IEEE 1989 ISSCC Digest of Tech. Papers, vol. 32, pp. 22-23, Feb. 1989.

[3] D. Nairn and C. Salama, "Ratio-Independent Current Mode Algorithmic Analog-to-Digital Converters", Proceedings, 1989 IEEE International Symposium on Circuits and Sys- tems, pp. 250-253, May, 1989.

[4] G. Wegmann and E. Vittoz, "Very Accurate Dynamic Current Mirrors", Electronics Letters, vol. 25, no. 10 , pp. 644-646, 11 May 1989.

[5] D. Vallancourt, "Programmable Integrated Analog Signal Processors with Transfer Function Coefficients Determined by Timing", Ph.D. Dissertation, Columbia University, NY, NY, 1987

.[6] S.J. Gould, "Bully for Brontosaurus", Natural History, pp. 16-24, Feb. 1990 .

[7] G. Wegmann and E. Vittoz, "Analysis and Improvements of Highly Accurate Dynamic Current Mirrors", Proceedings of the 1989 European Solid-State Circuits Conference

[8] S.J. Daubert and D. Vallancourt, "Operation and Analysis of Current Copier Circuits", to be published in IEE Proceedings, part G.

[9] W. Groeneveld, H. Schouwenaars, H.Termeer, C. Bastiaansen, "A Self-Calibration Technique for Monolithic High-Resolution D/A Converters", IEEE Journal of Solid-State Circuits, vol. 24, no. 6, pp. 1517-1522, Dec. 1989.

[10] D. Vallancourt, Y.P. Tsividis, and S.J. Daubert, "Sampled-Current Circuits", Proceedings, 1989 IEEE International Symposium on Circuits and Systems, pp. 1592-1595, May 1989.

[11] J. Robert, P.Deval, and G. Wegmann, "Very Accurate Current Divider", Electronics Letters, vol. 25, no.14, pp. 912-913, 6 July 1989.

[12] P.W. Li, M.J.Chin, P.R. Gray, and R. Castello, "A Ratio-Independent Algorithmic Analog-to-Digital Conversion Technique", IEEE Journal of Solid-State Circuits, vol. SC-19, no. 6, pp. 828-836, Dec. 1984.

[13] J. Robert, P. Deval, and G. Wegmann, "Novel CMOS Pipe- lined A/D Converter Architecture Using Current Mir- rors", Electronics Letters, vol. 25, no.11, pp. 691-692, 25 May 1989.

[14] G.C. Temes, "High-Accuracy Pipeline A/D Converter Con- figuration", Electronics Letters, vol. 21, no. 17, pp. 762-763, 15 Aug. 1985.[

15] P. Deval, G. Wegmann, and J. Robert, "CMOS Pipelined A/D Converter Using Current Divider", Electronics Letters, vol. 25, no. 20 , pp. 1341- 1343, 28 Sept. 1989.

[16] B.E. Boser, B.A. Wooley, "The Design of Sigma-Delta Modulation Analog-to-Digital Converters", IEEE Journal of Solid-State Circuits, vol. 23, no. 6, pp. 1298-130 8, Dec. 1988.

[17] H.J. Schouwenaars, D.W.J. Groeneveld, and H.A.H. Termeer, "A Low-Power Stereo 16-bit CMOS D/A Converter for Digital Audio", IEEE Journal of Solid-State Circuits, vol. 23, no. 6, pp.1290 -1297, Dec. 1988.

[18] J.B. Hughes, N.C. Bird, and I.C. MacBeth, "Switched Currents - A New Technique for Analog Sampled-Data Signal Processing", Proceedings of the 1989 International Symposium on Circuits and Systems, pp. 1584- 1587, May 1989.

[19] D. Vallancourt and Y.P. Tsividis, "A Fully Programmable Sampled-Data Analog CMOS Filter with Transfer-Function Coefficients Determined by Timing", IEEE Journal of Solid-State Circuits, vol. SC-22, no. 6, pp. 10 22- 10 30 , Dec. 1987

[20] Y.P Tsividis, M. Banu, and J. Khoury, "Continuous-Time MOSFET-C Filters in VLSI", IEEE Journal of Solid-State Circuits, vol. SC-21, no. 1, pp. 15-30 , Feb. 1986.

Chapter 15

Integrated Current Conveyor

Doug C.Wadsworth

15.1 Introduction

One of the most basic building blocks in the area of current-mode analogue signal processing is the current conveyor. As has been described in Chapters 3 and 4, an extensive literature on the application of current conveyors has evolved since the concept was originally disclosed in 1968 [1]. Over 100 references have demonstrated the universality of this element in the synthesis of almost all known active networks. Concurrently with these papers, a number of authors have outlined improved implementations designed to enhance the performance and utility of this circuit block. Although major progress has been made with these realizations, they were not entirely satisfactory, and it has been suggested as recently as 1988 [27] that the wide-spread acceptance of current-mode techniques by the working design engineer has been limited by the lack of standard integrated circuits available in the field (Chapter 4). This dearth of devices is in marked contrast to the situation in the voltage domain, where a wide range of high performance ICs, such as operational amplifiers, are commonplace. In this Chapter, a new current conveyor IC based on a novel bipolar topology will be described which attempts to address this disparity.

15.1.1 Previous Conveyor Realizations

In order to put this new current conveyor IC in perspective, it is informative to briefly examine a number of prior realizations. The original implementation of this idea [1], which was designated a first generation current conveyor or CCI, was relatively basic and had distortion and accuracy limitations due to base current errors and output impedance restrictions. Subsequently, a more accurate version of the CCI was presented [3] which utilized a Wilson current mirror [2], but it was not suitable for integration since both NPN and PNP transistors were required to match each other. The disclosure of the second generation current conveyor or CCII [4] augmented the CCI concept by adding a higher impedance input port, but a solution to the accuracy/integration problem was not given.
As a means of producing an accurate conveyor that was suitable for integration, a number of researchers [5,9,10,11] have reported on implementations using operational amplifiers alone or in combination with

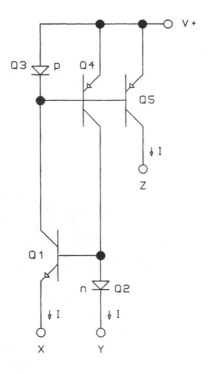

Figure 15.1   Basic Current Conveyor Figure

external transistors or operational transconductance amplifiers. However, these approaches were all subject to the frequency and transient response limitations of the op-amps. As described in Chapter 4, an innovative idea and a series of applications [17-24,26-30] were proposed which minimized these problems by sensing the output stage current in the op-amp and producing a current conveyor function with appropriately connected current mirrors. In practice, this was implemented by using the positive and negative supply pins, as these were the only suitable accessible leads on commercially available op-amps, but the indirect nature of this monitoring technique introduced errors due to the bias component. Recent efforts, as described in Chapters 3 and 4, have attempted to address this problem by designing an op-amp with a special output stage and utilizing CMOS technology.

Two other proposals for improved conveyor configurations [14-16,25] take an approach which is similar to the new IC to be described, in that they attempt to incorporate more accurate current mirrors into the original CCI implementation [1]. The particular mirrors that were used, however, had first order base current errors and potential transient response limitations.

15.1.2 New Conveyor Realization

This Chapter will describe a new current conveyor IC [33,34] that incorporates a topology [36] which is an evolution of the original CCI configuration [1]. This comprises an innovative patent pending connection of Wilson [2] current mirrors, an emitter degeneration compensation scheme to optimize the transient response and stability of these mirrors, and a novel output mirror arrangement which provides an enhanced output impedance that is twice that of the Wilson circuit. This enhancement matches the output impedance of a pure cascode configuration without appreciably disturbing the base current error cancellation scheme incorporated into the preceding mirrors. Also, this arrangement requires only that transistors of like polarity match each other, thereby maintaining compatibility with standard IC processing. These features, plus the direct nature of the implementation, which avoids the use of any major circuit blocks such as op-amps, results in a realization which offers significantly improved accuracy, frequency bandwidth, transient response, output impedance, and distortion.

15.2 Topology Evolution

The basic configuration of a first generation current conveyor or CCI, as described in the literature [1] is shown in Figure 15.1. Note that a diode symbol with the letter "n" or "p" represents an NPN or PNP diode-connected transistor respectively, where the base and collector terminals are shorted together. All transistors of like polarity are of identical size and construction to ensure matching. The CCI configuration consists basically of two cross- coupled simple current mirrors, one of NPN transistors (Q1,Q2), and the other of PNP transistors (Q3,Q4), with an additional output PNP transistor (Q5). If port Y is held at a reference potential, and current is sourced from port X, then to a first order, equal currents are sourced from ports Y and Z. Of particular interest is the fact that port X will be driven to the same potential as port Y, regardless of the current levels being sourced. Therefore, if port Y is referenced to ground, then port X becomes a virtual ground or low impedance input. Conversely, port Z becomes a high impedance output. In summary, current which is sourced from low impedance input port X is conveyed to high impedance output port Z , unaltered, except for impedance level. The main limitation of the simple CCI realization shown in Figure 15.1 is that the base currents of the transistors introduce a significant error in the accuracy with which current sourced from port X matches that sourced from ports Y and Z. This error is of the order of $100/\beta$ % where β is the forward current gain of the transistors. The magnitude of this deviation can easily reach 1 % or more, and results in a circuit which is unacceptable for most applications. In addition, the Early voltage or output conductance effect of Q5 causes a further degradation in performance.

As has been reported previously [2], the Wilson current mirror provides

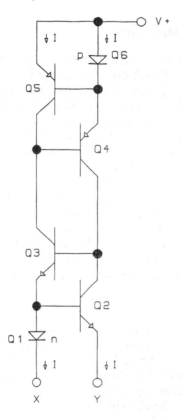

Figure   15.2   Partial Wilson Mirror Current Conveyor

much improved accuracy through a base current cancellation scheme which results in errors of the order of $200/\beta*\beta$ %. It is obvious that if the simple current mirrors of Figure 15.1 are replaced with Wilson mirrors, a much more accurate current conveyor will be obtained. Referring to Figure 15.2, the problem is that if an NPN Wilson mirror (Q1,Q2,Q3) is cross-coupled with a PNP Wilson current mirror (Q4,Q5,Q6), accurate conveying of current between ports X and Y is obtained, but it is not apparent how port Z can be implemented. Any attempt to add an output port in a manner similar to that of Q5 in Figure 15.1 will destroy the first order base current cancellation scheme.

15.2.1 Basic New Topology

The proposed new topology utilizes an innovative connection of Wilson current mirrors which solves the difficulties that have been outlined, and realizes a current conveyor of high accuracy. The basic concept is illustrated in Figure 15.3. Four Wilson mirrors are used to produce a first generation current conveyor or CCI. An NPN Wilson mirror (Q1,Q2,Q3)

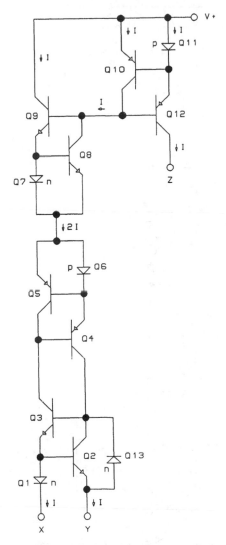

Figure 15.3   Basic New Current Conveyor Topology

and a PNP Wilson mirror (Q4,Q5,Q6) are connected as was shown in Figure 15.2. Referring to Figure 15.2, it should be noted that the current drawn from the supply rail, V+, is equal to twice the value of each of the equal currents sourced from ports X and Y respectively. Therefore, another NPN Wilson mirror (Q7,Q8,Q9) can be connected as shown in Figure 15.3 to perform a current splitting action. The current at the emitters of Q7 and Q8 is divided by this mirror such that half is shunted up through Q9 to the rail, and the remainder is delivered to an output PNP Wilson mirror (Q10,Q11,Q12). This results in a current being sourced from port Z which

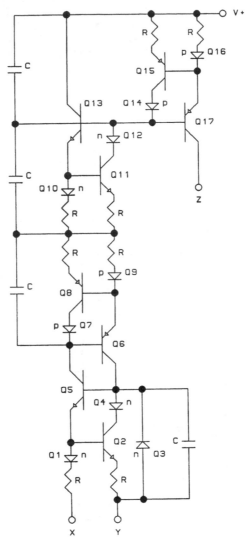

Figure 15.4   Improved New Current Conveyor Topology

is equal to that which is sourced from ports X and Y. Thus a basic CCI has been implemented which maintains the excellent accuracy of the Wilson current mirror.

There is a remote possibility that the two cross-coupled Wilson mirrors (Q1,Q2,Q3 and Q4,Q5,Q6) might not "start" when the circuit is first powered up. This situation is a result of the fact that Q3 generates its own base drive through the upper PNP mirror (Q4,Q5,Q6). If Q3 is off, it needs base drive to turn on, but it cannot produce base drive until it has turned on. Leakage currents would almost certainly overcome this contradiction, but the addition of device Q13 ensures that, if necessary, a turn on path would be established when port X fell three diode drops below port Y, thereby forcing current to flow through Q13 into the base of Q3. Once the mirrors start up, Q13 is reverse biassed and has no effect on the normal operation of the current conveyor.

15.2.2 Improved Topology

A number of refinements can be made to further improve this circuit. It has been shown previously [8] that an additional diode connected in series with the collector of one of the transistors in the Wilson mirror would improve its accuracy. This device is therefore added in to the collector leads of Q2, Q5, Q8, and Q10 in Figure 15.3. Another characteristic of the Wilson mirror that must be considered is that it tends to ring or overshoot on fast pulse waveform edges. This is of itself undesirable, but it becomes a serious problem with the two cross-coupled mirrors (Q1,Q2,Q3,Q4,Q5,Q6) in Figure 15.3. The overshoot is actually a result of the fact that the current gain of the Wilson mirror becomes greater than unity at high frequencies. Such a situation, along with parasitic capacitance on port X, can cause the loop gain around these two mirrors to exceed unity, with potential oscillation the result. The solution is to employ matching emitter de-generation resistors in each of the Wilson mirrors and then use capacitors to roll off the high frequency gain. The RC time constant is chosen to optimize the transient response to a critically damped characteristic, consistent with the parameters of the transistors utilized. These resistors have the added advantage of improving the accuracy of the mirrors. A schematic diagram of the current conveyor with these features added is shown in Figure 15.4.

15.2.3 Final Topology

Two final enhancements are required to complete the circuit. As has been documented in the literature [2,5,7], the Wilson current mirror has an output resistance that is much improved over a simple mirror, being approximately $\beta/2$ times greater. However, this is still only half that which can be obtained with a pure cascode circuit. As it is desirable to make the output resistance of port Z as high as possible, the proposed circuit incorporates a novel scheme whereby the performance of a pure cascode is obtained without introducing any significant errors into the Wilson base current cancellation arrangement. Referring to Figure 15.5, PNP transistors Q15 and Q20 have been added to effect this improvement. Q20 is the cascode transistor which raises the output resistance of the modified PNP Wilson mirror consisting of Q16, Q17, Q18, and Q19. The base of Q20 is biassed from the NPN current splitting mirror (Q11,Q12,Q13,Q14) in such a way that its base current is shunted up to the rail with the unused half of the split current. As a result, the accuracy of the NPN splitter mirror is not compromised. There is, however, a potential for a major inaccuracy in the output current, as base

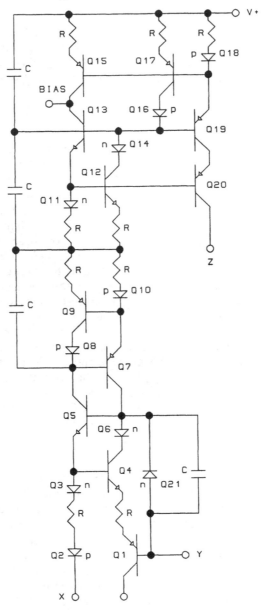

Figure 15.5   Final New Current Conveyor Topology

current is subtracted from it as it flows from the emitter to the collector of
Q20. This is reduced to a negligible error by introducing a balancing base
current with Q15. Since this transistor and its emitter resistor match Q17
and Q18, two base currents are injected into the mirror at Q18 to cancel the
first loss in Q19 and the second loss in Q20. The collector of Q15, through

which a "wasted" current flows, can be connected to the collector of Q13 which carries the unused half of the split current. These two extraneous currents are very nearly equal, and so virtually cancel. This allows the collectors of Q13 and Q15 to be tied to any convenient bias point about two diode drops below rail, even if it is relatively high impedance, as only a small difference current needs to be supported. The end result is that the output resistance of port Z is doubled at the expense of increasing the base current error from approximately $200/\beta*\beta$ % to $500/\beta*\beta$ %. This is negligible when weighed against the benefits of improved output resistance.

The addition of devices Q1 and Q2 to Figure 15.5 provides some flexibility in the application of the conveyor. If the collector and base of Q1 are shorted together, then a CCI is produced with port X at the emitter of Q2 and port Y at the connection of the base and collector of Q1. The operation of the current conveyor remains unchanged from previous descriptions, except that an additional diode drop has been added in series with ports X and Y respectively. If however, Q1 is utilized as a transistor, then a useful implementation of a second generation current conveyor or CCII [4] is realized. In this configuration, the collector of Q1 is returned to a separate more negative rail and the base becomes port Y. As with the CCI, the voltages at X and Y are forced to be equal and current sourced from port X is conveyed to port Z unchanged, except for impedance level. The current at port Y, however, no longer matches that at X and Z and is in fact reduced by the current gain of Q1. This approximates the terminal characteristics of a CCII which require that the current at port Y be zero. Even though the condition at port Y is non-ideal, this does not affect the utility and accuracy of this CCII implementation in many applications.

An example is shown in Figure 15.6 where a voltage to current convertor has been implemented using this CCII configuration. The voltage applied to port Y will be accurately duplicated at port X, since the currents through diode Q2 and the base-emitter junction of Q1 are essentially identical, even though the emitter current of Q1 subsequently splits between its base and collector. The magnitude of the current sourced from port Z is therefore equal to the input voltage at port Y divided by the value of resistor Ro at port X. An additional matching resistor between the collector of Q1 and the negative rail forces its collector to be at virtually the same potential as its base, within an accuracy of approximately $100/\beta$ %. This reduces the Early voltage error on Q1 to the level of the other second order errors in the Wilson mirror and an accurate voltage to current convertor results.

In summary, the topology presented in Figure 15.5 can be configured as a first generation current conveyor (CCI) of high accuracy, optimized transient response, and enhanced output impedance. Only transistors of like polarity are required to match to achieve this performance, thereby facilitating integrated circuit fabrication. In addition, a useful realization of a second generation current conveyor (CCII) can also be implemented

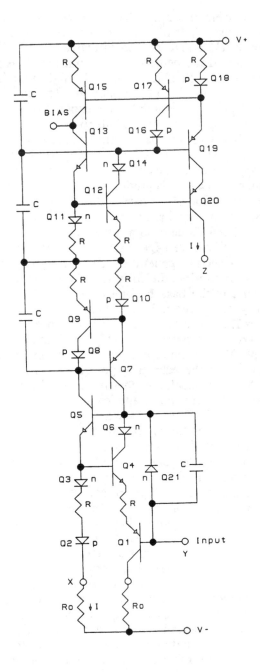

Figure 15.6   Voltage to Current Convertor

15.3 Process Characteristics

Although the new conveyor topology described in Section 15.2 is compatible with standard IC processing technology, it makes use of both NPN and PNP type current mirrors in the signal path. As a result, the overall performance depends in equal measure on the characteristics of both transistor polarity types, and this fact must be considered in producing a monolithic implementation.

15.3.1 Standard Bipolar Processing

Conventional bipolar IC fabrication requires that a lateral structure be used for PNP devices, while a much higher performance vertical arrangement is available for the NPN transistors. While an NPN device might exhibit a typical forward current gain (β) of 100 and a bandwidth (ft) of 350 MHz, the lateral PNPs would lag behind with a β and ft of 20 and 5 MHz respectively. As well, the current handling capability of lateral devices is poor when compared with vertical structures of similar size, thereby necessitating the use of additional chip area to compensate. Lateral PNP transistors would therefore be the limiting factor in conveyor performance with this type of processing technology.

15.3.2 Complementary Bipolar Processing

Fortunately, a newer complementary bipolar processing technology has been developed which incorporates vertical structures for both NPN and PNP transistors. Although this enhancement requires additional complexity and photomask layers to implement, the cost penalty has been steadily decreasing as more advanced processing techniques become standard. In fact, complementary bipolar processes are now available on a foundry basis which provide both NPN and PNP transistors with a typical β of 100 and an ft of 2.5 GHz. Therefore, it is logical to take advantage of these advances to produce an integrated version of this new conveyor configuration.

In IC fabrication, there is a direct trade-off between breakdown voltage (BVceo) and speed (ft). Processes with bandwidths in the GHz range are generally limited to a power supply of approximately 5 volts, due to the inherently thin base widths and compact device dimensions. As it is anticipated that a current conveyor IC would often be required to operate with op-amp supply rails, and in order to ensure adequate signal swing with the cascode output stage previously described, it is desirable to select a process with a +/-18 volt capability. A suitable complementary bipolar process meeting this voltage requirement offers both NPN and PNP transistors with a typical B of almost 100 and an ft of greater than 300 MHz. An outline of the specifications of this process is given in Table 15.1. This technology choice synergistically compleiments the new current conveyor topology.

Table 15.1   IC Process Characteristics

Process	Characteristics			
PARAMETER	CONDITIONS	NPN	PNP	UNITS
β	Ic = 1 mA	85	110	-
ft	Vce = 10 v	350	300	MHz
Early V	Ic = 500 μA	225	60	v
Vbe	Ic = 100 μA	743	748	mv
BVceo	Ic = 1 mA	38	47	v

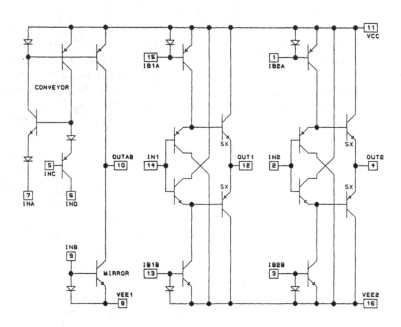

Figure 15.7   PA630 Current Conveyor IC Figure

15.4 Circuit description

Functional schematics of two versions of the new current conveyor
integrated circuit, designated the PA630 and PA630A, are shown in Figure
15.7 and Figure 15.8 respectively. The parts are fabricated on the
complementary NPN-PNP bipolar IC process outlined in Table 15.1. Each
circuit consists of two unity gain buffer amplifiers, a current mirror, and a
current conveyor block. While the PA630 is a 16 pin device, the PA630A
has two additional pins which provide some flexibility in interfacing.
Otherwise, the two ICs are electrically equivalent.   For clarity, the current

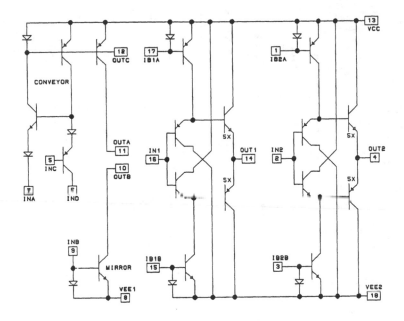

Figure 15.8   PA630A Current Conveyor IC

conveyor block and current mirror have been simplified in these diagrams. The actual on-chip circuitry incorporates the new conveyor topology and a Wilson mirror with the emitter degeneration compensation scheme, as described in Section 15.2.

15.4.1 Unity Gain Buffers

Each unity gain buffer amplifier consists of four emitter followers in complementary symmetry and two current sources for biassing, as shown in Figure 15.9. The quiescent operating point of each can be set independently with an external resistor (Rset) connected as indicated. As the current sources are actually Wilson mirrors (2 Vbe), and the output devices are five times larger than the input transistors, the quiescent current in the output stage can be calculated using Equation 15.1.

$$Io = 5 \frac{Vcc + Vee - 2.8v}{Rset}$$

$$(15.1)$$

These two follower amplifiers can obviously be used to buffer the input and output of the various conveyor block configurations.

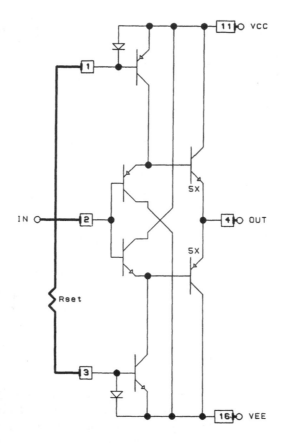

Figure 15.9   PA630/PA630A Buffer Amplifier

15.4.2 Inverting CCI Configuration

By combining the current conveyor block in CCI mode with the current mirror, an inverting current follower can be implemented as shown in Figure 15.10. Since pin 7 is forced to become a virtual ground, and the current mirror block is actually a Wilson source (2 Vbe), R3 sets up a quiescent current around the loop whose value is given by Equation 15.2.

$$Iq = \frac{Vee - 1.4v}{R3}$$

$$(15.2)$$

If VIN is open or connected to ground, the output at pin 10 will also be at ground, since all currents balance. The voltage gain of this configuration is - R2/R1,   and the input is bi-directional up to the level of current specified by

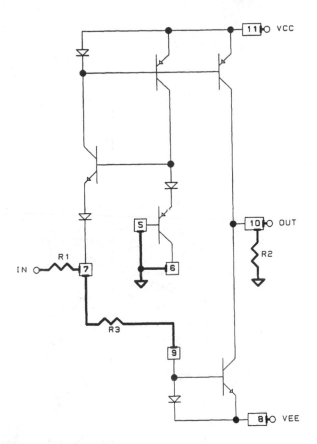

Figure 15.10   PA630 Inverting CCI Configuration

Equation 15.2. It should be noted that both the input and output parameters are currents which are converted to voltages by resistors R1 and R2 respectively. By eliminating one or both of these resistors, many current-mode applications can be addressed.

15.4.3 Non-Inverting CCII Configuration

If the conveyor is connected in CCII mode, as shown in Figure 15.11, a non-inverting block is produced with a higher impedance input on pin 5. The voltage gain is R2/R1, and R3 sets up the quiescent current as defined by Equation 15.2. With this arrangement, however, pin 5 swings with the input signal, and pin 7 is forced to follow. By making R4=R1 and R5=R3, the voltage on pin 6 will closely match that on pin 5 and pin 7, and the Early voltage (output conductance) error of the input transistor will be minimized. One of the on-chip buffers can be used to enhance the input impedance at pin 5, thereby implementing a CCII with more ideal terminal characteristics.

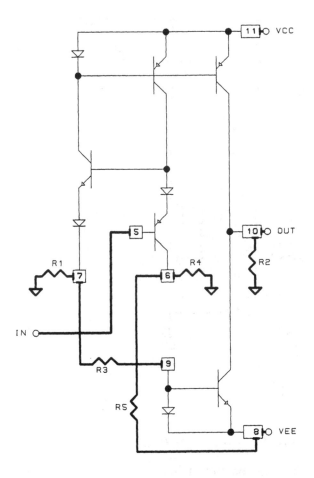

Figure 15.11   PA630 Non-Inverting CCII Configuration

15.4.4 JFET Output Configuration

Figure 15.12 demonstrates how the 18 pin PA630A can be interfaced with two external junction field effect transistors (JFETs). Suitable components would be a 2N5462 P-channel JFET and a 2N5459 N-channel JFET in the current source and sink paths respectively. This arrangement buffers the on-chip bipolar transistors [7], and provides extremely high output impedance, improved accuracy, and lower distortion, since the Early voltage errors in the output stage are virtually eliminated. With many applications such as gyrators [5], such an enhanced output impedance can be advantageous.

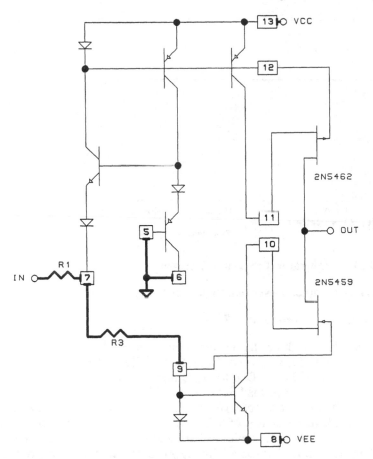

Figure 15.12   PA630A JFET Output

15.5 Performance

The new PA630/PA630A current conveyor ICs offer significantly improved performance due to both the innovative circuit topology, and the advanced fabrication process. In particular, the characteristics in the areas of output impedance, frequency bandwidth, transient response, distortion, and accuracy are noteworthy.

15.5.1 Conveyor Output Impedance

As a result of the cascode stage described in Section 15.2, the output impedance of the conveyor configurations shown in Figure 15.10 and Figure 15.11 is typically 3 MΩ, with a quiescent current of 2 mA in the output devices. With the same bias conditions, this parameter can be enhanced to over 100 MΩ by employing external JFETs, as shown in Figure 15.12. It is believed that, heretofore, this level of performance has been available only

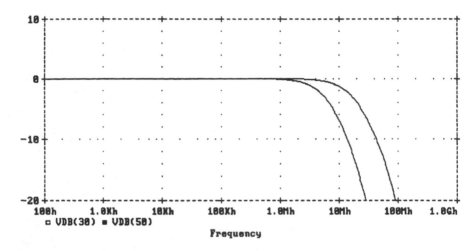

Figure 15.13   Conveyor Frequency Response

with relatively complex lower bandwidth circuitry.

15.5.2 Conveyor Frequency Response

The frequency bandwidth of the current conveyor block of the PA630 IC is presented in Figure 15.13. The configuration of Figure 15.10 was used with a quiescent current of 2 mA. Due to the nature of the current-mode output, a pole is introduced by the load resistance (R2), and the parasitic output capacitance. This effect is observed on the lower bandwidth curve in the figure, where a load resistance of 1 kΩ results in a -3 dB point of 6 MHz. With a 100 Ω load, the full conveyor bandwidth of 18 MHz is obtained, as indicated by the second trace. This represents more than an order of magnitude improvement over many prior realizations.

15.5.3 Conveyor Transient Response

The transient response of the conveyor with a +/-1 mA step change in current is displayed in Figure 15.14. The pulse width is 0.5 μs and the device was in the inverting CCI mode (Figure 15.10) with a quiescent current of 2 mA. The smooth pulse response that is evident results from the previously described emitter degeneration compensation scheme, and it is in marked contrast to the ringing and overshoot often experienced with other topologies.

15.5.4 Conveyor Harmonic Distortion

In Figure 15.15, the harmonic distortion performance of the inverting conveyor (Figure 15.10) is shown in Fourier plot format. A 1 kHz sine wave, with an amplitude of 1 volt peak, was used as the input. Both R1 and

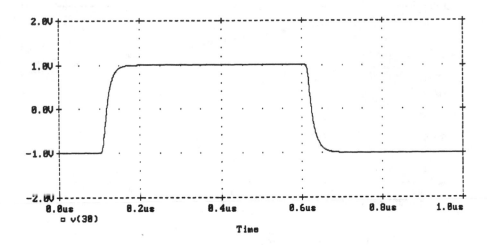

Figure 15.14   Conveyor Transient Response

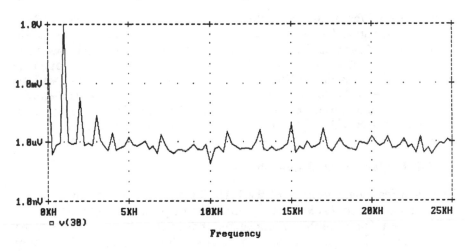

Figure 15.15   Conveyor Fourier Harmonics

R2 were equal to 1 kΩ, and R3 was used to set up a quiescent current of 2 mA. The Fourier harmonic components of the output are displayed with a log voltage scale on the Y axis. The first spike at 1 kHz is the 1 volt input signal, and the second harmonic may be read as 0.18 mv or 0.018% distortion. The third and higher order harmonics are below 50 μv or 0.005%. This low level of distortion makes possible a number of high performance uses in such areas as professional audio electronics.

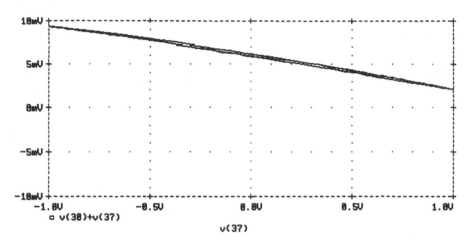

Figure 15.16   Conveyor Accuracy

15.5.5 Conveyor Accuracy

A measure of the accuracy of the inverting conveyor (Figure 15.10) is presented in Figure 15.16. With R1 and R2 equal to 1 kΩ and a quiescent current of 2 mA, a swept input of +/-1 volt, displayed on the X axis, was used to produce an output which should ideally be of equal value but opposite polarity. The absolute difference between these two signals will be representative of the accuracy of the circuit, and this deviation is therefore used as the Y axis parameter. As may be seen from the figure, the conveyor has a 6 mv offset with a deviation of +/-3.5 mv, for an incremental accuracy of 0.35%.

15.5.6 Performance Summary

The electrical characteristics of the PA630/PA630A current conveyor ICs are summarized in Table 15.2. Two items of particular interest are the enhancements to a number of conveyor parameters which result from the use of the JFET output, and the wide bandwidth (50 MHz) and high input resistance (11 MΩ) of the unity gain buffer amplifiers. Also, the noise performance, which is expressed as a current (5 nA), represents a 5 μv rms signal in the audio band, into a 1 kΩ load. As this is 106 dB below a 1 volt input level, the device is suitable for low noise applications such as professional audio electronics

15.6 Applications

The literature of current conveyor applications presents an extensive selection of implementations for many useful circuit functions, such as impedance convertors, oscillators, active filters, and gyrators. The

Table 15.2   PA630/PA630A Electrical Characteristics

Electrical Characteristics   Ta=25°C, Vcc=+15v, Vee=-15v, Buffer: Rset=68k, RL=10k
Conveyor: R1=1k, R2=1k, R3=6.8k unless otherwise specified

PARAMETER	CONDITIONS	BUFFERS Each Amp FIG 9			INVERTING CONVEYOR FIG 10			NON-INVERTING CONVEYOR FIG 11			JFET OUTPUT CONVEYOR FIG 12			UNITS
		MIN	TYP	MAX	MIN	TYP	MAX	MIN	TYP	MAX	MIN	TYP	MAX	
Input Bias Current		-10	+2	+10				-50	-15	-5				µA
Input Resistance	Incremental		11						0.1					Mohms
Virtual Ground	Incremental					2						2		ohms
Input Capacitance			8						3					pF
Input Common Mode Range	R1=10k													
Positive		+13.0	+13.3					+4.5	+5.2					V
Negative			-13.3	-13.0					-5.0	-4.5				V
Offset Voltage	Rsource=50 ohms	-40	-10	+40										mV
Virtual Ground					-5	-3	+5				-5	-3	+5	mV
Accuracy	Incremental		0.1			0.5			0.5			0.25		%
Output Resistance	Incremental					3			3			100		Mohms
Buffer	Incremental		9											ohms
Output Capacitance						11			11			7		pF
Output Voltage Swing	R2=10k													
Positive		+13.0	+13.3		+12.0	+12.8		+12.0	+12.8		+9.8	+10.8		V
Negative			-13.3	-13.0		-13.5	-13.0		-13.5	-13.0		-11.5	-10.5	V
Power Supply Current														
Vcc			3	3.5		6	6.5		6	6.5		6	6.5	mA
Vee		-3.5	-3		-4.5	-4		-6.5	-6		-4.5	-4		mA
Power Supply Rej Ratio	120 Hz													
Vcc			70			66			62			66		dB
Vee			64			75			70			75		dB
Bandwidth	3 dB point		50			6			6			8		MHz
Conveyor	R3=100 ohms					18			18			18		MHz
Rise Time			15			70			70			60		ns
Conveyor	R3=100 ohms					25			25			25		ns
Overshoot			0			0			0			0		%
Buffer	CL=20 pF		10											%
Slew Rate			150			50			50			50		V/µs
Harmonic Distortion	1 kHz; 0.5 Vrms		0.002			0.020			0.030			0.010		%
Buffer	RL=1 kohm		0.005											%
Output Noise Current	20 Hz - 20 kHz					5			5			5		nA

PA630/PA630A ICs can provide enhanced performance in the areas of frequency bandwidth, distortion, and transient response in many of these situations, and in addition the high output impedance offered can be advantageous with frequency selective networks. Specifically, limitations associated with the Q of many prior circuits will be greatly ameliorated. Other special attributes of these new devices make possible several particularly innovative uses, and these will be examined in the following sections.

15.6.1 Professional Audio Electronics

Due to the topology of the conveyor, the PA630/PA630A ICs can realize functions such as gain blocks, inverting followers, and virtual ground inputs without the use of the global negative feedback required by most other chips such as operational amplifiers. The parts are therefore inherently free of dynamically induced distortion mechanisms, and as such, can be used to advantage in the field of professional audio electronics [34]. This type of distortion [6], which includes transient intermodulation distortion (TIM) and slewing induced distortion (SID), can be produced when a fast rising complex waveform is introduced into a gain block with global negative feedback, such as an op-amp. Fortunately, audio signals are generally band limited and therefore may not contain these fast edges. Also, in a carefully designed system, global negative feedback is not necessarily problematic. Nevertheless, certain situations are particularly sensitive to these effects, and for example, the dynamic characteristics of any device used in the analogue stages of digital recording and compact disc equipment must be carefully considered to ensure distortion-free performance.

As a specific example, Figure 15.17 shows the circuit of the analogue section of a typical compact disc player. With this arrangement, the current output of a digital to analogue convertor (DAC), which must drive into a virtual ground, is opposed by a 2 mA current to establish the zero reference for the audio signal. An op-amp provides the virtual ground and performs the current to voltage conversion with a high frequency roll-off. This is followed by a Salen and Key second order linear phase filter based around an op-amp connected as a unity gain buffer. This configuration has the potential for the generation of dynamically induced distortion, as fast edges from the DAC can mix with the sinusoidal music signals in an amplifier with global negative feedback. Perhaps it is not surprising that a number of companies are retrofitting production CD players with higher slew rate op-amps and discrete components in an effort to eliminate what is perceived as a harshness in their musical reproduction.

Figure 15.18 indicates how the PA630 IC can replace the op-amps and perform the same function without global negative feedback. The current output from the DAC is fed directly into the virtual ground input of the current conveyor. The reference current is subtracted from the output of the conveyor by the opposing mirror and an equivalent output to Figure 15.17 is produced. One of the on-chip unity gain amplifiers buffers the output of the conveyor, while the other is used for the filter. This implementation completely eliminates the potential for TIM type distortion in the analogue stages of a CD player. Recent measurements and listening tests, where this configuration was evaluated with a state of the art digital audio system, have been very encouraging.

A more basic audio building block can be realized if the two on-chip follower amplifiers are used to buffer the input and output of the inverting

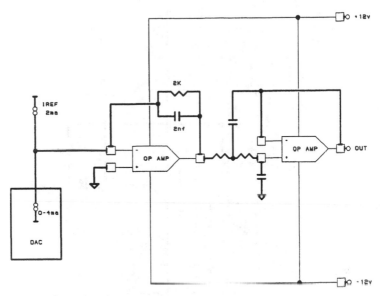

Figure 15.17   Analogue Section of Typical CD Player

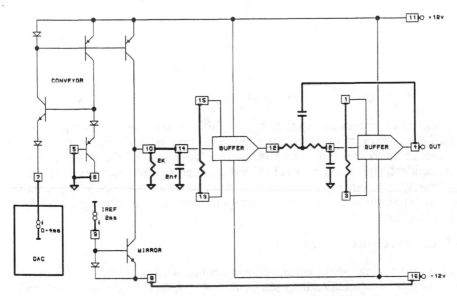

Figure 15.18   PA630 Analogue Section for CD Player

CCI configuration of Figure 15.10. A unity gain inverting amplifier using this approach would be extremely useful in professional audio systems, as signal inversion is often required. This implementation retains DC coupling and is inherently free of dynamically induced distortion.

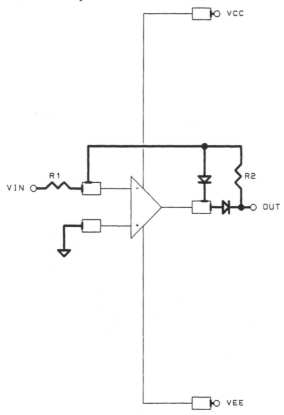

Figure 15.19   Opamp Precision Rectifier

One other interesting application of this current conveyor chip results from the nature of the high impedance current output. Since the input reference (port Y) and output reference (cold end of the load) can be independent, signals can be level-shifted between any two voltages within the common mode range. This property could be used to isolate two grounds within an audio system, thereby eliminating noise caused by ground loop circulating currents.

15.6.2 Precision Rectification

A very promising use for current conveyors has been suggested [24,26] in the area of precision rectification (see Chapter 4). With its wide bandwidth and optimized transient response, the PA630 can offer a major performance advantage over rectifiers utilizing operational amplifiers.

Figure 15.19 gives an example of such an application using a conventional op-amp. Since the diodes are within the feedback loop of the amplifier, a severe requirement is placed on the slew rate of the device as it attempts to slew through two diode drops at the input signal zero crossing. Using a

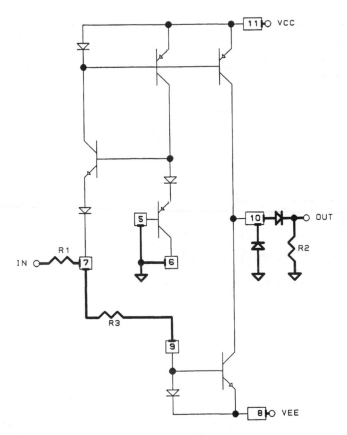

Figure 15.20   PA630 Precision Rectifier

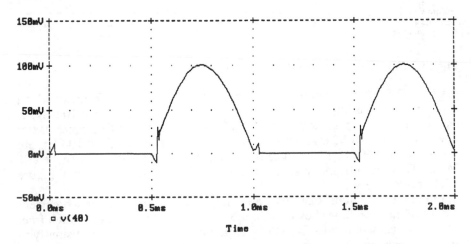

Figure 15.21   uA741 Rectifier - 1 kHz

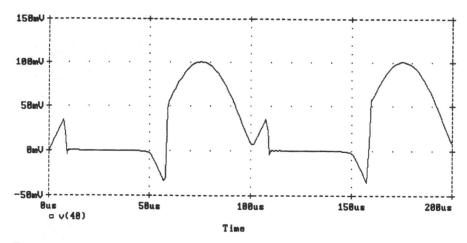

Figure 15.22   uA741 Rectifier - 10 kHz

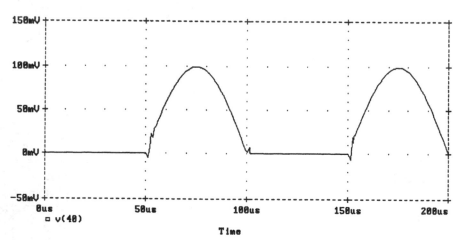

Figure 15.23   LM318 Rectifier - 10 kHz

μA741 (0.5 v/μs, 1 MHz) in the circuit of Figure 15.19 results in a waveform that has major glitches at 1 kHz and severe distortion at 10 kHz, as shown in Figure 15.21 and Figure 15.22 respectively. With a much faster device (30 v/μs, 20 MHz) such as the LM318, glitches show up at 10 kHz, are severe at 100 kHz, and culminate in a complete loss of waveform shape at 1 MHz. These results are presented in Figure 15.23, Figure 15.24, and Figure 15.25 respectively.

In contrast, a circuit that takes advantage of the high impedance current output of the PA630 is shown in Figure 15.20, and it is virtually glitch free. This improvement results from the fact that the diodes can be used simply to steer the conveyor output current either into the load or bypass it through a

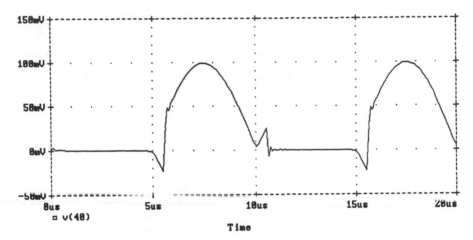

Figure 15.24   LM318 Rectifier - 100 kHz

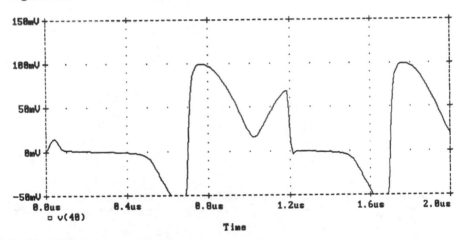

Figure 15.25   LM318 Rectifier - 1 MHz

second diode to ground. Since at the input zero crossing the output current is zero, the transition from one diode to the other can occur smoothly, without the instantaneous two diode slew required of op-amp circuits. This advantage is obvious when examining Figure 15.26, Figure 15.27, and Figure 15.28, which display the performance of the conveyor rectifier at 10 kHz, 100 kHz, and 1 MHz respectively. Although parasitic capacitance on the output eventually introduces distortion at 1 MHz, a potentially usable waveform is still available. With each of the precision rectifier waveforms presented in Figures 15.21 through 15.28, the input was a sine wave with the specified frequency and an amplitude of 100 mv peak.

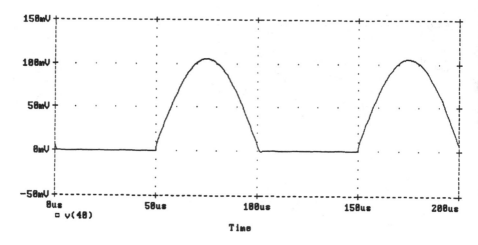

Figure 15.26   PA630 Rectifier - 10 kHz

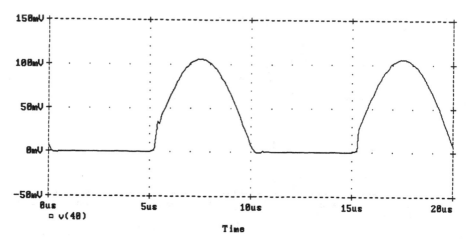

Figure 15.27   PA630 Rectifier - 100 kHz Figure

15.6.3 Current Feedback Amplifier

As has been described in Chapters 4 and 16, several major integrated circuit manufacturers have incorporated current-mode techniques into a new type of IC known variously as a current feedback amplifier, a current-mode feedback op-amp, or a transimpedance op-amp. A number of such devices are now available as standard products. The main advantage of these ICs, when compared with conventional op-amps, is that the bandwidth is relatively independent of the closed-loop gain. This not only improves the frequency response, but also results in more loop-gain being available at higher frequencies to reduce distortion and improve accuracy. Obviously,

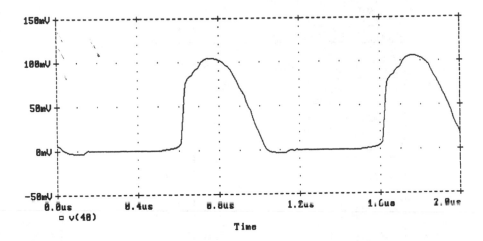

Figure 15.28   PA630 Rectifier - 1 MHz

these characteristics can be advantageous in many applications.

The data sheets generally describe these current feedback amplifiers as op-amps with a unity gain follower connected between the non-inverting and inverting inputs. The output current of the follower is sensed, amplified, and a fraction is applied as negative feedback to the inverting input. However, in the context of this book, it is instructive to define these devices in terms of standard current-mode building blocks. As such, it can be shown that these ICs basically consist of a current conveyor and a unity gain output buffer arranged as follows.

The inverting and non-inverting inputs correspond to port X and port Y of the conveyor respectively, and port Z is driven into a high impedance internal node to produce a large open-loop gain. This node is buffered to the output of the current feedback amplifier with the unity gain follower. The closed-loop gain is defined by feeding- back a current into the inverting input, which inherently exists as a virtual ground on port X of the current conveyor, independent of this feedback. The performance of the conveyor with regard to accuracy and distortion is relatively non-critical in this specific application, as the current-mode feedback will ameliorate such deficiencies.

The PA630 contains all the necessary elements to implement the current feedback amplifier function that has been outlined. By using the conveyor configuration of Figure 15.10, and employing one of the unity gain buffers, an inverting current feedback amplifier may be constructed as shown in Figure 15.29. The output of the conveyor (pin 10) drives the input impedance of the buffer (pin 14), resulting in a large open-loop gain. The closed-loop voltage gain is -Rf/Ri. The additional unity gain follower on the PA630 chip may be used as a buffer to increase the input impedance at Ri, or to isolate a capacitive load. A non-inverting current feedback amplifier may

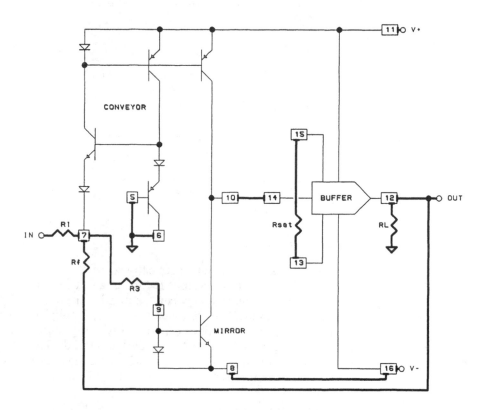

Figure 15.29   PA630 Current Feeedback Amplifier

also be implemented by utilizing the conveyor arrangement of Figure 15.11.

The frequency response performance of the PA630 IC, configured as an inverting current feedback amplifier (Figure 15.29), is shown in Figure 15.30. The quiescent current of the conveyor and the output current of the buffer were both set at 2 mA with R3 and R set respectively. The upper curve in the figure specifies the response with a closed-loop gain of -10 (Rf=1 kΩ, Ri=100 Ω), while the bottom trace represents a unity gain inverter (Rf=1 kΩ, Ri=1 kΩ). In each case, a load resistance (RL) of 10 kΩ was used. As is typical with this type of amplifier, the bandwidth may be observed to be relatively constant (10 MHz, 16 MHz) for the two different gains. The peaking in the response is due primarily to the interaction of the global negative feedback and parasitic capacitance on the virtual ground (inverting input) node. With this type of closed-loop topology, the stability of the system must be given careful consideration.

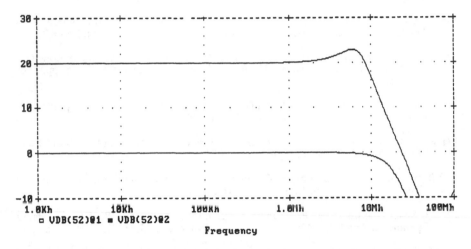

Figure 15.30   Current Feedback Amplifier Frequency Response.

Therefore, in this configuration, the PA630 exhibits the enhanced characteristics of a current feedback amplifier. Furthermore, it has the added flexibility of incorporating an accessible and accurate current conveyor, which may be employed to advantage as an open-loop current-mode building block in many other applications.

15.7 Conclusions

This chapter has presented a new integrable current conveyor topology which offers the promise of improved accuracy, wider frequency bandwidth, smoother transient response, and higher output impedance. This expectation has been confirmed by the performance of the PA630/PA630A ICs, which incorporate this topology and are fabricated on a complementary bipolar IC process. Based on these results, the characteristics of applications previously described in the literature will be enhanced, and new uses such as professional audio electronics are practicable. These new current conveyor ICs should stimulate the already growing interest in analogue current-mode signal processing, and assist in making these techniques more accessible to the general electronics community.

15.8 References

[1] K. C. Smith and A. Sedra, "The Current Conveyor: A New Circuit Building Block,Proc. IEEE, Vol. 56, pp. 1368-1369, Aug.1968.

[2] G. R. Wilson, "A Monolithic Junction FET-NPN Operational Amplifier," IEEE J. Solid-State Circuits, Vol. SC-3, pp.380- 387, Dec. 1968.

[3] K. C. Smith and A. Sedra, "A New Simple Wide-Band CurrentMeasuring Device," IEEE Trans. Inst. and Meas., Vol. IM-18, pp. 125-128, June 1969.

[4] K. C. Smith and A. Sedra, "A Second Generation Current Conveyor and its Applications," IEEE Trans. Circuit Theory, Vol. CT-17, pp. 132-134, Feb. 1970.

[5] G. G. A. Black, R. T. Friedmann, and A. Sedra, "Gyrator Implementation with Integrable Current Conveyors," IEEE J. Solid-State Circuits, Vol. SC-6, pp. 396-399, Dec. 1971.

[6] M. Otala, "Transient Intermodulation Distortion in Commercial Audio Amplifiers," J. Audio Engr. Soc., Vol. 22, pp.244-246, May 1974.

[7] R. C. Jaeger, "A High Output Resistance Current Source," IEEE J. Solid-State Circuits, Vol. SC-9, pp. 192-194, Aug. 1974.

[8] R. W. Barker and B. L. Hart, "D.C. Matching Errors in the Wilson Current Source," Electronics Letters, Vol. 12, pp. 389- 390, July 1976.

[9] M. Sharif-Bakhtiar and P. Aronhime, "A current Conveyor Realization Using Operational Amplifiers," Int. J. Electronics, Vol. 45, no. 3, pp.283-288, 1978.

[10] R. Senani, "Novel Circuit Implementation of Current Conveyors Using an O.A. and an O.T.A.," Electronics Letters, Vol. 16, pp. 2-3, Jan. 1980.

[11] J. L. Huertas, "Circuit Implementation of Current Conveyor," Electronics Letters, Vol. 16, pp. 225-226, 1980.

[12] U. Kumar, "Current Conveyors: A Review of the State of the Art, IEEE Circuits and Systems Mag., Vol.3, pp. 10-14, 1981.

[13] B. Wilson, "Low Distortion Feedback Voltage-Current Conversion Technique," Electronics Letters, Vol.17, pp.157-159, Feb.1981.

[14] A. Fabre, "Dual Translinear Voltage/Current Convertor," Electronics Letters, Vol. 19, pp. 1030-1031, Nov. 1983.

[15] A. Fabre, "Wideband Translinear Current Convertor," Electronics Letters, Vol. 20, pp. 241-242, March 1984.

[16] A. Fabre and P. Rochegude, "Ultra-Low-Distortion Current- Conversion Technique," Electronics Letters, Vol. 20, pp. 674- 676, Aug. 1984.

[17] B. Wilson, "High-Performance Current Conveyor Implementation," Electronics Letters, Vol. 20, pp. 990-991, Nov. 1984.

[18] C. Toumazou and F. J. Lidgey, "Accurate Current Follower," Electronics and Wireless World, Vol. 91, no. 1590, pp. 17-19, April 1985.

[19] C. Toumazou and F. J. Lidgey "Floating Impedance Convertors Using Current Conveyors," Electronics Letters, Vol. 21, pp. 640-642, July 1985.

[20] B.Wilson, "Floating FDNR Employing a New CCII Conveyor Implementation,"Electronics Letters, Vol. 21, pp. 996-997, Oct. 1985.

[21] B. Wilson, "Using Current Conveyors," Electronics and Wireless World, Vol. 92, pp. 28-32, April 1986.

[22] C. Toumazou and F. J. Lidgey, "Universal Active Filter using Current Conveyors," Electronics Letters, Vol. 22, pp. 662-664, June 1986.

[23] C. Toumazou and F. J. Lidgey, "Recent Developments in Current Conveyors," IEE Colloquium on Amplifiers, Oct. 1986

[24] C. Toumazou and F. J. Lidgey, "Wide-band Precision Rectification," IEE Proc. G, Electron. Circ. & Syst., Vol.134, pp. 7-15, Feb. 1987.

[25] R. W. J. Barker and B. L. Hart, "A Novel Integrable Voltage- Current Converter," IEEE J. Solid-State Circuits, Vol. SC-22, pp. 109-111, Feb. 1987

[26] C. Toumazou and F. J. Lidgey, "Fast Current-Mode Precision Rectifier," Electronics and Wireless World, Vol. 93, no. 1621, pp. 1115-1118, November 1987.

[27] C. Toumazou, F. J. Lidgey, and P.K Cheung, "Current Mode Signal Processing Circuits: A Review of Recent Developments ", Proc. IEEE Int. Symp. on Circ. and Syst., pp.1572-1579, May 1989.

[28] C. Toumazou and F. J. Lidgey, "Novel Current-Mode Instrumentation Amplifier," Electronics Letters, Vol. 25, pp. 228-230, Feb. 1989.

[29] B. Wilson, "Universal Conveyor Instrumentation Amplifier," Electronics Letters, Vol. 25, pp.470-471, March 1989.

[30] F J. Lidgey, C. Toumazou, and C. Makris, "High CMRR, No Matching," Electronics and Wireless World, Vol. 95, pp. 344- 345, April 1989.

[31] G. W. Roberts and A. S. Sedra, "All Current-Mode Frequency Selective Circuits," Electronics Letters, Vol. 25, no. 12, pp. 759-761, June 1989.

[32] C. Toumazou, F. J. Lidgey, and M. Yang, "Translinear Class AB Current Amplifier," Electronics Letters, Vol. 25, no. 13, pp. 873-874, June 1989.

[33] D. C. Wadsworth, "Accurate Current Conveyor Integrated Circuit," Electronics Letters, Vol. 25, no. 18, pp. 1251- 1253, Aug. 1989.

[34] D. C. Wadsworth, "A Professional Audio Integrated Circuit," presented at the Audio Engineering Society 87th Convention, New York, Oct. 1989.

[35] B. Wilson, "Performance Analysis of Current Conveyors," Electronics Letters, Vol. 25, no. 23, pp. 1596-1598, Nov. 1989.

[36] D. C. Wadsworth, "Accurate Current Conveyor Topology and Monolithic Implementation," IEE Proceedings (part G), Special Issue on Current-Mode Analogue Signal Processing, April 1990.

[37] D. C. Wadsworth, "The Performance and Applications of a New Current Conveyor Integrated Circuit," presented at the IEEE 1990 ISCAS conference, New Orleans, May 1990.

Chapter 16

Applying "Current Feedback" to Voltage Amplifiers

Derek F. Bowers

16.1 Introduction

The term "current feedback" when applied to voltage amplifiers is a source of considerable confusion and misunderstanding. When applied to operational amplifiers, the term is commonly used to describe a method of feedback relying on an inherently low impedance inverting input [1]. When applied to instrumentation amplifiers, the term has been utilised (by myself, unfortunately) to denote a differential feedback system relying upon the use of a linearised voltage to current converter as the feedback element [2]. Furthermore, the basic techniques described in reference [1] have their origins in instrumentation amplifier design [3,4,9]. The term "active feedback" has also been carelessly applied, and has frequently been perpetuated in the description of any feedback system using other than passive elements in a feedback loop. However, the term "active feedback" seems to have been initially used to describe the use of a voltage to current converter as an overall feedback element [5]. I have to admit that the latter usage is a useful and (potentially) non-ambiguous description for such a type of feedback, and from here on the term "current feedback" will be applied only in the foremost context.

Current and active feedback techniques are applicable to a wide range of analog functions, including nonlinear components (Chapter 2,12) and digital-to-analog converters (Chapter 13): however, for the purposes of definition and illustration, I believe it is sufficient to restrict the discussion to operational and instrumentation amplifiers. Unfortunately, the term "instrumentation amplifier" is also somewhat misunderstood and it is necessary to clarify the terminology used in the remainder of this Chapter. Please note that I am not attempting to propose a universal classification for amplifiers or feedback techniques (although there is a definite need for this), but rather to maintain consistency in my own analyses.

16.2 The Operational Amplifier

The term "operational amplifier" was first used in a paper by Ragazzini et al in May 1947 [6], though op-amps were (at least) in some use in the Second World War years; notably by Lovell and Parkinson in the development of the Western Electric (Mk.9) electronic anti-aircraft gun director. The classic

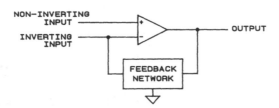

Figure 16.1 **The op-amp has evolved from an inverting-only to a fully-differential gain block.**

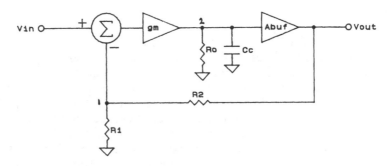

Figure 16.2 **analytical model for the conventional op-amp.**

op-amp (linear) feedback equations which have moulded the basis for much everyday analogue system design were both formulated and elaborated in the Ragazzini paper. Early op-amps were designed solely for analogue computing applications and were almost always used with the non-inverting input at ground potential. The modern op-amp is better regarded as a gain block with near infinite gain for voltages appearing between two high-impedance inputs [7] (see Figure 16.1).

Assuming linear operation, feedback from the output to the inverting input forces the latter to follow the non-inverting input closely, establishing a 'virtual low-impedance' at the inverting input. This steady-state condition holds as long as the amplifier gain is high enough. In a practical op-amp, the gain is a (mathematically) complex function and therefore exact analysis of AC and transient behaviour becomes very difficult. In amplifiers designed for high speed, of course, it is necessary to have an understanding of these effects, and hopefully this knowledge can be used to optimise the AC characteristics of a given amplifier.

A voltage-feedback operational amplifier can be modelled by the network shown in Figure 16.2. It contains a differential to single-ended converter, a transconductance amplifier, an RC compensation network and a unity gain output buffer. The resistor R_0 shown in the model is actually the effective parallel resistance (seen at the output of the transconductance stage) of all active or passive elements connected to that particular node. The dominant pole of the amplifier is set by the R_o-C_c time constant and the DC open-loop voltage gain is set by the product of R_o, g_m and A_{buf}, where A_{buf} is the gain of the output buffer and g_m is the transconductance of the transconductance stage. Feedback is applied around the loop, from the output to the inverting input, through the voltage divider formed by resistors R_1 and R_2. This leads to an overall transfer function described by:-

$$V_o = \left(V_{in} - \left(\frac{R_1}{R_1 + R_2} \right) V_o \right) \left(\frac{g_m R_o A_{buf}}{1 + s R_o C_c} \right)$$

(16.1)

Thus the overall closed-loop voltage gain over frequency can be derived as:-

$$\frac{V_o}{V_{in}} = \frac{\left(\dfrac{g_m R_o A_{buf}}{1 + s R_o C_c} \right)}{1 + \left(\dfrac{R_1}{R_1 + R_2} \right)\left(\dfrac{g_m R_o A_{buf}}{1 + s R_o C_c} \right)}$$

(16.2)

Some rearrangement leads to the following expression:-

$$\frac{V_o}{V_{in}} = \frac{1 + \dfrac{R_2}{R_1}}{\left(1 + \dfrac{(R_1 + R_2)/R_1}{g_m R_o A_{buf}} \right)\left(1 + s \left(\dfrac{R_o C_c}{1 + \dfrac{g_m R_o A_{buf}}{(R_1 + R_2)/R_1}} \right) \right)}$$

(16.3)

Thus, the circuit has a closed-loop gain approaching $1 + R_2/R_1$ (assuming that the product of g_m, R_o and A_{buf} is reasonably large) and a closed-loop pole of:-

$$f_{pole} = \frac{1 + g_m R_o A_{buf} \left(\dfrac{R_1}{R_1 + R_2} \right)}{2\pi R_o C_c}$$

(16.4)

Therefore, the closed-loop pole frequency is actually equal to the dominant open-loop amplifier pole multiplied by the loop-gain of the circuit, plus one. As the closed-loop gain is increased, the loop gain drops in inverse proportion and so does the closed-loop frequency. The concept of a finite gain-bandwidth product, almost ubiquitously perpetuated by op-amp manufacturers, can therefore be expressed as a simple equation if R_1 is set to infinity:-

$$ GBW \cong \frac{g_m A_{hf}}{2\pi C_c} $$

(16.5)

Assuming no relevant higher order poles (or any zeroes), this will, theoretically, remain constant for all closed-loop gains and will also be equal to the bandwidth of the amplifier when set to a gain of unity. In practice, few amplifiers can afford the luxury of the overcompensation necessary to guarantee even close to this ideal behaviour, and anomalies will exist causing the gain-bandwidth product to be an approximation at best. Nevertheless, the concept is characteristic, even if not totally descriptive, of voltage-feedback op-amps. This fundamental characteristic poses a limitation when attempting to achieve a high closed-loop gain and wide bandwidth simultaneously.

An additional limitation of voltage feedback amplifiers is that the slew rate [8] is restricted by the maximum output current from the input transconductance stage (usually equal to the 'tail' current of a differential input transistor pair) available to charge the compensation capacitor. This results in an absolute limitation on (what is commonly referred to) as 'large-signal bandwidth'. The large-signal bandwidth is somewhat loosely defined as the maximum frequency at which a given amplifier can deliver an output waveform of a given amplitude at its output when driven by a sinusoidal waveform. The classic equation relating slew-rate to large signal bandwidth is based upon equating the slew-rate to the maximum rate of change of such a sine waveform (which occurs at the zero crossing) and is given by:-

$$ FPBW = \frac{SR}{2\pi V_{peak}} $$

(16.6)

While it is certainly true that this is a limitation, there is actually no overall feedback whatsoever when the amplifier is at its slew-rate limit. This causes the full (open-loop) nonlinearity of the input stage to be present at this point. Therefore, the output can never be perfectly sinusoidal unless the input stage is perfectly linear (or the frequency is zero!). In practice, serious distortion will occur long before the large-signal bandwidth is exceeded and slew-rate limiting thus has a more calamitous impact on signal fidelity than might at first be assumed. Incidentally, the peak output value used for the large-signal bandwidth calculation is fairly well standardised as 10V for amplifiers capable of operating from symmetric 15V supply rails.

16.3   The Instrumentation Amplifier

The term "instrumentation amplifier" seems to have emerged in the late 1960's [9], although such amplifiers were earlier referred to as "data amplifiers"[10,11].   Aside from op-amps that have been called "instrumentation operational amplifiers" (because their DC characteristics were suitable for precision instrumentation applications) the term has generally denoted an amplifier performing differential to single-ended voltage conversion with a precisely defined gain. The 'precisely defined' gain is almost always set by a ratio of two resistors, and in the case of an integrated unit, one, both or neither resistor may be external. In the case of both being integrated, the gain is either fixed at a particular value, selectable by pin strapping, or selectable by a digital input word.   Therefore, the conceptual model for an instrumentation amplifier is actually simpler than for an operational amplifier, since the feedback conditions are well defined.

Unfortunately, attempting a more in-depth analysis of the instrumentation amplifier is somewhat perplexing owing to the plethora of different topologies involved.   Despite several grandiloquent claims, no single instrumentation amplifier topology has been universally successful and so some background information is necessary at this point.

Briefly, instrumentation amplifiers fall into two categories: those constructed using op-amps and resistors and those making more use of intrinsic transistor properties. The former class is more suited to discrete designs while the latter is more suitable for integrated implementation, though there are many exceptions to both these rules. The 'classic' op-amp implementation is shown in Figure 16.3.

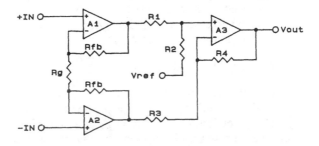

Figure 16.3   Triple op-amp instrumentation amplifier.

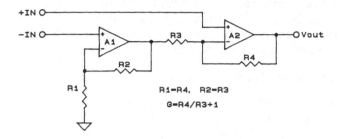

Figure 16.4   Dual op-amp instrumentation amplifier.

Essentially, it consists of a differential amplifier formed by A3 and R_1-R_4 (usually set to unity gain) preceeded by a differential input amplifier formed by the remaining components. V_{ref} is a voltage to which the output is referred, and is usually connected to ground. Assuming R_1-R_4 are all equal, the overall transfer function will be given by:-

$$\frac{V_o}{V_{ip} - V_{im}} = \left(1 + \frac{2R_{fb}}{R_g}\right)$$

(16.7)

Note that this is easily adjustable by means of one resistor, R_g. The DC errors of this configuration are mainly those of the op-amps themselves with one big exception: common mode rejection ratio. This term depends critically on the matching of R_1-R_4 and in unity gain matching to 0.01% is needed to obtain 80dB of CMRR. This is generally accomplished by trimming. AC errors are also a function of the op-amps used and the finite gain-bandwidth product of A1 and A2 cause the overall bandwidth to be inversely proportional to gain.

There are numerous other op-amp configurations, a two op-amp version being shown in Figure 16.4. They all have their relative advantages and disadvantages, but they all suffer from CMRR degradation with poor resistor matching and bandwidth reduction with increasing gain.

16.4   The 'Current Feedback' Operational Amplifier

Current feedback operational amplifiers were introduced primarily to overcome the bandwidth variation, inversely proportional to closed-loop gain, exhibited by voltage feedback amplifiers. In practice, current feedback op-amps have a relatively constant closed-loop bandwidth at low gains, and behave like voltage feedback amplifiers at high gains when a constant gain-bandwidth product eventually results. Another feature of the current feedback amplifier is the theoretical elimination of slew-rate limiting. In

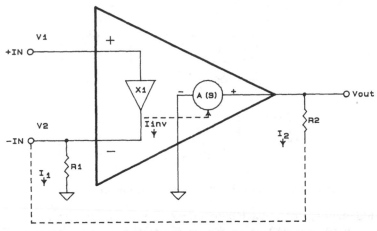

Figure 16.5 Simplified current feedback op-amp model

practice, component limitations do result in a maximum slew-rate, but this is usually very much higher (for a given bandwidth) than with voltage feedback amplifiers. The current feedback concept is illustrated in Figure 16.5.

The input stage is now a unity-gain buffer forcing the inverting input to follow the non-inverting input. Thus, unlike a conventional op-amp the latter input is at an inherently low (ideally zero) impedance.

Feedback is thus always treated as a current and because of the low impedance inverting terminal output, R_2 is always present, even at unity gain. Voltage imbalances at the inputs thus cause current to flow into or out of the inverting input buffer. These currents are sensed internally and transformed into an output voltage. The transfer function of this transimpedance amplifier is $A(s)$; the units are in ohms.

It can be shown that if $A(s)$ is high enough (like the open loop gain of a conventional op-amp) that very little current flows in the inverting input at balance. The overall closed loop transfer function now becomes:-

$$\frac{V_o}{V_{in}} = \left(1 + \frac{R_2}{R_1}\right)$$

$$(16.8)$$

which is the same as a conventional op-amp.

However, if the dominant pole is created by feeding the current imbalances into the compensation capacitor, the time constant will be set by the product of this capacitor (C_c) and the feedback resistor R_2. The closed loop bandwidth is now given by:-

$$BW = \frac{1}{2\pi R_2 C_c}$$

$$(16.9)$$

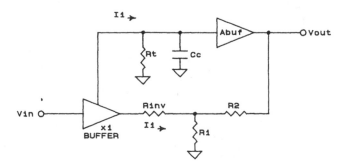

Figure 16.6   Analytical model for the current feedback op-amp.

and is independent of closed loop gain. A more complete mathematical analysis can be obtained using the representative model shown in Figure 16.6.   Notice that the input buffer has now been given a finite output impedance (R_{inv}) to model practically realisable buffers. The error current from the buffer is mirrored and fed into a transimpedance stage consisting of R_t and C_c, where current to voltage conversion takes place. The voltage generated here is buffered by another unity-gain stage and fed to the main amplifier output. Because the value of the small-signal transresistance, R_t, is very high (often in the MΩ range), only minute error currents are needed to change the voltage at node 2 by several volts. Consequently, the amount of current that must flow into or out of the inverting input terminal under steady state conditions is extremely small. The feedback network, even though it may be formed from quite low value resistors, therefore presents a very light effective load on the output of the input buffer. To derive an overall transfer function, Kirchoff's current law can be used at nodes 1 and 2. At node 1 we have:-

$$\frac{V_1}{R_1} + \frac{V_1 - V_o}{R_2} + \frac{V_1 - V_{in}}{R_{inv}} = 0$$

(16.10)

which leads to:-

$$V_1 = \frac{\dfrac{R_2}{R_{inv}} V_{in} + V_o}{1 + \dfrac{R_2}{R_1} + \dfrac{R_2}{R_{inv}}}$$

(16.11)

Doing a similar analysis at node 2 yields:-

$$V_2 = \frac{I_1 R_t}{1 + sR_t C_c}$$

$$(16.12)$$

Now, some expression is needed to relate the output voltage to the voltages at nodes 1 and 2. This is a simple task, since:-

$$I_1 = \frac{V_{in} - V_1}{R_{in}} = V_1 \left(\frac{1}{R_1} + \frac{1}{R_2} \right) - \frac{V_o}{R_2} \quad \text{and} \quad V_o = A_{bf} V_2$$

$$(16.13)$$

Thus, after the appropriate substitutions are performed to eliminate V1 and V2, we have:-

$$V_o = \left(\left(\left(\frac{\frac{R_2}{R_{in}} V_{in} + V_o}{1 + \frac{R_2}{R_1} + \frac{R_2}{R_{in}}} \right) \left(\frac{1}{R_1} + \frac{1}{R_2} \right) - \frac{V_o}{R_2} \right) \left(\frac{R_t A_{bf}}{1 + sR_t C_c} \right) \right)$$

$$(16.14)$$

It now remains to rearrange the last expression to obtain an explicit relationship between V_o and V_{in}. This yields the following result:-

$$\frac{V_o}{V_{in}} = \frac{1 + \frac{R_2}{R_1}}{1 + \frac{R_2 + \left(1 + \frac{R_2}{R_1} \right) R_{in}}{R_t A_{bf}} + s \frac{\left(R_2 + \left(1 + \frac{R_2}{R_1} \right) R_{in} \right) C_c}{A_{bf}}}$$

$$(16.15)$$

It is somewhat more instructive to perform further rearrangement in order to make this expression similar to that of the voltage feedback amplifier given in equation (16.). This leads to:-

$$\frac{V_o}{V_{in}} = \frac{1 + \frac{R_2}{R_1}}{\left(1 + \frac{R_2 + \left(1 + \frac{R_2}{R_1}\right)R_{inv}}{R_t A_{buf}}\right)\left(1 + s\left[\frac{\left(R_2 + \left(1 + \frac{R_2}{R_1}\right)R_{inv}\right)C_c}{A_{buf} + \frac{R_2 + \left(1 + \frac{R_2}{R_1}\right)R_{inv}}{R_t}}\right]\right)}$$

$$(16.16)$$

It can thus be seen that the DC term in the denominator approaches unity if the product of R_t and A_{buf} is made arbitrarily large. This is basically the same as saying that the closed loop gain approaches $1+R_2/R_1$ as the transconductance approaches infinity, which is the same result as a voltage feedback amplifier with large open-loop gain. The AC behaviour would appear to be somewhat less intuitive, however if the product of R_t and A_{buf} is large enough, the closed-loop pole frequency can be closely approximated by:-

$$f_{pole} \cong \frac{A_{buf}}{2\pi\left[R_2 + \left(1 + \frac{R_2}{R_1}\right)R_{inv}\right]C_c}$$

$$(16.17)$$

This result is indeed very different from the constant gain-bandwidth product of a voltage feedback amplifier. At low gains, when $R_2 \gg R_1$, and assuming $R_{buf} \ll R_2$ (which is always true in practice), the closed-loop pole frequency is predominantly dictated by the compensation capacitor and the feedback resistor, R_2, and is substantially independent of the exact gain setting. Thus the choice of feedback resistor is of somewhat more importance than in the case of voltage feedback amplifiers, and most manufacturers will state a suggested minimum to avoid oscillation problems.

At high gains (in practice 50 or more would be considered 'high') the R_{inv} term becomes dominant and the amplifier asymptotically assumes a constant gain-bandwidth product given by:-

$$GBW = \frac{A_{buf}}{2\pi R_{inv} C_c}$$

$$(16.18)$$

The gain of the output buffer, A_{buf}, also plays its part in determining the closed-loop pole frequency. As the main amplifier output is loaded, the gain drops below unity and causes a reduction in closed-loop bandwidth predicted by equation (16.17). This also tends to make practical amplifiers more stable when heavily loaded. With a constant load, the bandwidth reduction can be compensated by reducing the value of the feedback resistor.

This theory, although useful, is in practice (as always) compounded by second-order poles and zeroes, stray capacitive and parasitic inductive effects. Such limitations principally manifest themselves as peaking in the small signal response, overshoot and ringing in the transient response, and sensitivity of overall AC behaviour to loading conditions.

Other topologies of "constant bandwidth" instrumentation amplifiers, using current-conveyors, can be found in Chapter 4.

16.5 Large Signal Effects on the Current Feedback Op-Amp

As mentioned earlier, if a limit exists on the output rate-of-change of an op-amp (slew-rate), a large signal transient can cause the feedback loop to be temporarily broken. This results in severe nonlinear distortion. Such a limitation almost always occurs because of a restriction on the amount of current available to charge the compensation capacitor. Referring to Figure 16.6, the peak current flowing into C_C will be the peak current flowing into the buffer output. This in turn is theoretically the difference between the input and output voltage divided by the Thevenin equivalent of R_{inv}, R_1 and R_2. Since this is always proportional to the error voltage, the simplified model predicts no theoretical slew-rate limiting and a large-signal bandwidth equal to the small-signal one. This seems somewhat obvious considering that the model is totally linear, but the point to be made here is that such a model is in practice a much better approximation to a current feedback amplifier than is the case with the vast majority of voltage feedback amplifiers.

Practical current feedback op-amps do exhibit some slew-rate limiting. This can be due to a multiplicity of factors including slew-rate limiting of the input buffer, high current rolloff of gain in the buffer and current mirrors, and slew-rate limiting in the output stage. Even so, the slew-rate of available current feedback op-amps is often two orders of magnitude higher than voltage feedback amplifiers with comparable quiescent current and bandwidth.

16.6 Drawbacks of the Current Feedback Op-Amp

Theoretically, there is no fundamental reason why a current feedback op-amp, like a voltage feedback type, cannot be perfectly balanced and free from DC errors. A glance at the simplified schematic of almost any conventional op-amp, however, reveals an inherent input stage symmetry which is not present in the current feedback type with its need for a buffer between the two inputs. Furthermore, techniques to attain high speed (notably the use of

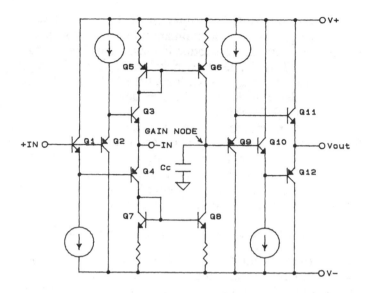

Figure 16.7 Basic design for a current feedback op-amp.

only one gain stage and an open-loop input buffer) have a deleterious effect on overall DC precision. Therefore, conventional op-amps generally have considerably higher precision than current feedback types, although it is not uncommon for current feedback types to outperform similar speed conventional amplifiers in high gain applications. This is because conventional high-speed op-amps usually have heavily degenerated input stages whose errors show up at the higher closed-loop gains.

The major areas where current feedback op-amps are deficient are common mode rejection, inverting input bias current and open-loop gain error (or transimpedance error, which amounts to the same thing in a given closed-loop gain environment). The reasons for this will be explored in the next section.

As far as AC performance is concerned, the bandwidth dependence on the value of the feedback resistor raises some interesting application problems. This is in fact a disguised form of external compensation and several manufacturers characterise their components using different values of feedback resistor for different closed-loop gains to obtain an optimum bandwidth. Reactive feedback, though, becomes an even bigger nightmare and direct connection of a large capacitor from the inverting input to the output (as in a conventional integrator or low pass filter) is almost guaranteed to produce oscillation. Such configurations can be stabilised by including an extra resistor in series with the inverting input, but at that point most of the advantages of the current feedback amplifier have been lost.

16.7 Design Techniques for Current Feedback Op-Amps

Most discussions of the design constraints concerning current feedback op-amps start with a configuration similar to that presented in Figure 16.7. The feedback loop is, of course, closed by the feedback resistor (R_{fb}) connected between the output and inverting input.

Q1-Q4 form the input buffer which forces the inverting input to the potential of the non-inverting one. Any imbalances in the collector currents of Q3 and Q4 are summed at the gain node, via current mirrors Q5-Q8, causing the output buffer, Q9-Q12, to move. Negative feedback via R_{fb} corrects the imbalance thus forcing a constant error current (ideally zero) into the the inverting input. Thus with appropriate gain setting resistors this configuration behaves as the model presented in Figure 16.6. The justification for starting with this configuration is because it is inherently simple, and therefore fast. In fact, the fastest current feedback op-amps still use little more circuitry than that shown in Figure 16.7. Computer simulation and reported results indicate that a low gain bandwidth of roughly Ft/5 can be obtained from this configuration, where Ft is the cutoff frequency of the slower type of transistor (usually the PNP). The availability of high-voltage discrete PNP's wth Ft's of 1GHz or more have enabled hybrid amplifiers with over 200MHz of bandwidth to be fabricated. However, the traditional bipolar monolithic IC process with its lateral PNP Ft of 3-10MHz has been totally unsuitable for this type of amplifier.

Many companies have developed complementary bipolar process (often replacing junction isolation with dielectric isolation) to produce PNP Ft's in the 200-600MHz range for 30V processes and in the 1-5Ghz range for low voltage processes. These processes have resulted in the appearance of many monolithic designs using current feedback.

The DC characteristics of the Figure 16.7 configuration leave much to be desired, however, and considerable effort has been focused on improving these deficiencies. Although the configuration at first glance seems symmetric, most of the symmetry depends on the matching of the PNP transistors to the NPN's. Firstly, the Vbe(on) at a given collector current is different because the Gummel number for the two types of transistor is independent. Secondly, beta control is also independent and in practice the beta's can differ by a factor of three. This leads to imperfect base current error symmetry, causing a net static current at the inverting input. Starting with the first problem, the buffer offset can be greatly reduced by inserting diode-connected PNP transistors in series with the emitters of Q1 and Q3, and similarly inserting diode-connected NPN transistors in series with the emitters of Q2 and Q4. This technique does have the drawback of doubling the output impedance of the buffer, and consequently bandwidth degradation will begin at a lower gain. Another technique for reducing the offset is to make the operating currents of the input transistors proportional to the ratio of their respective saturation currents [12]. This technique requires a very

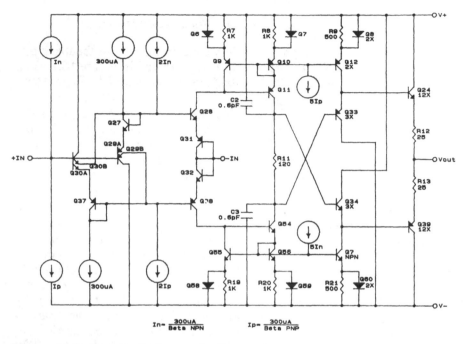

Figure 16.8 Improved current feedback op-amp design.

well controlled process to prevent the stage currents from varying widely with process fluctuations. A practical design addressing both problems is shown in Figure 16.8 [13].

Signal currents from the input stage are fed to two Wilson current mirrors (Q9-Q11 and Q54-Q56) before arriving at the gain node. These mirrors, of course, have a high output impedance which helps improve the overall transimpedance. Loading by the output stage is minimised by deriving the bias for Q33 and Q34 from the Wilson sources rather than from a separate bias line. To the extent that Early voltage is constant with varying collector-base voltage, the base currents of Q34 and Q57 (and their PNP counterparts) sum to a constant value independent of output voltage and therefore produce an offset term rather than a gain error. The offset term together with the alpha error of Q28 and Q38 is corrected by a cancellation scheme using currents which are dependent on transistor beta. The overall transimpedance is about 7 MΩ, which causes negligible error with the recommended 2.5 kΩ feedback resistor.

The input stage has the additional balancing diodes already mentioned (Q27, Q37, Q31 and Q32) to improve input offset voltage to around 2mV. Note also the addition of speed-up transistors, Q29b and Q30b. The latter are necessary to avoidslew degradation at low gains as a result of the relatively low stage currents used (this circuit is half of a dual amplifier and power dissipation was a major concern). The base current errors of Q27, Q28, Q37 and Q38 are corrected with currents from the previously mentioned cancellation scheme. This scheme is used yet again to provide bias current compensation for the non-inverting input, resulting in approximately 200nA of input current.

It has been mentioned that current feedback op-amps suffer from poor common mode rejection. This is because even with perfectly matched transistors the input buffer will only have zero offset at zero input (or more correctly the supply mid-point). The Early voltage effects of the input NPN and PNP transistors are additive, since increasing collector-base voltage on the PNP's translates to reducing voltages on the NPN's, and they have opposite contributions to offset voltage. It can be shown that the resulting CMRR will be:-

$$CMRR = 20\log\left(\frac{qV_{af}}{2kT}\right)$$

(16.19)

Where V_{af} is the transistor Early voltage. Thus for an Early voltage of 100 volts (a difficult figure to surpass for a high-speed process) CMRR will only be 65.7dB, which is very meagre compared to the 100dB or more obtainable with conventional op-amps. The obvious remedy of cascoding the input stage transistors tends to cause severe problems with input stage slew-rate, though such techniques have been applied to improve the DC precision of some amplifiers intended to work only in the inverting mode.

16.8 Practical Current Feedback Op-Amp Designs

The current feedback op-amp has been largely popularised by Comlinear Corporation in the mid 1980's, starting with a line of hybrid amplifiers. Typical of the performance obtainable in the hybrid format is the CLC205 amplifier. This amplifier can deliver bandwidths in excess of 200MHz at low gains and an impressive 80MHz at a gain of 50. Slew rate is typically 2400 V/uS which is particularly useful considering that this is an amplifier intended for operation on standard 15V supply rails. Additionally, a settling time of 22nS to 0.1% with a 10V output step can be achieved. While the DC performance of such parts does not rival conventional op-amps, the input offset voltage of 8mV maximum does not compare unfavourably to many other high speed op-amps. The common mode rejection ratio is typically 60dB and, as analysed earlier, is not a typical of this type of amplifier.

The first monolithic current feedback op-amp appears to have been the EL2020 from Elantec Inc. introduced in 1987. This amplifier is fabricated on a 30V complementary dielectrically-isolated bipolar process, similar to that pioneered by Harris Semiconductor. The EL2020 features a bandwidth of 50MHz and a slew rate of 500 V/us with a quiescent supply current of 9mA. Elantec also has very fast hybrid devices in its portfolio, and, at time of writing, some new ultra-fast monolithic amplifiers in development.

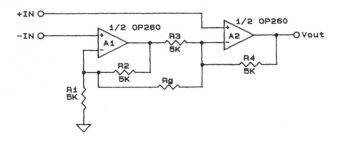

Figure 16.9 Instrumentation amplifier realisation using a matched dual current feedback op-amp.

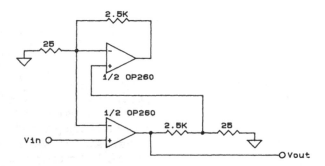

Figure 16.10 Low phase error amplifier.

Comlinear Corporation have recently introduced some monolithic current feedback op-amps built on a 12V high-speed junction-isolated complementary bipolar process, and these are intended for operation from dual 5V supply rails. The CLC400 is a good example, achieving a 200MHz bandwidth and 700 V/uS slew rate with 15mA of supply current.

Many of the DC errors of a current feedback design will cancel in differential applications provided that the amplifiers are matched. To take advantage of this fact, Precision Monolithics Inc. has introduced the monolithic dual op-amp, the OP-260, which is the production version of the circuit described in reference [13]. This device is built on a 30V complementary junction-isolated bipolar process, and features a 55MHz bandwidth and a 550 V/us slew rate with a supply current of 4.5mA per amplifier. Figure 16.9 shows the use of such a circuit to build an instrumentation amplifier with under 0.5mV of input offset voltage. The overall gain is set by resistor R_g according to the formula:-

$$\frac{V_o}{V_{in}} = \frac{10k\Omega}{R_g} + 2$$

(16.20)

By matching resistors R_1 to R_4, a CMRR in excess of 90dB (measured at 60Hz) is attainable, even though the CMRR of the individual amplifiers is only 62dB. Such techniques can be extended further to include AC, as well as DC, error correction [14]. Figure 16.10 shows an ultra-low phase error amplifier relying on the matched AC (and DC) characteristics of a dual amplifier.

Referring to the circuit, notice that each amplifier has the same feedback resistor network, corresponding to a gain of 100. Since the two matched amplifiers are set to equal gain, they will have a near identical frequency response. A pole in the feedback loop of an amplifier becomes a zero in the closed loop response. With one amplifier in the feedback loop of the other, the pole and zero are at the same frequency, thus cancelling and resulting in low phase error. Using the OP-260, the circuit of Figure 16.10 has only one degree of error at 1Mhz. For a single conventional op-amp to match this performance, it would require a gain-bandwidth product exceeding 10GHz.

16.9 'Current Feedback' Applied to Instrumentation Amplifiers.

It is a reasonable generalisation to say that current feedback applied to op-amps is advantageous from an AC standpoint and disadvantageous from a DC standpoint. Such a generalisation is not possible in the realm of instrumentation amplifiers. It is fair to say that current and active feedback both yield AC improvements, but active feedback can also benefit some DC parameters, notably common-mode rejection. The absence of any universal topology coupled with the nomenclature confusion mentioned earlier forces this discussion into an 'overview' format. At very least, the design principles are applicable to a wide range of circuitry. Instrumentation amplifiers can, of course, be made using current feedback op-amps in conventional configurations, such as that described in the previous section. These in turn can be analysed using theory already presented.

Current feedback (according to the nomenclature described in the introduction) can also be applied differentially in the case of an instrumentation amplifier, and has led to a whole class of commercially available amplifiers. This type of feedback has also been referred to as 'resistive feedback'. The second class of amplifiers to be discussed in this section involves those I have chosen to refer to as 'active feedback' types. These amplifiers rely on a differential voltage-to-current converter as the feedback element, rather than resistors or potential dividers.

Unfortunately, due to the multiplicity of different configurations in existence, it is not possible to present a very general analysis for either type of instrumentation amplifier. The approach taken here is to investigate representative topologies in a qualitative manner with some practical examples and references given for a more practical comprehension.

One implementation of the current feedback instrumentation amplifier is shown in Figure 16.11. Q1 and Q2 form a differential stage biased by I_1 and

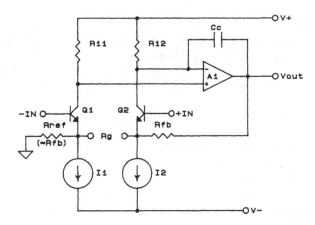

Figure 16.11   Basic current feedback instrumentation amplifier

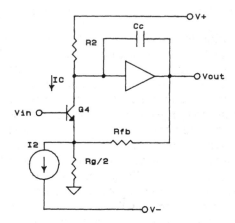

Figure 16.12   Half-circuit equivalent to Figure 16.11.

I_2. Differential inputs imbalance the symmetry of the pair due to current flow in R_{ref}, R_{fb} and R_g (if present). This asymmetry is corrected by A1 via R_{fb}, resulting in an overall output voltage given by:-

$$\frac{V_o}{V_{ip} - V_{im}} = \left(\frac{R_{ref} + R_{fb}}{R_g} + 1 \right)$$

$$(16.21)$$

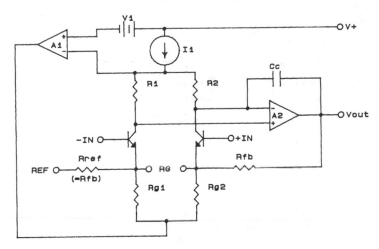

Figure 16.13 Improved topology for the current feedback instrumentation amplifier.

Common mode rejection is largely determined by the match of resistors R_{ref} and R_{fb}, and this can be trimmed by adjusting R_{ref}. From an AC standpoint, the circuit can be modelled by the half-circuit of Figure 16.12.

If Q2 is viewed as a buffer amplifier, the circuit is an exact parallel of the current feedback op-amp, except that the slew rate will be limited (in one direction at least) by I_2 and C_c. Thus, for values of R_g large compared to the output impedance of Q2 (given to a first order by kT/qI_2) the amplifier has a constant bandwidth. As R_g (or more correctly, $R_g/2$) approaches this output impedance, the bandwidth will reduce until ultimately a constant gain-bandwidth product is reached. The major drawback of this simple approach is that common mode voltage swings cause large variations in the operating currents of Q1 and Q2. This has a serious effect on AC and noise performance at different common mode input range levels. The latter drawback can be removed by adding an extra feedback loop, and one implementation of this is shown in Figure 16.13 [15].

The operating currents of Q1 and Q2 are now defined by a current source feeding their collectors rather than the emitters. The required emitter current is now provided by a nulling amplifier, A1, via resistors R_{g1} and R_{g2}. The overall gain equation now modifies to:-

$$\frac{V_o}{V_{ip} - V_{im}} = \left(\frac{R_{ref} + R_{fb}}{R_g} + \frac{R_{g1} + R_{g2}}{R_g} + 1 \right)$$

(16.22)

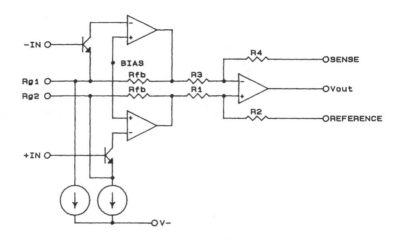

Figure 16.14 Triple amplifier type of current feedback instrumentation amplifier.

In order for A1 to provide the required emitter current there is a practical upper limit on R_{g1} and R_{g2}, and consequently this configuration is not usable below gains of about three. This configuration has the same AC analysis as the previous one, but has a slightly better noise performance due to the removal of the current sources in the transistor emitters, which are always noisier than resistors.

A version of the three op-amp instrumentation amplifier is also possible using a variation on this theme [3, 4] and is shown conceptually in Figure 16.14. The DC transfer equation is identical to the op-amp type given in equation (16.7), but the AC performance is similar to the configuration of Figure 16.11. Most of the other (advantageous and disadvantageous) features of the three op-amp type amplifier also apply to this configuration.

16.10 'Active feedback' applied to instrumentation amplifiers

The principle of active feedback can be illustrated by the circuit of Figure 16.15. The collector currents from two degenerated differential pairs (Q1-Q4) are added at the load resistors, R11 and R12. Amplifier A1 corrects for any imbalance in these currents by feedback returned to the base of Q4. If R_{g1} is equal to R_{g2} then any imbalance in Q1 and Q2 due to a differential input signal will be corrected by an identical differential voltage between the bases of Q3 and Q4. The circuit thus behaves as a unity-gain instrumentation amplifier. The important point to be made here is that common mode rejection ratio now no longer depends on resistor matching, but rather upon mismatches in the output impedances of the current sources, I_1-I_4. Using monolithic techniques, the absolute output impedance of these current sources can be made extremely high, so even relatively high mismatches have

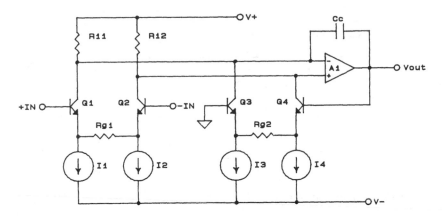

Figure 16.15 Basic active feedback concept.

minimal effect on common mode performance. This is one of the biggest reasons for using active feedback.

The simple configuration of Figure 16.15 can be made to have gain (or loss) by lowering (or raising) R_{g1}. Unfortunately, nonlinear behaviour in the transistors no longer cancels between the stages and substantial gain-error and distortion result. Gain could also be achieved by potentiometrically attenuating the feedback from A1 to Q4. This does work, but apart from exhibiting good CMRR has almost no redeeming features as an instrumentation amplifier. In particular, such an amplifier would become slow and noisy as the gain was increased, and excess DC errors due to the degeneration of the input stage would be multiplied by the overall gain.

This is a good point at which to get some confusion cleared up. Like op-amps, instrumentation amplifiers have input-referred DC error terms (such as input offset voltage) which appear at the output multiplied by the overall closed-loop gain. In addition, however, instrumentation amplifiers also have a fixed component which is present at the output regardless of gain setting.

For op-amp type configurations, the output referred error is largely caused by input offset currents dropped across the network resistors, and is usually very small. With current feedback types, mismatches in input stage operating currents are forced to flow through the feedback resistors causing much larger output referred errors. Active feedback amplifiers tend to be even worse offenders, because of the accumulated mismatch inherent to (often complicated) active V to I converters and current sources. However, the improvement in common mode rejection generally makes active feedback attractive (from a DC point at least) above gains of about 10-20. This (important) digression aside, it remains to find a good way of varying the gain of an active feedback instrumentation amplifier.

Using op-amps to correct for the transistor nonlinearities of the Figure 16.15 configuration leads to the rather cumbersome arrangement of Figure

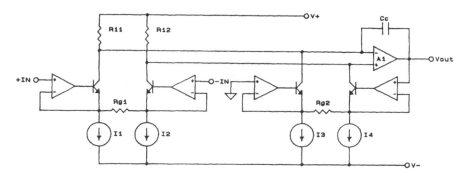

Figure 16.16   Linearised version of Figure 16.15.

16.16. Some merging of the op-amp functions is possible, and a design based on this idea has been integrated [16]. Assuming that the op-amps used for localised feedback have a bandwidth large in comparison to the overall feedback loop, this type of amplifier has a bandwidth which is theoretically totally independent of gain setting. In practice, the localised op-amps require no level shifting and can indeed be made very fast. This type of amplifier is thus limited more by DC problems than AC ones.

The main drawback here is the complexity of the input stage, which can never yield accuracy (or noise performance) close to that obtainable with a simple differential transistor pair. This is especially problematic when most of the main advantages of an active feedback amplifier are at high gains. The key to attaining a universally useful active feedback instrumentation amplifier is to find a configuration able to maintain an input stage with inherently high accuracy and low noise when used at high overall gains.

A conceptual solution, which has been realised in many different forms, is depicted in Figure 16.17. Neglecting biassing requirements, a precise and linear differential voltage to current (V to I) converter is employed as a feedback element. The V to I converter is driven directly by the nulling amplifier, A1, which serves to exactly balance the collector currents of Q1 and Q2. Since the latter transistor pair are always operating at equal currents, no nonlinearities are generated by them. A direct consequence of this is that the input differential voltage is forced across the gain setting resistor, Rg. Thus the V to I converter is supplying a differential output current equal to the differential input voltage divided by the gain setting resistor. The output voltage is thus proportional to the input voltage divided by the product of Rg and the transconductance of the V to I converter.

Assuming that the output impedence of the V to I converter can be kept high, this configuration will enjoy all the CMRR advantages of active feedback. A practical realisation of this technique is illustrated in Figure 16.18 [2 17]. The feedback V to I converter is composed of A2, A3, Q3 and Q4 with I_1 and I_2 providing emitter bias current for the input stage, Q1 and

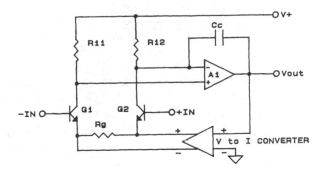

Figure 16.17 Conceptual active feedback amplifier with precision input characteristics.

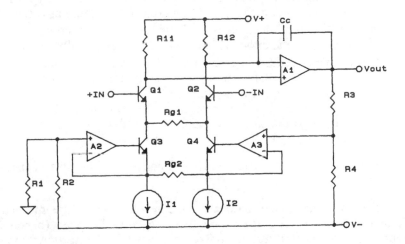

Figure 16.18 Precision active feedback instrumentation amplifier.

Q2. Unfortunately, the V to I converter cannot in this case be driven directly from the output of A1. This is because the static and also the signal dependent voltages at both inputs of the V to I converter need to allow for the full negative common mode excursion of the input stage, so some form of level shift is required. This is most easily provided by potential dividers comprising R_1-R_4 with the attenuation side referred to the negative supply rail. A2 and A3 can now be extremely simple since no level shift is required and any systematic offset will cancel between the two amplifiers. The effective transconductance of the V to I converter is the inverse of R_{g2}. The overall transfer function is now given by:-

$$\frac{V_o}{V_{ip} - V_{im}} = \left(\frac{R_4}{R_3 + R_4}\right)\left(\frac{R_{g2}}{R_{g1}}\right)$$

$$(16.23)$$

The gain can thus vary from zero to infinity (or more likely some upper practical limit) by scaling appropriate resistors, or with R_1-R_4 fixed and R_{g2} fixed, merely by adjustment of R_{g1}. Common mode rejection is once more very high with this circuit, but the fact that the V to I level shift network is returned to the negative supply will create power supply rejection (PSRR) problems if the ratio R_1/R_2 is not the same as R_3/R_4. The AC characteristics of this type of circuit are very similar to the previously discussed current feedback types. At low gains, R_{g1} is high compared to the dynamic impedance at the emitters of the input transistors, thus feedback from the V to I converter is relatively unattenuated before it reaches the nulling amplifier with its dominant pole (set by C_c or some equivalent network). At higher gains, the dynamic impedances of the input transistors combine with R_{g1} to form an AC "Π" attenutor, which attenutes the AC feedback causing the closed-loop pole to reduce in frequency. This "current feedback" type behaviour is, of course, still far superior in bandwidth performance to a conventional op-amp type implementation.

Several other techniques have been used for the realisation of the V to I converter, and although the fundamental ideas are the same, some of the performance tradeoffs may be interesting for further reading [18, 19].

16.11 Typical Current and Active Feedback Instrumentation Amplifiers

Two of the foremost companies pioneering the development of the integrated circuit instrumentation amplifier have been Analog Devices (based in Norwood, Massachusetts) and Burr-Brown (based in Tucson, Arizona). Both these companies have marketed examples of current feedback, active feedback and voltage feedback instrumentation amplifiers.

The Analog Devices AD524 is an excellent example of a monolithic current feedback instrumentation amplifier and follows the general topology of Figure 16.14, with the addition of input overvoltage protection (this amplifier is the commercial result of the paper presented in reference [4]). At unity gain, the bandwidth is 1MHz but it is still 150kHz at a gain of 100, illustrating the benefit of a current feedback topology. At a gain of 1000, the bandwidth reduces to 25kHz and this represents an essentially constant gain-bandwidth product of 25MHz at this and higher gains. Input referred offset and noise are less than 250µV and 1µV peak-to-peak respectively which is fairly typical of performance obtainable for both current feedback and active feedback instrumentation amplifiers with carefully designed bipolar input stages (voltage feedback configurations can also achieve similar specifications in this regard). Laser trimming achieves a minimum CMRR of 100dB at a gain of 100.

The INA110 from Burr-Brown uses a very similar topology but features junction field-effect-transistors in the input stage to reduce input currents to less than 100pA. The FET's also afford a bandwidth improvement, resulting in a low-gain bandwidth of 2.5MHz and a limiting gain-bandwidth product of 50MHz. The penalty to be paid for using a FET input stage is, of course, reduced input high-gain precision. The input offset voltage is about five times worse than a typical bipolar amplifier.

An amplifier using the active feedback technique of Figure 16.18 is the AMP-01 from Precision Monolithics, the outcome of the paper presented in reference [2]. This amplifier has the potential division ratio for the voltage to current converter internally fixed to 20, and for overall unity gain values of 200 kΩ and 10 kΩ are recommended for R_{g1} and R_{g2} respectively. As mentioned earlier, one advantage of this type of amplifier is that gains from zero to any practical maximum (limited by input stage precision and noise which is very similar to the AD524 discussed above) can be set by a suitable choice of these resistors. The other big advantage of active feedback is demonstrated by the 120dB minimum CMRR achieved at a gain of 100; no trimming is used to achieve this. However, at unity gain, the AMP-01 is roughly five times noisier than the AD524, most of the excess noise originating in the feedback V to I converter.

16.12 Conclusion

It should now be fairly clear that "current feedback" technology when applied to voltage amplifiers is not a sudden innovation but rather an evolutionary development of past concepts. The situation with the very high speed current feedback op-amps is such that design 'breakthroughs' are becoming less and less likely, whereas high-speed monolithic processing techniques are becoming the driving force in the quest for higher speed.

Throughout this Chapter, it should be apparent that the major obstacle preventing AC characteristics from being totally gain independent has, in theory, been the output impedance of the open-loop 'buffer' used as a feedback sensing element. The idea of using an active closed-loop buffer to ameliorate this situation is certainly not out of the question, but in the case of the current feedback op-amp the overall bandwidth of available amplifiers is already pushing the technology limitations. Inserting an additional loop within an already strained one is unlikely to result in any additional useable performance.

The case of instrumentation amplifiers, however, is totally different since they are not usually required to operate at such high speed. Indeed, the major advantage (from an AC viewpoint) that current feedback techniques have endowed upon instrumentation amplifiers is the ability to maintain a good AC performance at the very high gain settings (in excess of one thousand not being uncommon) required for precision work.

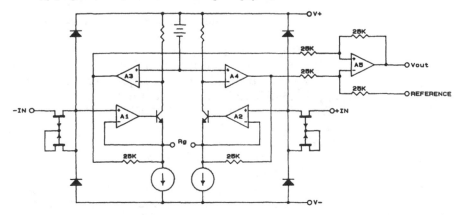

Figure 16.19   High-speed active feedback amplifier configuration.

Almost as a postscript, Figure 16.19 shows the simplified schematic for a precision instrumentation amplifier using the technique of Figure 16.14 coupled with high-speed closed loop input buffers, A1 and A2 [20]. The fully differential configuration enables simple high-speed (all NPN transistor) buffers to be used because of the inherent cancellation of systematic DC errors. This amplifier achieves a near constant bandwidth of 250kHz from low gains up to its maximum practical gain (>10,000).

To finalise, I firmly believe that all of the techniques discussed in this Chapter have actually or potentially been accessible for several decades (cathode feedback, a clear predecessor of current feedback, was used on many early vacuum tube amplifiers). It is only necessity which will drive the process technology, manufacturing and entrepreneurial infrastructure to respond to the theoretical possibilities.

16.13   Acknowledgment

I would like to acknowledge the help of my colleague, Mark Alexander, for his analytical contributions; particularly with the derivations associated with section (16.4).

16.14   References.

[1] "A New Approach to Op- Amp Design", Comlinear Corporation Application Note 300-1, March 1985.

[2] D. F. Bowers, "A Versatile Precision Instrumentation Amplifier", ESSCIRC'83 Digest of Technical Papers, September 1983.

[3] National Semiconductor LH0038 Data Sheet.

[4] S. A. Wurcer and L. counts, "A Programmable Instrumentation Amplifier for 12-Bit Resolution Systems", ISSCC Digest of Technical Papers, February 1982.

[5] B. Gilbert, "A High-Performance Monolithic Multiplier Using Active Feedback", IEEE Journal of Solid-State Circuits, volume SC-9, December 1974.

[6] J. R. Ragazzini, R. H. Randall and F. A. Russell, "Analysis of Problems in Dynamics by Electronic Circuits", Proc. I.R.E. May 1947.

[7] J. E. Solomon, "The Monolithic OP-Amp: A Tutorial Study", IEEE Journal of Solid-State Circuits, volume SC-9, December 1974.

[8] P. R. Gray and R. G. Meyer, "Recent Advances in Monolithic Operational Amplifier Design", IEEE Trans. Circuits and systems, vol CAS-21, May 1974.

[9] D. R. Breuer, "Some Techniques for Precision Monolithic Circuits Applied to an Instrumentation Amplifier", IEEE Journal of Solid-State Circuits, vol sc-3, no. 4, December 1968.

[10] J. Rose, "Straight Talk about Data Amplifiers", EDN Magazine, November 23 1966.

[11] M. H. Levin, "Advantages of Direct-Coupled Differential Data Amplifiers", Hewlett-Packard Journal, July 1967.

[12] I. A. Koullias, "A Wideband Low-Offset Current-Feedback Op-Amp Design", BCTM 89 Digest of Technical Papers, September 1989.

[13] D. F. Bowers, "A Precision Dual 'Current Feedback' Operational Amplifier", BCTM 88 Digest of Technical Papers, September 1988.

[14] J. Wong, "Active Feedback Improves Amplifier Phase Accuracy", EDN, September 1987.

[15] D. F. Bowers, "Un Ultra-Low-Noise Monolithic Microphone Preamplifier", presented at the October 1987 Audio Engineering Society Convention, preprint 2495 (K-2).

[16] R. J. Van De Plassche, "A Wide-Band Monolithic Instrumentation Amplifier", IEEE Journal of Solid-State Circuits, volume SC-10, December 1975.

[17] C. T. Nelson, "A 0.01% Linear Instrumentation Amplifier", ISSCC Digest of Technical Papers, February 1980.

[18] H. Krabbe, "A High Performancee Monolithic Instrumentation Amplifier", ISSCC Dig. Tech. Papers, February 1971.

[19] A. P. Brokaw and M. P. Timko, "An improved Monolithic Instrumentation Amplifier", IEEE Journal of Solid-State Circuits, volume SC-10, December 1975.

[20] "High Accuracy 8-Pin Instrumentation Amplifier", Precision Monolithics AMP-02 Data Sheet, August 1989.

Chapter 17

Neural Network Building Blocks
for Analog MOS VLSI

Steven Bibyk and Mohammed Ismail

17.1 Neural Networks and Analog VLSI

Artificial neural networks, or neurocomputers, provide an alternative form of computation that attempts to mimic the functionality of the human brain[1]. These networks seem to be better suited for information processing applications and tasks, such as optimization, pattern recognition and associative recall, than traditional digital computers.

Artificial neural networks have experienced significant growth in the last few years. However, only very large scale integration (VLSI) can realize the true computing potential of massively parallel neural networks. The realization of these neurocomputers, which are optimized to the computation of neural models, follows one of two approaches[2]:

1) general-purpose neurocomputers that consist of programmable processor arrays for emulating a range of neural network models, or,

2) special-purpose neurocomputers that are dedicated hardware implementations of a specific neural network model. However, any programmable neurocomputer is an order of magnitude slower than what could be achieved by directly fabricating a neural network in hardware. For this reason, far more dedicated neural hardware is being developed than programmable neural processors.

Technologies used in special purpose neural network implementation are broadly categorized into silicon[3-5], using analog or digital or mixed analog/digital integrated circuits, and optical or electro-optical[6-7].

Neural network designers could theoretically choose any neural model for implementation, although researchers favor the Kohonen[8] or the Hopfield[9] associative memory models because of their extreme simplicity. In fact, these simple models lend themselves naturally to analog electronic implementation. Furthermore, some of the traditional analog design requirements such as accurate absolute component values, device matching, precise time constants, etc. are not major concerns in neural networks. This is primarily because computation precision of individual neurons does not seem to be of paramount importance. For these reasons, analog VLSI has been identified as a major technology for future information processing and a large effort is devoted to neural network implementation in analog MOS VLSI[3,4,10].

Fortunately, the state of analog design tools has shown signs of dramatic changes. Design strategies and philosophies to bridge the gap between classical analog design and VLSI have been established. A large volume of activities on analog VLSI is currently underway[11-20]. The goal of these efforts is to develop efficient tools for synthesis at both circuit and layout levels, simulation, and testing of large-scale analog integrated circuits.

In our view, analog VLSI neural networks should make use of very simple building blocks with such features as reconfigurability, versatility and most importantly, simplicity. This results in a neural architecture that requires less design time, and makes effective use of VLSI computer-aided design (CAD) tools[17]. The more reconfigurable/versatile the analog circuit is, the more it becomes like a digital cell. We advocate the design of primitives or well-defined analog cells that are input-output compatible and can be interconnected to achieve different linear and/or nonlinear functions. The design of the analog cells needs to be done in parallel with the design of neural processing modules, and the modes of neural processing should be matched to the characteristics of the analog cells. Such analog cells[17] will ultimately bring analog VLSI design in general, and neural networks in particular, a step closer towards automation.

Although modern scaled MOS and Bi-CMOS technologies inherently possess increased high-speed analog capabilities, voltage signal handling is severely limited for analog applications. Therefore, more emphasis has recently been devoted to current-mode analog signal processing. This seems attractive for neural networks with increased complexity since many of the neural functions naturally involve currents rather than voltage. This chapter will present some interesting circuit approaches for the hardware implementation of neural network in analog MOS VLSI with emphasis on current-mode signal processing.

17.2 Neural Models

The two most popular artificial neural-network models (feedforward and feedback — see Figures 17.1 and Figure 17.2) share a neural-amplifier with an output that is a non-linear function of the adaptive weighted sum of the outputs of a number of other neurons. The D.C. output of a neuron in either network can be described by:

$$O_i = f\left(\Sigma_j(W_{ij}O_i) + \emptyset_i\right) \qquad (17.1)$$

where the function f is typically a sigmoidal non-linearity, W_{ij} is the weight from neuron j to neuron i, O_i, is the output of neuron j, and $\emptyset_i$ is the threshold of neuron i.

Neural networks "learn" by modifying the W_{ij} weights (synapses) according to some learning algorithm. These synapse weights typically must be able to take on a wide range of positive and negative values and must be alterable if the network is to learn. One of the most difficult requirements

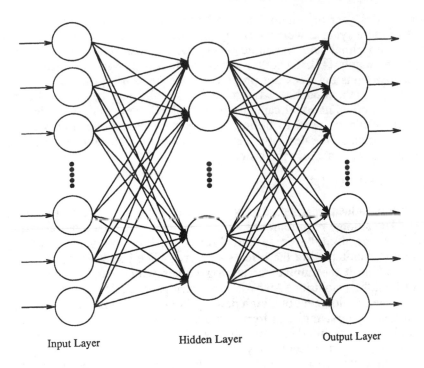

Figure 17.1. Feedforward network

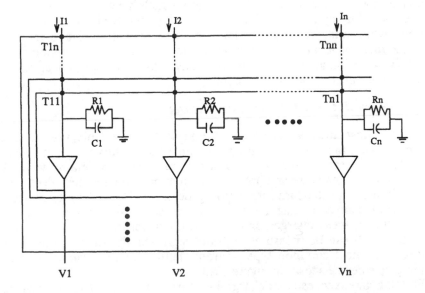

Figure 17.2. Feedback network

facing the designer of a neural network integrated circuit (NNIC) is the storage of these synapse weights. Many previous implementations of NNICs have utilized either non-adaptable [22,23] or non-continuously adaptable synaptic elements [24]. Another implementation provides continuously adaptable weights but does not provide means for storing their values on-chip [25]. These approaches, while possibly simplifying the circuit design, can limit the computational capacity of the NNIC. New integrated devices and circuits can be used to permit continuously adaptable, non-volatile synaptic elements. We describe the components of a current mode NNIC using non-volatile memory transistors.

17.3   Synaptic Weight Storage

If continuous valued weights are to be used, then there is a need to store these values. The storage requirement immediately causes a quantization effect, either by using digital storage and A/D converters, or by using analog storage and counteracting the effects of noise. The effective noise in the system will limit the number of analog values which can be stored and retrieved, ie. the resolution. An even more stringent requirement for weight storage is the development of a high density storage medium which is readily accessible in integrated circuit form. Digital storage (with D/As and A/Ds) may not have the required on chip density. Analog storage seems to require major breakthroughs in technology.

There are a variety of methods of producing analog storage. Capacitors and integrators are apparent methods, but the stored signal degrades with time too quickly. To prevent the degradation of signals, digital RAM uses a coded storage via cross coupled amplifiers which have noise margins and hence allow for signal restoration. Even in dynamic RAM, there are noise margins which allow for signal restoration. In pure analog storage, there are no noise margins and hence no possibility for restoration. An analog signal can only be maintained, with memory decay, and the design objective is to maintain the signal as long as necessary.

17.3.1 Nonvolatile Floating-gate Storage

One of the most attractive answers to the quest for an alterable, non-volatile, on-chip analog memory is the programmable-threshold voltage transistor. This device, based on the standard MOS transistor, has many desirable characteristics: small size and power consumption, slow memory decay and compatibility with standard fabrication processes. These devices typically have one or more extra gate layers which are used to store trapped charges. Once trapped, these charges produce a shift in the transistor's threshold voltage which can be measured without significantly altering the stored value. The most common type of nonvolatile memory transistor is the floating gate device shown in Figure 17.3.

Floating-gate devices store charge in a thin conducting layer sandwiched between two insulators. The original floating-gate memory device reported

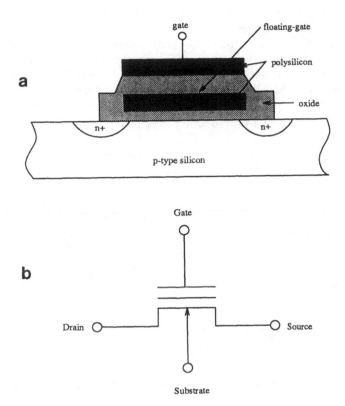

Figure 17.3. (a) Programmable floating gate transistor and (b), its symbol.

by Kahng and Sze had a gate that consisted of four layers: a tunneling oxide (< 50 Å), a floating metal gate, an insulator, and an active metal gate (MIMIS). This device required thin oxide to permit direct tunneling and was difficult to manufacture. This fabrication problem and poor retention prevented the MIMIS from being practical memory device[26].

Three FGMOS test transistors, with W/L ratios of 20/20, 40/20, and 80/20 μm, were included in the layout of our neural-network IC. The first layer of polysilicon serves as the floating-gate and has dimensions equal to that active area. The second layer of polysilicon serves as the active-gate and completely covers the floating-gate. The nominal oxide thickness between the substrate and the first layer polysilicon is 400 Å; first layer polysilicon thickness is 4000 Å; the oxide thickness between polysilicon layers is 750 Å; the second layer polysilicon thickness is 4000 Å. The devices share a common source and a common well but have separate gate and drain connections.

An HP4145B was used for measurement of threshold voltage. Threshold voltage was defined as the projected X-axis intercept of the linear portion of the $\sqrt{2I_d}$ vs. V_{gs} curve with $V_{ds} = 5$ v. $k = \mu_n C_{ox}$ W/L was taken as the square of the slope of this line.

The first series of tests utilized the HP4145B to generate positive and negative programming pulses. Unused gate and drain connections were grounded. The initial programming tests of NNIC1 used the HP4145B as a signal source to provide 20 s, 20 V gate pulses of either polarity to the test devices. Figure 17. 4 shows I_d vs. V_{ds} curves ($V_{gs} = 8$ V, 40/20 μm device) after each pulse; Table 17.1 provides a tabular listing of this data.

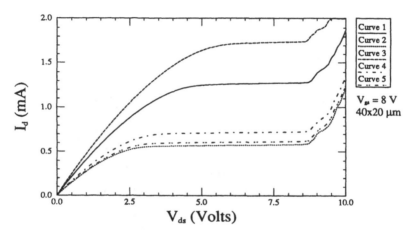

Figure 17.4. Programmed transistor curves of a 40/20μm floating gate transistor.

Table 17.1: FGMOS Threshold Shifts

Curve	Pulse Amplitude	V_t	ΔV_t
1	—	-6.00 V	—
2	-20 V	-1.36 V	7.36 V
3	+20 V	-2.53 V	5.80 V
4	-20 V	-2.53 V	5.80 V
5	-20 V	-1.70 V	0.83 V

17.4   Current Mode, Neural Network Building Blocks

Current mode signal processing offers several advantages when used in neural circuits. One of the most apparent advantages is that the summing of many signals is most readily accomplished when those signals are currents. Another advantage is the increased dynamic range of signals when MOS transistors are operated all the way from weak inversion to strong inversion. This dynamic range is especially critical for future scaled VLSI technologies

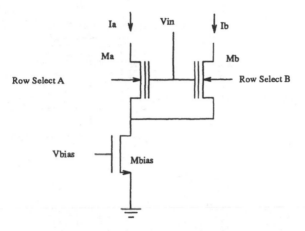

Figure 17.5. Programmable synaptic element.

which are expected to see a reduction in supply voltages. Another possible advantage is increased frequency of operation due to the use of low impedance internal nodes, minimizing capacitive charging and discharging. Finally, by representing signals as current differences at a virtual short, common mode nonlinear terms arising from nonlinear devices can be eliminated, allowing for increased linear range if needed.

17.4.1 Programmable Synaptic Element

A synapse in an NNIC must provide (a) storage of the synaptic weight (enhancing or inhibiting) and (b) an output that is a function of the weight times the input. The programmable synaptic element (PSE) used in NNIC1 provides both of these functions using only three transistors (Figure 17.5). The PSE can produce both positive and negative (enhancing and inhibiting) weights, simplifying interconnect requirements for the network. The PSE's compact design consists of a single-transistor current sink and two programmable transistors configured as a differential pair with the gates tied together. Each transistor is n-channel and is in its own p-well. In a circuit with non-programmable devices, this circuit configuration would result in the currents I_a and I_b being exactly equal (assuming perfect matching). In the PSE, however, the currents can be controlled by programming the devices with non-equal threshold voltages. The effective "weight" of the synapse is proportional to the difference in the currents I_a and I_b, which are in turn proportional to the difference in the threshold voltages of the two transistors M_a and M_b.

The differential transistors can be selectively programmed by proper biasing of the input voltage, the substrate (well) voltages, and the drain voltages of M_a and M_b. One programming technique is to provide a negative

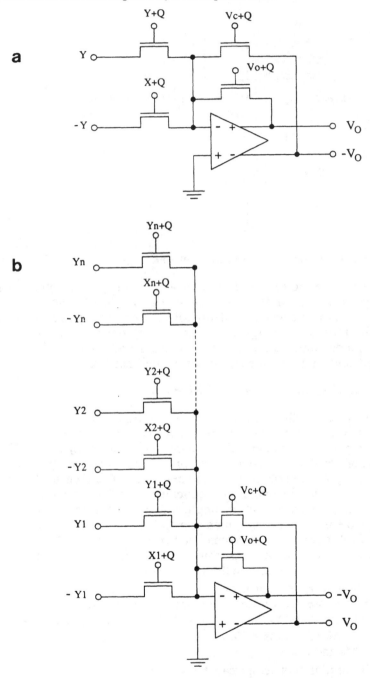

Figure 17.6. (a) A two variable multiplier and (b) its extension to a vector multiplier.

well bias equal to one half the reverse-breakdown voltage to the selected device (say M_a, with the well-bias provided by Row Select A) while grounding the substrate and the wells of M_b and M_{bias}. A positive pulse, equal to one half the reverse-breakdown voltage, is then applied to the drain of M_a, reverse biasing the drain of the selected device, causing avalanche breakdown and reverse current flow.   Charge will be trapped in M_a's floating gate via avalanche injection, resulting in a shift in the transistor's threshold voltage. The device's gate is given a positive bias (via V_{in}) to induce a positive V_t shift or a negative bias to to induce a negative V_t shift. The threshold-voltage of M_b will not be affected.

17.4.2 Inner Product Circuits.

As can be seen from Equation 17.1, a common analog computation for a neuron circuit is the inner product of two vectors, the weight vector and an input vector. A variety of circuits can be developed to form this product, with many of them using current summing. The circuits involve tradeoffs of complexity and density vs. flexibility and linearity. An example of flexibility is allowing for the use of negative weight values, and of wide dynamic range. The dynamic range can be further quantified as linear dynamic range vs. total, or nonlinear dynamic range.

A circuit that we have been developing for inner product computation, or what is referred to as vector multiplication for a set of neurons, is shown in Figure 17.6. This circuit has the main advantage of offering wide linear dynamic range, which is of importance in signal processing applications. In neural circuits, the extent of the linear dynamic range becomes a design parameter. In some cases, only a small signal range is needed, but in other cases, the wider the linear range the better. The circuit in Figure 17.6 is extremely well suited to advanced MOS VLSI technologies, since it uses only transistors. It offers the widest range of any transistor-only multiplier circuit, and has been described in detail elsewhere[27]. The linearity is developed by summing nonlinear currents in a transconductor into a virtual short. The common mode nonlinear terms are eliminated, while the difference current signals develop an inner product computation with wide linear range.

The design of the NNIC chip also included a vector multiplier, and used floating gate transistors for weight storage. The vector multiplication circuit is developed by modifying the application of the neural input as shown in Figure 17.7. This circuit forms the multiplication of Vin with the difference in threshold voltages of the differential pair. The differential pair allows for both positive and negative weight values. The linearity of this multiplier is shown via circuit simulation in Figure 17.8, for several different values of threshold voltage difference. The output of the multiplier cell is a current

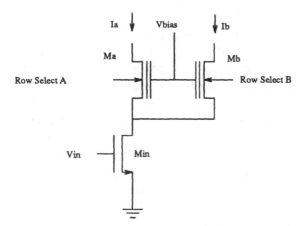

Figure 17.7. Programmable synaptic element for inner product computation.

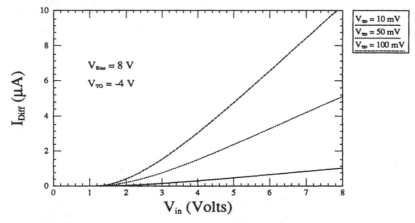

Figure 17.8. Output curves of the multiplier in figure 7.

difference, and the current differences of several cells are summed and converted to a voltage using a current thresholder. The multiplier is not as linear as the circuit in Figure 17.6, but is compact. The multiplier in Figure 17.6 could also use floating gate storage in the transconductor stage, and hence would be capable of storing weight values in the same manner as the NNIC chip.

Finally, a very compact method of multiplication for inner products uses a single transistor per cell, as shown in Figure 17.9. This inner product circuit has significant nonlinearity, which can limit the dynamic range of signal processing. Nevertheless, it has been demonstrated to be very useful for adaptive signal separation[28]. It can also use floating gate storage, but only for weight values of one sign. The limited dynamic range is due to the

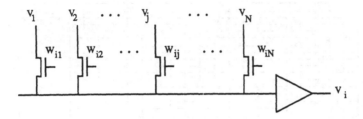

Figure 17.9. Compact inner product circuit.

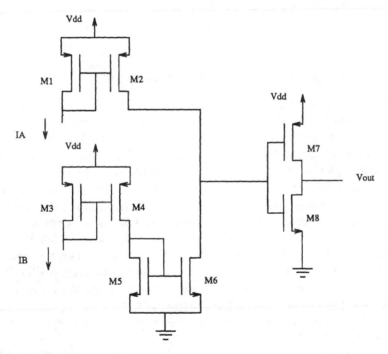

Figure 17.10. Current thresholder.

nonlinearities of the transistors as well as the need to operate the transistors in their triode regions.

17.4.3 Current Thresholding

The current thresholder shown in Figure 17.10 sums the difference currents generated by the attached PSEs and applies this sum to an amplifier with a

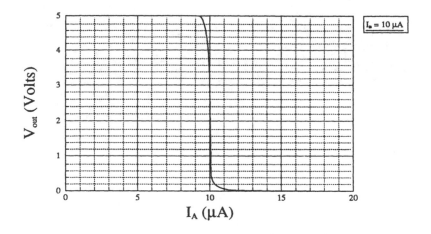

Figure 17.11. Current thresholder transfer characteristic

sigmoidal output function. The thresholder consists of two p-channel current mirrors, one n-channel current mirror and one CMOS inverter.

The current I_B (sum of all I_b currents from attached PSEs) is mirrored into the n-channel mirror formed by M5 and M6. This allows the mirrored I_A (sum of all I_a currents from attached PSEs), flowing out of M2, to be sunk by M6. If $I_A = I_B$, then M2 and M6 will attempt to source and sink, respectively, the same amount of current. Both transistors will remain saturated and their common drain voltage will be dependent on their respective transconductance parameters. If $I_A > I_B$, then M2 will try to source more current than M6 could sink. M6's drain voltage will rise to allow the current in both transistors to remain the same; V_{out} will be low. When $I_A < I_B$, a reverse situation occurs. M6 will attempt to sink more current than M2 can source; M6's drain voltage will decrease (i.e. M2's drain voltage will increase), allowing the drain currents to remain equal to each other; V_{out} will be high.

M1 and M3 are sized such that each will remain in saturation while supplying a current nI_{bias} to the PSEs, where n is the number of attached PSEs. M2, M6, M7, and M8 are sized to produce $V_{out} = \dfrac{V_{dd}}{2}$ when $I_A = I_B$ and optimized using SPICE.

The current thresholder was simulated by holding I_B constant while sweeping I_A. Figure 17.11 shows the transfer characteristic for an I_B current of 10 μA.

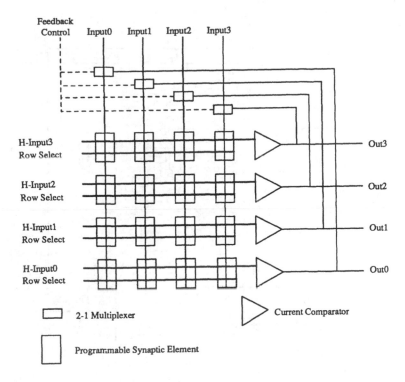

Figure 17.12. Network architecture.

17.4.4 Silicon Implementation

A four-neuron neural-network integrated circuit (NNIC1) utilizing the circuits described in the previous section was designed, fabricated, and tested. This section will discuss the architecture and design of this chip.

The architecture of the neural network (Figure 17.12) is reconfigurable such that it can be used either as a single layer in a multi-layer feed-forward network or as a Hopfield net (i.e. a single-layer network with full feedback). For an n neuron design there are n^2 synaptic connections, n feedforward inputs, $2n$ Hopfield-inputs (H-inputs), $2n$ row selects, and one feedback control line. Row selects are used for programming the synaptic elements. The feedback control line enables and disables feedback to the synapses through n 2-to-1 multiplexers. An extra column of PSEs with a separate input provides a programmable offset $\emptyset$. Figure 17.13 shows a two neuron implementation of this architecture.

NNIC1 was fabricated with the MOSIS $2\mu m$ double-poly, double-metal p-well process using a 40-pin ceramic DIP package. The limited number of available pins constrained the design to only four neurons.

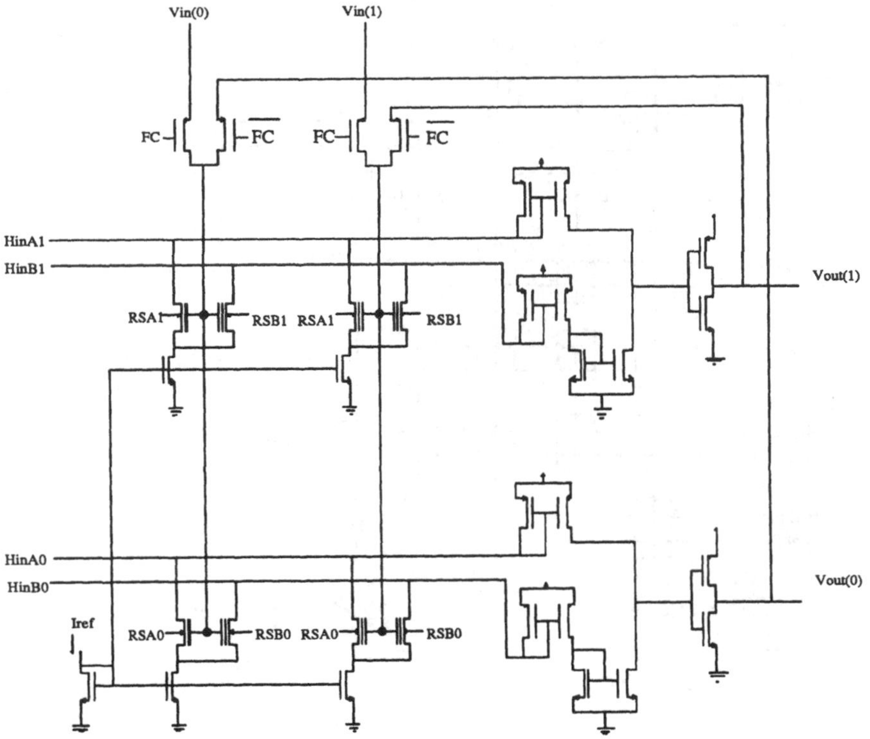

Figure 17.13. Two neuron circuit.

NNIC1 has two sets of inputs—four vertical single-ended voltage inputs (labeled **input**) and eight horizontal differential current inputs (labeled **H-input**). The inputs to the synapses are single-ended voltages and the outputs are differential currents. The differential currents of each row of synapses is summed in the current summing lines (H-inputs) and fed to the current thresholder. The current thresholder has a sigmoid-like output characteristic based on this differential current input.

17.5   Adaptive Filtering and Applications

Although neural network circuits hold great promise, the development of widely used applications has been minimal. Special purpose hardware exists to simulate many of the neural network algorithms that have been developed. Many of these algorithms are targeted for pattern recognition applications, and thus are appropriately categorized as signal processors, although searching large data bases for patterns and forming new patterns is also under the purview of artificial intelligence research.

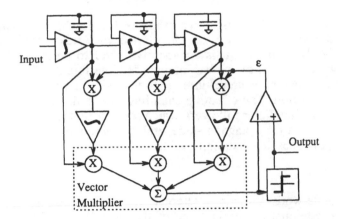

Figure 17.14. An analog implementation of Lucky's decision directed filter.

One application for neural network circuits in the AI field would be the development of realistic dataflow machine architectures[29]. Data flow machines are significantly different from conventional computer architectures, and have been more of a software pursuit than a hardware pursuit. With the potential development of effective associative memory hardware, data flow machines may be realized. Unfortunately, the use of neural network ideas for associative memories does not solve many of the difficulties with associative memories, which exist at the system level. These difficulties involve how to organize writing to the memory, how to readily expand the memory action across chip interfaces, and how to treat the case of identical data storage in different program contexts. Conventional memories store the same bit pattern at many different points in a program, and it is not clear how this style of programming should be changed when associative memories are used.

Another application of neural network circuits is in the signal processing area, and the development of VLSI hardware seems more practical in this endeavor. Many of the ideas developed in adaptive signal processing have their roots in early studies of neural systems.

Signal processing has become synonymous with discrete-time signal processing, which is usually implemented as a digital signal processing (DSP) system. In many cases, DSP systems are discrete forms of linear analog filtering. However, DSP has also evolved to incorporate a wide variety of nonlinear filtering, using prediction and estimation, which do not have obvious counterparts in the analog filtering domain. In a sense, the use of nonlinear analog neural networks to perform computations for signal processing are similar to the estimation and prediction techniques that have come to characterize certain DSP methods. The main difference with the neural circuits is that the signal processing is in continuous time, not discrete time.

An example of such a circuit is shown in Figure 17.14, which is an analog form of the decision directed filter developed by Lucky for channel equalization[30,31]. This circuit is an adaptive filter to perform channel equalization, which is used to combat intersymbol interference. This circuit is an example of using discrete-time nonlinear signal processing in a continuous time fashion. The effective sampling rate is determined by the time constant of the integrator followers which form the delay line. The time constant can be varied from stage to stage, and the input taps are an average sampled signal, as opposed to a discrete value.

17.6 Conclusions

Artificial neural networks provide an alternate form of computational signal processing that attempts to mimic some of the capabilities of the human brain. These networks seem better suited for certain information processing tasks, such as pattern recognition and optimization with limited computational hardware resources, than traditional digital computers. The utilization of nonlinearity, and continuous time vs. discrete time sampling, are used to increase signal processing power. Continuous time sampling strives to alleviate the bottlenecks of A/D conversion and sequential memory accesses in high speed, digital signal processing. Nonlinearity is still used to code signals, but memory storage in the digital sense is minimized, due to the memory I/O bottleneck and the fact that analog values cannot be stored in the same sense as digital values, since restoration is not used.

Dedicated neural hardware is being developed to achieve real-time signal processing and control. The development of continuous time neural networks requires the use of analog building blocks. These building blocks should have characteristics of reconfigurability, be input-output compatible, and be simple in nature. This allows for the better use of analog VLSI, and will bring it a step closer to automation. Some of the traditional analog design requirements, such as accurate absolute component values, device matching, precise time constants, and linearity, are relaxed somewhat for neural networks. This is primarily due to the use of nonlinearity and adaptation. Nonlinearity offers restoration of some signals, and adaptation is a form of feedback. The design requirements for neural networks in terms of precision fall in between the exacting criteria for analog design and the simplified criteria for digital design.

Current mode signal processing offers several advantages when used in neural circuits. One of the most apparent advantages is that the summing of many signals is most readily accomplished when those signals are currents. Other advantages are increased dynamic range in future VLSI technologies which are expected to see power supply reductions, high speed signalling at low impedance nodes due to minimal capacitive charging/discharging, and extending linear ranges in transistor circuits. The amount of linearity can be increased by representing signals as current differences in transistors and cancelling common mode nonlinear terms at virtual short inputs of operational transconductance amplifiers. Increased linearity is achieved using more complex cells.

Neural network circuits in signal processing applications are similar to nonlinear signal processing found in DSP. However, neural circuits need to use minimal amounts of analog storage, since analog storage is fundamentally different from digital storage. Also neural networks operate in continuous time, vs. sampled time. The design of signal processors which can take advantage of nonlinearity in a continuous time manner represents the challenge of developing neural network hardware applications. An initial design of a continuous time version of a decision directed filter, which uses a compact neural hardware structure, was presented.

17.7 Acknowledgements

The authors would like to thank the NASA Lewis Research Center, the National Science Foundation, and AT&T Technology Systems for their support. Parts of this research were supported under NASA grant NAG 3-1082 and NSF grant MIP-8896244. The advice and work provided by Ken Adkins, Tom Borgstrom, Scott Dupuie, Rich Kaul, Nabil Khachab, and Doug Yarrington is much appreciated.

17.8 References

[1] R.P. Lippman, "An Introduction to Computing with Neural Nets," *IEEE ASSP Magazine*, pp. 4-22, April 1987.

[2] P. Treleaven, M. Pacheco, and M. Vallasco, "VLSI Architecture for Neural Networks," *IEEE Micro*, pp. 8-27, December 1989.

[3] C. Mead, Analog VLSI and Neural Systems, Addison-Wesley, 1989.

[4] C. Mead and M. Ismail, Analog VLSI Implementation of Neural Systems, Kluwer Academic Publishers, 1989.

[5] D. Del Corso, K.E. Grosspietsch, and P. Treleaveng, "Silicon Neural Networks," special issue *IEEE Micro*, a collection of good papers on digital and analog artificial neural networks, December 1989.

[6] T.E. Bell, "Optical Computing: A Field in Flux," *IEEE Spectrum*, Vol. 23, No. 8, pp. 34-57, Aug. 1986.

[7] Y.S. Abu-Mostafa and D. Pslatis, "Optical Neural Computers," *Scientific American*, 256, 88-95, March 1987.

[8] T. Kohonen, "New Analog Associate Memories," International Joint Conference on Artificial Intelligence, pp. 1-8, Aug. 1973

[9] J.J. Hopfield, "Neural Networks and Physical Systems with Emergent Collective Computational Abilities," *Proc. National Academy of Science*, Vol. 79, pp. 2554-2558, 1982.

[10] S. Bibyk and M. Ismail, "Issues in Analog VLSI and MOS Techniques for Neural Computing," Ch. 5 in *Analog VLSI Implementation of Neural Systems*, Kluwer Academic Publishers, 1989.

[11] M. Ismail and J. Franka, Introduction to Analog VLSI Design Automation, Kluwer Academic Publishers, Boston, 1989.

[12] E. Habekotte, et al., "State-of-the-Art in the Analog CMOS Circuit Design," *Proc. IEEE*, Vol. 75, pp. 816-828, June 1987.

[13] Y. Tsividis, "Analog MOS Integrated Circuits: Certain New Ideas, Trends, and Obstacles," *IEEE J. Solid-State Circuits*, Vol. SC-22, pp. 317-321, June 1987.

[14] M. Ismail, "Continuous-time Analog Design for MOS VLSI," State-of-the-Art Review invited paper, *Proc. of the 30th Midwest Symp. on Circuits and Systems*, pp. 707-711, Elsevier Science Publishing Co., 1987.

[15] P.R. Gray, B. Wooley, and R.W. Broderson, Analog MOS Integrated Circuits, IEEE Press book, New York, 1989.

[16] P.E. Allen, "CAD for Analog VLSI," IEEE CAS Distinguished Lecturer Program, April 24, 1989.

[17] M. Ismail, "Reconfigurability, Versatility and Modularity in Analog IC Design," presented at the Semiconductor Research Corporation (SRC) Workshop on Analog Design Automation, December 2nd, 1988.

[18] J.L. Hilbert, SRC Private Communication, December 1988.

[19] L.R. Carley and R.A. Rutenbar, "How to Automate Analog IC Designs," *IEEE Spectrum*, pp. 26-30, August 1988.

[20] H.Y. Koh, C.H. Sequin, and P.R. Gray, "Auto Synthesis of Operational Amplifiers Based on Analytic Circuit Modes," *Proc. IEEE ICCAD*, pp. 502-505, November 1987.

[21] T. Borgstrom, M. Ismail, and S. Bibyk, "Programmable Current-Mode Neural Network for Implementation in Analog MOS VLSI," *Proc. IEE, Pt. G, Electronics Circuits and Systems*, April 1990.

[22] L.D. Jackel, H.P. Graf, and R.E. Howard, "Electronic neural network chips," *Appl. Opt.*, pp. 5077-5080, 1987.

[23] W. Hubbard, D. Schwartz, *et al*, "Electronic neural networks," in J. Denker (Ed.) *Proc. AIP Conference on Neural Networks for Computing*, pp. 227-234, 1986.

[24] M.A. Sivilotti, M.R. Emerling, and C.A. Mead, "VLSI architectures for implementation of neural networks," in J. Denker (Ed.) *Proc. AIP Conference on Neural Networks for Computing*, pp. 408-413, 1986.

[25] F.M.A. Salam, N. Khachab, M. Ismail, Y. Wang, "An analog MOS implementation of the synaptic weights for feedback neural nets," *Proc. IEEE International Symposium on Circuits and Systems*, pp. 1223-1226, 1989.

[26] J.J. Chang, "Nonvolatile Semiconductor Memory Devices," *Proceedings of the IEEE*, pp. 1039-1059, July 1976.

[27] N. Khachab and M. Ismail, "MOS Multiplier/Divider Cell for analogue VLSI," *Electronic Letters*, pp. 1550-1552, Nov. 1989.

[28] E.A. Vittoz and X. Arreguit, "CMOS Integration of Herault-Jutten Cells for Separation of Sources," in C. Mead and M. Ismail (Eds.) <u>Analog VLSI Implementation of Neural Systems</u>, chp. 3, 1989.

[29] J.B. Dennis, "Data Flow Supercomputers," *IEEE Computer*, Nov. 1980,pp. 48-56.

[30] R. Kaul, S. Bibyk, M. Ismail, and M. Andro, "Adaptive Filtering using Neural Network Integrated Circuits," in *Proc. IEEE International Symposium on Circuits and Systems*, May 1990.

[31] K. Adkins, M.J. Shalkhauser, and S. Bibyk, "Digital Compression Algorithms for HDTV transmission," in *Proc. IEEE International Symposium on Circuits and Systems*, May 1990.

Chapter 18

Future of Analogue Integrated Circuit Design

Phillip E. Allen

18.1 Present Status of Analogue Integrated Circuit Design

Analogue integrated circuit design has played an important role in the development of integrated circuit technology. As the level of integration increased in integrated circuit technology, digital circuit implementation became more desirable than analogue circuit implementation because of its robustness and simplicity of design. However, an all-digital implementation of complex integrated circuits is only found in certain types of applications such as memories and microprocessors and even in these integrated circuits, various types of analogue circuits are used. At the present time, a typical applications specific integrated circuit (ASIC) might contain 80% digital and 20% analogue circuits.

The focus of this book represents techniques for achieving higher performance in analogue circuits in the face of VLSI technologies. Current-mode analogue signal processing offers some important speed advantages over voltage signal processing. For example by nature, current is low impedance and therefore wider bandwidth. The various applications that have been discussed in the previous chapters give examples of how performance can be improved through the use of current signal processing techniques. One particularly important aspect of the current amplifier is found in the influence of feedback on the amplifier bandwidth. It has been shown that properly applied current feedback will cause the upper -3dB frequency to remain constant and independent of the closed loop gain [8]. This characteristic is extremely important in a circuit like a pipeline analogue-digital converter where the residue amplifier must have large bandwidth. A current amplifier having a gain of 2^N where N is the number of bits in the pipe would not lose bandwidth compared to a voltage amplifier. New design techniques will be developed to help improve the performance of analogue circuits as the technology places more constraints on performance. The trend will be to use digital circuits wherever possible along with highly optimized analogue circuits.

A system viewpoint of integrated circuit signal processing is shown in Figure 18.1. The key aspect of this system is the signal processing part which can be done by a digital processor or an analogue processor. Except in specific cases, digital signal processing is preferred because of reliability,

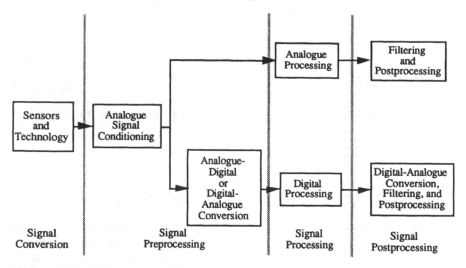

Figure 18.1   Illustration of a general signal processing system.

Table 18.1 - Comparison with analogue and digital design.

Characteristics	Analogue Design	Digital Design
Signal Amplitude	Continuum of values for amplitude (and time)	Two amplitude states
Design Methodology	Customized	Standardized
Component Values	Continuum of values	Components with fixed values
Model Requirements	Requires precise modeling capability	Only requires a precise large signal model
Programmability	Hard to change after design	Easily programmable by software
Design Level	Designed at the circuit level	Designed at the systems level
Use of CAD Tools	Difficult to use with CAD tools	Amenable to CAD tools

ease of design, programmability, and cost. However, it should be noted that analogue circuits are still employed in the general signal processor using digital signal processing.

It is important to compare analogue and digital design to understand the implications of the present status of analogue integrated circuit design. Table 18.1 compares analogue and digital design. The information found in Table 18.1 suggests that digital design is easier and cheaper than analogue design. This is confirmed in the fact that sophisticated CAD tools exist for digital circuits which permit the design at the systems level. This allows the use of integrated circuit technology by systems engineers and leads to a cycle/success ratio of near unity. In contrast, complex analogue integrated circuits require experienced circuit designers and have a cycle/success ratio that is between 2 and 3.

Why have analogue circuits continued to find use in large and very-large integrated circuits? The answer is found in the economic viewpoint. Many considerations make up the economic viewpoint. Technical considerations include such aspects as design time, probability of success, performance, and area. In general analogue integrated circuits require more design time and have a smaller probability of success compared with digital integrated circuits. Analogue circuits are only competitive with digital circuits in performance and area. Even if digital circuits could always outperform analogue circuits with smaller or equivalent area, analogue circuits would still be required. This is because the source and final destination of information is often in analogue form. Consider the case of digital audio illustrated in Figure 18.2. The source is in an analogue format and the final format of the music is also analogue. We see that conversion of analogue to digital and from digital to analogue is necessary even in an all-digital processing system.

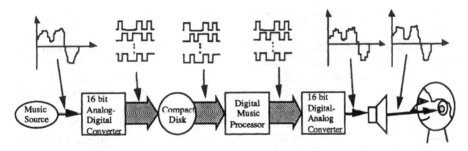

Figure 18.2   Example of a digital stereo processing system.

Analogue circuits can be divided into two classes of circuits. The first class is called the simple class and consists of analogue designs requiring a unique grouping and/or interconnection of ordinary analogue functions. The second class is called the complex class of analogue designs and consists of designs pushing the limits of precision and performance. The time to design each of these class varies widely. The simple class is suitable for design automation resulting in reduction of design time and a cycles/success ratio approaching that of digital circuits. The complex class of design can provide higher performance and less area but must be balanced against longer design time and a lower probability of cycles/success ratio.

One of the most important methods of characterizing signal processing is the bandwidth required. Figure 18.3 shows a graph of the bandwidths of a variety of signal. The bandwidths in this figure cover the enormous range of 11 orders of magnitude in frequency. At the low end are the seismic signals,

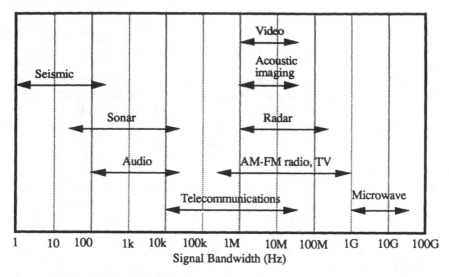

Figure 18.3 Bandwidths of signals used in signal processing applications.

which do not extend much below 1 Hz because of the absorption characteristics of the earth. At the high end are the microwave signals, which do not extend past 30 GHz because of the difficulties in performing even the simplest forms of signal processing at these frequencies. The higher the bandwidth, the larger the data rate that can be processed. Digital and analogue systems are compared in bandwidth by the sampling theorem. In theory, it is necessary to sample an analogue signal at least twice the rate of its frequency. In practice, it may be necessary to sample many times the highest frequency of the analogue signal. If the sampling rate is limited, then the equivalent bandwidth is limited by the sampling requirement. For example, suppose the maximum sampling rate is 10MHz but that at least 10 samples/cycle are necessary to preserve the integrity of the analogue signal. Thus the effective bandwidth of the digital system would be 1MHz. It is in the region from the digital signal processing bandwidths to the maximum sampling frequency that analogue finds the majority of its applications in signal processing.

The trends of analogue signal processing will be considering in the following sections. These trends will include, technology, design, simulation and modeling, design automation, and testing. The present status of these trends and their anticipated direction will be discussed. This chapter will close with an examination of the future expectations of analogue signal processing.

18.2 Technological Trends

The element of principle importance concerning analogue signal processing is the trend of technology. There are four viable integrated technologies for analogue circuits that exist today. These technologies are bipolar, CMOS, GaAs, and BiCMOS. A distinction should be made between standard integrated circuit technology and specialized integrated circuit technology. Standard IC technology as used here refers to processes which are commercially mature and are widely used. These technologies tend to limit the number of masks as much as possible. A specialized IC technology includes some aspect in the technology which is not found in the standard technologies. For example, high quality resistors or capacitors are examples. These elements are fabricated by extra processing steps. Other specialized technologies may include specially isolated devices which offer better performance at high frequencies.

Technology has always been a limiting factor in analogue circuit design. This has been caused by the fact that most technologies are optimized for digital. In the early days of IC bipolar technology, this prevented good quality passive components such as a capacitor. Consequently the analogue circuits implemented typically consisted of op amp, comparators, regulators, and similar circuits not requiring precise passive components. As different technologies became available, the analogue designer used good quality capacitors, which coupled with new design techniques, opened up new areas of integrated circuit analogue signal processing [1]. The contents of this book are a clear indication of the impact of technological advances on analogue circuit design. The "current-mode approach" is essentially a technology driven approach and many high performance circuits and systems have been developed to exploit the particular technology.

The application areas indicated for a given frequency range in Figure 18.3 can be related to the integrated circuit technology. Figure 18.4 shows a similar graph but with the technology bandwidths and applications designated. This figure shows that the highest signal processing capability is achieved by optical technology. GaAs has a bandwidth that ranges from 10GHz downward. The effective lower limit of GaAs is where silicon can out-perform it. The advantage of GaAs FETs over silicon bipolar is rapidly diminishing as bipolar silicon technologies continue to improve. Surface acoustic wave devices are implemented by charge-coupled devices fabricated in silicon and in materials such as lithium niobate. It is also seen that bipolar silicon can perform at higher frequencies than MOS silicon. This makes a strong case for BiCMOS technologies which combine the best aspects of both technologies into a single technology. Figure 18.4 also attempts to suggest a boundary between digital and analogue implementation occurring somewhere between 1 and 10 MHz. Note that analogue implementations are on the upper side of the technology bandwidth.

Bipolar technology has long been the mainstay of integrated circuit technology. Present bipolar technology has devices as small as 1 to 3 microns and uses trench isolation techniques to reduce the area required by isolation diffusions. A typical digital bipolar technology can have up to 4 levels of metal in order to simplify the routing of complex circuits. Capacitors are restricted to junction capacitors and are not of high quality for the standard

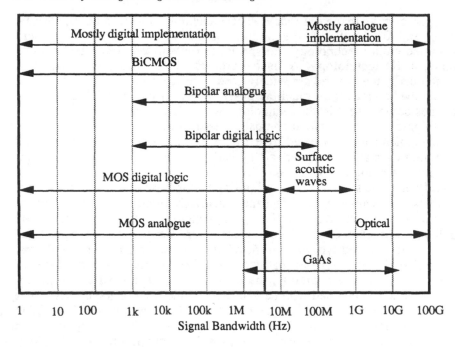

Figure 18.4   Bandwidths that can be processed by presentday technologies (1990).

bipolar process. Some of the advantages of the bipolar process are low noise, high transconductance, and fast switching. One of the important advantages of bipolar technology is that the structure and processing becomes simpler as the device is scaled down. This means that the performance of small bipolar devices will not be deteriorated compared with larger devices. This will become a significant advantage of bipolar technology as the resolution capability of lithography improves.

CMOS technology has become a dominant analogue technology primarily because of good quality capacitors and good switches. Typical CMOS technology varies from 1 to 2 microns in minimum channel length. Most CMOS technologies have several layers of metal plus one or more layers of polysilicon for interconnects. It is desirable but not crucial to have a double-polysilicon process in order to achieve high quality capacitance. The gate capacitance can be used as a high quality capacitance where necessary. Double metal capacitances using silicide dielectrics have shown good results. The advantages of CMOS technology are that it provides a good capacitor, a good switch, and dissipates low power. The disadvantages of CMOS include problems with scaling such as hot-electron effects and higher noise than bipolar (typically because of a higher 1/f corner). CMOS technology does not offer the promise of undegraded device performance as devices are made smaller.

BiCMOS technology combines both bipolar and CMOS technologies and obviously has the advantages of both. The combination of bipolar and CMOS also offer other advantages such as lightly-doped drains permitting increased breakdown voltage performance assuming punch-through is not a limit. BiCMOS offers the ability of low power dissipation using CMOS and high-speed performance using bipolar. Another advantage to the designer is the combination of bipolar and MOSFET devices to achieve performance not capable individually. For example consider a Darlington configuration. A combination of a MOSFET and bipolar will give high input resistance but large transconductance. It is expected that BiCMOS will become the dominant technology of the 1990s.

GaAs technology is quickly maturing and offers many possibilities as a niche technology. From an analogue viewpoint, GaAs, is well developed for microwave analogue but less developed for analogue signal processing [?] (Chapter 8). GaAs offers at present MESFETs but the highest performance promise is in the heterojunction bipolar transistors (HBTs). GaAs will not replace silicon, but instead will supplement it. The advances in bipolar silicon are giving GaAs MESFETs approximately a 5-to-1 speed advantage which may not be worth the extra problems in GaAs that the analogue designer must face. These problems include the lack of a good switch (the FET can become forward biased connecting the gate directly to the channel), poor models (the channel conductance becomes a function of frequency), substrate problems (the influence of devices in the proximity, the potential negative feedback of a device on itself causing the simulated gain to always be less than measured gain), high noise, and the lack of a p-channel device of equivalent performance. The advantages of GaAs MESFETs are a very high quality capacitor and resistor (even an inductor for that matter), the ability to interface with optical signals, and the technology is maturing very rapidly resulting in higher yields and more consistent device characteristics.

Figures 1.1,1.2 and 1.3 in the introductory Chapter 1 of this book describe the evolution of a new generation of analogue building blocks, circuit and sytems based upon current-mode analogue signal processing techniques. This volume has shown how many of these techniques have been employed to exploit the aforementioned technologies and obtain optimum analogue circuit performance.

18.3 Design Trends

It is clear that design techniques have a strong influence upon the performance of analogue circuits implemented in a given technology. This is particularly true for the analogue designer because the technologies have generally been optimized for digital circuits. This means that power supplies may be reduced, the transconductances may be reduced, noise is increased, and normal analogue models are not appropriate for the devices available.

A good example of how design techniques can influence the implementation of analogue circuits in an integrated technology is found in switched capacitor circuits [3]. In the early 1970s, analogue IC circuit design was restricted to requirements that did not need precise passive components. The concept of replacing a resistor by capacitors and periodically clocked switches suddenly allowed the design of circuits requiring precise time constants or ratios of passive components. Of course, new design techniques always cause the manifestation of new problems. In the case of switched capacitor circuits, there are several problems which appeared. One of these problems was the need to oversample the signal to be processed. This problem caused the digital approach to be equivalent in bandwidth to the analogue. However, the switched-capacitor resulted in much less area and power dissipation. Another problem is clock feedthough of the switches. Clever design techniques have minimized this problem but not eliminated it. A third problem that was unexpected was the simulation of such circuits. It was necessary to develop a discrete-time simulator that could predict the frequency response of switched-capacitor filters [4,5].

In Chapter 11 of this book a powerful alternative techique to switched capactors filters known as "switched current filters" is introduced that does not rely upon accurate capacitor ratios. Many other examples of clever design techniques such as dynamic current-mirrors and current-copier cells, current-mode ADCs, current-conveyors and current feedback op-amps, that improve the performance of analogue integrated circuits or result in analogue circuits compatible with an integrated technology can be found throughout this volume. In the literature, a special issue of IEE proceedings Part G (April 1990) is devoted entirely to current-mode analogue signal processing, and the December issues of the IEEE Journal of Solid-State Circuits highlight the general advances in analogue IC circuit and system design techniques.

Such advances in design techniques are a necessary part of the evolution of analogue IC design. As power supplies are reduced and smaller devices are used, the dynamic range of analogue circuits is being squeezed from both ends. In many signal processing applications, the need for high precision or large dynamic range is increasing. For example, stereo audio requires analogue-digital converters approaching 16-bit accuracy.with bandwidths up to 20-40kHz. 16-bits corresponds to a dynamic range of 96 dB which is extremely difficult to achieve even in present technology. Other problems occur when switches are used. Switches inject a noise called kT/C. This noise is inherent in any circuit which charges a capacitor through a switch. This relationship will be developed as an example of the problems facing the analogue designer. Assume that the maximum clock frequency of the switch is f_c and is given as

$$f_c = \frac{1}{t_s} = \frac{1}{10RC} \qquad (18.1)$$

where t_s is the settling time caused by the resistance of the switch (R) and the capacitor (C) being charged (or discharged). The dynamic range, DR, can be defined as the ratio of the reference voltage, V_{ref}, to the rms value of the kT/C noise and is expressed as

$$DR = \frac{V_{ref}}{Noise} = \frac{V_{ref}}{\sqrt{kT/C}} = 2^N \qquad (18.2)$$

where N is the number of bits required. Solving for C in Eq. (18.1) and substituting into Eq. (18.2) gives

$$1/2^N = \frac{\sqrt{kT}\sqrt{10f_cR}}{V_{ref}} \qquad (18.3)$$

or

$$2^N\sqrt{f_c} = \frac{V_{ref}}{\sqrt{kT10R}} \qquad (18.4)$$

Finally, taking the $\log_{10}$ of both sides of Eq. (18.4) results in

$$N\log_{10}(2) + 0.5\log_{10}(f_c) = \log_{10}(V_{ref}) - 0.5\log_{10}(kT10R) \qquad (18.5)$$

Eq. (18.5) is plotted in Figure 18.5 for several values of V_{ref} and switch resistances . It is assumed that V_{ref} is equal to or less than the power supply. Figure 18.5 shows that the fundamental limit of the precision (bits) of sampled systems reduces as the sampling rate increases and as the power supply decreases. At a sampling rate of 10MHz, a reference voltage of 5V, and a switch resistance of 10kΩ, the precision is limited to 14 bits assuming all else is ideal.

Fundamental limits such as kT/C noise will place a severe challenge on the creativity of analogue designers. New design trends which are being explored, as discussed in this book, may lead to much-needed solutions for analogue circuits to remain competitive with digital circuits. One interesting design technique is the implementation of neural network concepts in analogue circuit design. Concepts of redundancy, parallelism, and repetition should be explored as potential ways to maintain circuit performance. Neural concepts are being used via parallelism and liberal use of digital memory to accomplish higher precision analogue signal processing [6]. These techniques may result in circuits which can train themselves to meet the required precision. Unfortunately, the design effort and expertise required is high.

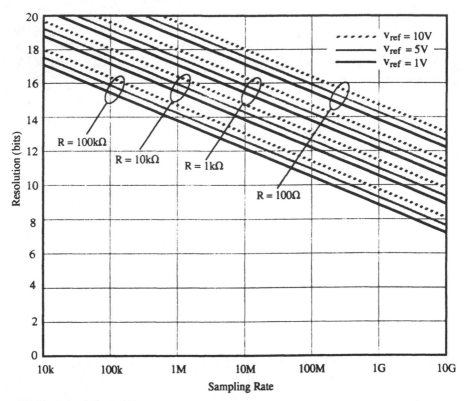

Figure 18.5   Relationship between resolution, sampling rate, and kT/C noise.

Another recent example of increasing the precision of analogue circuits is the oversampled analogue-digital converter. This example is significant in that it is an obvious tradeoff of analogue for digital in order to improve the precision. Only the sigma-delta modulator is implemented in analogue circuits. This permits the designer to enhance the performance of this simpler circuit and to use digital circuits to decimate and filter the modulator output resulting in a high-precision analogue-to-digital converter. The oversampled analogue-digital converter achieves precision by repetition or oversampling. Typical oversampling ratios are in the range of 50 to 100 depending upon the characteristics of the modulator. The oversampled analogue-digital converter is one of the highest precision analogue-digital converters techniques available today [7].

A number of analogue circuit applications employing current-mode signal processing techniques to improve precision can be found throughout this book.

Table 18.2 - Analogue circuit hierarchy from a simulation viewpoint.

ANALOGUE MODELING CATEGORY	MODELING PRIMITIVES	STRUCTURAL IMPLICATIONS
Higher Level	Linear and nonlinear mathematical equations, tables, formulas, etc.	Spatially unrelated to the original circuit, KVL and KCL need not be globally satisfied
Macromodel	Linear and nonlinear mathematical equations, tables, formulas, SPICE primitives, etc.	Spatially unrelated to the original circuit, KVL and KCL must be globally satisfied
Circuits	SPICE primitives	One-for-one relation to the actual circuit
Components	Physical equations, geometrical aspects, processing constants	Related to the physical and processing aspects of the device

18.4 Simulation and Modeling Trends

Simulation and modeling is one of the key areas determining the success of analogue integrated circuit design. If the designer is unable to satisfactorily predict the performance of analogue circuits implemented in an IC technology, clever design techniques and wide bandwidth technologies are useless. The simulation and model step of analogue design is the verification approach which used to be implemented in breadboards before the advent of integrated circuit technology. Without an adequate verification, one can only use designs which have been proven in the past providing the technology is constant.

In order to understand the simulation and modeling issues in analogue integrated circuit design, one must understand hierarchy in analogue circuits. Table 18.2 shows one possible hierarchy of analogue circuits from a simulation viewpoint. The modeling level is important in defining the simulation level of an analogue circuit. For example, a table model may contain the I-V data of the large signal model of a device, the nonlinear function of a controlled source, or the nonlinear voltage transfer function of a circuit. Thus the model mechanism can serve for different levels of hierarchy. The circuit model level is well-understood because of its universal usage in SPICE and SPICE-like simulators. The modeling elements are called SPICE primitives and include passive components, active components, and linear and nonlinear controlled sources. The macromodel

Simulation Error

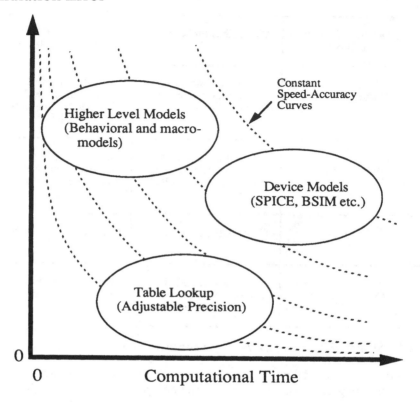

Figure 18.6 Illustration of the accuracyspeed tradeoffs for analogue models.

level has been used extensively to result in efficient models for circuits containing passive components and active components. The macromodel may not be physically related to the actual circuit but captures the terminal performance with various levels of accuracy. Higher level models are similar to macromodels but have some minor differences. One of these differences is KCL or KVL only needs to be satisfied on a local basis, i.e. at a node or loop.

One of the requirements of analogue modeling is to be able to select the desired accuracy and speed. Unfortunately, the accuracy-speed product of a model is almost constant so that a tradeoff must be made. Figure 18.6 illustrates this tradeoff and suggests that different level models can achieve different regions in the accuracy-speed characteristics. The ultimate model would perform near the origin where the error is small and the computational time is small.

Analogue modeling and simulation is facing a serious problem as technology continues to shrink. This problem is not having models with sufficient precision to undertake successful analogue design efficiently. The present trend is for the analogue devices in a VLSI circuit to have channel lengths which are 3 to 5 times larger than the minimum. The reason for this circumstance is to permit the present analogue device models to be accurate enough and to avoid the increasing noise of short-channel devices. It is necessary to develop models that will permit accurate and efficient simulation of short-channel devices for analogue applications. If the designer can use short-channel devices, higher bandwidths (and higher noise) can be obtained. The best solution to this problem is the table look-up model [9]. The table look-up model has out-performed formula-based models in both accuracy and speed. The table model requires a large amount of memory (approximately 10 KBytes for a typical MOSFET) but can be quickly updated and has the feature of permitting a tradeoff between accuracy and computational time.

The presence of both analogue and digital circuits in ASICs require a simulation capability that can be applied to both analogue and digital circuits. Three basic approaches have been taken toward mixed analogue-digital simulation [10,11]. The first method uses an analogue simulator to perform both analogue and digital simulation. The second uses a digital simulator to perform both digital and analogue simulation. The third method couples together an analogue and a digital simulator. In the first method, the digital elements are usually analyzed by the same mechanisms as the analogue ones, which results in a simulation which is too accurate and too inefficient. The second approach extends the digital (discrete) methods to analogue elements. This usually provides a simulation which is quite efficient but rather inaccurate. Only the coupled approach combines the advantages of analogue (accuracy) and digital (efficiency) simulations. However, this approach depends upon the level of coupling. Loosely coupled simulators basically execute two independent simulation programs (analogue and digital) that "communicate" whenever they need information from the other part of the circuit. Loosely coupled simulators are relatively simple to design but do not perform simulation of mixed analogue-digital circuits with much accuracy or efficiency. Tightly coupled or integrated simulators "synchronize" the two simulation mechanisms at the level of internal timesteps and time event controls so they can easily avoid any redundant evaluations without any loss of accuracy.

Any implementation of integrated mixed-mode simulation must solve two basic questions. The first is the conversion of analogue to digital and digital to analogue information at interfaces of analogue and digital components. The second is the synchronization of the (usually variable) timesteps of the analogue simulation with the event list that drives the (event-driven) digital simulation. The analogue-to-digital conversion can be handled by establishing voltage thresholds and corresponding digital signals. The digital-to-analogue conversion is more difficult because it must generate a continuous analogue waveform on the basis of discrete digital values. Two

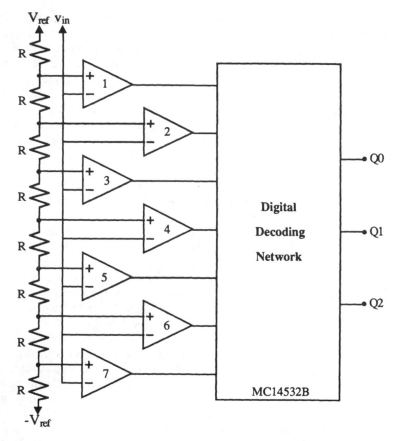

Figure 18.7   3bit flash analoguetodigital converter used as a simulation benchmark.

popular solutions assume that the converted waveforms are piecewise linear or piecewise exponential.

Several mixed analogue-digital simulators have been applied to an analogue-digital circuit in order to determine their capabilities [12]. The benchmark circuit is a 3-bit flash analogue-digital converter shown in Figure 18.7. A public domain mixed analogue-digital simulator using SPICE-PAC [13] was compared with two commercial mixed analogue-digital simulators, PSPICE [14] and SABER [15]. The results are shown in Tables 18.3 and 18.4. Table 18.3 is the time-domain analysis for a slow ramp through all possible states. Table18.4 is the time-domain analysis for a step response from the 000 state to just past the MSB by 10mV. The simulation was performed at two levels of modeling the comparators. The first is at the circuit level and the second was at the macromodel level. All of the simulators were run on a SUN 3/60.

Table 18.3 - Comparison of SPICE-PAC, PSPICE, and SABER for a slow-ramp applied to a 3-bit, flash, analogue-digital converter.

	Comparator Modeling Level	
Simulator	Circuit	Macro
SPICE-PAC Conversion parameter = 3	1554.02 seconds	213.70 seconds
Conversion parameter = 7	2812.16 seconds	362.24 seconds
PSPICE	2030.62 seconds	537.23 seconds
SABER	5390.00 seconds	365.00 seconds

Table 18.4 - Comparison of PSPICE-PAC, PSPICE, and SABER for a step-response applied to a 3-bit, flash, analogue-digital converter.

	Comparator Modeling Level	
Simulator	Circuit	Macro
SPICE-PAC Conversion parameter = 3	1343.26seconds	300.12 seconds
Conversion parameter = 7	1573.70seconds	318.88 seconds
PSPICE	1062.40 seconds	210.33 seconds
SABER	4350.00 seconds	144.00 seconds

Simulation and modeling trends are moving in the direction of being able to simulate complex analogue and digital circuits efficiently with the ability to adjust the model levels as needed. Modeling techniques must be flexible and be able to adapt to changing technology and new requirements. An adequate simulation and modeling capability are important in many other activities of design such as design automation and testability.

18.5 Design Automation

Design automation is another important factor that is becoming a strong influence on analogue circuits and their application. Design automation is being applied to both the simple and complex class of analogue circuit design. Design automation in the simple class is providing synthesis and automated layout tools which extend the design of these circuits to individuals without design experience. These tools are maturing quickly and are being commercially supported. Design automation applied to the complex class of analogue circuits takes the form of designer assistance. These tools fall in the category of helping the expert designer to be more efficient and to extend the designers capability. These tools tend to have a great deal of interaction with the designer.

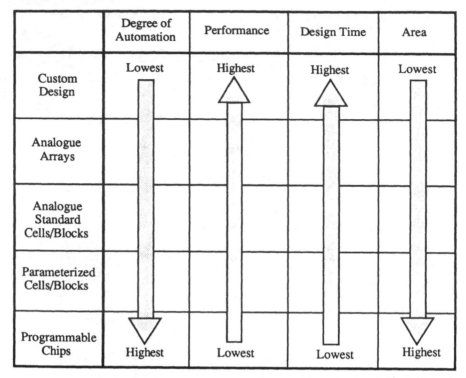

	Degree of Automation	Performance	Design Time	Area
Custom Design	Lowest	Highest	Highest	Lowest
Analogue Arrays				
Analogue Standard Cells/Blocks				
Parameterized Cells/Blocks				
Programmable Chips	Highest	Lowest	Lowest	Highest

Figure 18.8   Illustration of the influence of design methodologies on circuit performance.

Design automation is based on several methods/techniques. These include knowledge or experience accumulation generally in the form of libraries, rules, or guidelines, equation-based techniques which use algorithmic approaches, and artificial intelligence which is capable of learning. It is interesting to note that artificial intelligence has not had much impact on analogue design automation. Other tools that go into the design automation approach include modeling, simulation, and strong use of hierarchy.

The design methodologies used for the simple class of circuits can be categorized into (1) analogue arrays, (2) analogue standard cells, (3) analogue parameterized cells, and (4) analogue programmable chips [16]. Analogue arrays are preprocessed integrated circuits containing unconnected components and groups of connected components which are programmed by defining interconnections on one or more mask layers. Analogue standard cells are predesigned circuits which reside in a software database and can be used to implement the design of an analogue integrated circuit. Analogue parameterized cells are partially predesigned circuits which reside in a software database and can be programmed or parameterized at the time of the design of the integrated circuit. Analogue programmable chips are completely fabricated chips which are capable of programming by electrical or some other means. The influence of these methodologies on the performance of analogue circuit is illustrated in Figure 18.8. The tradeoffs between various aspects of performance versus design methodologies are shown.

The application of design automation techniques to the simple class of analogue circuits has resulted in various types of analogue compilers. An analogue compiler is a computer implemented program which accepts a circuit or system specification and synthesizes a circuit or system design in a given technology and generates a physical description of that circuit or system in a form ready for fabrication. Compilers generally make strong use of hierarchy and are classified by how they use hierarchy. Successive approximation compilers decompose the design from a high-level description down through a hierarchy of intermediate levels to the lowest level units in a top-down approach. Incremental refinement compilers are a generalization of the traditional block approach and make use of parameterized blocks to implement a given specification in a bottom-up approach. The design automation methodologies of standard cell and parameterized cell are frequently used in compiler implementation. The goals of compilers are to generate analogue circuits or systems from high level specifications, emulate the design practices of human circuit designers, keep the databases as small as possible, provide for expanding or changing the database, and allow for designer interaction. The advantages of compilers are that they do not require design expertise to use, circuits are correct by construction, and design time is reduced. Disadvantage of compilers are that their performance may be limited, compilers are difficult to design, topologies may be limited, and the databases may not be easy to change.

Many analogue compilers have been developed for various applications. The performance of three different analogue compilers designed for circuits such as an op amp have been compared and show different capabilities [17]. These compilers use the equation based approach based on the data stored in the library. One of the advantages of analogue compilers is that the design time is minimized and permits more design exploration or optimization. The analogue compiler usually consists of two parts. These parts are the synthesis part and the layout part. Analogue layout compilers must take into account the specialized requirements for analogue circuit performance.

Circuits level analogue compilers have been used to create systems level analogue compilers through the application of hierarchy. Typically, the application such compilers is limited to an area such as switched capacitor filters or analogue-digital conversion. Systems level compilers can be found in both the academic and commercial environment. One of the problems with analogue system compilers is being able to verify the performance of the system through simulation. Correctness by construction may not be valid as the possibilities for the system topology configuration increase. A typical systems level compiler flow chart is shown in Figure 18.9. This compiler was designed to implement successive approximation analogue-to-digital

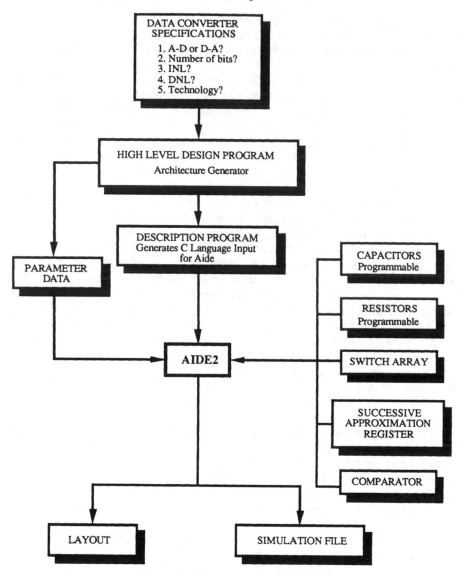

Figure 18.9   Flow chart for analogue systems compiler for successive approximation analoguedigital converters.

converters [18]. It used the parameterized methodology imbedded in a CAD platform called AIDE2 [19]. The compiler output consisted of a physical layout and a simulation input file. The weakness in this compiler was the inability to predict the actual performance through simulation.

Analogue compilers are beginning to make an impact on analogue integrated circuit design. The largest impact is placing the design of simple analogue circuit in the hands of those without design expertise. These compilers are primarily suited only to specific application areas. Analogue compilers need better simulation and modeling capabilities for the design of complex circuits and systems. Hierarchy coupled with equation-based methods has resulted in compilers with practical capabilities which are presently found in commercially supported CAD systems.

18.6 Testing Trends

Analogue testability is an area which is becoming more important as the size of analogue and digital integrated circuits increases The motivation for testability is found in the economics of the situation. The cost and time to test some integrated circuits is a major part of the total cost of the IC. Further motivation is found in the success with which testing has been used in digital circuits. The possibilities of increased design automation and better modeling and simulation promise to be opportunities that may help analogue testability evolve to the level equivalent of digital circuit testability.

Testability is a measure of how accurately the deviation of a component/circuit from its nominal value/performance can be determined by measurements. The objective of analogue testability is to detect or predict a malfunction in the operation of an integrated circuit. The definitions and terms used in analogue testability can be found in the various references [20-23].

Types of analogue tests include functional, parametric, static, and dynamic. Analogue faults can come from several sources. These sources include design errors, fabrication/process errors, packaging/manufacturing errors, and errors in the test procedure itself. Analogue testing focuses on which parameters should be tested and how accurately they can be tested. Digital testing does not focus on if the testing can be done, but rather can it be done efficiently, reliably, and with ease of use. The implications of scaling down the device size has a direct bearing on the cost of testing. The test cost is proportional to the square of the number of transistors. As a transistor is scaled down by a factor of two, the number of transistors increases by 4 and the test cost increases by a factor of 16. This brings up some important issues. Since the cost of testing is proportional to the amount of testing, the customer may want to perform part of the testing in exchange for a lower purchase price.

The activities in analogue testability can be categorized into approximately 6 areas. These areas are accessibility, product testing, simulation before testing (SBT), simulation after testing (SAT), design for testability, and self testing/trainable design. Accessibility deals with the ability to access nodes/parts of a circuit for the purposes of analogue testing. Product testing is used to determine the defect level of a circuit after fabrication in a production environment. SBT is the simulation of the circuit/system before testing/fabrication. It is used to predict the testability of a circuit/system. SAT is the simulation of the circuit/system after testing/fabrication to identify, locate, and diagnose analogue faults. Design for testability is the

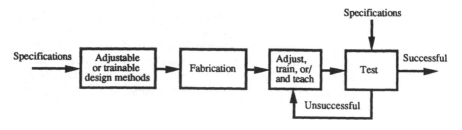

Figure 18.10   Block diagram illustrating self testing/trainable design methodology.

incorporation of testability considerations into the design phase of the IC development. Self testing/trainable analogue circuits/systems are the design of analogue circuits/systems that are capable of post-fabrication adjustment that enables the performance to meet the specifications of the design.

Progress is being made in the area of applying design for testability methods to ASICs. Emphasis is placed on techniques necessary for proper isolation, control, and observation of each partition of the circuit. The methods used are based on the standard and parameterized cell design methodologies. The principles are to isolate and control cell inputs and outputs and be able to observe the cell output nets. A method of incorporating accessibility has been applied to use digital inputs to establish the test mode. The test strategy is designed into the cell along with other documentation at a local level. Global considerations are incorporated when the cells are assembled into a system level implementation.

Self testing/trainable analogue circuits are closely related to design techniques. The objective is to develop design methodologies which use an interactive test and training phase to guarantee that analogue circuits either meet design specification or are nonfunctional (unable to meet specifications). Figure 18.10 shows a block diagram illustration of this approach to analogue testability. This methodology solves the testability problem by requiring that operational characteristics be taught to circuits and systems rather than designed. The functionality of a circuit or system is determined by whether or not it can be successfully taught to operate as desired. This approach uses techniques of massive parallelism, the ability to learn during an interactive test and training period, and self calibration. Two important questions of how will the circuit be tested and who will design the test are solved by this approach. In this method, the problem of the time required to train and test should not be significant. Circuits should be designed in such a manner as to only require the fine tuning of an almost operational circuit. This is important since the time taken to refine an almost operational network is much shorter than the time taken to teach a completely unformed network.

The importance of analogue testability has been recognized and many different approaches are being taken to solve this problem. Accomplishments include the incorporation of design for testability methods in analogue design methodologies, the use of mixed analogue-digital and multilevel modeling techniques for fault simulation, and a much better understanding of the issues, problems, and potential solutions.

18.7 Future Trends and Expectations

The future of analogue signal processing and of analogue integrated circuit design holds many exciting challenges and important advances. This section will conclude this chapter with a projection of what can be expected to occur in analogue signal processing and analogue integrated circuit design during the next 10 years.

In the area of technology, BiCMOS will become the dominant technology. This technology will be optimized for digital circuits, but will remain useful for analogue as long as reasonably good capacitors are available. Niche technologies such as GaAs will develop and serve important but specific application areas particularly in the optical-electronic area. Some niche technologies may be combined with BiCMOS on the same chip to more closely couple the niche technologies with silicon. New technologies such as superconducting transistors may appear and make major changes in design methods. Another technological aspect that will have significant influence is floating-gate technology. There are many applications that require the ability to adjust a parameter with high resolution in a minimum area. If such a technology becomes available it will drive a whole new set of design methods based primarily on neural circuit concepts.

While design techniques are often driven by advances in technology, new and clever design techniques will always be developed that will permit the more efficient use of existing technology. The next 10 years may see a reversal of this technology-design relationship where the design requirements may drive the technology in certain high performance areas. In the short term, neural network and neural concepts will have a significant impact on design techniques. Techniques such as current signal processing, the subject of this book, should provide improved performance over voltage signal processing. Self-training and self-tuning design methods will decrease the analogue cycle/success ratio further reducing an important disadvantage of analogue circuits.

The largest changes in analogue IC design are likely to be in the area of modeling and simulation. There are two factors which contribute to advances in simulation and modeling. The first is that truly integrated simulators and models are just beginning to be developed. These simulators, as pictured in Figure 18.11, attempt to allow specialized simulation and modeling to occur within a uniform database and architecture. Such simulators can be tuned to the problem at hand and will permit modeling tradeoffs between speed and accuracy. Another advance in simulation and modeling will be the coupling of the simulator/solvers of Figure 18.11 with process simulators. The combination of a process simulator with a table

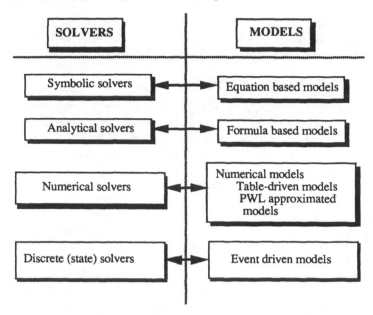

Figure 18.11 Vertical integration viewpoint of mixed analguedigital simulators a▮ models.

look-up model will permit quick and precise modeling of complex devices. Process shifts can readily be included in the model.

The second factor that will have an impact on modeling and simulation is the influence of new hardware. Computers will be faster, have more memory, and be more accessible to the designer. If no improvements in modeling and simulation were made, simulation would still advance through the increased capability of the hardware. Fortunately, both new hardware and new simulation and modeling techniques will join to provide some of the most significant advances in analogue circuit design that will be seen in the next 10 years.

Design automation will allow the simple class of analogue circuits to be designed by individuals outside the field. This in turn will create a demand for better design automation and more design capabilities. This field is where CAD companies in analogue will probably see a profit. Artificial intelligence should begin to make inroads into design automation but it probably will not compete with equation-based methods and will only be useful in specific application areas. Improvements in analogue compilers will be in the area of robustness, versatility, and better circuit/system performance.

Analogue testability is another area where major improvements will occur. New CAD methods for testability will be developed. These will include methods for identifying, locating, and diagnosing analogue faults, determining the appropriate place to test in a production environment, and practical methods of determining the testability of a circuit. Testability will be integrated with design and new design methodologies based on self-testing and self-trainable circuits and systems will be developed.

The future of analogue signal processing is very dynamic and holds many opportunities and challenges. The ideas and concepts presented in the previous chapters are part of the foundation for these advancements and accomplishments that will be witnessed over the next 10 years.

18.8   References

[1]   D.A. Hodges, P.R. Gray, and R.W. Brodersen: "Potential of MOS Technologies for Analog Integrated CIrcuits," IEEE J. of Solid-State Circuits, vol. SC-13, no. 3, pp. 285-294, June 1978.

[2]   L.E. Larson, K.A. Martin, and G.C. Temes, "GaAs Switched Capacitor Circuits forHigh-Speed Signal Processing," IEEE J. of Solid-State Circuits, vol. SC-22, no. 6, pp. 971-981, Dec. 1987.

[3].   R.W. Brodersen, P.R. Gray, and D.A. Hodges: "MOS Switched-Capacitor Filters,"Proc. of the IEEE, vol. 67, no. 1, pp. 61-75, January 1979.

[4]   S.F. Fang, Y.P. Tsividis, and O. Wing, "SWITCAP: A Switched-Capacitor NetworkAnalysis Program - Part I: Basic Features," IEEE Circuits and Systems Magazine, vol. 5, no. 5, pp. 4-10, Dec. 1983.

[5]   S.F. Fang, Y.P. Tsividis, and O. Wing, "SWITCAP: A Switched-Capacitor Network Analysis Program - Part II: Advanced Applications," IEEE Circuits and Systems Magazine, vol. 5, no. 6, pp. 41-46, Sept. 1983.

[6].   T.L. Sculley and M.A. Brooke, "A Neural Approach to High Performance Analog Circuit Design," Proc. of 32 Midwest Sym. on Circuits and Systems, Aug. 1989, Champaign-Urbana, IL.

[7].   B.E. Boser and B.A. Wooley, "The Design of Sigma-Delta Modulation Analog-to-Digital Converters," IEEE J. of Solid-State Circuits, vol. 23, no. 6, pp. 1298-1308, Dec. 1988.

[8].   P.E. Allen and M.B. Terry, "The Use of Current Amplifiers for High Performance Voltage Applications," IEEE J. of Solid-State Circuits," vol. SC-15, no. 2, pp. 155-162, April 1980.

[9].   P.E. Allen and K.S. Yoon, "A Table Look-Up MOSFET Model for Analog Applications," "IEEE Inter. Conf. on Computer-Aided Design, Santa Clara, CA, pp. 124-127, Nov. 1988.

[10]. R. Goering, "A Full Range of Solutions Emerge to Handle Mixed-Mode Simulation," Computer Design, vol. 27, no. 3, pp. 57-65, 1988.

[11]. T. Tormey, "Mixed-Signal Simulator Eases System Integration," Computer Design, vol. 28, no. 9, pp. 103-106, 1989.

[12]. P.E. Allen, B.P. Lum Shue Chan, and W.M. Zuberek, "Comparison of Mixed Analog-Digital Simulators," Proc. of 1990 Inter. Sym. on Circuits and Systems, New Orleans, LA, 1990.

[13]. W.M Zuberek, "SPICE-PAC2G6c - User's Guide", Technical Report #8902, Dept. of Computer Science, Memorial Univer. of Newfoundland, St. John's, Canada A1C-5S7, 1989.

[14]. M. Wimbrow, "Simulating Mixed Analog-Digital Systems," Paper DT-4.2, ATE & Instrumentation Conf. West, Jan. 1988, Anaheim, CA.

[15]. Electronics Staff, "SABER cuts SPICE out of Analog Simulation," Electronics Magazine, vol. 59, no. 34, pp. 80-92, 1986.

[16]. P.E. Allen, "Computer aided design of analogue integrated circuits," Journal of Semicustom ICs, vol. 4, no. 2, pp. 22-32, Dec. 1987.

[17]. L. R. Carley and R.A. Rutenbar, "How to Automate Analog IC Designs," IEEE Spectrum, pp. 26-30, August 1988.

[18]. P.E. Allen and P.R. Barton, "A Silicon Compiler for Successive Approximation A/D and D/A Converters," Proc. of 1986 Custom Integrated Circuits Conf., Rochester, NY, pp. 552-555, May 1986.

[19]. P.E. Allen, E.R. Macaluso, S.F. Bily, and A.P. Nedungadi, "AIDE2: An Automated Analog IC Design System," Proc. of 1985 Custom Integrated Circuits Conf., Portland, OR, pp. 489-501, May 1985.

[20]. C.F. Hawkins, H.T. Nagle, R.R. Fritzemeir, and J.R. Guth, "The VLSI Circuit Test Problem - A Tutorial," IEEE Trans. on Industrial Electronics, vol. 36, no. 2, pp. 111-116, May 1989.

[21]. R.R. Fritzemeir, H.T. Nagle, and C.F. Hawkins, "Fundamentals of Testability - A Tutorial," IEEE Trans. on Industrial Electronics, vol. 36, no. 2, pp. 117-128, May 1989.

[22]. P.P. Fasang, "Analog/Digital ASIC Design for Testability, " IEEE Trans. on Industrial Electronics, vol. 36, no. 2, pp. 219-226, May 1989.

[23]. K.D. Wagner and T.W. Williams, "Design for Testability of Analog/Digital Networks,", IEEE Trans. on Industrial Electronics, vol. 36, no. 2, pp. 227-230, May 1989

Index